SWITCHED-CURRENTS
an analogue technique
for digital technology

Edited by
C. Toumazou, J. B. Hughes
& N. C. Battersby

Supported by the IEEE Circuits and Systems Society
Technical Committee on Analog Signal Processing

Peter Peregrinus Ltd. on behalf of the Institution of Electrical Engineers

Published by: Peter Peregrinus Ltd., on behalf of the
Institution of Electrical Engineers, London, United Kingdom

Peter Peregrinus Ltd.,
The Institution of Electrical Engineers,
Michael Faraday House,
Six Hills Way, Stevenage,
Herts. SG1 2AY, United Kingdom

British Library Cataloguing in Publication Data

A CIP catalogue record for this book
is available from the British Library

ISBN 0 86341 294 7

Printed in England by Short Run Press Ltd., Exeter

To Gina Toumazou,

Patricia Hughes, and

the late Marjorie Battersby

Foreword

In this, the fifth volume in the IEE Circuits and Systems series, the editors have compiled an excellent record of the state of the art of this field. The book is essentially part of a deliberate policy by the IEE for the extremely rapid publication of highly topical material in niche areas of electronics.

Switched-current circuit techniques have already proved their great potential in analogue and mixed-signal circuit design. The field is still developing and this volume brings together contributions from the most advanced researchers and practitioners of this art from Europe and North America.

As with other books in this series, the present work is being launched in conjunction with an international conference. This practice provides a very sharp focus and effective means of exchange and discussion of ideas in particular fields. The present work will be launched at the IEEE International Symposium on Circuits and Systems in Chicago in May 1993 and will accompany a special session on the topic. The need to meet this date imposed a time scale of around five months for the editors and authors to provide camera ready material. In the event the wide coverage and depth to which the material is treated is testimony to their great effort and the importance which is attached to this exciting field.

The work is of immense importance and the breadth, depth and style of treatment of the various constituent topics will be found to be of value and appeal to advanced undergraduate and postgraduate students and particularly to practising engineers in industry.

Randeep Singh Soin
David Haigh
Fareham, April 1993

Preface

"This is not the end. It is not even the beginning of the end. But it is perhaps the end of the beginning."
Winston Churchill.

The text-book "Analogue IC design: the current-mode approach" edited by Chris Toumazou, John Lidgey and David Haigh, and the second volume in the IEE series on Circuits and Systems, is acclaimed as the first international compilation of the important area of current-mode analogue signal processing. The current volume, which has arisen from this text, is devoted specifically to switched-current sampled-data analogue processing. We have seen over the past two years a phenomenal rise in both industrial and academic research interest in the switched-current technique. At many IEEE conferences including the 1992 and 1993 Custom Integrated Circuit Design Conference and the 1989 - 1993 International Symposium on Circuits and Systems (ISCAS) many sessions and papers have been devoted to the switched current area. In fact at the 1993 IEEE ISCAS this year, Professor Kenneth Laker special session chairman approved a special session organised by one of the editors, Chris Toumazou, on switched-capacitors versus switched-currents for future VLSI , and it is because of these events that the idea for this volume arose.

From book conception to final camera ready copy took less than five months. Meeting the tight production schedule coupled with the wide diversity of subject material covered in the book is due to tough demands on the authors, and the special efforts they made in getting the material to us on time. These 'super-human' efforts will be remembered for a very long time. Through the editing of the book we have witnessed very impressive developments from the basic switched current concept to current state-of-the-art in switched current systems.

The switched current technique is set to have a tremendous impact on future mixed signal VLSI.

Chris Toumazou
John Hughes
Nicholas Battersby
London, April 1993

List of Contributors

Corné Bastiaansen
Philips Research Laboratories.
PO Box 80000
5600 JA Eindhoven
The Netherlands.

Nicholas C. Battersby
Dept. of Electrical & Electronic Eng.
Imperial College
Exhibition Road
London SW7 2BT
England.

Ian Bell
Dept. of Electronic Engineering
University of Hull
Cottingham Road
Hull, EHU6 7RX
England

Neil Bird
Philips Research Laboratories
Cross Oak Lane
Redhill
Surrey, RH1 5HA
England.

Philip J. Crawley
Microelectronics and Computer systems
Laboratory
Dept. of Electrical Engineering.
McGill University
3480 University Street
Montreal, H3A 2A7.
Quebec, Canada

Steven Daubert
Room 3P-232
AT & T Bell Laboratories.
200 Laurel Avenue
Middletown
NJ 07748
USA

Philippe Deval
Electronics Laboratories.
Swiss Federal Institute of Technology
Ecole Polytechnique Federale de
Lausanne
(EPFL)
EL- Ecublens
CH-1015 Lausanne,
Switzerland

Rafael Dominguez-Castro
Dept. de Diseno Analogico
Centro Nacional de Microelectronica
University de Sevilla (Edifico CICA)
Avd. Reina Mercedes s/n
41012-Sevilla
Spain.

Servando Espejo
Dept. de Diseno Analogico
Centro Nacional de Microelectronica
University de Sevilla (Edifico CICA)
Avd. Reina Mercedes s/n
41012-Sevilla
Spain.

Wouter J. Groeneveld
Philips Research Laboratories.
PO Box 80000
5600 JA Eindhoven
The Netherlands.

José L. Huertas
Dept. de Diseno Analogico
Centro Nacional de Microelectronica
University de Sevilla (Edifico CICA)
Avd. Reina Mercedes s/n
41012-Sevilla
Spain

John B. Hughes
Philips Research Laboratories
Cross Oak Lane
Redhill
Surrey, RH1 5HA
England.

Kenneth W. Moulding
Philips Research Laboratories
Cross Oak Lane
Redhill
Surrey, RH1 5HA
England.

Ian Macbeth
Pilkington Micro-electronics Limited
Sherwood House Gadbrook Business
Centre
Rudheath Northwich
Cheshire CW9 7TN

Antônio Carlos Moreirão de Queiroz
Departamento de Engenharia Eletrica
COPPE/EE
Universidade Federal do Rio de Janeiro
CP-68504 CEP 21945 Rio de Janeiro
Brazil

David Nairn
Dept. of Electrical Engineering
Queen's University at Kingston
Kingston
Ontario
Canada K7L 3N6

Douglas M. Pattullo
Pilkington Micro-electronics Limited
Sherwood House Gadbrook Business
Centre
Rudheath Northwich
Cheshire CW9 7TN

William Redman-White
Department of Electrical Engineering
University of Southhampton
England

Gordon W. Roberts
Microelectronics and Computer systems
Laboratory
Dept. of Electrical Engineering.
McGill University
3480 University Street
Montreal, H3A 2A7.
Quebec, Canada

Angel Rodríguez-Vázquez
Dept. de Diseno Analogico
Centro Nacional de Microelectronica
University de Sevilla (Edifico CICA)
Avd. Reina Mercedes s/n
41012-Sevilla
Spain

Adoración Rueda
Dept. de Diseno Analogico
Centro Nacional de Microelectronica
University de Sevilla (Edifico CICA)
Avd. Reina Mercedes s/n
41012-Sevilla
Spain

Hans Schouwenaars
Philips Research Laboratories.
PO Box 80000
5600 JA Eindhoven
The Netherlands.

Adel S Sedra
Dept. of Electrical Engineering
University of Toronto
Ontario, M5S 1A4
Canada

Gaynor Taylor
Dept. of Electronic Engineering
University of Hull
Cottingham Road
Hull, EHU6 7RX
England

Henk Termeer
Philips Research Laboratories.
PO Box 80000
5600 JA Eindhoven
The Netherlands.

Christofer Toumazou
Dept. of Electrical & Electronic Eng.
Imperial College
Exhibition Road
London SW7 2BT
England.

George Wegmann
Centre de Conception de Circuits
Integres
Ecole Polytechnique Federale de
Lausanne
EPFL C3i-LEG
EL- Ecublens
CH-1015 Lausanne,
Switzerland

Paul Wrighton
Dept. of Electronic Engineering
University of Hull
Cottingham Road
Hull, EHU6 7RX
England

Alberto Yúfera
Dept. de Diseno Analogico
Centro Nacional de Microelectronica
University de Sevilla (Edifico CICA)
Avd. Reina Mercedes s/n
41012-Sevilla
Spain

Acknowledgements

We would like to thank Professor Robert Spence of Imperial College and Professor John Lidgey of Oxford Brookes University for the use of photocopying and printing facilities.

Wiesia Hssisen at Imperial college has provided outstanding secretarial and moral support in getting to the final version. We would also like to thank the staff at the IEE for encouragement and support. In particular the invaluable help of Fiona MacDonald for meticulously proof reading of many of the chapters, Robin Mellors for administrative support and for initiating a very quick printing schedule, in fact only two weeks! Support from secretarial staff of Philips Research Labs (Redhill) is also gratefully acknowledged.

As editors, we were privileged to have had such an eminent group of contributing authors, their efforts and dedication, in writing their chapters, is very gratefully acknowledged.

Finally, we would like to thank our families for their support, especially Ann Toumazou.

The Editors

Table of Contents

Filters

Data Converters

Other Applications

Analysis, Simulation and Test

Future Directions

Chapter 1

Introduction

Chris Toumazou, John B. Hughes and Nicholas C. Battersby

1.1 VLSI Technology

The recent phenomenal advances in computer and communications technology have been made possible by the progress of integrated circuit processing technology, that has allowed ever more complex systems to be integrated onto a single monolithic integrated circuit. Applications which, just a few years ago, required dozens of separate generic analogue and digital integrated circuits can now be realised using just one or two ASICs (Application Specific Integrated Circuits). This progress looks set to continue in the next few years as processing technology advances yet further.

The development of integrated circuit processing technology has now reached the stage where it is possible to integrate several million transistors onto a single silicon chip. The availability of such vast numbers of transistors has given a tremendous impetus to digital signal processing and computational circuits, by enabling complex systems to be integrated onto a single monolithic integrated circuit. The dominant processing technology used to realise these very densely packed integrated circuits is currently CMOS (Complementary Metal Oxide Semiconductor). Although, the combination of CMOS and bipolar technologies (BiCMOS) has recently begun to emerge as a competitor to CMOS for some applications [1].

1.2 The Analogue Dilemma

The emergence of complex digital integrated circuits has steadily displaced analogue solutions from many applications. Often, the advantages of digital implementations are overwhelming; they offer programmability, flexibility, additional product functionality, short design cycles and they exhibit good immunity to both noise and manufacturing process tolerances. However, for a digital system to interact effectively with an inherently analogue world, analogue signal conditioning and data conversion circuits are still required. The role of analogue integrated circuits is therefore changing. Most often complex systems consist of a core of digital signal processing and computation circuits, buffered from an analogue external environment by a layer of analogue interface circuits. Complete analogue systems will still continue to be required in some applications: those in which the frequency of operation is too high for digital implementation, where low complexity does not justify a digital implementation and in very low power applications.

In recent years, the quest for ever smaller and cheaper electronic systems has led manufacturers to integrate entire systems onto a single chip. It is now becoming common to find that a single mixed analogue and digital (mixed-mode) integrated circuit contains both a digital signal processor and all the analogue interface circuits required to interact with its external analogue transducers and sensors as shown in Figure 1.1 below

Whilst these mixed-mode integrated circuits are advantageous from both economic and systems design viewpoints, combining both analogue and digital circuits on a single chip makes the circuit design and simulation process considerably more complex.

As a typical mixed-mode integrated circuit contains primarily digital circuits, it is natural that the processing technology be tailored to optimise digital performance. In practice, this means that CMOS is the most commonly employed technology. Analogue interface circuits that are fully compatible with digital CMOS processing technology are therefore required if mixed-mode integrated circuits are to be manufactured economically.

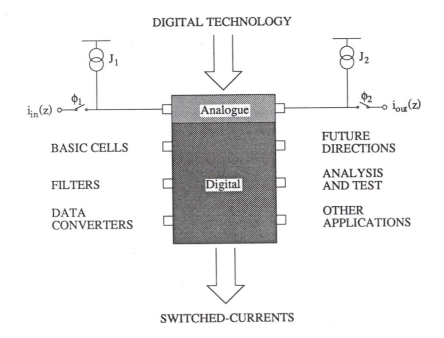

Figure 1.1 Mixed-signal ASIC.

1.3 Switched-Currents

State-of-the-art analogue circuit design is receiving a tremendous boost from the application of current-mode analogue signal processing and this is apparent from the widespread academic and industrial interest in the recent text-book 'Analogue IC design: the current-mode approach' [2]. The term current-mode is somewhat misinterpreted in the literature. General statements such as current-mode designs allow for gain independent of bandwidth, give lower noise, wider dynamic range, higher speeds, are not strictly correct, since these performances are very technology and application specific. Circuit architecture's such as common-gate or common-base transistor configurations, current-mirrors and current-amplifiers, current-feedback circuit configurations and circuits which present devices with 'virtual ground' low impedance nodes are now more generally becoming labelled as current-mode circuits. In the context of this book what we mean by a switched current-mode system is a system in which signals are represented by current samples. To be clear on this point, any circuit which handles signal currents must of course develop internal voltage swings. In [2] the contents are arranged to cover continuous-time and sampled data current-mode processing. Whilst a tremendous amount of research and development has been expended in the continuous-time current-mode area, the seeds of a new sampled-data current-mode analogue signal processing technique, switched-currents, has evolved and a compilation of references to its foundations can be found in [2]. This book builds on these foundations and develops them into practical switched-current analogue circuits and systems on a pure digital technology.

1.4 An Analogue Technique For Digital Technology

Traditionally the switched-capacitor technique [3,4] has been employed extensively in the analogue interface portion of mixed-mode designs. However, switched-capacitors are not fully compatible with digital CMOS processing technology and, as the technology advances further, the drawbacks of switched-capacitors are becoming more significant. Switched-capacitors traditionally require high quality linear capacitors, which are usually implemented using two layers of polysilicon. The second layer of polysilicon used by switched-capacitors is not needed by purely digital circuits and may become unavailable as process dimensions shrink to the deep submicron range. The trend towards submicron processes is also leading to a reduction in supply voltages, directly reducing the maximum voltage swing available to switched-capacitors and consequently reducing their maximum achievable dynamic range. With lower supply voltages the realisation of high speed high gain operational amplifiers will become more difficult.

The difficulties faced by switched-capacitors, and other 'voltage-mode' analogue interface circuits, in coping with the advance of digital

CMOS processing technology has revived interest in 'current-mode' techniques [2]. The switched-current technique is a relatively new analogue sampled-data signal processing technique which fully exploits digital CMOS technology [5]. Switched-current circuits do not require linear floating capacitors or operational amplifiers. The switched-current technique is heralding a new era in analogue sampled data signal processing and is giving a renewed impetus to mixed signal VLSI on standard digital technology. This has been acknowledged by many analogue designers from industry and academia all over the world who have made a notable contribution to this very timely first book entirely devoted to switched-current analogue signal processing.

The switched-current technique is a prime example of technology driven analogue IC design, and as such represents an analogue design philosophy which has a very close integration between system level blocks and process technology. In order to portray this close bond between technique and technology we felt it very appropriate to represent the contents of this book via schematics of integrated circuits as shown in the following figures.

1.5 Book Style

This book has been arranged in the six key sections illustrated in Figure 1.1. The schematic also illustrates the disproportionate amount of analogue circuitry on a mixed-mode ASIC, and an input of pure digital technology out of which comes switched-currents. It must be stressed that this book does not consider the design of high quality, linear V to I and I to V converters which will ultimately be required in a complete switched-current system if it is to be to interfaced to the voltage domain. While the performance of these interface circuits will certainly affect overall system performance, this book only considers switched-current processors with current in and current out as shown in Figure 1.1. The design of a variety of linear tuneable V to I and I to V converters can be found in [2].

This book introduces the basic switched-current technique, reviews the state-of-the-art and presents practical chip examples, in some cases with performances set to out-perform switched-capacitor techniques in particular application areas. Throughout the text numerous application areas are described ranging from filters, data converters to cellular neural networks for image processing applications.

Chapters have been organised in the general areas shown in Figure 1.1. More specific reference to the chapters is shown in the chip photographs of Figures 1.2 and 1.3. Chapter contents are illustrated by pad labels on each IC. Referring back to Figure 1.1 we begin with **Basic cells,** which traces the evolution of the switched-current technique (Chapter 2), introduces the basic switched-current memory cell (Chapter 3), identifies major limitations and non-idealities within the basic cell (Chapter 4), gives a comprehensive analysis of noise performance (Chapter 5), proposes a number of novel cell enhancements to improve cell performance such as the use of fully differential structures (Chapter 6), and considers low-power class AB cells (Chapter 7).

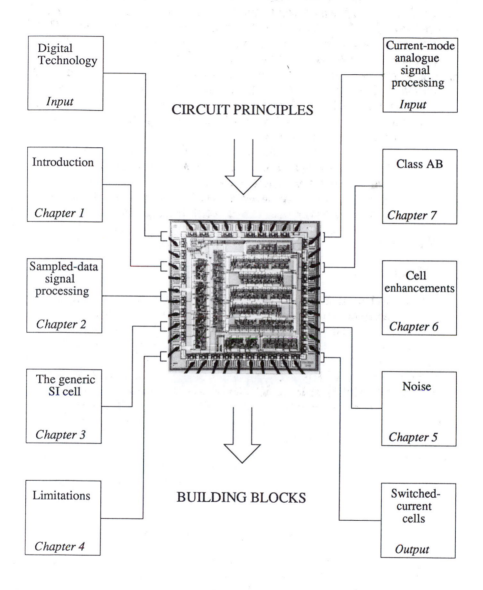

Figure 1.2   Switched-current building blocks.

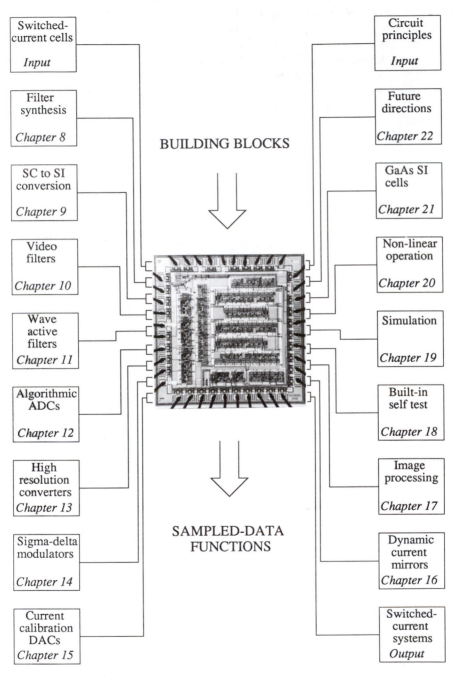

Figure 1.3   Switched-current systems.

Filters, covers the basic rules for the synthesis of switched-current filters (Chapters 8 and 9), introducing key filter building blocks such as switched-current delay cells, integrators, differentiators (Chapters 8 and 10), programmable wave active filter adaptor elements (Chapter 11) and employing these blocks to realise encouraging filter performance for audio (Chapter 8 and 11) and video frequency operation (Chapter 10). Filter topologies ranging from standard two pole biquads to complex high order Elliptic filters, demonstrate the flexibility of the technique (Chapter 8).

Data Converters, covering switched-current algorithmic and pipelined data converters (Chapter 12), high resolution algorithmic A/D converters based on dynamic current memories (Chapter 13), building blocks for switched-current sigma delta modulators (Chapter 14), current calibration D/A converters (Chapter 15), are described all with state-of-the-art test chip results.

Other Applications, covering topics from dynamic current-mirrors and their application (Chapter 16), to switched-current cellular neural networks for image processing applications (Chapter 17).

Test, Simulation and Analysis, introducing techniques for the important area of built-in self test of switched-current circuits (Chapter 18), a frequency domain algorithm for analysis and simulation of transistor level switched-current cells (Chapter 19), and finally a detailed analysis of the non-linear behaviour of switched-current systems (Chapter 20).

Future directions, a look into the future at some promising research avenues, in particular attempts to realise Gallium Arsenide Switched-Current Cells for front-end RF applications, with GHz clock-rates, are presented (Chapter 21). A concluding chapter (Chapter 22), summarises all state of the art material in this book and presents promising future prospects of the technique

1.6 Conclusion

It has not been the intention of this book to make explicit comparisons throughout with performance achievable with switched-capacitor techniques but rather to identify the state-of-the-art of switched-current techniques. It is acknowledged that besides the general advantages of switched-current techniques allowing the implementation of mixed analogue/digital systems on digital technology, the feasibility of realising switched-current structures as good as current state-of-the-art switched-capacitor techniques is still a challenge. However, great strides have already been made to accommodate analogue defects of basic memory cells and in Chapters 11-15 and 22 even greater promise is expressed for future switched-current signal processing.

Where will the switched-current technique find its niche? Is it in low-power or low-voltage or high speed or high resolution applications or a combination of all, or as with switched-capacitor techniques will it be optimised in performances specific to a whole variety of applications? The editors feel that it is too soon to be sure that the great variety of techniques given in this book may guide the reader to his/her own conclusion.

The switched-current technique represents a powerful future direction for mixed analogue and digital VLSI, and as with the early days of switched capacitors both innovation and creativity will be paramount in ensuring the acceptance and maturity of the technique. This book has attempted to give a full treatment of the fundamental basics of switched-current processing whilst at the same time also capturing state-of-the-art developments in the area. Hopefully, this book should stimulate interest in making, 'switched-currents: an analogue technique for digital technology' more accessible to the industry.

References

[1]   A. A. Iranmanesh and B. Bastani, "BiCMOS emerges for gate-arrays and memories," *IEEE Circuits and Devices Mag.*, vol. 8, pp. 14-17, March 1992.

[2]   C. Toumazou, F. J. Lidgey and D. G. Haigh, Eds., *Analogue IC Design: the current-mode approach*, London: Peter Peregrinus Ltd., 1990.

[3]   G. S. Moschytz, Ed., *MOS Switched-Capacitor Filters: Analysis and Design*, New York: IEEE Press, 1984.

[4]   R. Gregorian and G. C. Temes, *Analog MOS Integrated Circuits for Signal Processing*, New York: John Wiley & Sons, 1986.

[5]   J. B. Hughes, N. C. Bird and I. C. Macbeth, "Switched-currents - A new technique for analog sampled-data signal processing," *Proc. IEEE International Symp. Circuits Syst.*, pp. 1584-1587, May 1989.

Chapter 2

The Evolution of Analogue Sampled-Data Signal Processing

Nicholas C. Battersby and Chris Toumazou

2.1 Introduction

This chapter attempts to provide a historical perspective on the development of switched-current circuits by surveying the evolution, role and current trends in the development of analogue sampled-data signal processing systems.

The practical application of analogue sampled-data signal processing systems originated some forty or so years ago [1] and since then changing system requirements and available technology have resulted in the development of a variety of techniques, each tailored to the circumstances of their era. This chapter charts this development and the challenges currently posed by the continuing advance of integrated circuit processing technology. However, it is not the aim of this chapter to provide a detailed discussion of particular techniques, for this the reader is referred to appropriate references.

This chapter commences by reviewing the evolution of integrated circuit based techniques, culminating in the ubiquitous switched-capacitor technique. Then the role that analogue sampled-data systems play within the context of a general information processing environment is examined and the required characteristics are identified. Trends in silicon integrated circuit processing technology are summarised and their impact on sampled analogue circuits, specifically the switched-capacitor technique, is identified. Finally, the application of analogue current-mode techniques, and in particular switched-currents, to overcome some of the limitations of previous voltage based approaches, is discussed.

2.2 The Evolution of Analogue Sampled-Data Techniques

In the forty or so years since the introduction of practical analogue sampled-data signal processing systems [1], several different techniques have evolved. Most of these developments occurred in the late sixties and early seventies when MOS (Metal Oxide Semiconductor) integrated circuit technology became well established and the realisation of compact integrated sampled-data analogue systems became possible. In this section the evolution and applications of the various techniques are reviewed.

2.2.1 Early work

The application of circuits comprising linear invariant elements and periodically operated switches has played a significant role in communications engineering for a considerable time, specifically in the realisation of modulators for telephone systems. In 1939 *Caruthers* provided what seems to have been the first analysis of these modulators [2] and then during the fifties their analysis received further attention [3-5].

In 1952 *Janssen* introduced what appears to have been the first realisation of a true analogue sampled-data signal processing system [1]. His system was a simple analogue delay line, consisting of storage capacitors interconnected by switches and buffer amplifiers, and the motivation behind this development was the need to realise an analogue delay line free of both the linear and non-linear distortion observed in *LC* delay lines [1].

During the sixties a number of new analogue sampled-data systems were proposed [6-10]. Amongst these were the first realisations of switched-capacitor networks [9] and the first system to be realised as an integrated circuit, the bucket brigade device [10].

2.2.2 Bucket brigade devices

The bucket brigade device (BBD) was initially proposed as an integrated circuit technique by *Sangster and Teer* in 1969 [10], although a similar discrete technique had been proposed in 1967 by *Krause* [6]. A bucket brigade device is essentially an analogue shift register, in which information is conveyed as charge packets passed along a line of storage capacitors under the control of interconnecting switches implemented by transistors. Three stages of the simplest bucket brigade circuit, using a two phase complementary clock, are illustrated in Figure 2.1.

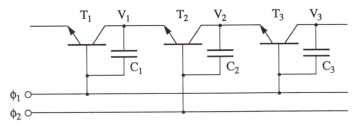

Figure 2.1 Three stages of the simplest bucket brigade circuit.

Referring to Figure 2.1, it can be seen that each stage of the bucket brigade circuit consists of an npn bipolar transistor and a storage capacitor. The capacitors are all of equal value C and the stages are clocked alternately by

ϕ_1 and ϕ_2. The incoming signal voltage samples U_k consist of an AC signal and a DC bias voltage, such that all the sample voltages are positive but less than a fixed reference voltage $+U$.

To understand the operation of the circuit, consider the case in which C_1 contains a signal sample U_k whilst the bias voltage $+U$ appears across C_2 and C_3. To transfer the information from C_1 to C_2 the base of T_2 is raised to $+U$ whilst those of T_1 and T_3 remain at ground. Since $U > U_k$ the base-emitter junction of T_2 is forward biased and it starts to conduct. Charge now flows from C_2 into C_1 until the voltage across C_1 is equal to U (less the V_{be} of T_2) leaving the signal voltage U_k stored across C_2. If the base voltage of T_3 is now raised to $+U$, whilst those of T_1 and T_2 are at ground, the signal voltage will be shifted onto C_3. By continuing in this manner, the signal voltage U_k will be shifted along through a series of storage capacitors, forming an analogue shift register. Furthermore, by connecting alternate stages to clock lines ϕ_1 and ϕ_2, and switching the clocks repetitively between ground and $+U$ in a complementary fashion, a stream of signal samples (one sample per two stages) can be processed by the circuit simultaneously.

Clearly, to realise a practical circuit, input and output circuitry is required, as are amplifiers to compensate for signal loss due to the finite base current and V_{be} of the bipolar transistor. Furthermore, it was quickly realised that an MOS transistor would make a more suitable switch than the bipolar transistor and circuits based on MOS technology emerged rapidly [11].

There were three main areas of application for which bucket brigade devices were considered useful [11]. The first of these was as simple delay lines, in which the delay time could be varied simply be changing the clock frequency. The second application was that of transversal filtering, in which the stored samples were weighted and summed to realise an FIR (Finite Impulse Response) filter. The third application was that of imaging arrays, in which the incidence of light on the capacitors caused the stored charge to dissipate at a rate determined by the light intensity. Therefore the charge left on each capacitor, after a period of time, indicated the total amount of light incident on that particular capacitor. The image stored on the bucket brigade array was then read by sequentially clocking out the charge samples stored on each capacitor.

When they were first developed, bucket brigade devices showed considerable promise; they could be integrated easily, had good packing density and had a number of useful areas of application. However, bucket brigade devices were relatively short lived as they were soon superseded by charge coupled devices.

2.2.3 *Charge coupled devices*

Charge coupled devices (CCDs) were introduced in 1970 by *Boyle and Smith* [12] and they performed very similar sampled analogue signal processing functions to their predecessor, the bucket brigade device.

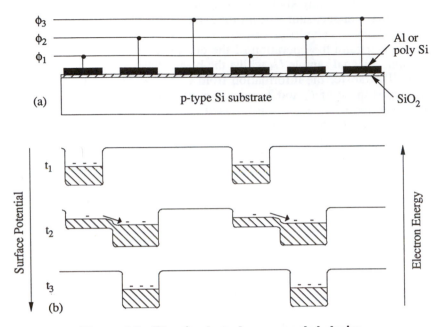

Figure 2.2   The simplest charge coupled device.

The primary difference between the two techniques was that CCDs stored charge samples in potential wells rather than in the explicit capacitors used by bucket brigade devices. This allowed CCDs to integrate the capacitance, MOS switch and MOS sensing device together, such that a smaller silicon area was used and higher performance could be achieved. These advantages allowed CCDs to rapidly displace bucket brigade devices as the implementation for all potential applications.

The simplest implementation of a CCD is the three clock phase device shown in Figure 2.2(a). In this device a p-type substrate is used so that the charge packets are conveyed in the form of electrons, by using an n-type substrate and applying negative voltages to the electrodes, holes can instead be used to convey the charge.

A brief description of the operation of the CCD (Figure 2.2) follows. The application of a positive voltage step to one of the electrodes causes a depletion region to form in the p-type silicon below it. This causes an electron energy minimum to form at the Si-SiO$_2$ interface below the electrode. If electrons are now introduced into the well, they will remain there because of the local electron energy minimum. However, thermally generated electrons will eventually fill the well and destroy the signal, therefore charge can only be stored temporarily in each well. The process of transferring charge from one well to another is illustrated in Figure 2.2(b). The first stage in the process is to raise the potential of a neighbouring electrode to be the same as that of the first electrode. The potential wells formed under both electrodes will then merge and the stored electrons will share between them, such that their energy is minimised. The potential of the first electrode is now gradually reduced, thus increasing the energy of the electrons remaining in its well. As this process continues all the electrons in the first well will transfer to the second well so that their energy is minimised. Eventually the second well will contain all the charge and the first well will have disappeared. By repeating this process a packet of charge can be shifted through a number of stages, forming an analogue shift register.

Although the basic principle underlying the operation of CCDs has been explained above, it is clearly necessary to use improved device structures and to employ interface circuits in order to build a practical device. Numerous papers have been published on the design and analysis of CCDs and some of the most significant early work is included in [13].

CCDs have been applied to three major system applications; self-scanned imaging arrays [14], analogue transversal filters [15] and digital memories [16]. Of these applications, that of self-scanned imaging arrays has proved to be by far the most successful and has evolved into the dominant imaging technique of today. CCD analogue transversal filters enjoyed a limited degree of success, but today are only used in special applications, often in support of CCD imaging arrays [17]. Their lack of flexibility compared with the switched-capacitor technique, specifically their difficulties in realising IIR (Infinite Impulse Response) filters, has limited their use in general analogue signal processing systems. Despite CCD digital memories showing much early promise [18], their inherently serial mode of access meant that they were unable to compete with MOS RAM (Random Access Memory).

2.2.4   Switched-capacitors

The switched-capacitor (SC) technique is based on the idea that a periodically switched capacitor can be used to simulate a resistor (provided that the switching frequency is much higher than the frequencies of interest). This idea was first proposed by *Maxwell* in 1891 [19], although at the time there was no practical application.

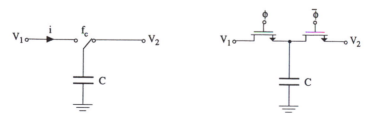

(a) Ideal circuit   (b) MOS implementation

Figure 2.3 Switched-capacitor resistance simulation.

The basic concept of how a switched-capacitor simulates a resistance is illustrated by Figure 2.3(a) and an MOS implementation of this circuit is shown in Figure 2.3(b). The switch alternately connects the capacitor C to voltage sources V_1 and V_2 with frequency f_c. When the capacitor is connected to V_1 the charge that it stores, q, is CV_1. When the switch is thrown, the capacitor is connected to V_2 and the net change flow of charge from the capacitor to V_2 is $C(V_1 - V_2)$. Since this process occurs f_c times per second the net flow of current i from V_1 to V_2 is $C(V_1 - V_2)f_c$. The equivalent resistance, R_{eq}, of the circuit is thus given by:

$$R_{eq} = \frac{V_1 - V_2}{i} = \frac{1}{f_c C} \tag{2.1}$$

Although originally proposed in 1891, it was not until 1967 that a practical application of the switched-capacitor concept was found, when *Baker* [9] proposed combining switched-capacitor resistance simulations with unswitched capacitors to create an RC time constant dependent only on a capacitance ratio and the clock frequency. This early work on switched-capacitors [9,20] used the resistance simulation, along with capacitors and buffering amplifiers to synthesise passive RC filters that could be tuned by simply varying the clock frequency. However, the main advantages of the technique only became readily apparent with the independent introduction in 1977 by *Caves* et al and *Hosticka* et al of the first monolithic switched-capacitor equivalents of active-RC filters [21, 22].

Active-RC filters had been used extensively for many years, but had proved resistant to realisation in monolithic form. The fundamental problem was that an accurate RC time constant could not be formed on an integrated circuit, where the absolute values of the resistors and capacitors could easily vary by as much as ±20%. Non-standard processing options, such as thin film resistors and trimming, could be used to improve the situation, but associated with these solutions was a significant cost penalty. Furthermore, even if good accuracy could be achieved at manufacture this would not be maintained because the ageing characteristics and temperature coefficients of the resistors and capacitors differ, resulting in time constant drift.

The introduction of switched-capacitor filters overcame these difficulties and enabled accurate compact analogue filters to be realised in monolithic form for the first time. The primary advantages that switched-capacitor filters offered were:

- Monolithic switched-capacitor filters were much more compact and potentially cheaper than discrete active-RC filters.

- In principle, time constants were determined solely by capacitor ratios and the clock frequency. Consequently they were more accurate and suffered less drift than realisations relying on absolute values or ratios between different types of component [23].

- Special processing options required to form precision resistors were not required.

- Active-*RC* filter structures could be used as a basis for switched-capacitor designs, thus avoiding the need to redesign from scratch.

Since the late seventies the switched-capacitor technique has evolved from just being a simple replacement for active-*RC* filters to a versatile analogue sampled-data signal processing technique used in a wide range of filtering and data conversion applications. This development has generated a vast number of publications and several books [24-26] which chart the progress of the technique.

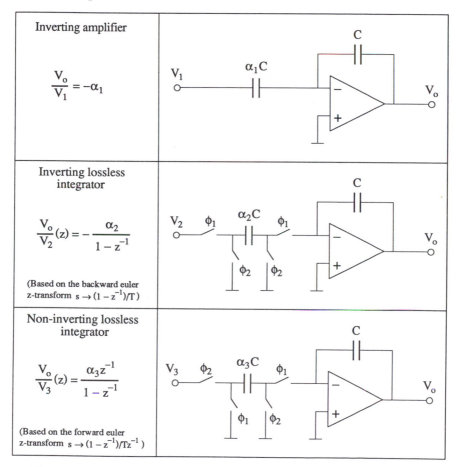

Inverting amplifier	
$\dfrac{V_o}{V_1} = -\alpha_1$	

Inverting lossless integrator	
$\dfrac{V_o}{V_2}(z) = -\dfrac{\alpha_2}{1 - z^{-1}}$ (Based on the backward euler z-transform $s \to (1 - z^{-1})/T$)	

Non-inverting lossless integrator	
$\dfrac{V_o}{V_3}(z) = \dfrac{\alpha_3 z^{-1}}{1 - z^{-1}}$ (Based on the forward euler z-transform $s \to (1 - z^{-1})/Tz^{-1}$)	

Figure 2.4 Parasitic insensitive switched-capacitor building blocks.

Rather than describe the switched-capacitor technique in detail, some basic 'parasitic insensitive' switched-capacitor building blocks are shown in Figure 2.4 to illustrate the technique. These building blocks are an inverting amplifier and forward and backward-Euler integrators, which can be combined together to synthesise many different filtering functions [26]. In Chapter 3 it will be shown how the switched-current technique can be used to realise functionally identical building blocks and thus utilise well established switched-capacitor design techniques.

Considering the implementation of switched-capacitor filters in MOS technology, two very important characteristics can be identified from the building blocks shown in Figure 2.4. Firstly, they use just three basic components; linear floating capacitors (usually realised using two polysilicon layers), switches (realised by MOS transistors) and operational amplifiers. Secondly, circuit transfer functions are determined solely by capacitor ratios and the clock frequency (assuming ideal circuit behaviour). These two characteristics made switched-capacitors very amenable to MOS monolithic realisation and contributed greatly to the success that the technique has enjoyed.

2.3   The Role of Analogue Sampled-Data Signal Processing

The evolution of analogue sampled-data signal processing has been discussed in the previous section and the emergence of the switched-capacitor technique as the dominant general purpose approach detailed. This section considers how the role of analogue sampled-data signal processing, in the context of a general signal processing environment, has changed and how it is likely to develop.

The role of early switched-capacitor filters was to replace active-RC filters, particularly in telecommunication applications [27]. As the technique matured it was used to realise more and more complex signal processing functions and at its peak switched-capacitors were used to realise complete systems [28]. Since the mid-eighties the role of switched-capacitors has been eroded by the emergence of digital signal processing (DSP) techniques. The move towards DSP solutions began around the mid-eighties when digital VLSI (Very Large Scale Integration) processing technology had evolved to the stage where complete DSP systems could be realised economically on a single chip. The digital approach offered many advantages over switched-capacitors; in particular, flexibility, programmability, ease of design and test, and freedom from noise and parasitic effects. However, since the outside world, with which any general signal processing system must at some stage interface, is inherently analogue in nature, a DSP system requires data converters and analogue signal conditioning circuits at its periphery.

The two approaches (analogue and digital) to the realisation of a general sampled-data signal processing system are illustrated in Figure 2.5. Referring to Figure 2.5, it can seen that both analogue and digital sampled-data systems share a common requirement for continuous time analogue circuits to act as buffers and bandlimiting filters at their input and output interfaces. The bandlimiting filters are needed to limit the bandwidth of the continuous time analogue signals to no more than half the sampling frequency of the filter (the Nyquist frequency) and their complexity is determined by system requirements and the ratio of the sampling frequency to the signal frequencies of interest.

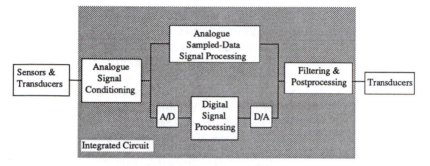

Figure 2.5 General sampled-data signal processing systems.

The numerous advantages of a DSP based signal processing system over one based on sampled analogue techniques are overwhelming. However, sampled analogue circuits still have a vital role to play. The requirement to realise both a DSP core and the associated analogue interface circuits in a single system, and the economic advantage to be gained from a single chip solution, has led to the emergence of mixed-mode signal processing chips. It is in the data conversion portions of these mixed-mode chips that analogue sampled-data circuits are now predominantly found.

Looking to the future, it seems likely that highly integrated mixed-mode solutions will displace generic analogue building blocks in all but a few areas of application. Semicustom design techniques will evolve, and will enable mixed-mode ASICs (Application Specific Integrated Circuits) to be applied economically in the relatively low volume applications which have hitherto been the preserve of multi-chip solutions. The main area in which analogue signal processing will maintain a competitive edge is in the area of very high speed filtering, such as is found in the front-end circuitry of radio receivers. Progress towards the realisation of very high speed sampled analogue filters has recently been made with the demonstration by *Haigh et al* of a 500MHz clock rate switched-capacitor filter using GaAs MESFET technology [29]. However, for very high speed applications the inherent simplicity and speed offered by continuous time filters has many advantages. Specifically, the tunable transconductance-C filter shows much promise for very high speed applications [30,31], although this area is outside the scope of this book.

Overall, although general purpose analogue sampled-data signal processing techniques have been applied to many filtering problems in the past, their future role will be primarily in the realisation of data converters and signal conditioning circuits at the interface between the digital and analogue domains.

2.4 Processing Trends and their Impact on Analogue Circuits

In the previous section the role of general purpose analogue sampled-data systems was examined and the areas of data converters and signal conditioning were identified as the primary applications.

Single chip mixed-mode systems are emerging as the best choice for many signal processing systems requiring analogue interfaces. But since the sampled analogue circuits only occupy a small proportion of the total silicon area, the use of a process tuned for analogue performance cannot be justified economically. Sampled analogue interface circuits will therefore have to be fabricated on a process that is tuned for digital performance. To gain an insight into the impact of processing advances on sampled analogue circuits it is therefore appropriate to consider processing trends for digitally oriented CMOS processes.

When implementing sampled analogue circuits on a mainstream digital process the first things that will be lost are the special analogue process options. The low proportion of the silicon area that the analogue circuits use makes the provision of these options an uneconomic luxury, although it may still be possible to provide these options at extra cost [32]. The analogue process options that look likely to disappear are, high quality resistors of various resistivities, double polysilicon (for high quality floating capacitors) and extra threshold voltage implants.

The driving force behind the dramatic advance of microelectronics in recent years has been the ever shrinking size of devices, which has both improved the speed and packing density of integrated circuits. Progress has been such that the realisation of a 64Mb DRAM (Dynamic Random Access Memory) on a single chip is now possible [33] and MOS transistors with gate lengths approaching $0.1\mu m$ have been fabricated [34,35].

Before considering specific process trends it may be useful to consider a fundamental limit on the signal processing capability of a transistor, which is determined by its speed of operation and dynamic range. In 1965 *Johnson* showed, by ideal analysis [36], that the fundamental performance limit of a transistor is set by:

$$V_m f_T = \frac{E_m v_s}{2\pi} \qquad (2.2)$$

where V_m is the maximum voltage that can be applied to the transistor, f_T is its cut-off frequency, E_m is the material breakdown field and v_s is the saturated drift velocity of carriers in the material. Since V_m and f_T indicate the dynamic range and speed of the device respectively, $V_m f_T$ is a measure of the signal processing capability of the device. As $V_m f_T$ is determined by the physical characteristics of the material, there is a limit to signal processing capability of a transistor made from a specific material; for a silicon transistor this limit is 200 V.GHz. Although derived in 1965, recent work [37], verified that (2.2) still holds true. The implication of (2.2) is that there is a fundamental trade-off between the speed of a device and its dynamic range, therefore as device speeds increase the voltages applied to them must be reduced.

The design of submicron devices requires the careful balancing of many conflicting requirements, i.e. reliability, speed and power consumption. In the past, the use of scaling theory has proved to be a useful tool in predicting device performance. The most natural of these theories is constant electric field (CE) scaling, which can achieve both high-performance and reliability. However, in the generation of devices with channel lengths greater than ~1μm, constant voltage (CV) scaling has been used in order to achieve higher performance and maintain TTL compatibility. For submicron devices it is no longer possible to use simple CE or CV scaling; CE scaling results in too great a reduction in power supply voltage for optimum speed, whilst CV scaling results in reliability problems. A more general scaling theory, which allows linear dimensions and potentials to be scaled independently by factors λ and κ respectively, is required. Such a generalised theory was proposed by *Baccarani et al* in 1984 [38] and the scaling factors associated with the most important parameters are shown in Table 2.1.

Parameter	Expression	Scaling Factor		
Linear dimensions	W, L, T_{ox}, x_j		$1/\lambda$	
Potentials	V_G, V_S, V_D		$1/\kappa$	
Doping concentration	N_A, N_D		λ^2/κ	
Electric field	E		λ/κ	
Capacitances	AC_{ox}, AC_j		$1/\lambda$	
Current	$(W/L)\mu C_{ox}(V_{GS} - V_T)V_{DS}$		λ/κ^2	
Current (vel. sat.)	$k_s W C_{ox}(V_{GS} - V_T)V_{sat}$		$1/\kappa$	
Power	$I_D V_{DD}$	$1/\kappa^2$		λ/κ^3
Power density	$I_D V_{DD}/A$	λ^3/κ^3		λ^2/κ^2
Gate delay	$C_g V_{DD}/I_D$	$1/\lambda$		κ/λ^2
Power-delay product	$I_D V_{DD} t_d$		$1/\lambda\kappa^2$	

Table 2.1 Generalised MOS scaling rules.

In Table 2.1 scaling factors are included for the case of carrier velocity saturation (leftmost column) as well as quadratic *I-V* relation (rightmost column). The use of these generalised scaling rules allows the voltages to be reduced independently of the linear dimensions so that the best compromise between device reliability, power consumption and speed can be made.

To determine the optimum choice of voltage scaling factor κ for a reduction in linear dimensions λ requires the consideration of many device and circuit constraints, which is beyond the scope of this work [40-48]. However, a brief summary of the most important factors is given below.

System considerations

For the last few years developments in CMOS technology have remained compatible with 5V power supplies and interfaces, but the development of submicron processes has now led to the adoption of a 3.3V standard [39]. However, the forthcoming generation of deep-submicron processes (<0.5μm gate length) will require even lower supply and interface voltages, possibly 2.5V or even 1.5V [40]. The considerable advantages to be gained from using well established interface standards has led to the development of on-chip voltage down converters. This has enabled internal circuitry to operate at a lower voltage than the supply and thus allowed compatibility with existing standards to be maintained as process geometries have shrunk [41].

The use of portable computers and communications equipment has experienced explosive growth over the last few years. With this has come the requirement for high performance integrated circuits which run from low supply voltages and consume little power. These requirements could lead to supply voltages dropping to below 1.5V, with parallelism being used to boost performance if necessary [42].

Reliability

As MOS transistors are scaled into the deep-submicron region, various design trade-offs between reducing the channel length L, the gate oxide thickness T_{ox} and the power supply voltage must be made [43,44]. These design trade-offs must include device reliability, which is limited by hot-electron degradation and time-dependent dielectric breakdown [43], and sets the maximum supply voltage (for a minimum lifetime of 10 years) for a given L and T_{ox}.

Modifications to the structure of MOS transistors have been found to result in improved reliability [34,45,46] and innovative circuit techniques are being employed to avoid exposing individual devices to the full supply voltage [41]. These techniques will allow the use of higher supply voltages than would otherwise have been possible.

Power dissipation

The impact of scaling on power dissipation can be seen from Table 2.1. Assuming that the carrier velocity is not saturated, the power density increases as λ^3/κ^3 and consequently if the supply voltage is not scaled to the same extent as linear dimensions the power dissipation will increase very rapidly. The management of power dissipation will therefore become a major problem in scaled processes unless the supply voltages are reduced sufficiently.

Carrier velocity saturation

As the electric field in the channel of a MOS transistor is increased the velocity of the charge carriers will also increase until the electric field reaches a critical value, E_c at which the carrier velocity saturates. As a result of velocity saturation the current becomes a linear function of the gate voltage rather than a quadratic one, the current drive is therefore significantly reduced. This implies that there is little speed advantage to be gained from operating above a certain critical voltage V_c. The value of V_c was derived by *Kakumu* and *Kinugawa* [47] and is defined as $V_c = 1.1E_cL$. For a 0.3µm process V_c was found to be 2.43V.

Threshold voltage scaling

As supply voltages are scaled down with submicron processes the threshold voltages will also be reduced. The limit as to how low the threshold voltages can be made is mainly determined by the subthreshold leakage current, which can result in significant static power dissipation [42].

The above summary has provided an overview of some of the issues governing the scaling of devices into the deep-submicron region. Clearly the trade-offs required in the design of deep submicron processes are very complex and at the present time a wide variety of approaches are being explored. Furthermore there are no standards in place for the reduction of supply voltages beyond 3.3V [39] that will be required by deep-submicron processes. To illustrate some of the approaches to submicron process design Table 2.2 summarises key process parameters for some recently published experimental submicron processes.

Effective channel length (µm)	Oxide thickness (nm)	Supply voltage (V)	Threshold voltages (V)	Reference
0.35 (n), 0.45 (p)	6.5	3.6	±0.4	[41]
0.25	7	2.5	±0.4	[48]
0.15	4	1.5	0.45 (n only)	[35]

Table 2.2 Key parameters for some recently reported MOS processes.

Although the data shown in Table 2.2 comes from experimental processes, it is clear that power supply voltages will continue to reduce significantly and that some reduction in threshold voltages will be seen.

Aside from process shrinkage, the use of the mixed CMOS and bipolar technology, BiCMOS, is emerging as an alternative to CMOS for demanding applications [49]. The combination of the CMOS and bipolar technologies offers advantages over both pure CMOS and bipolar technologies. For CMOS technology, the addition of a bipolar device brings significant speed advantages when large loads have to be driven, such as in gate arrays, bus drivers and interfaces [50]. For bipolar, the addition of CMOS devices allows increased packing density and reduced power consumption without sacrificing speed [51].

The main disadvantage of BiCMOS is that the extra process steps required to form the bipolar transistor inevitably increase the cost. A further disadvantage is emerging as deep-submicron processes are developed; that is that the V_{be} required to turn the bipolar device on does not scale. Consequently it has been found that conventional BiCMOS logic gates loose their advantage over CMOS gates at low supply voltages [52]. However, new circuit techniques and scaling rules are being investigated to try and overcome this problem [52,53,54].

To the analogue interface designer, BiCMOS brings an added degree of design freedom and the potential of exploiting the bipolar device to obtain higher speed than is possible with a pure CMOS technology. It also brings the challenge of using the technology effectively, particularly at low voltages.

The discussion of processing trends has so far concentrated on the requirements of digital circuits and the subsequent trends in processing technology. The direct impact of process scaling on analogue sampled-data circuit parameters is now considered.

Parameter	CE Scaling	Generalised Scaling		
		Full Scaling	Constant Area	
			Const. Power	Const. Current
Linear dimensions (W, L)	$1/k$	$1/\lambda$	$1/\lambda$	$1/\lambda$
Voltages (V_D, V_S, V_G)	$1/k$	$1/\kappa$	$1/\kappa$	$1/\kappa$
Currents (I_{DS})	$1/k$	λ/κ^2	λ^3/κ^2	1
Capacitances (AC_{ox}, AC_j)	$1/k$	$1/\lambda$	λ	λ
Transconductance (g_m)	1	λ/κ	λ^3/κ	$\sqrt{\lambda^3}$
Max. transconductance $(g_{m(m)})$	$1/k$	λ/κ^2	λ^3/κ^2	1
Dynamic range (DR)	$1/k$	$\sqrt{\lambda/\kappa^2}$	$\sqrt{\lambda^3/\kappa^3}$	$\lambda^{3/4}/\kappa$
Max. dynamic range $(DR_{(m)})$	$1/\sqrt{k^3}$	$\sqrt{\lambda}/\kappa^2$	$\sqrt{\lambda^3}/\kappa^2$	$1/\kappa$
Surface mismatch [1] (σ_{ms})	$1/k$	$1/\lambda$	1	1
Edge Mismatch [2] (σ_{me})	$\sqrt{k}$	$\sqrt{\lambda}$	$1/\lambda$	$1/\lambda$
1/f Noise [3] $(V_{1/f})$	$k^{1-m/2}$	$\lambda^{1-m/2}$	$\lambda^{-m/2}$	$\lambda^{-m/2}$
Charge injection[4] $(\Delta V/V_D)$	1	1	κ/λ^2	κ/λ^2

Table 2.3 Impact of process scaling on analogue circuit parameters.

---

[1] Surface mismatch due to random parameter fluctuations over the area of the device.
[2] Edge mismatch due to lithographic accuracy.
[3] m is a constant ranging between 1 and 2.
[4] Values for constant area assume operation at constant speed.

In 1990 *Vittoz* considered the impact of CE scaling on analogue circuit parameters [55]. However, CE scaling does not accurately reflect current processing trends and so in Table 2.3 the work of *Vittoz* is extended to include the generalised scaling theory of Table 2.1 [38]. The values shown in Table 2.3 have been calculated according to the simplifying approximations derived in [55]. The second column shows the impact of CE scaling, which reduces the area of the analogue circuits by k^2 but also degrades most of the key parameters. The third column shows the effects of applying generalised scaling, where linear dimensions and voltages are scaled by different factors λ and κ, respectively. As processes are scaled to the deep-submicron region, voltages will be scaled down, but not by as much as CE scaling would require, thus $\lambda > \kappa$. Using this it can be seen that most of the key parameters are still degraded by scaling, but not by as much as when CE scaling is applied.

Clearly, full digital process scaling degrades analogue circuit parameters. However, analogue interface circuits are only expected to occupy a small proportion of the total silicon area and so in many instances it will be acceptable to keep the area of the analogue circuits constant. The last two columns of the table show the effect of scaling when the area is kept constant and either the power consumption or currents are also constant. If constant currents are maintained, most of the key parameters are improved, the notable exception being dynamic range. However, if constant power consumption is acceptable the dynamic range can also be enhanced.

Under circumstances where it is acceptable to maintain analogue circuit area and power consumption, whilst digital circuits are scaled, key analogue performance parameters can actually be improved by scaling.

Overall, enhanced analogue circuit performance on scaled digital CMOS processes is possible if appropriate circuit techniques are developed. On the negative side, the circuit designer can no longer rely on analogue process options to provide good quality passive components and must design circuits to operate with less voltage headroom.

2.5   Current-Mode Analogue Signal Processing

Current-mode analogue signal processing can loosely be defined as analogue signal processing in which current rather than voltage is the primary, although not necessarily exclusive, information-carrying medium.

In the previous section the trend towards ever higher levels of integration and of combining analogue and digital signal processing on a single silicon chip was discussed. The performance of the analogue interface circuits tends to set the dynamic range and speed limits of the overall system. This is because the small feature size of submicron processes makes possible the realisation of fast digital systems with almost arbitrarily long word lengths. On the other hand, traditional voltage based analogue interface circuits are limited by the available voltage swing and the need to drive load capacitances. In this environment, where capacitance dominates and low supply voltages must be used, the use of current-mode circuit techniques may be favourable from both dynamic range and speed viewpoints.

This quest for ever higher performance and the need for co-existence with digital processing technology has led to a considerable revival in current-mode analogue signal processing techniques [56]. At the present time there is considerable interest in the development of current-mode circuits because they promise to overcome some of the bottlenecks encountered by voltage-mode techniques when attempting to realise very high speed circuits in the digital environment. Some of the current-mode building blocks being developed include translinear circuits [56: Chapter 2], current conveyors [56: Chapters 3, 15], current-mode amplifiers [56: Chapters 4,16], transconductors [56: Chapters 5, 8, 9, 10, 12], dynamic current mirrors [56: Chapter 7], switched-current filters [56: Chapter 11], and neurons [56: Chapter 17]. However, some of these ideas, such as the current conveyor, are far from new [56: Chapter 3] but the process technology required to realise high quality devices has only recently become available [56: Chapter 15].

The wide range of applications in which current-mode techniques can be applied to advantage is documented in [56] and it can be seen that the application of current-mode techniques brings performance enhancements to many areas of analogue design.

In the field of general voltage domain analogue sampled-data signal processing it has been seen that the switched-capacitor technique dominates. However, the advancement of digital process technology will result in degraded switched-capacitor performance. The two major difficulties facing switched-capacitors are the loss of two layers of polysilicon and the continuing trend towards reduced power supply voltages.

Double polysilicon has traditionally been used by switched-capacitors to form high-quality linear floating capacitors. It is, of course, possible to realise these capacitors using the inter-metal capacitance. However, the specific capacitance is much lower than for double polysilicon capacitors and therefore more area is required and hence more costs incurred. An alternative technique has recently been proposed, by *Bermudez* [57], which allows most of the linear capacitors to be replaced by non-linear capacitors. Whilst this technique shows promise, it has yet to be fully proven by integrated results.

The reduction in power supply voltages also poses problems for switched-capacitors. Whilst it is possible to design low-voltage switched-capacitor filters [58,59], achieving high speed and dynamic range is difficult. Specifically, it is difficult to design low voltage high speed operational amplifiers. Furthermore, it has been shown [60] that the fundamental dynamic range limit of a switched-capacitor integrator can only be maintained with reduced voltages if the size of the integrating capacitor is increased with the square of the voltage reduction. Thus the area of high dynamic range switched-capacitor filters could become prohibitively large, particularly if double polysilicon capacitors are not available.

The concept of processing sampled currents instead of voltages offers to alleviate some of the difficulties that digital process scaling has created for switched-capacitors. The new generation of sampled current techniques are all based on the principle that, by storing the gate voltage of a field effect transistor, the current flowing through it can be memorised. The first application of this principle appears to have been that of storing the current generated by a photodiode and was proposed in 1972 by *Matsuzaki* and *Kondo* [61]. In 1987 *Bird* proposed the application of the sampled current principle to sampled current filters [62,63]. Then in 1988 *Daubert*, *Vallancourt* and *Tsividis* proposed the current copier [64] which was followed in 1989 by independent work for different applications, self-calibrating D/A converters by *Groeneveld*, *Schouwenaars* and *Termeer* [65], current copiers by *Vallancourt* and *Tsividis* [66], current-mode A/D converters by *Nairn* and *Salama* [67], dynamic current mirrors by *Wegmann* and *Vittoz* [68] and switched-current filters by *Hughes*, *Bird* and *Macbeth* [69].

This book develops the concept of switched-current circuits, discusses enhancements to the technique for improved performance, a number of promising areas of application and early work on CAD tools.

2.6 Summary

In this chapter the evolution of analogue sampled-data techniques has been reviewed. The switched-capacitor technique has been seen to dominate in general purpose applications, whilst CCDs are being used extensively in imaging arrays.

The emergence of mixed-mode analogue and digital signal processing and the subsequent need to integrate analogue sampled-data circuits on a digital VLSI process was identified. It was then seen that switched-capacitors are not fully compatible with digital VLSI technology and that advances in technology will degrade their performance. Finally, the use of switched-current circuits, as a digital process compatible analogue sampled-data signal processing technique, has been identified and this forms the basis of this text.

References

[1] J. M. L. Janssen, "Discontinuous low-frequency delay line with continuously variable delay," *Nature*, vol. 169, pp. 148-149, 26 Jan. 1952.

[2] R. S. Caruthers, "Copper oxide modulators in carrier telephone systems," *Bell Syst. Tech. J.*, vol. 18, pp. 315-337, April 1939.

[3] H. B. Hååd and C. G. Svala, "Means for detecting and/or generating pulses," *US Patent 2718621*, 20 Sept. 1955, First filed in Sweden 12 March 1952.

[4]   W. R. Bennett, "Steady-state transmission through networks containing periodically operated switches," *IRE Trans. Circuit Theory*, vol. CT-2, pp. 17-21, March 1955.

[5]   A. Fettweis, "Steady-state analysis of circuits containing a periodically-operated switch," *IRE Trans. Circuit Theory*, vol. CT-6, pp. 252-260, Sept. 1960.

[6]   G. Krause, "Analog-speicherkette: eine neuartige schaltung zum speichern und verzögern von signalen," *Electronics Letters*, vol. 3, pp. 544-546, Dec. 1967.

[7]   W. Kuntz, "A new sample-and-hold device and its application to the realization of digital filters," *Proc. IEEE*, vol. 56, pp. 2092-2093, Nov. 1968.

[8]   R. Boite and J. P. V. Thiran, "Synthesis of filters with capacitances, switches, and regenerating devices," *IEEE Trans. Circuit Theory*, vol. CT-15, pp. 447-454, Dec. 1968.

[9]   L. Baker, "Dynamic transfer networks," *US Patent No. 3469213*, Sept. 23 1969, Filed May 16 1967.

[10]   F. L. J. Sangster and K. Teer, "Bucket-brigade electronics - new possibilities for delay, time-axis conversion and scanning," *IEEE J. Solid-State Circuits*, vol. SC-4, pp. 131-136, June 1969.

[11]   F. L. J. Sangster, "The "bucket-brigade delay line", a shift register for analogue signals," *Philips Technical Review*, vol. 31, pp. 97-110, 1970.

[12]   W. S. Boyle and G. E. Smith, "Charge coupled semiconductor devices," *Bell Syst. Tech. J.*, vol. 49, pp. 587-593, April 1970.

[13]   R. Melen and D. Buss, Eds., *"Charge-Coupled Devices: Technology and Applications,"* New York: IEEE Press, 1977.

[14]   D. F. Barbe, "Imaging devices using the charge-coupled concept," *Proc. IEEE*, vol. 63, pp. 38-67, Jan. 1975.

[15]   R. W. Brodersen, C. R. Hewes and D. D. Buss, "A 500-stage CCD transversal filter for spectral analysis," *IEEE Trans. Electron Devices*, vol. ED-23, pp. 143-152, Feb. 1976.

[16]   L. M. Terman and L. G. Heller, "Overview of CCD memory," *IEEE Trans. Electron Devices*, vol. ED-23, pp. 72-78, Feb. 1976.

[17]   A. M. Chiang and M. L. Chuang, "A CCD programmable image processor and its neural network applications," *IEEE J. Solid State-Circuits*, vol. SC-26, pp. 1894-1901, Dec. 1991.

[18]   "16,384 Bit CCD Serial Memory," Intel Data Catalog, pp. 4-15-5-2, 1976.

[19]   J. C. Maxwell, *A Treatise on Electricity and Magnetism*, Oxford: Clarendon Press, pp. 420-425, 1891.

[20]   D. L. Fried, "Analog sample-data filters," *IEEE J. Solid-State Circuits*, vol. SC-7, pp. 302-304, Aug. 1972.

[21]   J. T. Caves, M. A. Copeland, C. F. Rahim and S. D. Rosenbaum, "Sampled analog filtering using switched capacitors as resistor equivalents," *IEEE J. Solid-State Circuits*, vol. SC-12, pp. 592-599, Dec. 1977.

[22] B. J. Hosticka, R. W. Brodersen and P. R. Gray, "MOS sampled data recursive filters using switched capacitor integrators," *IEEE J. Solid-State Circuits*, vol. SC-12, pp. 600-608, Dec. 1977.

[23] J. L. McCreary and P. R. Gray, "All-MOS charge redistribution analog-to-digital conversion techniques-part 1," *IEEE J. Solid-State Circuits*, vol. SC-10, pp. 371-379, Dec. 1975.

[24] G. S. Moschytz, Ed., *"MOS Switched-Capacitor Filters: Analysis and Design,"* New York: IEEE Press, 1984.

[25] M. S. Ghausi and K. R. Laker, *"Modern Filter Design: Active RC and Switched Capacitor,"* Englewood Cliffs, NJ: Prentice-Hall Inc., 1981.

[26] R. Gregorian and G. C. Temes, *"Analog MOS Integrated Circuits for Signal Processing,"* New York: John Wiley & Sons, 1986.

[27] P. R. Gray, D. Senderowicz, H. Ohara and B. M. Warren, "A single-chip NMOS dual channel filter for PCM telephony applications," *IEEE J. Solid-State Circuits*, vol. SC-14, pp. 981-991, Dec. 1979.

[28] J. H. Fischer et al, "Line and receiver interface circuit for high-speed voice-band modems," *IEEE J. Solid-State Circuits*, vol. SC-22, pp. 982-989, Dec. 1987.

[29] D. G. Haigh, C. Toumazou, S. J. Harrold, K. Steptoe, J. I. Sewell and R. Bayruns, "Design optimization and testing of a GaAs switched-capacitor filter," *IEEE Trans. Circuits Syst.*, vol. 38, pp. 825-837, Aug. 1991.

[30] B. Nauta, "A CMOS transconductance-C filter technique for very high frequencies," *IEEE J. Solid-State Circuits*, vol. SC-27, pp. 142-153, Feb. 1992.

[31] W. M. Snelgrove and A. Shoval, "A balanced 0.9-μm transconductance-C filter tunable over the VHF range," *IEEE J. Solid-State Circuits*, vol. SC-27, March 1992.

[32] R. W. Gregor, K. J. O'Brien, G. R. Wesley, W. H. Stinebaugh Jr., H. Chew and C. W. Leung, "A submicron analog CMOS technology," *Proc. IEEE Custom Integrated Circuits Conf.*, pp. 18.5.1-18.5.4, 1989.

[33] Y. Nakagome et al, "An experimental 1.5V 64-Mb DRAM," *IEEE J. Solid-State Circuits*, vol. SC-26, pp. 465-472, April 1991.

[34] M. Aoki et al, "Design and performance of 0.1μm CMOS devices using low-impurity-channel transistors (LICT's)," *IEEE Electron Device Letters*, vol. 13, pp. 50-52, Jan. 1992.

[35] R.-H. Yan et al, "89-GHz f_T room temperature silicon MOSFET's," *IEEE Electron Device Letters*, vol. 13, pp. 256-258, May 1992.

[36] E. O. Johnson, "Physical limitations on frequency and power parameters of transistors," *RCA Review*, vol. 26, pp. 163-177, June 1965.

[37] M. Nagata, "Limitations, innovations, and challenges of circuits and devices into a half micrometer and beyond," *IEEE J. of Solid-State Circuits*, vol. SC-27, pp. 465-472, April 1992.

[38] G. Baccarani, M. R. Wordeman and R. H. Dennard, "Generalized scaling theory and its application to a 1/4 micrometer MOSFET design," *IEEE Trans. Electron Devices*, vol. ED-31, pp. 452-461, April 1984.

[39] Solid State Products Engineering Council, "*JEDEC standard no. 8, Standard for reduced operating voltages and interface levels for integrated circuits,*" Electronic Industries Association, Washington D.C., Dec. 1984.

[40] R. H. Dennard, "Power-supply considerations for future scaled CMOS systems," *Proc. IEEE Symp. VLSI Tech. Syst. Applications*, pp. 188-192, 1989.

[41] Y. Nakagome et al, "Circuit techniques for 1.5-3.6V battery-operated 64Mb DRAM," *IEEE J. Solid-State Circuits*, vol. SC-26, pp. 1003-1010, July 1991.

[42] A. P. Chandrakasan, S. Sheng and R. W. Brodersen, " Low power CMOS digital design," *IEEE J. of Solid-State Circuits*, vol. SC-27, pp. 473-484, April 1992.

[43] J. E. Chung, M.-C. Jeng, J. E. Poon, P.-K. Ko and C. Hu, "Performance and reliability design issues for deep-submicrometer MOSFET's," *IEEE Trans. Electron Devices*, vol. 38, pp. 545-554, March 1991.

[44] M. Kakumu, M. Kinugawa and K. Hashimoto, "Choice of power-supply voltage for half-micrometer and lower submicrometer CMOS devices," *IEEE Trans. Electron Devices*, vol. 37, pp. 1334-1342, May 1990.

[45] R. Izawa, T. Kure and E. Takeda, "Impact of the gate-drain overlapped device (GOLD) for deep submicrometer VLSI, " *IEEE Trans. Electron Devices*, vol. 35, pp. 2088-2093, Dec. 1988.

[46] D. Hisamoto, T. Kaga, Y. Kawamoto and E. Takeda, "A fully depleted lean-channel transistor (DELTA) - A novel vertical ultrathin SOI MOSFET," *IEEE Electron Device Letters*, vol. 11, pp. 36-38, Jan. 1990.

[47] M. Kakumu and M. Kinugawa, "Power-supply voltage impact on circuit performance for half and lower submicrometer CMOS LSI," *IEEE Trans. Electron Devices*, vol. 37, pp. 1902-1908, Aug. 1990.

[48] B. Davari et al, "A high performance 0.25μm CMOS technology," *Proc. IEEE International Electron Devices Meeting*, pp. 56-59, 1988.

[49] A. Iranmanesh and B. Bastani, "BiCMOS emerges for gate-arrays and memories," *IEEE Circuits and Devices Mag.*, vol. 8, pp. 14-17, March 1992.

[50] A. A. Iranmanesh, V. Iiderem, M. Biswal and B. Bastani, "A 0.8-μm advanced single-poly BiCMOS technology for high-density and high-performance applications," *IEEE J. Solid-State Circuits*, vol. SC-26, pp. 422-426, March 1991.

[51] K. Yamaguchi, "A 1.5-ns access time, 78-μm^2 memory-cell size, 64-kb ECL-CMOS SRAM, " *IEEE J. Solid-State Circuits*, vol. SC-27, pp. 167-174, Feb. 1992.

[52] H. Momose, Y. Unno and T. Maeda, "Supply voltage design tradeoffs between speed and MOSFET reliability of half-micrometer BiCMOS gates," *IEEE Trans. Electron Devices*, vol. 38, pp. 566-572, March 1991.

[53] C.-L. Chen, "2.5-V bipolar/CMOS circuits for 0.25-μm BiCMOS technology," *IEEE J. Solid-State Circuits*, vol. 27, pp. 485-491, April 1992.

[54] G. P. Rosseel and R. W. Dutton, "Scaling rules for bipolar transistors in BiCMOS circuits," *Proc. IEEE International Electron Devices Meeting*, pp. 33.2.1-33.2.4, 1989.

[55] E. A. Vittoz, "Future of analog in the VLSI environment," *Proc. IEEE International Symp. Circuits and Syst.*, pp. 1372-1375, 1990.

[56] C. Toumazou, F. J. Lidgey and D. G. Haigh, Eds., "*Analogue IC design: the current-mode approach,*" London: Peter Peregrinus Ltd., 1990.

[57] J. C. M. Bermudez, M. C. Schneider and C. G. Montoro, "Linearity of switched capacitor filters employing nonlinear capacitors," *Proc. IEEE International Symp. Circuits and Syst.*, pp. 1211-1214, 1992.

[58] T. Adachi, A. Ishikawa, A. Barlow and K. Takasuka, "A 1.4V switched capacitor filter," *Proc. IEEE Custom Integrated Circuits Conf.*, pp. 8.2.1-8.2.4, 1990.

[59] R. Castello and L. Tomasini, "1.5V high-performance SC filters in BiCMOS technology," *IEEE J. Solid-State Circuits*, vol. SC-26, pp. 930-936, July 1991.

[60] R. Castello and P. R. Gray, "Performance limitations in switched-capacitor filters," *IEEE Trans. Circuits Syst.*, vol. CAS-32, pp. 865-876, Sept. 1985.

[61] S. Matsuzaki and I. Kondo, "Information holding apparatus," *UK Patent 1359105*, 10 July 1974, Filed 6 July 1972.

[62] N. C. Bird, "Storing sampled analogue electrical currents," *UK Patent 2209895*, 24 May 1989, Filed 16 Sept. 1987.

[63] N. C. Bird, "A method of and a circuit arrangement for processing sampled analogue electrical signals," *UK Patent 2213011*, 2 Aug. 1989, Filed 16 Sept. 1987.

[64] S. J. Daubert, D. Vallancourt and Y. P. Tsividis, "Current copier cells," *Electronics Letters*, vol. 24, pp. 1560-1562, 8 Dec. 1988.

[65] W. Groeneveld, H. Schouwenaars and H. Termeer, "A self-calibration technique for high resolution D/A converters," *Proc. IEEE International Solid-State Circuits Conf.*, pp. 22-23, 1989.

[66] D. Vallancourt and Y. P. Tsividis, "Sampled current circuits," *Proc. IEEE International Symp. Circuits Syst.*, pp. 1592-1595, May 1989.

[67] D. G. Nairn and C. A. T. Salama, "Current mode analogue to digital converters," *Proc. IEEE International Symp. Circuits Syst.*, pp. 1588-1591, May 1989.

[68] G. Wegmann and E. A. Vittoz, " Very accurate dynamic current mirrors," *Electronics Letters*, vol. 25, pp. 644-646, 11 May 1989.

[69] J. B. Hughes, N. C. Bird and I. C. Macbeth, "Switched-currents - A new technique for analog sampled-data signal processing," *Proc. IEEE International Symp. Circuits Syst.*, pp. 1584-1587, May 1989.

Chapter 3

Switched-Current Architectures and Algorithms

John B. Hughes, Neil C. Bird and Ian C. Macbeth

3.1   Introduction

The switched-current technique is a current-mode signal processing technique which utilises the ability of an MOS transistor to maintain its drain current, when its gate is open-circuited, through the charge stored on its gate oxide capacitance. Although early attempts to exploit this property were made as early as 1972 [1], it was not until the late 1980s that it was revived independently by a number of researchers [2,3,4,5,6]. Of these, switched-currents [6] aimed to provide a technique for sampled data filtering, like switched-capacitors, but requiring only baseline VLSI CMOS processing.

Initially, switched-currents was described in terms of so-called 'first generation' circuit modules [7] which included delay cells and integrators for use in state-variable and active ladder filters. The modules were based on a memory cell developed from the simple current mirror but suffered inevitable errors resulting from transistor mismatch. Although proving suitable only for low-Q filters, it nevertheless established the technique and its architectures which are still used today. The introduction of the current-copier [8] enabled this shortcoming to be overcome and the so-called 'second generation' switched-current circuits were developed [9,10,11]. It is this generation of circuits which is the subject of this chapter.

In this chapter, the scene is set with a very simple review of switched-capacitor circuits highlighting the principal differences with a switched-current system. The memory cell is described and then used to develop delay lines, integrators and differentiators as filter building blocks. Switched-current filter synthesis is demonstrated with the design of a low-pass filter based on cascaded biquadratic sections (see Chapters 8,9,10,11 for other filter synthesis approaches).

3.2   Switched-Capacitor Background

A good starting point for introducing a technique such as switched-currents is to review the state-of-the-art alternatives. In the case of a mature technique such as switched-capacitors a complete review is not needed as many books and review papers are available [12,13,14]. The intention here is merely to highlight the major features of the technique so that switched-currents may be viewed from the right perspective. Most switched-capacitor filter structures have resulted from the substitution of an active-RC filter's continuous-time integrators by switched-capacitor counterparts. This approach has been applied to state-variable filters [15,16] and to filters which simulate the nodal voltages of lossless ladder prototypes [17]. Three of the principal switched-capacitor integrator building blocks are shown in Figure 3.1.

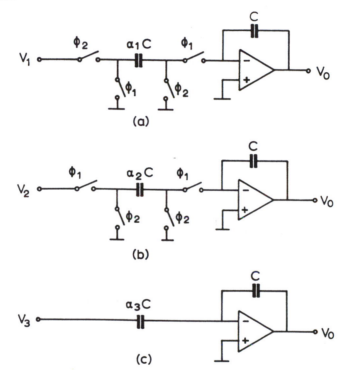

Figure 3.1 Switched-capacitor circuits (a) non-inverting lossless integrator (b) inverting lossless integrator (c) inverting amplifier.

3.2.1   *Non-inverting integrator*

The non-inverting integrator (Figure 3.1(a)) operates as follows. On phase ϕ_2 of the non-overlapping clock period $(n-1)$, the charge on capacitor C holds the output voltage at $v_0(n-1)$ while capacitor α_1C is charged to $v_1(n-1)$. The next clock phase is ϕ_1 of period (n), and capacitor α_1C is discharged into capacitor C causing the output voltage to charge to $v_0(n)$. It is easily shown that

$$v_0(n) = v_0(n-1) + \alpha_1 v_1(n-1) \qquad (3.1)$$

which gives the z-domain transfer function

$$H_1(z) = \frac{v_0(z)}{v_1(z)} = \frac{\alpha_1 z^{-1}}{1 - z^{-1}} \tag{3.2}$$

This is the Forward Euler z-transform $(s \rightarrow (1 - z^{-1})/Tz^{-1})$ of a non-inverting continuous-time lossless integrator $(H(s) = 1/sRC)$ where $\alpha_1 = T/RC$.

3.2.2 Inverting integrator

The inverting integrator shown in Figure 3.1(b) operates as follows. On clock phase ϕ_2 of period $(n-1)$, capacitor C holds the output voltage at $v_0(n-1)$ while capacitor $\alpha_2 C$ is discharged. On the next phase ϕ_1 of period (n), capacitor $\alpha_2 C$ is charged to $v_2(n)$ and capacitor C charges to $v_0(n)$. It is easily shown that

$$v_0(n) = v_0(n-1) - \alpha_2 v_2(n) \tag{3.3}$$

which gives the z-domain transfer function

$$H_2(z) = \frac{v_0(z)}{v_2(z)} = -\frac{\alpha_2}{1 - z^{-1}} \tag{3.4}$$

This is the Backward Euler z-transform $(s \rightarrow (1 - z^{-1})/T)$ of an inverting integrator $(H(s) = -1/sRC)$ where $\alpha_2 = T/RC$.

3.2.3 Inverting amplifier

The inverting amplifier shown in Figure 3.1(c) operates as follows. On clock phase ϕ_2 of period $(n-1)$, capacitor C holds the output at $v_0(n-1)$ and capacitor $\alpha_3 C$ charges to $v_3(n-1)$. On the next ϕ_1 clock phase of period (n), capacitor $\alpha_3 C$ charges to $v_3(n)$ and in so doing transfers a charge of magnitude $(v_3(n) - v_3(n-1))\alpha_3 C$ onto capacitor C causing the output voltage to change to $v_0(n)$. It is easily shown that

$$v_0(n) = v_0(n-1) - \alpha_3[v_3(n) - v_3(n-1)] \tag{3.5}$$

which gives the z-domain transfer function

$$H_3(z) = \frac{v_0(z)}{v_3(z)} = -\alpha_3 \tag{3.6}$$

This amplification is the result of delivering a current, proportional to the derivative of the input signal, to the integrator loop formed by capacitor C and the op-amp. It should be noted that this amplifier would respond to continuous-time inputs and so sample and held signals would normally be supplied. All the circuits shown in Figure 3.1 have infinite DC gain, as indeed has the continuous-time integrator, and would never be used in isolation.

3.2.4 Generalised integrator

The virtual earth of the op-amp is a current summing node and this enables any number of the input branches of the circuits of Figure 3.1 to be combined.

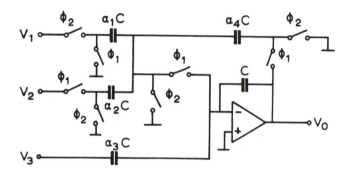

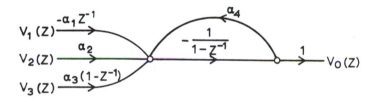

Figure 3.2 Switched-capacitor generalised integrator
(a) circuit topology (b) z-domain SFG

This is shown in the generalised integrator of Figure 3.2(a). In this arrangement, the three input branches previously described are combined with a fourth branch $(\alpha_4 C)$ which is driven from the output, v_0, having the same switch phasing as the second branch $(\alpha_2 C)$. By superposition,

$$v_0(n) = v_0(n-1) + \alpha_1 v_1(n-1) - \alpha_2 v_2(n) - \alpha_3[v_3(n) - v_3(n-1)] - \alpha_4 v_0(n) \quad (3.7)$$

which, after some manipulation and z-transformation, gives

$$v_0(z) = \frac{\dfrac{\alpha_1}{1+\alpha_4} z^{-1}}{1 - \dfrac{1}{1+\alpha_4} z^{-1}} v_1(z) - \frac{\dfrac{\alpha_2}{1+\alpha_4}}{1 - \dfrac{1}{1+\alpha_4} z^{-1}} v_2(z) - \frac{\dfrac{\alpha_3}{1+\alpha_4}(1 - z^{-1})}{1 - \dfrac{1}{1+\alpha_4} z^{-1}} v_3(z) \quad (3.8)$$

The signal flow graph (SFG) is shown in Figure 3.2(b). Clearly, the extra feedback branch $(\alpha_4 C)$ has introduced damping into the integrator. The generalised integrator is a frequently used building block for even-order state-variable filters. Each biquadratic section comprises a combination of lossless or damped integrators in a feedback loop, and can be made to execute the bilinear z-transform $(s \rightarrow 2(1 - z^{-1})/(1 + z^{-1})T)$ which enables 'exact' filter design.

3.2.5 Subtracting integrator

A second useful building block is the subtracting integrator shown in Figure 3.3(a). This operates as follows. On clock phase ϕ_2 of period $(n-1)$, capacitor $\alpha_1 C$ is charged to $v_1(n-1)$ while capacitor C holds the output voltage at $v_0(n-1)$. On phase ϕ_1 of the next clock period (n), capacitor $\alpha_1 C$ charges to $v_2(n)$ and in so doing transfers a charge with magnitude $\alpha_1 C(v_1(n-1) - v_2(n))$ to capacitor C. The branch $\alpha_3 C$ produces damping as described earlier. The resulting output voltage is given by

$$v_0(n) = v_0(n-1) + \alpha_1[v_1(n-1) - v_2(n)] - \alpha_3[v_3(n) - v_3(n-1)] \quad (3.9)$$

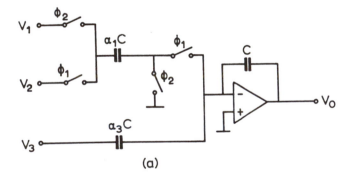

(a)

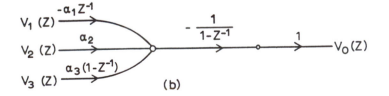

(b)

Figure 3.3 Switched-capacitor subtracting integrator
(a) circuit topology (b) z-domain SFG

which, after z-transformation, gives

$$v_0(z) = \frac{\alpha_1}{1 - z^{-1}} \left[v_1(z)z^{-1} - v_2(z) \right] - \alpha_3 v_3(z) \tag{3.10}$$

This is expressed in the SFG shown in Figure 3.3(b). The circuit may be regarded as a special case of the generalised integrator of Figure 3.2(a) with $\alpha_4 = 0$ and $\alpha_2 = 1$ but avoiding mismatch errors resulting from separate branches ($\alpha_1 C$ and $\alpha_2 C$). It finds particular use in filters designed by simulation of lossless ladder prototypes.

3.2.6 Switched-capacitor characteristics

In summary, the switched-capacitor technique may be characterised as follows:

- Switched-capacitor systems are usually made by substitution of switched-capacitor integrators in well-established active-RC circuit structures. This enables them to retain their modularity and low sensitivity to component spreads.

- The integrators execute algorithms, defined by their difference equations, involving the manipulation of past and present voltage samples. Low supply voltage operation necessarily implies degraded performance.

- Manipulation of voltage samples is accomplished by transferring charge between floating capacitors. For linear operation these capacitors should also be linear (although some recent work [18] indicates that non-linear capacitors may be used internally) and so switched-capacitor circuits are not totally compatible with digital VLSI processes.

- Building blocks have been defined which are versatile (generalised integrator), self-contained (only capacitors within the integrator require critical matching) and which have low design interaction (loading of a module by other modules does not significantly change the module's performance). These are necessary attributes for hierarchical design leading to design-automation.

3.3 Switched-Current Systems

A switched-current system may be defined as a system using analogue sampled-data circuits in which signals are represented by current samples. This is in contrast with switched-capacitor circuits in which signals are represented by voltage samples. The applications for switched-current systems will be much the same as for switched-capacitors viz. filters, A/D and D/A converters, general signal processing etc. but it is a prime aim that switched-current circuits should be implemented using a standard VLSI CMOS process. Linear floating capacitors are not needed in switched-current circuits and, in principle if not in practice, voltage swings need not be large as signals are represented by currents.

To be clear on this last point, any circuit which handles signal currents must, of course, develop internal voltage swings. However, the voltages developed within switched-current circuits need be neither large nor linear for correct algorithmic operation and this gives switched-currents the potential for low voltage operation. It is practical considerations (precision, dynamic range, linearity) which dictate the use of larger internal voltage swings and these aspects are covered in Chapter 4. For the moment, switched-current circuits will be treated as ideal circuits which perform algorithms with signal currents.

A switched-current counterpart of the generalised switched-capacitor integrator shown in Figure 3.2(a) would have the following difference equation:

$$i_0(n) = i_0(n-1) + \alpha_1 i_1(n-1) - \alpha_2 i_2(n) - \alpha_3 [i_3(n) - i_3(n-1)] - \alpha_4 i_0(n) \quad (3.11)$$

Taking z-transforms gives

$$i_0(z) = \frac{\dfrac{\alpha_1}{1+\alpha_4} z^{-1}}{1 - \dfrac{1}{1+\alpha_4} z^{-1}} i_1(z) - \frac{\dfrac{\alpha_2}{1+\alpha_4}}{1 - \dfrac{1}{1+\alpha_4} z^{-1}} i_2(z) - \frac{\dfrac{\alpha_3}{1+\alpha_4} (1 - z^{-1})}{1 - \dfrac{1}{1+\alpha_4} z^{-1}} i_3(z) \quad (3.12)$$

A switched-current generalised integrator described by the algorithm above could be used directly to make switched-current systems employing structures which are duals of those already used in active-RC and switched-capacitor systems. Such a switched-current system should inherit the desirable qualities of switched-capacitor systems (modularity, component insensitivity), just as switched-capacitors did from earlier active-RC circuits.

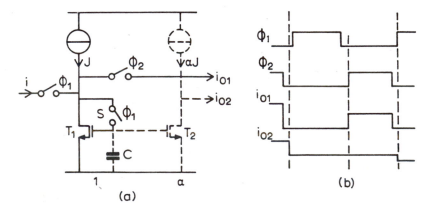

Figure 3.4 Single transistor current memory cell
(a) circuit topology (b) clock waveforms.

A switched-current circuit capable of exercising the above algorithm ((3.11) and (3.12)) must possess the following attributes: it must be capable of algebraic manipulation of currents: summation (or subtraction) can be simply achieved in current mode circuits by feeding the currents (or their inverse) into a low impedance circuit node; scaling of currents (i.e. multiplication by a fixed coefficient) can be achieved by current mirror action. It must be capable of "memorising" currents. This required a new technique and this is explained in the next section.

3.4   Delay Modules

The basis of the delay modules is the single-transistor current memory or current copier [8].

3.4.1   *Current memory cell*

The arrangement in Figure 3.4(a) can achieve current memory within a single transistor, T_1. It is driven by the clock waveforms shown in Figure 3.4(b) and operates as follows. On phase ϕ_1, switch S closes as does the switch supplying input current, i. This input current adds to the bias current J and the current $J+i$ flows initially into the discharged gate-source capacitor C. As C charges, the gate-source voltage V_{gs} rises and when it exceeds the threshold voltage, T_1 conducts. Eventually, when C is fully charged, the whole of current $J+i$ flows in the drain of T_1.

During phase ϕ_2, switch S is opened and the value of V_{gs} at the end of phase ϕ_1 is held on capacitor C and sustains current $J+i$ in the drain of T_1. With the input switch now open and the output switch closed, the imbalance between the bias current, J, and the drain current, $J+i$, forces an output current, $i_{01} = -i$, to flow throughout phase ϕ_2. The output current, i_{01}, is a memory of the input current i. Of course, this is achieved by virtue of the charge retained on capacitor C. The current memory operates without the need for linear floating capacitors and so retains the important advantage of being compatible with digital VLSI processes.

It is noted that the output current, i_{01}, is not available during phase ϕ_1. When an output current is needed throughout the whole clock period this may be achieved by adding transistor T_2 and its associated bias current. Output current, i_{02}, flows during the sampling phase (ϕ_1) and the holding phase (ϕ_2) by current mirror action. If the aspect ratio (*W/L*) of T_2 is α times that of T_1, then $i_{02} = -\alpha i$.

In practice, the ϕ_1 drive to switch S needs to be slightly in advance of that driving the input switch to ensure that "memorising" of the input sample has occurred before the input current is interrupted. Also, the memory transistor, T_1, and its switch, S, (also an MOS transistor) are non-ideal and this leads to analogue errors. These considerations are covered in Chapter 4 and for the rest of this chapter ideal transistor operation is assumed.

3.4.2 Delay cell

Next, consider the delay cell shown in Figure 3.5. It comprises two cascaded current memories and an optional extra output stage for a full clock period output signal. It operates as follows.

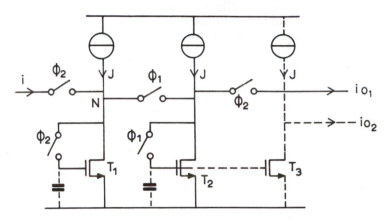

Figure 3.5 Delay cell.

On phase ϕ_2 of a clock period, arbitrarily called period *(n-1)*, the input signal current *i(n-1)* sums with the first bias current in transistor T_1. On the next phase, ϕ_1, which falls in the next clock period *(n)*, T_1 holds current *J+i(n-1)* and the output current *-i(n-1)* is sampled in the second current memory *(T$_2$)*. On the next phase ϕ_2 (still in clock period *(n)*), T_2 holds current *J-i(n-1)* and the output current $i_{01}(n) = i(n-1)$. The optional output, i_{02}, maintains the value *i(n-1)* throughout phases ϕ_1 and ϕ_2 of clock period *(n)*.

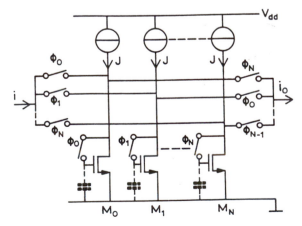

Figure 3.6 Delay line.

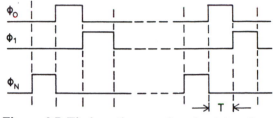

Figure 3.7 Timing diagram for the delay line.

3.4.3 Delay line

A delay line could be generated by cascading delay cells but this would increase the transmission error and degrade analogue performance. A better approach is shown in Figure 3.6. The circuit is an array of $N+1$ paralleled memory cells and is operated from the clock sequence shown in Figure 3.7. On clock phase ϕ_0, memory cell M_0 receives an input signal while cell M_1 delivers its output. Similarly, on clock phase ϕ_1, cell M_1 receives input signal while cell M_2 delivers its output. This continues until after cell M_N has received its input signal and cell M_0 has delivered its output signal, and then the cycle repeats. Clearly, each cell delivers its output signal current immediately before its next input and N periods (NT) after its previous output phase. With $N=1$, the delay line is an inverting unit delay cell or, with a continuous input signal, it is a sample and hold circuit. Note that each memory cell is neither receiving nor delivering signals for N-1 clock phases of the cycle. At these times the voltages at the drains of the memory transistors change to force a match between each bias current (in practice, generated by a PMOS transistor) and the current held in its associated memory transistor. The z-domain transfer characteristic is given by

$$H(z) = -z^{-N} \qquad (3.13)$$

3.5 Integrator Modules

3.5.1 Non-inverting integrator

Figure 3.8(a) shows the delay cell with output current i_f (equivalent to i_{01} in Figure 3.5) fed back to the input summing node. As with the delay cell, $i_f(n) = i(n-1)$, and the fed back signal $i_f(n)$ sums with input sample $i(n)$. The two current memories form a feedback loop which captures input samples and adds them sequentially to the running sum of samples stored in the loop. Intuitively, the circuit is operating as a lossless integrator. The circuit can be simplified by inspection to that shown in Figure 3.8(b) because the ϕ_1 and ϕ_2 switches in the integrator loop are in parallel and may be replaced by a short circuit.

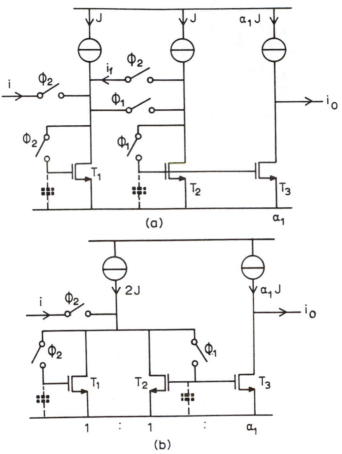

Figure 3.8 Non-inverting lossless integrator
(a) delay cell with feedback (b) simplified structure.

Formally, the transfer characteristic of the integrator (Figure 3.8(b)) can be derived as follows. On phase ϕ_2 of clock period $(n\text{-}1)$, transistor T_1 is diode-connected and receives current from the input, i(n-1), from the bias, $2J$, and from T_2, $-(J - i_0(n\text{-}1)/\alpha_1)$. So, the current in T_1 is I_1 where

$$I_1 = J + i(n\text{-}1) + \frac{i_0(n\text{-}1)}{\alpha_1} \tag{3.14}$$

On the next phase ϕ_1 of clock period (n), transistor T_2 is diode-connected and passes current I_2 where

$$I_2 = 2J - I_1 = J - i(n - 1) - \frac{i_0(n\text{-}1)}{\alpha_1} \tag{3.15}$$

$$i_0(n) = \alpha_1(J - I_1) = i_0(n\text{-}1) + \alpha_1 i(n\text{-}1) \tag{3.16}$$

Taking z-transforms gives

$$H(z) = \frac{\alpha_1 z^{-1}}{1 - z^{-1}} \tag{3.17}$$

The exact performance with physical frequencies is found by multiplying numerator and denominator by $z^{1/2}$ and substituting $z = e^{j\omega T}$. After some algebraic manipulation it can be shown that the transfer function for the non-inverting lossless integrator, $H(e^{j\omega T})$, is given by

$$H(e^{j\omega T}) = \left[\frac{\alpha_1 / T}{j\omega} \right] \left[\frac{\frac{\omega T}{2}}{sin\frac{\omega T}{2}} \right] e^{-j\omega T/2} \tag{3.18}$$

Equation (3.18) has three terms: the first term corresponds to the frequency response of a lossless continuous-time non-inverting integrator with a gain-constant α_1 / T; the second term, $(\omega T/2)/sin(\omega T/2)$, is the deviation of the sampled-data integrator from the ideal response; and the third term, $e^{-\omega T/2}$, is an excess phase lag.

3.5.2 Non-inverting damped integrator

This integrator may be damped as shown in Figure 3.9. It contains an extra feedback stage (T_4 and current source) which is weighted α_4, and the output stage is weighted α_1. On phase ϕ_1, T_2 and T_4 are connected in parallel and the current they receive is shared between them. On phase ϕ_2, these currents are stored but only that current stored in T_2 is fed-back to the summing node. Formally, the transfer characteristic can be found as follows. On phase ϕ_2 of clock period $(n\text{-}1)$, transistor T_1 is diode-connected and passes current I_1 where

$$I_1 = J + i(n\text{-}1) + \frac{i_0(n\text{-}1)}{\alpha_1} \tag{3.19}$$

On the next phase, ϕ_1 of clock period (n), transistors T_2 and T_4 are diode-connected in parallel, and the current in T_2 is I_2 where

$$I_2 = [(2 + \alpha_4)J - I_1] \frac{1}{1 + \alpha_4} \tag{3.20}$$

$$= J - \frac{i(n\text{-}1)}{1 + \alpha_4} - \frac{i_0(n\text{-}1)}{\alpha_1(1 + \alpha_4)} \tag{3.21}$$

$$i_0(n) = \alpha_1(J - I_1)$$

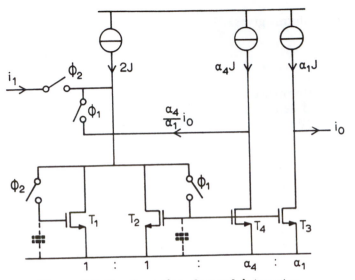

Figure 3.9 Non-inverting damped integrator.

$$= \frac{\alpha_1}{1 + \alpha_4} i(n\text{-}1) + \frac{i_0(n\text{-}1)}{1 + \alpha_4} \tag{3.22}$$

Taking z-transforms and rearranging gives

$$H_1(z) = \frac{i_0(z)}{i(z)} = \frac{\dfrac{\alpha_1}{1 + \alpha_4} z^{-1}}{1 - \dfrac{1}{1 + \alpha_4} z^{-1}} \tag{3.23}$$

The performance with physical frequencies, following the earlier procedure for $\omega T \ll 1$ is given by

$$H_1(e^{j\omega T}) = \frac{\left(\dfrac{\alpha_1}{\alpha_4}\right) e^{-j\omega T/2}}{1 + j \left[\dfrac{2}{T} \dfrac{\omega}{2 + \alpha_4} \right]} \tag{3.24}$$

This is clearly equivalent to a damped continuous time integrator response except for the excess phase lag term $e^{-j\omega T/2}$. The low-frequency gain a_0 and the cut-off frequency ω_0 are given by

$$a_0 = \frac{\alpha_1}{\alpha_4} \tag{3.25}$$

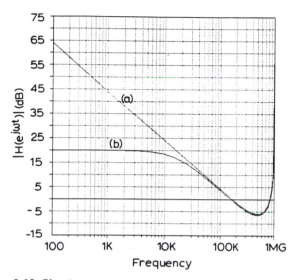

Figure 3.10 Simulated response of a non-inverting integrator
(a) lossless ($\alpha_1 = 1$, $\alpha_4 = 0$) (b) damped ($\alpha_1 = 1$, $\alpha_4 = 0.1$).

$$\omega_0 = \frac{2}{T}\left(\frac{\alpha_4}{2 + \alpha_4}\right)$$
(3.26)

Figure 3.10 shows the frequency response of a non-inverting integrator simulated using the CAD package SCALP2. The clock frequency was 1MHz ($T = 1\mu s$) and the integrator model was ideal, i.e. the parasitics were negligible. The deviation from the ideal response at high frequency is shown. The effect of introducing damping ($\alpha_4 = 0.1$) is to limit the low-frequency gain to $a_0 = 10$ and produce a -3dB cut-off frequency $f_0 = 15.15$kHz in agreement with (3.25) and (3.26).

3.5.3 *Sensitivity*

The sensitivities of a_0 and ω_0 to variations in the coefficient α_4 are given by

$$S^{a_0}_{\alpha_4} = \frac{\alpha_4}{a_0}\frac{\partial a_0}{\partial \alpha_4} = -1$$
(3.27)

and

$$S^{\omega_0}_{\alpha_4} = \frac{\alpha_4}{\omega_0}\frac{\partial \omega_0}{\partial \alpha_4} = \frac{2}{2 + \alpha_4}$$
(3.28)

The sensitivities to MOS transistor aspect ratio mismatch in these circuits are identical to the sensitivities to capacitor ratio mismatch in switched-capacitor circuits.

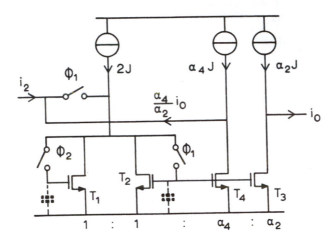

Figure 3.11 Inverting damped integrator.

3.5.4   *Inverting damped integrator*

Figure 3.11 shows an inverting damped integrator which differs from the non-inverting damped integrator just described only in that the input current is sampled on phase ϕ_1 instead of phase ϕ_2. On phase ϕ_1, the input current enters T_2 (being diode-connected) and the output current, i_0, is mirrored immediately. Analysis of this circuit gives the following z-domain transfer characteristic.

$$H_2(z) = \frac{i_0(z)}{i_2(z)} = -\frac{\dfrac{\alpha_2}{1 + \alpha_4}}{1 - \dfrac{1}{1 + \alpha_4} z^{-1}} \tag{3.29}$$

The performance with physical frequencies, following the earlier procedure for $\omega T \ll 1$, is given by

$$H_2(e^{j\omega T}) = -\frac{\left(\dfrac{\alpha_2}{\alpha_4}\right) e^{-j\omega T/2}}{1 + j\left[\dfrac{2}{T} \dfrac{\omega}{\dfrac{\alpha_4}{2 + \alpha_4}}\right]} \tag{3.30}$$

This is clearly equivalent to a damped continuous time integrator response except for the excess phase lag term $e^{-j\omega T/2}$, where the low-frequency gain a_0 and the cut-off frequency ω_0 are given by

$$a_0 = -\frac{\alpha_2}{\alpha_4} \tag{3.31}$$

$$\omega_0 = \frac{2}{T}\left(\frac{\alpha_4}{2 + \alpha_4}\right) \tag{3.32}$$

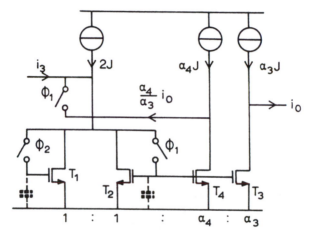

Figure 3.12 Inverting damped amplifier.

3.5.5 *Inverting damped amplifier*

Figure 3.12 shows an inverting damped amplifier which is identical to the integrators except that the input current, i_3, is fed directly, without being switched, to the summing node. On phase ϕ_2, the input current has a value of $i_3(n-1)$ and flows in T_1, while on phase ϕ_1 it has a value of $i_3(n)$ and flows in T_2. Of course, on phase ϕ_1, the stored current from T_1 also flows into T_2 giving an effective input signal in T_2 of $i_3(n) - i_3(n-1)$ and so the damped integrator is effectively driven by the derivative of the input signal. Analysis gives the z-domain transfer response

$$H_3(z) = \frac{i_0(z)}{i_3(z)} = -\frac{\dfrac{\alpha_3}{1 + \alpha_4}(1 - z^{-1})}{1 - \dfrac{1}{1 + \alpha_4}z^{-1}} \tag{3.33}$$

The performance with physical frequencies, following the earlier procedure for $\omega T \ll 1$, is given by

$$H_3(e^{j\omega T}) = -(j\omega T)\,\frac{\left(\dfrac{\alpha_3}{\alpha_4}\right)e^{-j\omega T/2}}{1 + j\left[\dfrac{2}{T}\dfrac{\alpha_4}{2+\alpha_4}\right]} \tag{3.34}$$

This is clearly equivalent to a damped continuous-time inverting amplifier response except for the excess phase lag term $e^{-j\omega T/2}$ and has resulted from the cascade of a differentiator ($j\omega T$ term) and an inverting damped integrator. The low-frequency gain a_0 and the cut-off frequency ω_0 are given by

$$a_0 = -\frac{\alpha_3}{\alpha_4} \tag{3.35}$$

$$\omega_0 = \frac{2}{T}\left(\frac{\alpha_4}{2+\alpha_4}\right) \tag{3.36}$$

Note that with $\alpha_4 = 0$ (lossless), the response is simply that of an inverting amplifier.

3.5.6 Generalised integrator

Figure 3.13(a) shows a generalised integrator configuration made from the superposition of non-inverting, inverting and amplifying inputs. The input currents are weighted α_1, α_2 and α_3, accomplished by scaling the weights (W/L ratios) of the output stages supplying these currents, and the output stage has unit weight. This achieves an independent gain constant for each of the inputs. By superposition of (3.23), (3.29) and (3.33), the z-domain output current is given by

$$i_0(z) = \frac{\dfrac{\alpha_1}{1+\alpha_4}z^{-1}}{1 - \dfrac{1}{1+\alpha_4}z^{-1}}i_1(z) - \frac{\dfrac{\alpha_2}{1+\alpha_4}}{1 - \dfrac{1}{1+\alpha_4}z^{-1}}i_2(z) - \frac{\dfrac{\alpha_3}{1+\alpha_4}(1-z^{-1})}{1 - \dfrac{1}{1+\alpha_4}z^{-1}}i_3(z)$$
$$\tag{3.37}$$

The z-domain SFG is shown in Figure 3.13(b).

Figure 3.14 shows the frequency responses of the generalised lossless ($\alpha_4 = 0$) integrator simulated with SCALP2. The clock frequency is 1MHz and the integrator is ideal (negligible parasitics). The non-inverting and inverting integrator responses are indistinguishable.

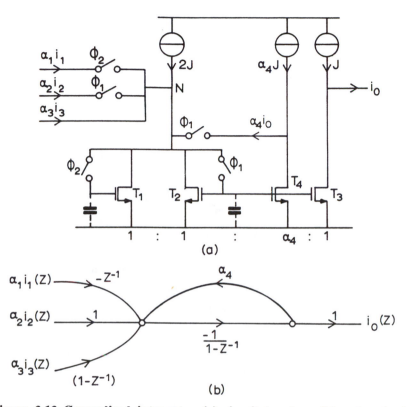

(a)

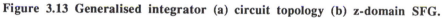

(b)

Figure 3.13 Generalised integrator (a) circuit topology (b) z-domain SFG.

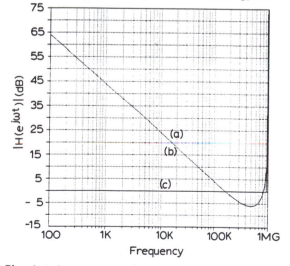

Figure 3.14 Simulated response of the generalised integrator ($\alpha_4 = 0$) (a) non-inverting ($\alpha_1 = 1$) (b) inverting ($\alpha_2 = 1$) (c) amplifying ($\alpha_3 = 1$).

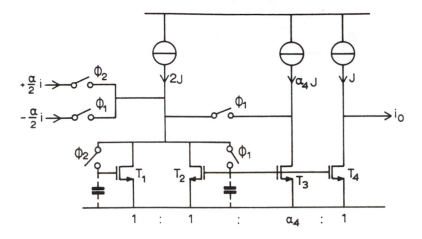

Figure 3.15 Bilinear z-transform integrator.

3.5.7 *Bilinear z-transform integrator*

A bilinear z-transform integrator may be formed by setting $i_1(z) = -i_2(z) = i(z)/2$, $i_3(z) = 0$ and $\alpha_1 = \alpha_2 = \alpha$ (in (3.37)) as shown in Figure 3.15. The z-domain transfer response then becomes

$$H(z) = \frac{i_0(z)}{i(z)} = \frac{\frac{\alpha}{2}(1 + z^{-1})}{1 - \dfrac{1}{1 + \alpha_4} z^{-1}} \tag{3.38}$$

The performance with physical frequencies, following the earlier procedure, is given by

$$H(e^{j\omega T}) = \frac{\alpha\left(\dfrac{1 + \alpha_4}{\alpha_4}\right)}{1 + j\left(\dfrac{2 + \alpha_4}{\alpha_4}\right)\tan\dfrac{\omega T}{2}} \tag{3.39}$$

This again is equivalent to a damped continuous time integrator response and, as expected for a bilinear integrator, there is frequency warping but neither excess phase nor amplitude distortion.

An elliptic filter design, based on the application of bilinear z-transform integrators, is discussed in Chapter 8.

3.5.8 Comparison with the switched-capacitor integrator

- In switched-capacitor circuits, integration is performed by transferring charge packets from an input sampling capacitor onto a Miller integrating capacitor. In switched-currents, an equivalent function is formed by a current integrator loop comprising two current memory cells. Since no charge transfer is involved, parasitic grounded capacitors suffice.

- Switched-capacitor integrators are rendered parasitic insensitive by the input switching arrangement which sequentially discharges the parasitics. The switched-current integrator employs the parasitic gate-oxide capacitance and so obviates the problem.

- Switched-capacitor integrators reset their sampling capacitors to zero voltage during each cycle of operation. Similarly, the switched-current integrator resets its input current to zero during each period.

3.5.9 Integrator-based biquadratic section

The generalised integrator of Figure 3.13 may be used to implement filters based on known switched-capacitor topologies. Mapping is straightforward as there is a direct correspondence between capacitor ratios in the switched-capacitor filter and W/L ratios in the switched-current counterpart.

As an example, a biquadratic section is now derived from a known switched-capacitor biquad. A similar approach may be adopted for other biquadratic sections or for the active ladder approach. The method is as follows:

1. Choose a suitable switched-capacitor configuration

2. Identify the integrators and their type (Forward Euler etc)

3. Define equivalent switched-current integrators

4. Construct switched-current filter configuration

The particular biquadratic section chosen for this example is well-known [13] but is described here for the sake of completeness. The general s-domain biquadratic transfer function is as follows

$$H(s) = \frac{x_0(s)}{x(s)} = \frac{k_2 s^2 + k_1 s + k_0}{s^2 + \dfrac{\omega_0}{Q} s + \omega_0^2} \tag{3.40}$$

where $x(s)$ and $x_0(s)$ are the input and output signals (voltage, current etc.), k_0, k_1 and k_2 are constants and ω_0 and Q are the pole frequency and quality factor. Cross-multiplying and rearranging (3.40) gives

$$x_0(s) = -\frac{1}{s}\left\{(k_1 + k_2s)x(s) + \frac{\omega_0}{Q}x_0(s) - \omega_0x_1(s)\right\}$$

where

$$x_1(s) = -\frac{1}{s}\left\{\frac{k_0}{\omega_0}x(s) + \omega_0x_0(s)\right\} \tag{3.41}$$

There are many other ways of partitioning (3.40) and each results in a particular circuit structure. The chosen partition, expressed in (3.41), results in the system shown as a SFG in Figure 3.16. It can be seen that the first integrator is fed from weighted input and output signals $x(s)$ and $x_0(s)$ with an overall inversion of both signals. The second integrator is fed from weighted input, output and intermediate signals $x(s)$, $x_0(s)$ and $x_1(s)$ to give a mixture of inverted and non-inverted integration and inverted amplification.

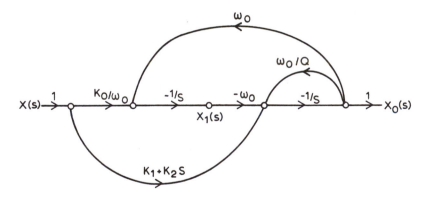

Figure 3.16 s-domain SFG of a biquadratic function.

A z-domain SFG may be constructed with corresponding integrators displaying the same features and this is shown in Figure 3.17. It is readily shown that the transfer function is given by

$$H(z) = \frac{x_0(z)}{x(z)} = -\frac{(a_5 + a_6)z^2 + (a_1a_3 - 2a_6 - a_5)z + a_6}{(1 + a_4)z^2 + (a_2a_3 - a_4 - 2)z + 1} \tag{3.42}$$

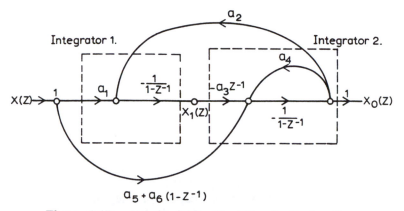

Figure 3.17 z-domain SFG of a biquadratic function.

Now, if (3.42) can be made to match the bilinear z-transform of (3.40), then the resulting transfer function will be 'exact', i.e. it will have neither amplitude distortion nor excess phase errors. To achieve this we first apply the bilinear-z transform, $s \rightarrow 2(1 - z^{-1})/T(1 + z^{-1})$, to (3.40). This, after some algebraic manipulation, gives

$$H(z) = \frac{\left[\dfrac{4k_2+2k_1T+k_0T^2}{X}\right]z^2 + \left[\dfrac{2k_0T^2-8k_2}{X}\right]z + \left[\dfrac{4k_2-2k_1T+k_0T^2}{X}\right]}{\left[\dfrac{\omega_0T^2 + \dfrac{2\omega_0T}{Q} + 4}{X}\right]z^2 + \left[\dfrac{2\omega_0T^2 - 8}{X}\right]z + 1} \qquad (3.43)$$

where

$$X = \omega_0T^2 - \frac{2\omega_0T}{Q} + 4 \qquad (3.44)$$

Comparing coefficients with (3.42) gives

$$a_6 = \frac{4k_2 - 2k_1T + k_0T^2}{X} \qquad (3.45)$$

$$a_5 = \frac{4k_1T}{X} \qquad (3.46)$$

$$a_4 = \frac{4\omega_0T}{QX} \qquad (3.47)$$

$$a_2a_3 = \frac{4\omega_0^2T^2}{X} \qquad (3.48)$$

$$a_1 a_3 = \frac{4k_0 T^2}{X} \qquad (3.49)$$

So, if the coefficients are chosen according to (3.44)-(3.49) then, bilinear-z mapping of the biquadratic function is achieved even though bilinear-z integrators were not employed. In fact, an identical transfer function could have been implemented using bilinear-z integrators but a more complex structure would have resulted as they require both polarities of input signal.

Circuit implementation of the biquadratic SFG (Figure 3.17) is achieved by substitution of integrators with appropriate input type (inverting or non-inverting integration, inverting amplification). The well-known switched-capacitor biquadratic section shown in Figure 3.18 results from substitution of switched-capacitor integrators (Figure 3.2(a)). To implement a switched-current biquadratic section integrators with suitable input type (Figure 3.13) are substituted in just the same manner. The resulting switched-current biquadratic section is shown in Figure 3.19. Of course, given the switched-capacitor circuit the switched-current circuit can be derived simply by direct replacement of integrators.

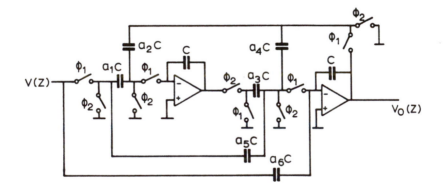

Figure 3.18 Switched-capacitor implementation of the biquadratic function.

The s-domain transfer function corresponding to the z-domain transfer function given in (3.42) can be found by performing an inverse bilinear z-transformation:

$$z \rightarrow \frac{2 + sT}{2 - sT} \qquad (3.50)$$

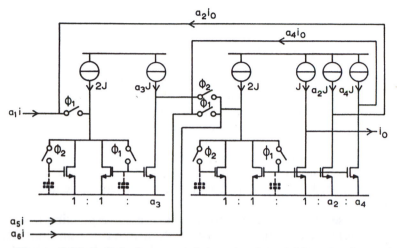

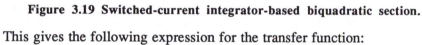

Figure 3.19 Switched-current integrator-based biquadratic section.

This gives the following expression for the transfer function:

$$H(s) = \frac{\left[\dfrac{4a_6 + 2a_5 - a_1a_3}{D}\right]s^2 + \left[\dfrac{4}{T}\dfrac{a_5}{D}\right]s + \left[\dfrac{4}{T^2}\dfrac{a_1a_3}{D}\right]}{s^2 + \left[\dfrac{4}{T}\dfrac{a_4}{D}\right]s + \left[\dfrac{4}{T^2}\dfrac{a_2a_3}{D}\right]}$$

(3.51)

where

$$D = 2a_4 - a_2a_3 + 4$$

(3.52)

Comparing (3.51) and (3.40) gives

$$k_0 = \frac{4}{T^2}\frac{a_1a_3}{D}$$

(3.53)

$$k_1 = \frac{4}{T}\frac{a_5}{D}$$

(3.54)

$$k_2 = \frac{4a_6 + 2a_5 - a_1a_3}{D}$$

(3.55)

$$\omega_0 = \frac{2}{T}\sqrt{\frac{a_2a_3}{D}}$$

(3.56)

$$Q = \frac{1}{2a_4}\sqrt{a_2a_3D}$$

(3.57)

Equations (3.52)-(3.57) give the inverse relationships to those in (3.44)-(3.49).

3.6 Differentiator Modules

Differentiators have not received the same attention as integrators as modules for modelling the differential equations of state-variable or ladder filters. This is because, in the classical active-RC implementation, circuit performance was severely degraded by high-frequency noise. Recently, a family of switched-capacitor differentiators was proposed [19] for the realisation of high-pass and band-pass filters. These circuits had good noise rejection properties [20] and avoided low-frequency instability problems associated with certain ladder circuits when implemented with integrators.

The purpose of this section is to show that a corresponding family of switched-current differentiators exists. Non-inverting and inverting differentiators are described and these are developed into generalised structures suitable for use in state-variable and ladder filters. A bilinear z-transform differentiator is also described.

3.6.1 Inverting differentiator

An inverting differentiator module is shown in Figure 3.20.

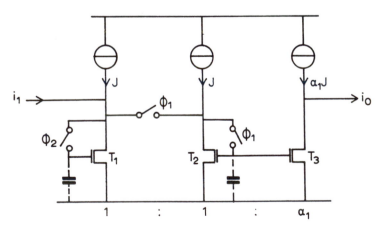

Figure 3.20 Inverting differentiator.

The operation is as follows. During period *(n-1)* on clock phase ϕ_2, current memory T_1 is diode-connected and receives input current $J +i_1(n-1)$. On the next phase, ϕ_1 of clock period *(n)*, current memory T_2 is diode-connected and receives current from the input, $i_1(n)$, from the bias $2J$ and from the current stored in T_1, $-(J + i_1(n-1))$. The resulting current in T_2 is

$$I_2 = i_1(n) + 2J - [J + i_1(n-1)] = J + i_1(n) - i(n-1) \qquad (3.58)$$

Current I_2 is mirrored into T_3 giving an output current $i_0(n)$ where

$$i_0(n) = \alpha_1(J - I_2) = -\alpha_1[i_1(n) - i_1(n-1)] \qquad (3.59)$$

Transforming to the z-domain gives

$$i_0(z) = -\alpha_1 i(z)(1 - z^{-1})$$

$$H_1(z) = \frac{i_0(z)}{i(z)} = -\alpha_1(1 - z^{-1}) \qquad (3.60)$$

Equation (3.60) is the Backward Euler z-transform $(s \rightarrow (1 - z^{-1})/T)$ of an inverting differentiator $(H(s) = -s\tau)$ where

$$\alpha_1 = \frac{\tau}{T} \qquad (3.61)$$

The exact performance for physical frequencies is given by re-arranging (3.60) and substituting $z = e^{j\omega T}$.

$$H_1(z) = -\alpha_1(z^{1/2} - z^{-1/2})z^{-1/2}$$

$$H_1(e^{j\omega T}) = -j2\alpha_1 \sin\frac{\omega T}{2}\, e^{-j\omega T/2}$$

$$= -j\omega[\alpha_1 T]\left[\frac{\sin\frac{\omega T}{2}}{\frac{\omega T}{2}}\right]e^{-j\omega T/2} \qquad (3.62)$$

where $\alpha_1 T$ is the gain constant, $\sin(\omega T/2)(\omega T/2)$ is the deviation of the sampled-data differentiator response from ideal and $e^{-j\omega T/2}$ indicates an excess phase lag. For $\omega \ll 1/T$

$$H_1(e^{j\omega T}) = -j\omega[\alpha_1 T]e^{-j\omega T/2} \qquad (3.63)$$

Figure 3.21 shows the simulated response with $\alpha_1 = 1$ and $T = 1\mu s$.

3.6.2 Generalised inverting differentiator

Figure 3.22(a) shows a generalised inverting differentiator in which the coefficients α_1, α_2 and α_3 are defined by the weights of the input currents. For the input $\alpha_1 i_1$ (i.e. i_2, $i_3 = 0$), the response is the same as for the inverting differentiator described above.

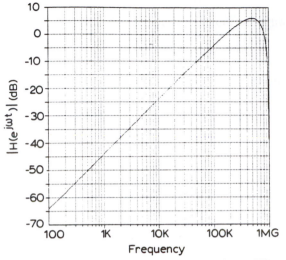

Figure 3.21 Simulated response of an inverting differentiator
$(\alpha_1=1, T=1\mu s)$.

With $i_1 = i_3 = 0$, the input current is $\alpha_2 i_2$ and the circuit operates as follows. On phase ϕ_2 of period $(n-1)$, current memory T_1 is diode-connected and conducts current $J + \alpha_2 i_2(n-1)$. On the next phase ϕ_1 of period (n), T_2 is diode-connected and conducts current $J - \alpha_2 i_2(n-1)$ and the output current is $i_0(n)$ where

$$i_0(n) = \alpha_2 i_2(n-1) \tag{3.64}$$

The output is merely the input current delayed by one clock period.

With $i_1 = i_2 = 0$ the input current is $\alpha_3 i_3$ and the circuit operates as follows. On phase ϕ_1 of period (n), the current memory T_2 is diode-connected and conducts current $J + \alpha_3 i_3(n)$. This is mirrored into T_3 to give an output current $i_0(n)$ where

$$i_0(n) = -\alpha_3 i_3(n) \tag{3.65}$$

The combined response is given by the superposition of (3.59),(3.64) and (3.65):

$$i_0(n) = -\alpha_1[i_1(n) - i_1(n-1)] + \alpha_2 i_2(n-1) - \alpha_3 i_3(n) \tag{3.66}$$

Taking z-transforms gives

$$i_0(z) = -\alpha_1 i_1(z)(1 - z^{-1}) + \alpha_2 i_2(z)z^{-1} - \alpha_3 i_3(z) \tag{3.67}$$

This equation is represented in the z-domain SFG shown in Figure 3.22(b).

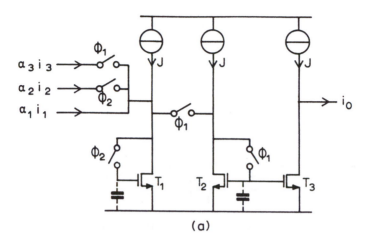

(a)

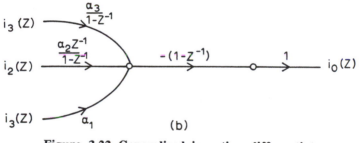

(b)

Figure 3.22 Generalised inverting differentiator
(a) circuit topology (b) z-domain SFG.

3.6.3   Non-inverting differentiator

Figure 3.23 shows a non-inverting differentiator. It comprises an inverting differentiator, the output of which is inverted and fed back to the input summing node. The total loop therefore has positive feedback which makes the module unstable. It may only be used in conjunction with other circuits which render it stable by applying negative feedback (e.g. in a biquad arrangement).

The operation of the circuit is as follows. On phase ϕ_2 of period $(n-1)$, the signal current fed to the input summing node is $i_1(n-1) + i_0(n-1)/\alpha_1$. The current in T_1 is I_1 where

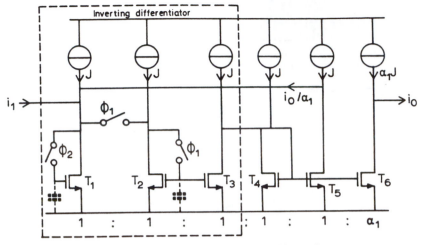

Figure 3.23 Non-inverting differentiator.

$$I_1 = J + i_1(n-1) + \frac{i_0(n-1)}{\alpha_1} \qquad (3.68)$$

On phase ϕ_1 of period (n), T_1 stores current I_1 and T_2 is diode-connected and passes I_2 where

$$I_2 = 2J + i_1(n) + \frac{i_0(n)}{\alpha_1} - I_1$$

$$= J + i_1(n) - i_1(n-1) + \frac{1}{\alpha_1}\left[i_0(n) - i_0(n-1)\right] \qquad (3.69)$$

Further,

$$\frac{i_0(n)}{\alpha_1} = J - I_5 = J - I_4 = I_2 - J$$

$$= i_1(n) - i_1(n-1) + \frac{1}{\alpha_1}\left[i_0(n) - i_0(n-1)\right]$$

$$i_0(n-1) = \alpha_1\left[i_1(n) - i_1(n-1)\right] \qquad (3.70)$$

Taking z-transforms:

$$H_1(z) = \frac{i_0(z)}{i(z)} = \alpha_1\frac{(1 - z^{-1})}{z^{-1}} \qquad (3.71)$$

The exact performance for physical frequencies is given by substituting $z = e^{j\omega T}$

$$H_1(e^{j\omega T}) = j\omega \, [\alpha_1 T] \left[\frac{sin\dfrac{\omega T}{2}}{\dfrac{\omega T}{2}} \right] e^{j\omega T/2} \qquad (3.72)$$

where $\alpha_1 T$ is the gain constant, $sin(\omega T/2)/(\omega T/2)$ is the deviation of the sampled-data differentiator's response from ideal and $e^{j\omega T/2}$ indicates an excess phase lead. For $\omega \ll 1/T$

$$H_1(e^{j\omega T}) = j\omega[\alpha_1 T] \, e^{j\omega T/2} \qquad (3.73)$$

3.6.4   Generalised non-inverting differentiator

Figure 3.24(a) shows a generalised non-inverting differentiator in which coefficients α_1, α_2 and α_3 are defined by the weights of the input currents. With $i_2 = i_3 = 0$, the input current is $\alpha_1 i_1$ and the response is the same as that of the non-inverting differentiator described above.

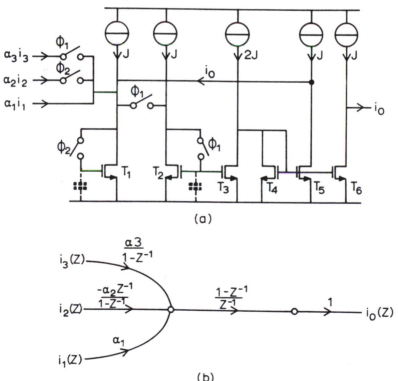

(a)

(b)

**Figure 3.24 Generalised non-inverting differentiator
(a) circuit topology (b) z-domain SFG.**

With $i_1 = i_3 = 0$, the input current is $\alpha_2 i_2$ and the circuit operates as follows. On phase ϕ_2 of period $(n-1)$, T_1 is diode-connected and its current I_1 is given by

$$I_1 = J + \alpha_2 i_2(n-1) + i_0(n-1) \tag{3.74}$$

On phase ϕ_1 of period (n), T_2 is diode-connected and its current I_2 is

$$I_2 = 2J + i_0(n) - I_1$$

$$= J + i_0(n) - i_0(n-1) - \alpha_2 i_2(n-1) \tag{3.75}$$

Further

$$i_0(n) = I_2 - J = i_0(n) - i_0(n-1) - \alpha_2 i_2(n-1)$$

$$i_0(n-1) = -\alpha_2 i_2(n-1) \tag{3.76}$$

The output is merely the inverse of the input current.

With $i_1 = i_2 = 0$, the input current is $\alpha_3 i_3$ and the circuit operates as follows. On phase ϕ_2 of period $(n-1)$, T_1 is diode-connected and

$$I_1 = J + i_0(n-1) \tag{3.77}$$

On phase ϕ_1 of period (n), T_2 is diode-connected and

$$I_2 = 2J + i_0(n) + \alpha_3 i_3(n) - I_1$$

$$= J + i_0(n) - i_0(n-1) + \alpha_3 i_3(n) \tag{3.78}$$

Further,

$$i_0(n) = I_2 - J = i_0(n) - i_0(n-1) + \alpha_3 i_3(n)$$

$$i_0(n-1) = \alpha_3 i_3(n) \tag{3.79}$$

Equation (3.79) is clearly non-causal since it states that the output current is equal to the input current from the next period.

The combined response is given by the superposition of (3.70), (3.76) and (3.79):

$$i_0(n-1) = \alpha_1 [i_1(n) - i_1(n-1)] - \alpha_2 i_2(n-1) + \alpha_3 i_3(n) \tag{3.80}$$

Taking z-transforms gives

$$i_0(z) = \alpha_1 i_1(z) \frac{1 - z^{-1}}{z^{-1}} - \alpha_2 i_2(z) + \frac{\alpha_3 i_3(z)}{z^{-1}} \tag{3.81}$$

This equation is represented in the z-domain SFG shown in Figure 3.24(b)

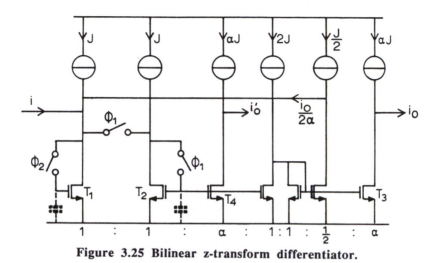

Figure 3.25 Bilinear z-transform differentiator.

3.6.5 *Bilinear z-transform differentiator*

Although the alternate connection of inverting and non-inverting differentiators produced bilinear z-transform mapping in biquads (see Section 3.6.6), adoption of this strategy in ladder filters results in unrealisable terminations [21]. A bilinear z-transform differentiator does not suffer this drawback and may be used in either biquad or ladder filter designs.

Figure 3.25 shows the configuration of the switched-current bilinear differentiator. It is similar in structure to the non-inverting differentiator except that the positive feedback is halved and an additional output stage (T_4) is used. The operation is as follows. On phase ϕ_2 of period $(n-1)$, the current in T_1 is I_1 where

$$I_1 = J + i(n-1) + \frac{i_0(n-1)}{2\alpha} \tag{3.82}$$

On the next phase, ϕ_1 of period (n), the current in T_2 is I_2 where

$$I_2 = 2J + i(n) + \frac{i_0(n)}{2\alpha} - I_1$$

$$= J + i(n) - i(n-1) + \frac{i_0(n)}{2\alpha} - \frac{i_0(n-1)}{2\alpha} \tag{3.83}$$

Also,

$$i_0(n) = \alpha I_2 - \alpha J = \alpha[i(n) - i(n-1)] + \frac{[i_0(n) - i_0(n-1)]}{2} \qquad (3.84)$$

Rearranging (3.84) and taking z-transforms gives

$$H(z) = \frac{i_0(z)}{i(z)} = 2\alpha \frac{1 - z^{-1}}{1 + z^{-1}} \qquad (3.85)$$

Now considering the mirroring in the output stages,

$$i_0 = \alpha J - I_3 = \alpha I_2 - \alpha J = I_4 - \alpha J$$

and

$$i_0' = \alpha J - I_4 = \alpha J - (i_0 + \alpha J) = -i_0 \qquad (3.86)$$

So, the outputs i_0 and i_0' are complementary and

$$H'(z) = \frac{i_0'(z)}{i(z)} = -2\alpha \frac{1 - z^{-1}}{1 + z^{-1}} \qquad (3.87)$$

The bilinear differentiator may be used to implement the inverting and non-inverting differentiators in either biquad structures or ladder filters.

3.6.6 *Differentiator-based biquadratic section*

Following the approach adopted for switched-capacitors [22], the generalised biquadratic function

$$H(s) = \frac{i_0(s)}{i(s)} = - \frac{k_2 s^2 + k_1 s + k_0}{s^2 + \dfrac{\omega_0}{Q} s + \omega_0^2} \qquad (3.88)$$

can be arranged as

$$i_0(s) = -s \left[\frac{k_0}{\omega_0^2} \frac{i(s)}{s} + \frac{i_0(s)}{Q\omega_0} + \frac{i_1(s)}{\omega_0} \right] \qquad (3.89)$$

where

$$i_1(s) = s \left[\left(\frac{k_2}{\omega_0} + \frac{k_1}{\omega_0 s} \right) i(s) + \frac{i_0(s)}{\omega_0} \right] \qquad (3.90)$$

Equations (3.89) and (3.90) are expressed as a SFG in Figure 3.26(a).

Using the generalised inverting and non-inverting differentiator SFG's (Figures 3.22 and 3.24), a corresponding z-domain SFG is constructed and is shown in Figure 3.26(b). Developing the equations implicit in Figure 3.26(b) gives the z-domain transfer function

$$H(z) = -\frac{(a_2a_3 + a_1a_3)z^2 + (a_0 - a_1a_3 - 2a_2a_3)z + a_2a_3}{a_3a_5z^2 + (1 + a_4 - 2a_3a_5)z + (a_3a_5 - a_4)} \qquad (3.91)$$

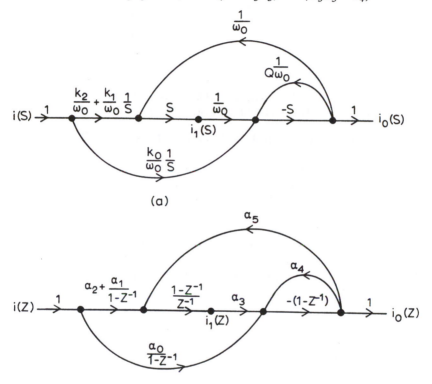

(a)

(b)

Figure 3.26 Differentiator-based biquadratic SFG's
(a) s-domain (b) z-domain.

Coefficient	Value
a_0	k_0/ω_0^2
a_1a_3	$k_1/T\omega_0^2$
a_2a_3	$(k_0 - 2k_1/T + 4k_2/T^2)/4\omega_0^2$
a_4	$1/QT\omega_0$
a_5a_3	$(4/T^2 + 2\omega_0/QT + \omega_0^2)/4\omega_0^2$

Table 3.1 Coefficients for use with differentiator-based biquad section.
T is the clock period.

We wish to derive a biquad with a bilinear z-mapping $(s \rightarrow 2(1 - z^{-1})/T(1 + z^{-1}))$. If (3.88) is transformed to the z-domain using the bilinear z-transformation, then the transfer function is

$$H(z) = \frac{\left[\dfrac{4k_2+2k_1T+k_0T^2}{D}\right]z^2 + \left[\dfrac{2k_0T^2-8k_2}{D}\right]z + \left[\dfrac{4k_2-2k_1T+k_0T^2}{D}\right]}{\left[\dfrac{\omega_0T^2 + \dfrac{2\omega_0T}{Q} + 4}{D}\right]z^2 + \left[\dfrac{2\omega_0T^2 - 8}{D}\right]z + 1} \qquad (3.92)$$

where $D = \omega_0T^2 - 2\omega_0T/Q + 4$. The required mapping is produced merely by setting the coefficients, $a_0 \ldots a_5$, to make (3.91) and (3.92) identical, and these are given in Table 3.1. The switched-current circuit which performs the SFG of Figure 3.26(b) is shown in Figure 3.27.

3.7 Filter Synthesis Example 6th Order Lowpass Filter

Having established a switched-current integrator-based biquadratic section in Section 3.5.9, this will now be used in the design of a 6th order low-pass filter with a Chebyshev response (0.5dB equiripple) and a -3dB cut-off frequency of 5MHz using a 20MHz clock frequency. The purpose of this exercise is to verify the theoretical treatment of earlier sections by using the structures and equations to perform the design and then to check the response by simulation. It is not the intention to make an optimised filter design: no consideration is given to silicon area, dynamic range or power consumption optimisation.

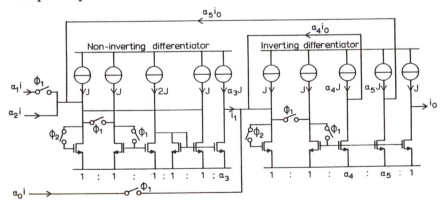

Figure 3.27 Differentiator-based biquadratic section.

The first step in the numerical design is to select the polynomial coefficients for a 6th order Chebyshev (0.5dB ripple) low-pass filter (1rads^{-1}) from design tables [23]. They are expressed in factorised form as

$$H(s) = \frac{n_1}{(s^2 + m_1 s + n_1)} \frac{n_2}{(s^2 + m_2 s + n_2)} \frac{n_3}{(s^2 + m_3 s + n_3)} \qquad (3.93)$$

The coefficients are given in Table 3.2.

	First Section	Second Section	Third Section
m	0.155300	0.424288	0.579588
n	1.023022	0.590010	0.156997

Table 3.2 Coefficients m and n of the quadratic factors (s^2+ms+n) of a 6th order 0.5dB ripple Chebyshev low-pass function ($\omega_c = 1.041029$rads).

The response produces peaks at +0.5dB and a -3dB cut-off radian frequency of 1.041029 rads^{-1}. To engineer a filter with a cut-off frequency of 5MHz, the coefficients of the biquadratic factors must be scaled to a prewarped cut-off frequency defined by

$$\omega_p = \frac{2}{T} \tan \frac{\omega T}{2} \qquad (3.94)$$

due to the frequency warping produced by the bilinear z-transformation. In our example $\omega = 10\pi 10^6 s^{-1}$ (corresponding to 5MHz) and $T = 0.05 \ 10^{-6}s$ (corresponding to 20MHz clock frequency) giving $\omega_p = 40 \ 10^6 s^{-1}$ (corresponding to 6.3662MHz). The coefficients are then scaled according to the following:

$$m_p = \left[\frac{\omega_p}{1.041029}\right] m \qquad (3.95)$$

$$n_p = \left[\frac{\omega_p}{1.041029}\right]^2 n \qquad (3.96)$$

where m_p and n_p are the scaled coefficients corresponding to $\omega_0 = Q$ and ω_0^2, respectively, of a prewarped prototype to be used to derive the switched-current filter. Further, to produce a response peaking at 0dB, the gain of the first stage is set to -0.5dB (0.9406). The scaled parameters are given in Table 3.3.

	First Section	Second Section	Third Section
$m_p = \omega_0/Q$	5.9673 10^6	16.3268 10^6	22.2711 10^6
$n_p = \omega_0^2$	1510.4 10^{12}	871.07 10^{12}	231.179 10^{12}
k_0	1425.9 10^{12}	871.07 10^{12}	231.179 10^{12}

Table 3.3 Scaled and prewarped filter parameters (s^{-1}).

The sampled-data filter coefficients are then calculated using (3.52) to (3.57). These are given in the Table 3.4.

	First Section	Second Section	Third Section
a_1a_3	1.984504	1.913393	0.984467
a_2a_3	2.102105	1.913393	0.984467
a_4	0.165887	0.716175	1.894220
a_5	0.000000	0.000000	0.000000
a_6	0.496126	0.478348	0.246117

Table 3.4 Sampled-data filter coefficients.

Using this table and Figure 3.19, the complete switched-current filter is defined. To verify the design, the filter was modelled and the small-signal response was simulated on SCALP2. The model for a memory cell, including parasitics (shown dotted), is shown in Figure 3.28, although for the purpose of verifying the circuit structures and theory the parasitics are omitted. The transistors T_1 and T_2 are simply modelled as ideal transconductors. Transconductor g_{m1} is arbitrary for ideal modelling and may be set to unity, and g_{m2} is chosen to give the required coefficient value (Table 3.4). Of course, parasitic resistors and capacitors may be added to the model for non-ideal simulation.

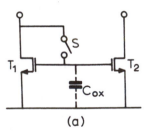

(a)

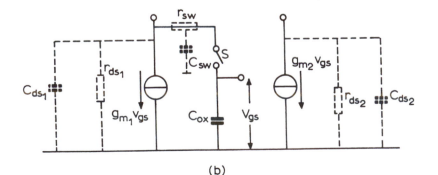

(b)

**Figure 3.28 Small signal model of current memory cell
(a) memory cell (b) model.**

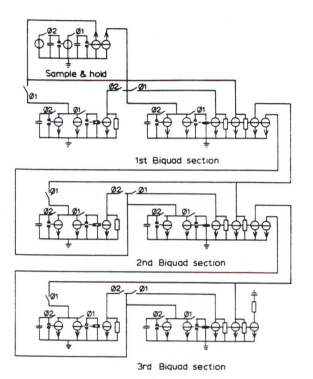

Figure 3.29 Small signal model of 6th order Chebyshev low-pass filter.

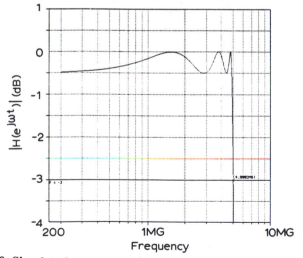

Figure 3.30 Simulated response of 6th order Chebyshev low-pass filter
(f_{C0} = 5MHz, equiripple = 0.5dB).

Using the ideal model of the memory cell, a complete model of the filter (including a sample and hold circuit) was constructed and the filter was simulated. The model is shown in Figure 3.29 and the resulting ideal amplitude response is shown in Figure 3.30. Note that the amplitude response was adjusted by the CAD package to remove *sinx/x* distortion resulting from the sampling process. It can be seen that an ideal Chebyshev response with 0.5dB ripple and 5MHz cut-off frequency has resulted. This confirms the theoretical background to switched-current circuits established earlier.

3.8 Summary

Switched-current circuits have been presented which perform sampled-data signal processing in the current domain. Current memory is achieved by virtue of the parasitic gate oxide capacitance of the memory transistor, making the technique ideally suited for mixed analogue/digital ICs using the most basic VLSI processes. The current memory or delay cell may be used directly to build delay lines / FIR filters or as elements of integrators or differentiators for synthesising state-variable or active ladder filters. Whereas filter coefficients of switched-capacitor circuits are defined by capacitor ratios, they may be defined by the ratios of transistor widths in switched-current circuits.

First, a family of integrator configurations capable of non-inverting and inverting integration and inverting amplification was developed from the current memory cell. The sensitivity of the damped integrator's low-frequency gain and cut-off frequency to coefficient errors was as good as that resulting from capacitor ratio errors in switched-capacitor circuits. A generalised integrator with non-inverting, inverting and amplifying input terminals was given which maps directly to a well-known switched-capacitor counterpart. A bilinear z-transform integrator was also presented. Using the generalised integrator, most known switched-capacitor filters can be transposed into switched-current counterparts.

Next, a family of differentiator circuits was developed from the current memory cell. Generalised inverting and non-inverting differentiator structures were presented which were suitable for implementing biquads capable of bilinear z-transformation. A bilinear z-transform differentiator was given which could give an alternative biquad implementation or be used directly in ladder filters. It is expected that, like their switched-capacitor counterparts, differentiator-based solutions will be best suited to high-pass or band-pass filters. Since the switched-current structures are duals of their switched-capacitor counterparts they will possess similar sensitivity to coefficient errors.

To demonstrate the theory developed in this chapter, a switched-current biquad was developed from a known switched-capacitor biquad and was used to design a 6th order Chebyshev lowpass filter with a 5MHz cut-off frequency.

References

[1]   S Matsuzaki and I Kondo, *UK Patent No. 1 359 105*, July 1972

[2]   D. Vallancourt and Y. P. Tsividis, "Sampled current circuits," in *Proc. IEEE International Symposium on Circuits and Systems*, pp. 1592-1595, May 1989.

[3]   W. Groeneveld, H. Schouwenaars and H. Termeer, "A self-calibration technique for high-resolution D/A converters," *IEEE International Solid-State Circuits Conference*, pp. 22-23, 1989,

[4]   D. G. Nairn and C. A. T. Salama, "Current mode analogue to digital converters," in *Proc. IEEE International Symposium on Circuits and Systems*, pp. 1588-1591, 1989.

[5]   G. Wegmann and E. A. Vittoz, "Very accurate dynamic current mirrors," *Electronics Letters*, vol. 25, pp. 644-646, 11th May 1989.

[6]   J. B. Hughes, N. C. Bird and I. C. Macbeth, "Switched-currents, A new technique for analogue sampled-data signal processing," in *Proc. IEEE International Symposium on Circuits and Systems*, pp. 1584-1587, 1989.

[7]   J. B. Hughes, I. C. Macbeth and D. M. Pattullo, "Switched-current filters," *Proc. IEE*, Pt. G, vol. 137, pp. 156-162, April 1990.

[8]   S. J. Daubert, D. Vallancourt and Y. Tsividis, "Current copier cells," *Electronics Letters*, vol. 24, pp. 1560-1562, 8th Dec. 1988.

[9]   J. B. Hughes, I. C. Macbeth and D. M. Pattullo, "Second generation switched-current circuits" in *Proc. IEEE International Symposium on Circuits and Systems*, pp. 2805-2808, 1990.

[10]   J. B. Hughes, I. C. Macbeth and D. M. Pattullo, "Switched-current system cells," in *Proc. IEEE International Symposium on Circuits and Systems*, pp. 303-306, 1990.

[11]   J. B. Hughes, I. C. Macbeth and D. M. Pattullo, "New switched-current integrator," *Electronics Letters*, vol. 26, pp. 694-695, 24 May 1990.

[12]   G. C. Temes, ed.,"Special section on switched-capacitor circuits", *Proc. IEEE*, vol. 71, pp. 926-1005, Aug. 1983.

[13]   R. Gregorian and G. C. Temes, *Analog MOS Integrated Circuits for Signal Processing*, John Wiley & Sons, pp. 280-284.

[14]   A. S. Sedra, "Switched-capacitor filter synthesis", Chapter 9 in, *Design of MOS VLSI circuits for Telecommunications*, Prentice Hall, 1985.

[15]   K. Martin and A. S. Sedra, "Exact design of switched-capacitor bandpass filters using coupled-biquad structures," *IEEE Trans. Circuits Syst.*, vol. CAS-27, pp. 469-475, June 1980.

[16]   J. C. M. Bermudez and B. B. Bhattacharyya, "A Systematic Procedure for the generation and design of parasitic insensitive SC biquads," *IEEE Trans. Circuits Syst.*, vol. CAS-32, pp. 767-783, Aug. 1985,.

[17]   G. M. Jacobs, D. J. Allstot, R. W. Broderson and P. R. Gray, "Design techniques for MOS switched-capacitor ladder filters," *IEEE Trans. Circuits Syst.*, vol. CAS-25, pp. 1014-1021, Dec. 1978.

[18]   J. Carlos, M. Bermudez, M. Schneider and C. Montoro, "Linearity of switched-capacitor filters employing non-linear capacitors," in *Proc. IEEE International Symposium on Circuits and Systems*, pp. 1211-1224, 1992.

[19]   C.-Y. Wu and T.-C. Yu, "The design of high-pass and band-pass ladder filters using novel SC differentiators," in *Proc. International Symposium on Circuits and Systems*, pp. 1463-1466, May 1989.

[20]   C.-Y. Wu, T.-C. Yu and S.-S. Chang, "New monolithic switched-capacitor differentiators with good noise rejection," *IEEE J. Solid-State Circuits*, vol. SC-24, Feb. 1989.

[21]   T.-C. Yu, C.-H. Hsu, C.-Y. Wu, "The bilinear-mapping SC differentiators and their applications in the design of biquad and ladder filters", in *Proc. IEEE International Symposium on Circuits and Systems*, pp. 2189-2192, 1990.

[22]   C.-Y. Wu, private communication

[23]   F. W. Stephenson, *RC Active Filter Design Handbook*, John Wiley and Sons.

Chapter 4

Switched-Current Limitations and Non-Ideal Behaviour

John B. Hughes and William Redman-White

4.1 Introduction

This chapter describes the non-ideal behaviour of switched-current circuits resulting from imperfect MOS transistor operation. Just as with the switched-capacitor technique, MOS transistor imperfections result in deviations from the ideal performance described by the algorithmic properties of its signal processing modules (see Chapter 3). The chapter is divided into five parts according to the type of MOS transistor imperfection: Section 4.2 analyses the effects of transistor gain and threshold voltage mismatch, Section 4.3, the effects of transistor drain conductance, Section 4.4, the effects of transistor bandwidth, Section 4.5, the effects of switch charge injection and Section 4.6, the effects of noise. In each case, the analysis derives the resulting errors in the frequency response of the switched-current delay cell and integrator. Circuit techniques for controlling these non-ideal effects and enhancing performance are discussed in Chapter 6.

4.2 Mismatch Errors

Mismatch errors are those errors resulting from the small but inevitable differences in the DC characteristics of the memory cell's transistors. Figure 4.1(a) shows the basic memory cell. As only one transistor is used to both sink and source current, no mismatch errors can arise. However, the integrator uses current mirrors to define loss and to facilitate a sampled and held output current and mismatch errors in these mirrors lead to non-ideal operation.

4.2.1 Unity gain current memory

A current memory with mirrored output is shown in Figure 4.1(b) and the mismatch errors of this cell will now be analysed. With switch S closed the circuit is a simple current mirror and if T_1 and T_2 are identical, then the input current I_1 is faithfully reproduced as I_2. In practice, T_1 and T_2 will not be identical due to small local variations in the IC processing which

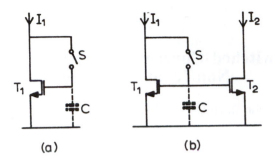

Figure 4.1 Current memory cells
(a) basic cell (b) cell with mirrored output.

give T_2 slightly different DC characteristics from T_1 with the result that I_2 contains an error current. When the switch S opens, the current I_2 is 'stored', including its error current which is propagated into other memory cells connected to its output with resulting degradation of performance. With T_1 and T_2 both in saturation their drain currents are given, in the usual notation, by

$$I_1 = \frac{\beta_1}{2}(V_{gs1} - V_{T1})^2(1 + \lambda_1 V_{ds1}) \tag{4.1}$$

where

$$\beta_1 = \mu_1 C_{OX1}\left[\frac{W}{L}\right]_1 \tag{4.2}$$

and

$$I_2 = \frac{\beta_2}{2}(V_{gs2} - V_{T2})^2(1 + \lambda_2 V_{ds2}) \tag{4.3}$$

where

$$\beta_2 = \mu_2 C_{OX2}\left[\frac{W}{L}\right]_2 \tag{4.4}$$

From (4.1)

$$V_{gs1} = V_{T1} + \sqrt{\frac{2I_1}{\beta_1(1 + \lambda_1 V_{ds1})}} \tag{4.5}$$

With the switch S closed, V_{gs2} is forced to the value for V_{gs1} in (4.5) and is held substantially at that value when the switch is opened. So, putting V_{gs2} = V_{gs1} in (4.3) gives

$$I_2 = \frac{\beta_2}{2}(V_{gs1} - V_{T2})^2(1 + \lambda_2 V_{ds2})$$

$$= \frac{\beta_2}{2}(V_{gs1}^2 + V_{T2}^2 - 2V_{gs1}V_{T2})(1 + \lambda_2 V_{ds2}) \qquad (4.6)$$

Substituting (4.5) and putting $V_{T1} \approx V_{T2} \approx V_T$ gives, after some simplification,

$$I_2 = \frac{\beta_2}{2}(1 + \lambda_2 V_{ds2})\left[\delta V_T^2 + 2\delta V_T(V_{gs} - V_T)\right] + \frac{\beta_2}{\beta_1}\left[\frac{1 + \lambda_2 V_{ds2}}{1 + \lambda_1 V_{ds1}}\right] \qquad (4.7)$$

where $\delta V_T = V_{T1} - V_{T2}$. Dividing by I_1 (from (4.1)) gives

$$\frac{I_2}{I_1} = \frac{\beta_2}{\beta_1}\left[\frac{1 + \lambda_2 V_{ds2}}{1 + \lambda_1 V_{ds1}}\right]\left[1 + \left(\frac{\delta V_T}{V_{gs} - V_T}\right)^2 + \frac{2\delta V_T}{V_{gs} - V_T}\right] \qquad (4.8)$$

$$\approx \frac{\beta_2}{\beta_1}\left[\frac{1 + \lambda_2 V_{ds2}}{1 + \lambda_1 V_{ds1}}\right]\left[1 + \frac{2\delta V_T}{V_{gs} - V_T}\right] \qquad (4.9)$$

It is noted that the influence of drain-source voltage may be minimised by forcing $V_{ds2} = V_{ds1} = V_{ds}$ (which can be achieved with circuit enhancements). Then, with β redefined as the gain at a drain-source voltage V_{ds}, the channel shortening factors $(1 + \lambda V_{ds})$ may be dropped. This gives

$$\frac{I_2}{I_1} = \frac{\beta_2}{\beta_1}\left[1 + \frac{2\delta V_T}{V_{gs} - V_T}\right] \qquad (4.10)$$

For a 1:1 memory, the mismatch error, ε_M, is given by

$$\varepsilon_M = \frac{I_2 - I_1}{I_1} \approx \frac{\delta\beta}{\beta} + \frac{2\delta V_T}{V_{gs} - V_T} \qquad (4.11)$$

where $\delta\beta = \beta_2 - \beta_1$. The above result expresses the matching error for specific errors in gain $(\delta\beta)$ and threshold voltage (δV_T).

Consider now the effect of the mismatch error, ε_M, on integrator performance. A lossless integrator has a mirrored output and so the non-ideal integrator response is simply

$$H(e^{j\omega T}) = (1 + \varepsilon_M)H_i(e^{j\omega T}) \qquad (4.12)$$

where $H_i(e^{j\omega T})$ is the response of an ideal lossless integrator. Clearly, the error of the integrator is just the same as that of the mirror. Now, the output current of a non-ideal 1:1 current mirror with bias currents J at both input and output and with a signal current, i, is given by

$$i_0 = -i(1 + \varepsilon_M) - J\varepsilon_M \qquad (4.13)$$

For $i = \hat{I}sin\omega T$, it can be shown that

$$i_0 \approx -i - J\varepsilon_M - \left(\frac{\delta\beta}{\beta} + \frac{\delta V_T}{V_{gs} - V_T}\right)\hat{I}sin\omega t$$

$$-\left[\frac{\hat{I}}{J}\right]\frac{\delta V_T}{8(V_{gs} - V_T)}\hat{I}cos2\omega t + \left[\frac{\hat{I}}{J}\right]^2\frac{\delta V_T}{32(V_{gs} - V_T)}\hat{I}sin3\omega t \qquad (4.14)$$

This indicates that an integrator with a non-ideal output current mirror will have not only an error in its gain (as indicated by amplitude of the fundamental signal) but also it will have an offset error and harmonic distortion.

Returning to the current mirror, the mismatch errors, in practice, are random in nature and have been researched extensively in recent years. *Shyu* [1] reported in 1984 on the analysis and measurement of random errors in current sources (and MOS capacitors) in a 3.5µm NMOS process. In 1986, further studies by *Lakshmikumar* [2] for a 3µm p-well CMOS process, quantified current ratio errors in terms of threshold and gain errors. Most recently (1989), *Pelgrom* [3] reported on current mirror matching errors for N-well CMOS processes with minimum feature sizes ranging from 2.5µm to 1.6µm. The physical causes of random errors include random local variations in surface charges (both due to ion-implanted charges and surface states), in the oxide (both thickness and permittivity) and in carrier mobility. Although there is some reported disagreement in the relative significance of these physical causes, perhaps due to the development of technology since 1984, the following general points may be drawn from these studies for close pairs of MOS transistors:

1   There is very low correlation between errors in V_T and β.

2   The variances of V_T and β mismatch errors are inversely proportional to transistor area.

3   As CMOS processes have scaled to smaller dimensions, the variance of V_T mismatch errors has varied with the square of the oxide thickness but the variance of β mismatch errors has remained constant.

4   The variance of current ratio is nearly as small as that of capacitor ratio in a switched-capacitor process.

Returning to (4.11), the static current ratio error of the memory cell with equal drain-source voltages has two terms: current gain error, $\delta\beta/\beta$, and threshold voltage error, $2\delta V_T/(V_{gs}-V_T)$. As there is very low correlation between these randomly varying mismatch errors (point 1) the variance of the current ratio mismatch error is formed by the mean square sum

$$\frac{\sigma^2(I)}{I^2} = \frac{\sigma^2(\beta)}{\beta^2} + \frac{4\sigma^2(V_T)}{(V_{gs}-V_T)^2} \tag{4.15}$$

From point 2 above

$$\frac{\sigma^2(\beta)}{\beta^2} = \frac{A_\beta^2}{WL} \tag{4.16}$$

$$\sigma^2(V_T) = \frac{A_{V_T}^2}{WL} \tag{4.17}$$

where A_β and A_{V_T} are process dependent constants. Substituting (4.16) and (4.17) in (4.15)

$$\frac{\sigma^2(I)}{I^2} = \frac{K_\sigma}{WL} \tag{4.18}$$

$$K_\sigma = A_\beta^2 + \frac{4A_{V_T}^2}{(V_{gs}-V_T)^2} \tag{4.19}$$

For a current CMOS process, the N-channel transistors have the following parameters [3]:

$$A_\beta = .02\mu m \ , \ A_{V_T} = 20mV\mu m \tag{4.20}$$

The contributions due to β mismatch and V_T mismatch are equal when

$$(V_{gs}-V_T)_{opt} = \frac{2A_{V_T}}{A_\beta} \tag{4.21}$$

For the numerical values given in (4.20)

$$(V_{gs}-V_T)_{opt} = 2V \tag{4.22}$$

This is somewhat higher than desired for low supply voltage operation (5V or 3.3V). So designing with lower $(V_{gs}-V_T)$ will give increased $\sigma(I)/I$ which will be dominated by V_T mismatch errors. Nevertheless, a pair of transistors with, say, $WL = 1000\mu m$ and $(V_{gs}-V_T) = 1V$ gives $\sigma(I)/I = 0.14\%$ which is comparable with achievable matching errors for two similar sized capacitors [1].

4.2.2 Non-unity gain current memory

So far the discussion has centred on the accuracy of unity gain current memory cells. Since mirror current-ratios are to be used to define coefficient values, they will in practice be non-unity and most probably non-integer. In switched-capacitor circuits non-integer capacitor ratios are made from arrays of unit capacitors with an adjustment to the geometry of one unit to trim the value to give the required ratio. A similar approach is possible with arrays of unit transistors to achieve an accurate current ratio.

Consider again a current memory cell of the type shown in Figure 4.1(b). To implement a non-unity current ratio *M:N*, T_1 would be made from *M* unit transistors and T_2 from *N* unit transistors, each set being connected in parallel. If each unit transistor is assumed to have an independent, normally distributed, current spread with variance σ_i^2 in its average current, *I*, then the relative variance is σ_i^2/I^2. A pair of such unit transistors connected to form a unity gain current mirror *(M/N = 1)*, would have a relative variance in the error between their currents given by

$$\frac{\sigma^2(I)}{I^2} = \frac{2\sigma^2 i}{I^2} \qquad (4.23)$$

An array of *M* unit transistors, each carrying an average current *I*, would pass an average total current *MI*. The variance in this current *MI* is

$$\sigma^2_{(MI)} = M\sigma^2 i \qquad (4.24)$$

Similarly, an array of *N* unit transistors would pass an average total current of *NI* with variance

$$\sigma^2_{(NI)} = N\sigma^2 i \qquad (4.25)$$

The variance in the current ratio, *R = MI/NI* can be found from

$$\sigma^2(R) \approx \left[\frac{\partial R}{\partial (MI)}\right]^2 \sigma^2_{(MI)} + \left[\frac{\partial R}{\partial (NI)}\right]^2 \sigma^2_{(NI)} \qquad (4.26)$$

$$\approx \frac{\sigma^2 i}{I^2}\left[\frac{M}{N^2} + \frac{M^2}{N^3}\right] \qquad (4.27)$$

The average value of the ratio *R* is given by

$$\overline{R} = \frac{M}{N} \qquad (4.28)$$

and so the relative variance in *R* is given by

$$\frac{\sigma^2(R)}{R^2} = \frac{\sigma^2 i}{I^2}\left[\frac{1}{M} + \frac{1}{N}\right] \tag{4.29}$$

Substituting from (4.23) and (4.18) gives

$$\frac{\sigma^2(R)}{R^2} = \frac{K_\sigma}{2WL}\left[\frac{1}{M} + \frac{1}{N}\right] \tag{4.30}$$

$$\approx \frac{K_\sigma}{2}\left[\frac{1}{A_1} + \frac{1}{A_2}\right] \tag{4.31}$$

where K_σ is as defined in (4.19) and A_1 and A_2 are the areas of transistors T_1 and T_2 comprising M and N unit transistors each of area WL.

Equation (4.31) also holds when T_1 and T_2 are just two transistors with different aspect ratios (W/Ls) rather than arrays of unit transistors. Such an arrangement has extra current ratio errors due to systematic errors in the two transistors' widths and lengths due to spreads in processing and lithography. If the transistors' dimensions are nominally W_1/L_1 and W_2/L_2 but with systematic errors δW and δL, respectively, then the relative systematic error in the current ratio is given by

$$\varepsilon_{sys} = \frac{\dfrac{W_2 + \delta W}{L_2 + \delta L}\dfrac{L_1 + \delta L}{W_1 + \delta W}}{\dfrac{W_2}{L_2}\dfrac{W_1}{L_1}} \tag{4.32}$$

$$\approx \delta W\left|\frac{1}{W_2} - \frac{1}{W_1}\right| + \delta L\left|\frac{1}{L_1} - \frac{1}{L_2}\right| \tag{4.33}$$

This systematic error is in addition to the random error given by (4.31).

Consider for example a 5:1 current mirror ratio using the process constants described in (4.20). For $V_{gs} - V_T = 1V$ as before, $K_\sigma = 0.002\mu m^2$. First, using 5:1 arrays of unit transistors with $W = 20\mu m$ and $L = 10\mu m$, the relative standard deviation in the current ratio is $\sigma(R)/\overline{R} = 0.23\%$. Next, using $W_1 = 100\mu m$, $W_2 = 20\mu m$, $L_1 = L_2 = 10\mu m$, the relative standard deviation in the current ratio is again 0.23% (since the areas are the same as those of the arrays) but there is an additional systematic error (for $\delta W = 0.25\mu m$) of $\varepsilon_{sys} = 1\%$. This clearly demonstrates the benefit of using arrays of unit transistors.

4.3 Output-Input Conductance Ratio Errors

On the sampling phase of the basic memory cell (ϕ_1, Figure 4.1), the memory transistor's drain voltage is forced to the same value as that at its gate. On the holding phase, when the cell delivers output current to its load, the drain voltage may change and this causes an error in the current flowing at the drain due to two main effects. Firstly, channel shortening effects as expressed in the familiar square-law equation for the MOS transistor in saturation:

$$I_{ds} = \frac{\beta}{2}(V_{gs} - V_T)^2(1 + \lambda V_{ds})$$

where λ is the channel shortening parameter, gives the memory transistor a drain conductance g_{ds} given by

$$g_{ds} = \frac{\partial I_{ds}}{\partial V_{ds}} = \lambda I_{ds} \qquad (4.34)$$

Secondly, with the gate of the memory transistor held open, changes in the drain voltage cause charge to flow through the drain-gate overlap capacitance C_{dg} into the memory capacitance C. This disturbs the gate-source voltage held on C and produces an error in the drain current.

Together, these two effects produce an error current δI_{ds} resulting from a change in the drain voltage δV_{ds} given by

$$\delta I_{ds} = \delta V_{ds}\left(g_{ds} + \frac{C_{dg}}{C + C_{dg}}g_m\right)$$

The memory cell behaves as though it had an output conductance g_{OM} given by

$$g_{OM} = g_{ds} + \frac{C_{dg}}{C + C_{dg}}g_m \qquad (4.35)$$

A small signal transmission error ε_G occurs in the basic memory cell when g_{OM} is not negligibly small compared with the small signal input conductance (g_m) of the diode-connected memory transistor. This error and the effect on the memory cell's frequency response is analysed in the following section.

4.3.1 *Memory cell response with conductance ratio errors*

Consider cascaded memory cells as shown in Figure 4.2. The memory cell is modelled as an ideal memory transistor T with a conductance g_0 connected between the drain and source where

$$g_0 = g_{ds} + g_{ds(J)} + \frac{C_{dg}}{C + C_{dg}}g_m \qquad (4.36)$$

where $g_{ds(J)}$ is the drain conductance of the transistor acting as the current source J. On phase ϕ_1, T_1 is diode connected and conducts the sum of the bias current J and the input signal i. For small signals, $V_{ds1} = V_{gs1} = V_{gs} + i(n-1)/g_m$ where V_{gs} and g_m are at $I_{ds} = J$ and I_1 is given by

$$I_1 = J + i(n-1) - I_{g01} = J + i(n-1)\left(1 - \frac{g_0}{g_m}\right) - V_{gs}g_0$$

On phase ϕ_2, I_1 is held at the above value and the first memory cell outputs current $i_0(n-1/2)$ to the second memory cell, T_2. The drain voltage of T_1 is determined by the gate voltage of T_2, i.e. $V_{ds1} = V_{gs2} \approx V_{gs} + i_0(n-1/2)/g_m$. Again, it is readily shown that $i_0(n-1/2)$ is given by

$$i_0(n-1/2) = J - I_1 - I_{g02} = -i(n-1)\left(1 - \frac{2g_0}{g_m}\right) \approx -\frac{i(n-1)}{1 + \frac{2g_0}{g_m}}$$

Taking z-transforms:

$$i_0 z^{-1/2} = -\frac{i(z)z^{-1}}{1 + \frac{2g_0}{g_m}}$$

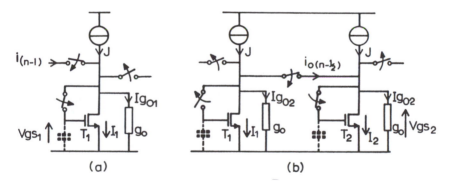

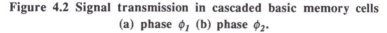

(a)   (b)

Figure 4.2 Signal transmission in cascaded basic memory cells
(a) phase ϕ_1 (b) phase ϕ_2.

$$H(z) = \frac{i_0(z)}{i(z)} = \frac{-z^{-1/2}}{1 + \frac{2g_0}{g_m}} = \frac{H_i(z)}{1 + \frac{2g_0}{g_m}} \tag{4.37}$$

where $H_i(z)$ is ideal value of $H(z)$ with $g_0/g_m = 0$ and corresponds to a half-clock period delay $(T/2)$ with signal inversion.

For physical frequencies, $z = e^{j\omega T}$ and the frequency response is given by

$$H(e^{j\omega T}) = \frac{H_i(e^{j\omega T})}{1 + \dfrac{2g_0}{g_m}} \tag{4.38}$$

where $H_i(e^{j\omega T}) = -e^{-j\omega T/2}$ is the frequency response of an ideal memory cell. A useful approximation for any network relating actual and ideal frequency response for small errors is as follows [5]:

$$H(e^{j\omega T}) = \frac{H_i(e^{j\omega T})}{(1 - m(\omega))e^{-j\theta(\omega)}} \approx \frac{H_i(e^{j\omega T})}{1 - m(\omega) - j\theta(\omega)} \tag{4.39}$$

where $m(\omega)$ and $\theta(\omega)$ are the magnitude and phase errors. Comparing (4.38) and (4.39) gives

$$m(\omega) = -\frac{2g_0}{g_m} = \varepsilon_G$$

$$\theta(\omega) = 0 \tag{4.40}$$

So, the transmission error ε_G is a magnitude loss of $2g_0/g_m$ with no phase error. This may be expressed as follows:

$$\varepsilon_G = -\frac{2g_0}{g_m} = -2\left(\frac{g_{ds} + g_{ds(J)}}{g_m} + \frac{C_{dg}}{C + C_{gd}}\right) \tag{4.41}$$

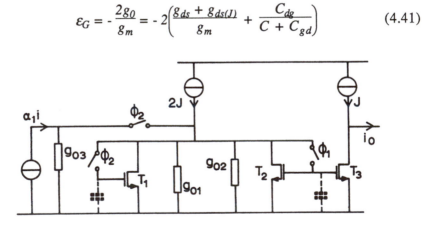

Figure 4.3 Non-inverting lossless integrator with non-zero output conductance.

In the basic memory cell, the error ε_G can be a few percent and while it can be lowered by using large values of channel length, L, this gives larger chip area and reduced bandwidth. So alternative solutions employing circuit enhancements may be more appropriate and these are discussed in Chapter 6.

4.3.2 Integrator response with conductance ratio errors

We will now analyse the effect of this error on the integrator's small signal performance. The approach to be used is to re-establish the difference equations taking account of the error ε_G. Consider a non-inverting lossless integrator which is ideal in all respects except for the finite transconductances, gm, and the effects of channel shortening and capacitive feedback as modelled by conductances, g_{01} and g_{02}, of its memory cells as shown in Figure 4.3. If the transistors T_1 and T_2 are identical then, for small signals, $g_{01} = g_{02} = g_0$ where g_0 is the memory cell output conductance with $I_{ds} = J$.

On phase ϕ_2 of period *(n-1)*, the integrator is configured as shown in Figure 4.4(a). The conductance at the summing node is Σg_0 where

$$\Sigma g_0 = g_{01} + g_{02} = 2g_0 \tag{4.42}$$

The voltage at the current summing node is $V_{gs1} = V_{gs} + i_1/g_m$ where V_{gs} and g_m are for $I_{ds} = J$ and i_1 is the small signal component of I_1 i.e. $i_1 = I_1 - J$. The current in Σg_0 is I_{g0} where

$$I_{g0} = \left(V_{gs} + \frac{i_1}{g_m}\right)\Sigma g_0 = \left(V_{gs} + \frac{i_1}{g_m}\right)2g_0 \tag{4.43}$$

Also,

$$I_2 = I_3 = J - i_0(n-1) \tag{4.44}$$

We have,

$$I_1 = 2J - I_{g0} - I_2 + \alpha_1 i(n-1) = J + i_1 \tag{4.45}$$

So,

$$i_1 = \alpha_1 i(n-1) + i_0(n-1) - \left(V_{gs} + \frac{i_1}{g_m}\right)2g_0 \tag{4.46}$$

Rearranging (4.46),

$$i_1 = \frac{\alpha_1 i(n-1) + i_0(n-1) - V_{gs}2g_0}{1 + \dfrac{2g_0}{g_m}} \tag{4.47}$$

On phase ϕ_1 of period *(n)* the integrator is configured as shown in Figure 4.4(b). The conductance at the summing node is again $2g_0$. The current I_2 is given by

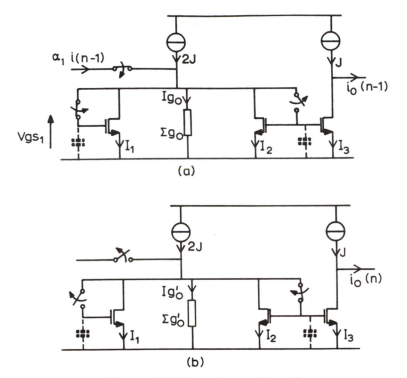

Figure 4.4 Integrator configurations
(a) on phase ϕ_2 of period $(n\text{-}1)$ (b) on phase ϕ_1 of period (n).

$$I_2 = 2J - I_1 - I'_{g_0} = J + i_2 \qquad (4.48)$$

where i_2 is the small signal component of I_2 and

$$I'_{g_0} = \left(V_{gs} + \frac{i_2}{g_m}\right)(2g_0)$$

$$i_2 = \frac{-i_1 - V_{gs}(2g_0)}{1 + \dfrac{2g_0}{g_m}} = -i_0(n) \qquad (4.49)$$

Substituting (4.47) in (4.49) then it can be shown that for $g_0 \ll g_m$

$$i_0(n)\left(1 + 4\frac{g_0}{g_m}\right) \approx \alpha_1 i(n\text{-}1) + i_0(n\text{-}1) \qquad (4.50)$$

Taking z-transforms gives

$$H(z) = \frac{i_0(z)}{i(z)} = \frac{\alpha_1 z^{-1}}{1 + 4\dfrac{g_0}{g_m} - z^{-1}} \qquad (4.51)$$

Equation (4.51) has the same form as that of an ideal lossy non-inverting integrator where the loss coefficient α_4 is equivalent to $4g_0/g_m$. So, non-zero g_0/g_m turns the ideal integrator into a first-order lowpass section with low-frequency gain and a cut-off frequency given by

$$a_0 = \frac{g_m}{4g_0}$$

$$\omega_0 = \frac{2\alpha_1}{T} \frac{2}{2 + \dfrac{g_m}{g_0}} \qquad (4.52)$$

Putting $\varepsilon_G = -2g_0/g_m$ in (4.51) and multiplying by $z^{1/2}$ gives

$$H(z) = \frac{\alpha_1 z^{-1/2}}{z^{1/2} - z^{-1/2} - 2\varepsilon_G z^{1/2}}$$

For physical frequencies, $z = e^{j\omega T}$,

$$H(e^{j\omega T}) = \frac{\alpha_1 e^{-j\omega T/2}}{j(2 - 2\varepsilon_G)\sin\dfrac{\omega T}{2} - 2\varepsilon_G \cos\dfrac{\omega T}{2}}$$

When $\varepsilon_G = 0$, the ideal response results where

$$H_i(e^{j\omega T}) = \frac{\alpha_1 e^{-j\omega T/2}}{j2\sin\dfrac{\omega T}{2}} \qquad (4.53)$$

$$H(e^{j\omega T}) = \frac{H_i(e^{j\omega T})}{1 - \varepsilon_G - j\dfrac{\varepsilon_G}{\tan\dfrac{\omega T}{2}}}$$

Using the approximation given in (4.39), the magnitude and phase errors are given by

$$m(\omega) \approx \varepsilon_G$$

$$\theta(\omega) \approx \frac{\varepsilon_G}{\tan\dfrac{\omega T}{2}} \qquad (4.54)$$

It can be seen from (4.53) that the ideal integrator's unity gain frequency is given by

$$\omega_1 = \frac{2}{T} sin^{-1}\left(\frac{\alpha_1}{2}\right)$$

and the magnitude and phase errors at the unity gain frequency for $\omega_1 T \ll 1$ are given by

$$m(\omega_1) \approx \varepsilon_G$$

$$\theta(\omega_1) \approx \frac{2\varepsilon_G}{\alpha_1} \qquad (4.55)$$

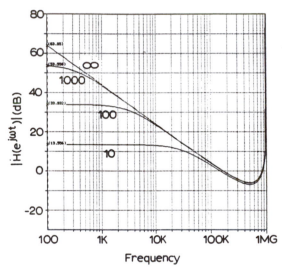

Figure 4.5 **Simulated frequency response of a non-inverting integrator with $g_m/\Sigma g_{ds}$ = ∞, 1000, 100, 10 and C_{dg} = 0 (T = 1μs, α_1 = 1).**

4.3.3 Comparison with switched-capacitor integrator

The above result is analogous to the effect of finite operational amplifier gain, A_0, on the performance of an active-RC integrator [4] or a switched-capacitor integrator [5]. It is interesting to note that the effect of finite g_m/g_0 in switched-current integrators is to produce incomplete transfer of current between memory cells, while in switched-capacitor integrators it produces finite low-frequency voltage gain in their op-amps and hence incomplete transfer of charge. Perhaps not surprisingly, it results in finite low-frequency gain in both types of integrator.

Just as switched-capacitor filters use op-amps with a low-frequency gain of 70-80dB to make errors negligible, switched-current filters should be designed with $\varepsilon_G \approx$ -70dB. This is clearly beyond the capability of the simple current memory which can achieve $\varepsilon_G \approx$ -30dB, indicating that more complex circuit structures are needed. Circuit techniques for increasing the memory cell's ratio of input to output conductance involve the use of negative feedback either to reduce the output conductance or increase the input conductance. Such techniques are described in Chapter 6.

4.3.4 Simulations

The above theory was checked by the simulation of a lossless non-inverting integrator which was ideal except for g_0 and g_m using an in-house switched-capacitor analysis program (SCALP2). First, g_0 was defined with resistors connected between the memories' drains and sources to model g_{ds}. They were set to give ratios of $g_m/\Sigma g_{ds}$ of infinity, 1000, 100 and 10 and the resulting transfer characteristics were compared with an ideal integrator ($\alpha_1 = 1$). The simulated performance is shown in Figure 4.5. The low-frequency gains and cut-off frequencies agree exactly with those computed from (4.52). Next, g_0 was defined by capacitors connected between the memory transistors' drains and gates to model C_{dg}. Again, the values were set to give ratios of C/C_{dg} of infinity, 1000, 100 and 10. The simulations are shown in Figure 4.6 and show exact agreement with that given in (4.52).

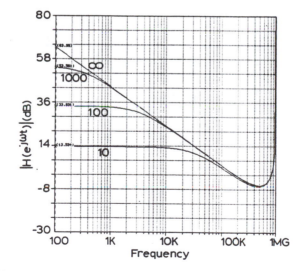

Figure 4.6 Simulated frequency response of a non-inverting integrator with $C/C_{dg} = 1$, 1000, 100, 10 and $\Sigma g_{ds} = 0$(T = 1µs, $\alpha_1 = 1$).

4.4 Settling Errors

The operation of the switched-current memory cell involves the charging of its gate capacitor to the gate-source voltage of a diode-connected input transistor (see Figure 4.1). If this charging is not completed during the time interval in which switch S is closed *(T/2)*, a residual error voltage results. At the end of this interval, switch S opens, the error voltage is stored, and results in an error in the memory cell's output current.

The electrical behaviour of the memory cell shown in Figure 4.1(b) will now be analysed. The small signal equivalent circuit is shown in Figure 4.7. Note that C is the combined oxide capacitance of T_1 and T_2. It can be shown that the transfer function is given by

$$H(s) = \frac{i_0(s)}{i(s)} = -\frac{1}{1 + s\left[\dfrac{C + C_d}{g_m}\right] + s^2\left[\dfrac{CC_d}{g_m g_s}\right]} \tag{4.56}$$

This is a second-order low-pass response with pole frequency and Q-factor given by

$$\omega_0 = \sqrt{\frac{g_m g_s}{CC_d}} \tag{4.57}$$

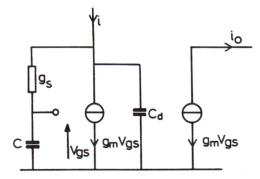

Figure 4.7 s-domain equivalent circuit of memory cell with switch S closed.

$$Q = \frac{\sqrt{\dfrac{g_m}{g_s} CC_d}}{C + C_d} \tag{4.58}$$

The response can be underdamped (for $Q > 0.5$), critically damped (for $Q = 0.5$) or overdamped (for $Q < 0.5$).

4.4.1 Underdamped response

The time-domain response to a unit input current step may be derived from (4.56) by applying the inverse Laplace transform. For the under-damped case $(Q > 0.5)$ this gives

$$i_0(t) = 1 - \frac{2Q}{\sqrt{4Q^2 - 1}} e^{\frac{\omega_o}{2Q}t} \sin\left[\frac{\sqrt{4Q^2 - 1}}{2Q}\omega_0 t + \phi\right]$$

where

$$\phi = tan^{-1} \sqrt{4Q^2 - 1}$$

which describes an overshoot followed by an exponentially damped sinusoid. The value of $i_0(t)$ at $t = T/2$ is stored for the subsequent half clock period, resulting in a settling error for the underdamped case, $\varepsilon_S(u)$, of

$$\varepsilon_S(u) = 1 - i_0(T/2)$$

$$\varepsilon_S(u) = \frac{2Q}{\sqrt{4Q^2 - 1}} e^{\frac{\omega_o}{2Q}t} \sin\left[\frac{\sqrt{4Q^2 - 1}}{4Q}\omega_0 t + \phi\right] \qquad (4.59)$$

It can be shown (see Figure 4.8) that the overshoot occurs at a time t_{OS}, and has an amplitude, ε_{OS}, given by

$$t_{OS} = \frac{2\pi Q}{\omega_0\sqrt{4Q^2 - 1}} \qquad (4.60)$$

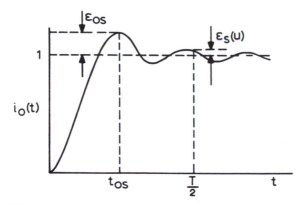

Figure 4.8 Underdamped second-order response.

$$\varepsilon_{OS} = exp\left(- \frac{\pi}{\sqrt{4Q^2 - 1}}\right) \qquad (4.61)$$

The minimum settling time, t_S min, results when the specified error, ε, equals ε_{OS} and this occurs when Q is optimised to the value Q_{opt} where

$$Q_{opt} = \frac{1}{2}\sqrt{1 + \left(\frac{\pi}{\ln\varepsilon}\right)^2} \qquad (4.62)$$

From (4.57), (4.58) and (4.60) this gives

$$t_S\ min = \frac{2\pi Q_{opt}^2}{\sqrt{4Q_{opt}^2 - 1}}\tau \qquad (4.63)$$

where

$$\tau = \frac{(C + C_d)}{g_m} \qquad (4.64)$$

Substituting (4.62) gives

$$t_S\ min = \frac{1}{2}\left[1 + \left(\frac{\pi}{\ln\varepsilon}\right)^2\right]\tau \ln\frac{1}{\varepsilon} \qquad (4.65)$$

4.4.2 Critically damped response

For the critically damped case ($Q = 0.5$), the time-domain response is given by

$$i_0(t) = 1 - (1 + \omega_0 t)e^{-\omega_0 t}$$

This is an exponential response and so has no overshoot. The settling error for the critically damped case, $\varepsilon_S(c)$, is given by

$$\varepsilon_S(c) = 1 - i_0(T/2) = \left(1 + \frac{\omega_0 T}{2}\right)e^{-\omega_0 T/2} \qquad (4.66)$$

The settling time $t_S(c)$ for a specified settling error ε is defined implicitly by

$$\varepsilon = (1 + \omega_0 t_S(c))e^{-\omega_0 t_S(c)} \qquad (4.67)$$

4.4.3 Overdamped response

For the overdamped case ($Q < 0.5$) the time domain response is given by

$$i_0(t) = 1 - \frac{1}{2\sqrt{1-4Q^2}}\left[(1+\sqrt{1-4Q^2})\exp\left(-\frac{1-\sqrt{1-4Q^2}}{2Q}\omega_0 t\right)\right.$$

$$- \left(1-\sqrt{1-4Q^2}\right)\exp\left(-\frac{1+\sqrt{1-4Q^2}}{2Q}\omega_0 t\right)\bigg]$$

which is exponential with no overshoot. The settling error for the overdamped case is

$$\varepsilon_S(o) = 1 - i_0(T/2)$$

$$\varepsilon_S(o) = \frac{1}{2\sqrt{1-4Q^2}}\bigg[\left(1+\sqrt{1-4Q^2}\right)\exp\left(-\frac{1-\sqrt{1-4Q^2}}{4Q}\omega_0 T\right)$$

$$- \left(1-\sqrt{1-4Q^2}\right)\exp\left(-\frac{1+\sqrt{1-4Q^2}}{4Q}\omega_0 T\right)\bigg] \quad (4.68)$$

The settling time, $t_S(o)$, for a specified settling error ε is defined implicitly by

$$\varepsilon = \frac{1}{2\sqrt{1-4Q^2}}\bigg[\left(1+\sqrt{1-4Q^2}\right)\exp\left(-\frac{1-\sqrt{1-4Q^2}}{2Q}\omega_0 t_S(o)\right)$$

$$- \left(1-\sqrt{1-4Q^2}\right)\exp\left(-\frac{1+\sqrt{1-4Q^2}}{2Q}\omega_0 t_S(o)\right)\bigg] \quad (4.69)$$

Figure 4.9 shows simulations of the current memory's response to a unit step input current. The current memory is simply modelled as the second-order system shown in Figure 4.7 with the following parameters: $g_m = 1S$, $C = 1F$, $C_d = 1F$. The three responses are for $r_s = 1/g_s = 0$ ($Q = 0$), $r_s = 1\Omega$ ($Q = 0.5$, $\omega_0 = 1Hz$), $r_s = 2\Omega$ ($Q = 0.707$, $\omega_0 = 1.414Hz$) corresponding to extreme over-damping, critical damping and underdamping, respectively. Universal curves of the step response of second-order systems can be found in textbooks [6].

It can be seen from the previous analysis that the underdamped error alternates in polarity (4.59). If the sampled error is negative, the gain becomes greater than unity and the system may become unstable. This situation cannot arise with overdamping or critical damping as their errors are always positive ((4.66) and (4.68)). So, although the underdamped response can in principle be tuned to give the shortest settling time (4.65), greater safety results from the overdamped response. The pole frequency and Q-factor ((4.57) and (4.58)) are in part determined by parasitic components and thus are subject to large variations due to processing spreads and bias conditions. No doubt an overdamped response must be chosen and the degree of overdamping (i.e. the Q-factor) will be a matter of engineering judgement which inevitably will involve a settling time penalty. No explicit solution to the overdamped settling time can be found ((4.68) gives it implicitly) but

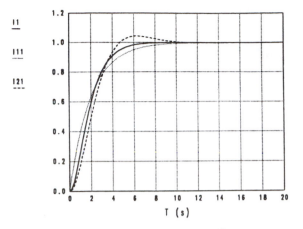

Figure 4.9 Basic cell simulated settling responses
I1...Critically damped, $C = C_d = 1F$, $g_m = g_s = 1S$ I11..Overdamped,
$g_s = 1S$ I21..Underdamped, $g_s = 0.5S$

a feel for the penalty can be obtained by comparing the settling time for extreme overdamping ($Q = 0$) with t_S *min*. Setting $Q = 0$ in (4.69) and substituting in (4.65) gives

$$t_S(Q=0) = \tau \ln\left(\frac{1}{\varepsilon}\right) = \frac{2t_S \ min}{1 + \left(\frac{\pi}{\ln\varepsilon}\right)^2} \qquad (4.70)$$

So, even with $Q = 0$, the settling time is less than twice the theoretical minimum. As an example, consider a current memory with the following parameters: $g_m = 165\mu S$, $C = 3pF$, $C_d = 0.3pF$, $\varepsilon = 0.001$, τ is 20ns (4.64) and Q_{opt} is 0.5493. This is achieved by setting g_s to $45\mu S(r_s \approx 22K\Omega)$ (4.58). This gives a minimum settling time of 83ns (4.65). For extreme overdamping ($Q = 0$), $g_s = 1$ and the settling time is 137ns (4.70). In practice, an intermediate value of Q would be chosen to give a more practical switch design while still giving a safety margin from the underdamped condition.

4.4.4 *Memory cell response with settling errors*

We now consider the effect of settling error ε_S on the frequency response of the current memory. In the following analysis it is assumed that the memory cell is ideal except for its settling error. The current memory and its waveforms are shown in Figure 4.10. During clock phase ϕ_1, the memory transistor is diode-connected and the drain current I increases from its previous level of $I(n-1)$ towards a new level of $\hat{I}(n) = J + i(n-1/2)$. For small signals ($i \ll J$) it is fair to assume the network is linear and so I reaches a final value $I(n)$ given by

$$I(n) = I(n-1) + [\hat{I}(n) - I(n-1)](1-\varepsilon_S) \tag{4.71}$$

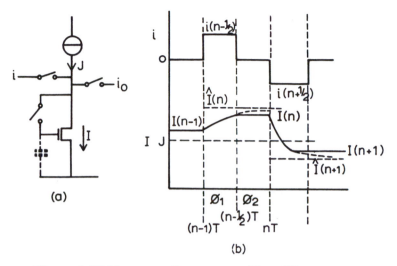

Figure 4.10 Memory cell response with settling error
(a) memory cell (b) waveforms.

During the next phase ϕ_2, the drain current is held at $I(n)$ and produces an output current $i_0(n) = J - I(n)$. Similarly, during the previous phase ϕ_2, the drain current was $I(n-1) = J - i_0(n-1)$. Substituting in (4.71) gives

$$i_0(n) = \varepsilon_S i_0(n-1) - (1-\varepsilon_S)i(n-1/2) \tag{4.72}$$

This shows that the settling error produces not only an attenuation of the input signal but also a memory of signals from earlier clock cycles. Taking z-transforms of (4.72) gives

$$i_0(z)(1 - \varepsilon_S z^{-1}) = -(1 - \varepsilon_S)i(z)z^{-1/2}$$

$$H(z) = \frac{i_0(z)}{i(z)} = -z^{-1/2}\frac{1 - \varepsilon_S}{1 - \varepsilon_S z^{-1}} = \frac{H_i(z)}{\dfrac{1}{1 - \varepsilon_S} - \dfrac{\varepsilon_S z^{-1}}{1 - \varepsilon_S}} \tag{4.73}$$

where $H_i(z)$ is the response of an ideal memory cell. For physical frequencies, $z = e^{j\omega T}$

$$H(e^{j\omega T}) = \frac{H_i(e^{j\omega T})}{\dfrac{1}{1-\varepsilon_S} - \dfrac{\varepsilon_S}{1-\varepsilon_S}\cos\omega T + j\dfrac{\varepsilon_S}{1-\varepsilon_S}\sin\omega T} \tag{4.74}$$

where $H_i(e^{j\omega T}) = -e^{-j\omega T/2}$. For $\varepsilon_S \ll 1$, this simplifies to

$$H(e^{j\omega T}) \approx \cfrac{H_i(e^{j\omega T})}{1 + \cfrac{\varepsilon_S}{1-\varepsilon_S}\cfrac{\sin^2\omega T}{2} + j\cfrac{\varepsilon_S}{1-\varepsilon_S}\sin\omega T}$$

Using the approximation in (4.39) and for $\omega T \ll 1$, the magnitude and phase errors are given by

$$m(\omega) \approx -\left(\frac{\varepsilon_S}{1-\varepsilon_S}\right)\frac{\omega^2 T^2}{2}$$

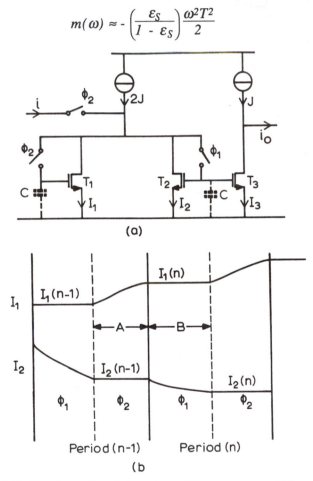

(a)

(b

Figure 4.11 **Non-inverting integrator with non-zero settling time**
(a) circuit configuration (b) current waveforms.

$$\theta(\omega) \approx -\left(\frac{\varepsilon_S}{1-\varepsilon_S}\right)\omega T \qquad (4.75)$$

So the effect of non-zero settling error is to increase the ideal delay of $T/2$ by $\varepsilon_S T/(1 - \varepsilon_S)$ with very little attenuation for low-frequency signals.

4.4.5 *Integrator response with settling errors*

We now consider the impact of the settling error on the transfer characteristic of the integrator. The method to be adopted is to re-establish the difference equations of the integrator taking account of the settling error. A non-inverting lossless integrator (with $\alpha_1 = 1$) is shown in Figure 4.11(a) and the current waveforms are shown in Figure 4.11(b).

First consider the circuit operation during phase ϕ_2 of period $(n-1)$, i.e. during period A. The circuit and current waveforms are shown in Figure 4.12. With the input switch closed, T_1 is diode-connected (with the equivalent circuit shown in Figure 4.7) and is driven by an input current step equal to the input signal current, $i(n-1)$, plus the bias current, $2J$, less the current stored in T_2, $J - i_0(n-1)$. This net current is $\hat{I}$ where

$$\hat{I}_1 = 2J + i(n-1) - (J - i_0(n-1))$$

$$= J + i(n-1) + i_0(n-1) \tag{4.76}$$

At the start of period A, the current in T_1 is $I_1(n-1)$ and the 'excess' current, $\hat{I}_1 - I_1(n-1)$, flows into C. As C charges, I_1 increases from its starting value, $I_1(n-1)$, towards a final value, $\hat{I}_1$. It never reaches this value because after a time $T/2$ has elapsed, period A is terminated. I_1 at this moment has reached the value $I_1(n)$ which is held throughout period B. It is readily shown that

$$I_1(n) = \varepsilon_S I_1(n-1) + (1 - \varepsilon_S)\hat{I}_1 \tag{4.77}$$

For practical reasons an overdamped response will be chosen giving $\varepsilon_S = \varepsilon_S(o)$ as defined by (4.69).

$$I_1(n) = \varepsilon_S I_1(n-1) + (1 - \varepsilon_S)(J + i(n-1) + i_0(n-1)) \tag{4.78}$$

Next we consider the circuit operation during phase ϕ_1 of period (n), i.e. period B. The circuit and current waveforms are shown in Figure 4.13. T_2 is now diode-connected and T_1 stores the current $I_1(n)$ given in (4.78). The diode-connected transistor T_2 receives a current step equal to the bias current $2J$ less the current stored in T_1, $I_1(n)$. This net current is $\hat{I}_2$ where

$$\hat{I}_2 = 2J - I_1(n) \tag{4.79}$$

As before, I_2 increases from its starting current, $I_2(n-1)$, towards $\hat{I}_2$ and the rise is terminated after time $T/2$ to give a stored current in T_2 of $I_2(n)$ where

$$I_2(n) = \varepsilon_S I_2(n-1) + (1 - \varepsilon_S)\hat{I}_2$$

$$= \varepsilon_S(J - i_0(n - 1)) + (1 - \varepsilon_S)(2J - I_1(n)) \tag{4.80}$$

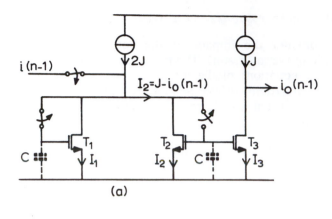

(a)

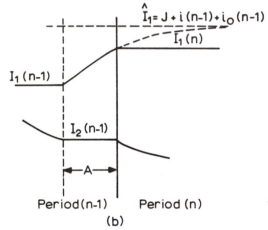

(b)

Figure 4.12 Integrator waveforms during period *A*
(a) circuit configuration (b) current waveforms.

Since $I_2(n)$ is mirrored into T_3 to produce $i_0(n)$, we also have

$$I_2(n) = J - i_0(n) \qquad (4.81)$$

Combining (4.80) and (4.81) gives

$$I_1(n) = J - \frac{\varepsilon_S}{1 - \varepsilon_S} i_0(n - 1) + \frac{1}{1 - \varepsilon_S} i_0(n) \qquad (4.82)$$

Since the operation during the previous clock period was similar to that just described we can also write

$$I_1(n - 1) = J - \frac{\varepsilon_S}{1 - \varepsilon_S} i_0(n-2) + \frac{1}{1 - \varepsilon_S} i_0(n-1) \qquad (4.83)$$

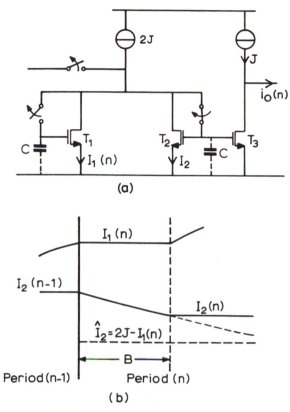

(a)

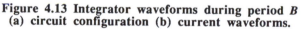

(b)

**Figure 4.13 Integrator waveforms during period B
(a) circuit configuration (b) current waveforms.**

If (4.82) and (4.83) are substituted in (4.78) we have

$$i_0(n) - (1 + \varepsilon_S^2)\, i_0(n-1) + \varepsilon_S^2\, i_0(n-2) = (1 - \varepsilon_S)2i(n-1)$$

Taking z-transforms gives

$$i_0(z)(1 - (1 + \varepsilon_S^2)z^{-1} + \varepsilon_S^2 z^{-2}) = (1 - \varepsilon_S)^2 z^{-1} i(z)$$

$$H(z) = \frac{i_0(z)}{i(z)} = \frac{(1 - \varepsilon_S)^2 z^{-1}}{(1 - z^{-1})(1 - \varepsilon_S^2 z^{-1})} \qquad (4.84)$$

Equation (4.84) gives the z-domain transfer characteristic of the non-inverting lossless integrator with non-zero settling error. Note that as $\varepsilon_S \to 0$

$$H(z) \to H_i(z) = \frac{z^{-1}}{1 - z^{-1}}$$

which is the transfer characteristic of an ideal non-inverting (Forward Euler) lossless integrator. Equation (4.84) may be re-written as follows:

$$H(z) = \gamma(z)H_i(z) \tag{4.85}$$

where

$$\gamma(z) = \frac{(1 - \varepsilon_S)^2}{1 - \varepsilon_S^2 z^{-1}} \tag{4.86}$$

$\gamma(z)$ represents the error in the integrator response introduced by the settling error of the memory cell. To find the response of this error function to physical frequencies we first manipulate (4.86) to give

$$\gamma(z) = \frac{z^{1/2}(1 - \varepsilon_S)^2}{z^{1/2} - z^{-1/2} + (1 - \varepsilon_S^2)z^{-1/2}}$$

Putting $z = e^{j\omega T}$ gives

$$\gamma(e^{j\omega T}) = \frac{e^{j\omega T/2}(1 - \varepsilon_S)^2}{j(1 + \varepsilon_S^2)\sin\frac{\omega T}{2} + (1 - \varepsilon_S^2)\cos\frac{\omega T}{2}} \tag{4.87}$$

For $\omega T \ll 1$

$$\gamma(e^{j\omega T}) = \frac{\left[\dfrac{1-\varepsilon_S}{1+\varepsilon_S}\right]e^{j\omega T/2}}{1 + j\left[\dfrac{\omega T(1+\varepsilon_S^2)}{2(1-\varepsilon_S^2)}\right]} = \frac{a_0(\gamma)e^{j\omega T/2}}{1 + j\dfrac{\omega}{\omega_0(\gamma)}} \tag{4.88}$$

We see that the error function is a first-order low-pass response with low-frequency gain given by

$$a_0(\gamma) = \frac{1 - \varepsilon_S}{1 + \varepsilon_S} \tag{4.89}$$

and a -3dB cut-off frequency given by

$$\omega_0(\gamma) = \frac{2}{T}\frac{1 - \varepsilon_S^2}{1 + \varepsilon_S^2} \tag{4.90}$$

In practice, a low value of ε_S would need to be chosen to minimise errors in the integrators' magnitude and phase response. More useful expressions for these errors can be obtained when the errors are small. From (4.85) and (4.86), putting $z = e^{j\omega T}$ gives the response

$$H(e^{j\omega T}) = \frac{(1 - \varepsilon_S)^2 H_i(e^{j\omega T})}{1 - \varepsilon_S^2(\cos\omega T - j\sin\omega T)}$$

$$= \frac{H_i(e^{j\omega T})}{\left[\dfrac{1}{1-\varepsilon_S}\right]^2 - \left[\dfrac{\varepsilon_S}{1-\varepsilon_S}\right]^2 \cos\omega T - j\left[\dfrac{1}{1-\varepsilon_S}\right]^2 \sin\omega T} \tag{4.91}$$

For small errors, $\varepsilon_S \ll 1$ and

$$H(e^{j\omega T}) \approx \frac{H_i(e^{j\omega T})}{1 + 2\varepsilon_S - \left[\dfrac{\varepsilon_S}{1-\varepsilon_S}\right]^2 \cos\omega T - j\left[\dfrac{1}{1-\varepsilon_S}\right]^2 \sin\omega T} \tag{4.92}$$

Using the approximation in (4.39) gives the magnitude and phase errors $m(\omega)$ and $\theta(\omega)$:

$$m(\omega) = -2\varepsilon_S + \left[\frac{\varepsilon_S}{1 - \varepsilon_S}\right]^2 \cos\omega T \approx -2\varepsilon_S$$

and

$$\theta(\omega) = \frac{\sin\omega T}{(1 - \varepsilon_S)^2} \tag{4.93}$$

For $\omega T \ll 1$,

$$m(\omega) \approx -2\varepsilon_S$$

$$\theta(\omega) \approx 0 \tag{4.94}$$

So, the effect of non-zero settling error on the non-inverting integrator's low-frequency response is to introduce an attenuation with magnitude $2\varepsilon_S$ but very little phase error.

If the analysis is repeated for the inverting (Backward Euler) integrator it can be shown that the response with finite settling time is given by

$$H'(z) = -\frac{(1 - \varepsilon_S)(1 - \varepsilon_S z^{-1})}{(1 - z^{-1})(1 - \varepsilon_S^2 z^{-1})} \tag{4.95}$$

As $\varepsilon_S \to 0$,

$$H'(z) \to H_i'(z) = -\frac{1}{1 - z^{-1}} \tag{4.96}$$

which is the z-domain response of an ideal inverting lossless integrator. Equation (4.95) may be re-written

$$H'(z) = \gamma(z)H_i'(z)$$

where

$$\gamma(z) = \frac{(1 - \varepsilon_S)(1 - \varepsilon_S z^{-1})}{(1 - \varepsilon_S^2 z^{-1})} \tag{4.97}$$

Putting $z = e^{j\omega T}$ in (4.95) gives the error response for physical frequencies:

$$H'(e^{j\omega T}) = \frac{(1-\varepsilon_S)(1 - \varepsilon_S(\cos\omega T - j\sin\omega T))}{(1 - \varepsilon_S^2(\cos\omega T - j\sin\omega T))} H_i'(e^{j\omega T}) \tag{4.98}$$

For small errors ($\varepsilon_S \ll 1$)

$$H'(e^{j\omega T}) \approx (1 - \varepsilon_S)(1 - \varepsilon_S\cos\omega T + j\sin\omega T)H_i'(z)(e^{j\omega T}) \tag{4.99}$$

$$\approx \frac{H_i'(e^{j\omega T})}{1 + \varepsilon_S(1+\cos\omega T) - j\varepsilon_S\sin\omega T} \tag{4.100}$$

$$= \frac{H_i'(e^{j\omega T})}{1 - m'(\omega) - j\theta'(\omega)}$$

where $m'(\omega)$ and $\theta'(\omega)$ are the magnitude and phase errors, respectively:

$$m'(\omega) = -\varepsilon_S(1 + \cos\omega T) \tag{4.101}$$

$$\theta'(\omega) = \varepsilon_S\sin\omega T \tag{4.102}$$

For low frequencies, $\omega T \ll 1$

$$m'(\omega) \approx -2\varepsilon_S \tag{4.103}$$

$$\theta'(\omega) \approx \varepsilon_S\omega T \tag{4.104}$$

4.4.6 Comparison with switched-capacitor integrator

The effect of non-zero settling time on the inverting integrator low-frequency response is to introduce an attenuation with identical magnitude to that of the non-inverting integrator (compare (4.94) and (4.103)) but, unlike the non-inverting integrator, with phase error $\varepsilon_S\omega T$. Analysis of equivalent switched-capacitor integrators [5] gives an almost identical result. In this case, the non-zero settling time is introduced by the finite unity gain bandwidth of the operational amplifier given by

$$GB = \frac{g_m}{C_c} \qquad (4.105)$$

where g_m is the transconductance of the differential input stage and C_c is the compensation capacitance. These are having much the same effect on the switched-capacitor integrator's performance as that of the memory cell's transconductance and gate capacitance on the performance of the switched-current integrator.

4.4.7 Simulations

To verify the analysis, a lossless non-inverting integrator was simulated using an in-house switched-capacitor analysis program (SCALP2). First, the simulation was performed with $T = 1\mu s$ and ideal transistors ($C/g_m = C_d/g_s \ll T$) and then it was repeated for non-ideal transistors ($C/g_m = T = 1\mu s$, $C_d/g_s \ll T$). The results are shown in Figure 4.14(a). The ideal response shows the expected -6dB/octave roll-off. The non-ideal response shows lower gain, brought about by the value of $a_0(\gamma)$, and a transition to -12dB/octave roll-off, due to the extra pole at frequency $\omega(\gamma)$. The calculated values for $a_0(\gamma)$ and $\omega_0(\gamma)$ for $T = C/g_m = 1\mu s$ are -12.22dB and 147kHz, respectively, and these agree exactly with the simulated results. Figure 14(b) shows the effect of non-ideal C/g_m on a lossy integrator ($\alpha_4 = 0.1$). As expected, the low-frequency gain is reduced to 20dB for both ideal and non-ideal integrators and the cut-off frequency of the non-ideal integrator is decreased due to the same fall in gain constant experienced by the lossless integrator.

4.5 Charge Injection Errors

The operation of the current memory has so far been described in terms of a current sampling phase and a current holding phase. In the current sampling phase, the switch S of Figure 4.15(a) closes and allows the charging of the gate-source capacitance to a voltage sufficient to support the input current, I, at the drain. In the holding phase, switch S is opened and ideally the gate-source voltage is left at the value established during the sampling phase so that the output current matches the sampled input current exactly. In practice the switch is a MOS transistor S (Figure 4.15(b)) which is driven between its linear (closed state) and cut-off (open state) regions. In the turn-off process, the MOS switch's mobile charges formed in the switch's inversion layer when it was rendered conducting, flow out of its drain, source and substrate. Further, the rapidly changing gate voltage causes charges to flow from the gate diffusion overlap capacitances and out of the MOS switch's source and drain. A fraction of this total charge, q, enters the memory transistor's gate source capacitance, C, and leaves an error voltage δV on capacitor C, resulting in an error in the cell's output current, δI.

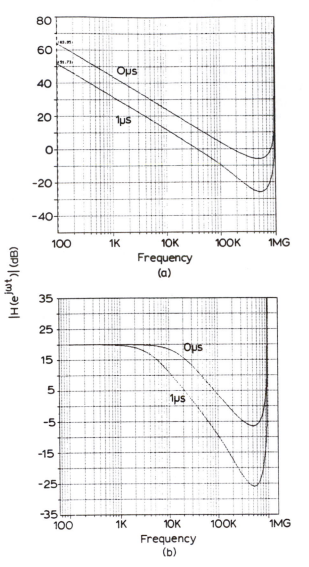

Figure 4.14 Simulated frequency responses of a non-inverting integrator with $C/g_m = 0$ (ideal) and 1µs with T = 1µs,

(a) lossless integrator (b) damped integrator ($\alpha_4 = 0.1$).

4.5.1 Switched-capacitor charge injection.

The problem of switch charge injection was identified in the early stages of switched-capacitor circuit development [7,8,9]. Since then there have been various attempts to model charge injection. In 1983, *MacQuigg* [10] offered empirical model equations, supported by measurements, for a simple voltage driven sample and hold circuit. In 1984, *Sheu* [11]

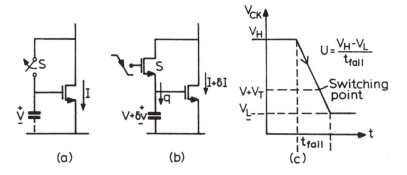

**Figure 4.15 Switch turn-off in current memory
(a) ideal circuit (b) practical circuit (c) clock waveform.**

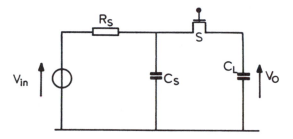

Figure 4.16 Basic voltage sample and hold.

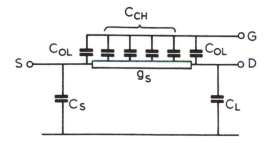

Figure 4.17 Model of MOS transistor's channel.

developed an analytical expression for the error voltage in the basic voltage driven sample and hold circuit and from this predicted its dependence on gate falling rate, signal voltage level and process parameters. In 1985, *Wilson* [12] presented more extensive measurements and observations of the basic sample and hold circuit, and supported this with computer simulations. Analyses of a more general sample and hold circuit with non-zero source resistance R_S and capacitance C_S (see Figure 4.16) were presented by *Shieh* [13] in 1987 and by *Wegmann* [14], and more recently by Eichenberger [15,16]. The analysis they presented had no closed-form solution for the general sample and hold so instead a numerical solution was found and this enabled a normalised diagram to be computed which showed the fraction of charge injected into the sampling capacitor C_L as a function of a switching parameter, $B = (V_H - V_T)\sqrt{\beta/UC_L}$, where U is the dV/dt of the clock transitions, and the capacitor ratio C_L/C_S. The following observations were made:

- For small values of B, corresponding to short fall-time of the switch gate voltage, the channel charge splits equally between source and drain independent of the ratio C_L/C_S.

- For large values of B, corresponding to long fall-time of the switch gate voltage, the channel charge splits according to the ratio of C_L/C_S.

- For intermediate values of B, the charge splitting depends on the value of B.

- For $C_L/C_S = 1$, the channel charge splits equally regardless of the value of B.

A qualitative explanation was offered (by *Shieh*) in terms of the distributed nature of the switch model (Figure 4.17) which gives the channel a characteristic delay T_0 given by

$$T_0 \propto \frac{L_S^2}{(V_H - V_T)} \qquad (4.106)$$

where L_S is the effective length of the switch channel.

For large values of B (corresponding to large values of the ratio t_{fall}/T_0) the channel behaves homogeneously and its conductance, g_S, has time to equalise the voltage at the switch source and drain. This makes the charge split proportional to C_L/C_S. For small values of B (corresponding to small values of the ratio t_{fall}/T_0) the channel is quickly pinched-off before the channel conductance has time to equalise the switch source and drain voltages. Providing negligible charge flows to the substrate (usually the case for short-channel switches) equal channel charge flows from the switch source and drain.

During the turn-off process, charge also flows from the diffusion side of the gate-diffusion overlap capacitance, C_{OL}, (Figure 4.17), Even when the channel has been destroyed, charge continues to flow from C_{OL} into C.

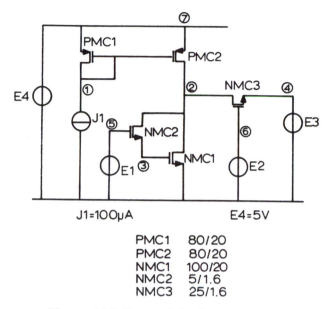

Figure 4.18 Charge injection test circuit.

4.5.2 *Current memory charge injection*

The dynamics of charge injection in a simple current memory are now considered. To help this discussion the test circuit shown in Figure 4.18 was simulated. Transistor NMC1 is the memory transistor ($g_m = 240\mu\text{S}$, $C_{gs} = 2.08\text{pF}$), transistor NMC2 is the memory switch ($r_S \approx 1.8\text{k}\Omega$, $C_{CH} = 7\text{fF}$, $C_{OL} = 1.6\text{fF}$) and the sum of the diffusion capacitances at node 2, $C_d = 0.3\text{pF}$. In the test, switch NMC2 is first closed with switch NMC3 open and current J_1 (100μA) is mirrored via PMC1 and PMC2 into the diode-connected memory transistor NMC1. After 1ns, the memory switch's gate (node 5) is driven from +5V to 0V in 5ns ($U = 1\text{V/ns}$) to cut-off NMC2 and leave NMC1 holding the input current. After a further 5ns (i.e. at 11ns) the gate of NMC3 is driven from 0V to +5V to turn-on NMC3 and connect node 2 to voltage generator E3. As the current from PMC2 is maintained throughout the test, only the error current in NMC1 flows into E3. The value of E3 was chosen to equal the voltage occurring at node 2 when switch NMC2 was closed so that, apart from a small voltage drop across NMC3 (1mV), node 2 settles at a value of 1.652V on both the sample and hold phases. This eliminates errors arising from the output conductances of transistors NMC1 and PMC2. The error current flowing in E3 is the result of charge injection of NMC2 only. First, a comment about the models employed by the simulator. The earlier discussion of reported work on charge injection effects in the basic sample and hold circuit assumed the distributed model shown in Figure 4.17 for the switch transistor. Such sophistication is unnecessary in the test circuit of Figure

4.18 because the switch time constant $(r_S C_{CH} = 12.6\text{ps})$ is very much shorter than the gate fall-time (5ns). Instead, the channel capacitance is lumped with $C_{CH}/2$ at the drain and source when NMC2 is in the linear region or set to zero when the channel is pinched-off.

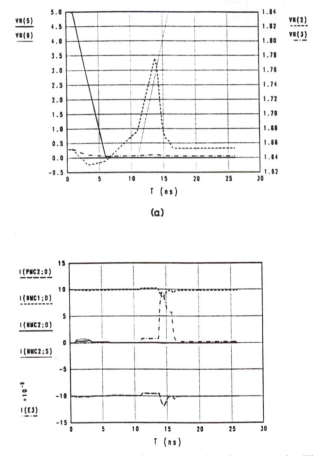

(a)

Figure 4.19 Simulation results of the test circuit shown in Figure 4.18.

Now consider the simulation results shown in Figure 4.19. First we concentrate on the turn-off transients of switch NMC2, i.e. the time interval 1ns to 6ns during which the gate voltage of NMC2, VN(5), is driven from +5V to 0V. Pinch-off occurs when VN(5) reaches 2.8V at time 3.2ns. In the interval from 1 ns to 3.2ns the channel is present and the equivalent circuit is as shown in Figure 4.20(a). Initially, nearly equal currents, i_d and i_s, flow in the drain and source of switch NMC2 equal to $U(C_{CH}/2 + C_{OL}) = 5.1\mu\text{A}$. As C_d is much smaller than C, the voltage tends to fall faster at node 2 than at node 3. This creates a discharge current flowing through the channel resistance r_S from node 3 to node 2 with a time constant of approximately $r_S C_d$ (= 0.54ns initially). The current i_s increases and the current i_d decreases, but as VN(5) approaches the level at which pinch-off occurs ($\approx V_T$ above VN(3)) r_S increases towards infinity and the current through r_S diminishes, and i_d and i_s return towards their initial values. The current i_s flowing into C causes the voltage at node 3 to fall below the value of V_{gs} required to sustain the input current. The resulting fall in the memory transistor's drain current causes an extra current to flow in r_S in the opposite sense to the discharge current and tends to reduce the charge injection error (classical negative feedback). However, the dominant loop time constant is $C/g_m = 8.7$ns, so the effect is not very marked in this test circuit. Memory circuits with much lower values of C/g_m or much longer fall times could enjoy considerably reduced charge injection errors due to the loop negative feedback.

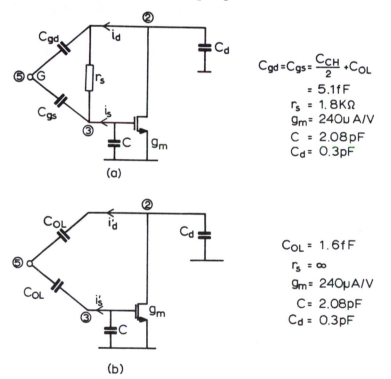

$$C_{gd} = C_{gs} = \frac{C_{CH}}{2} + C_{OL}$$
$$= 5.1 \text{fF}$$
$$r_s = 1.8 \text{K}\Omega$$
$$g_m = 240u \text{ A/V}$$
$$C = 2.08 \text{pF}$$
$$C_d = 0.3 \text{pF}$$

(a)

$$C_{OL} = 1.6 \text{fF}$$
$$r_s = \infty$$
$$g_m = 240\mu \text{A/V}$$
$$C = 2.08 \text{pF}$$
$$C_d = 0.3 \text{pF}$$

(b)

Figure 4.20 Equivalent circuit of current memory
(a) before pinch-off (b) after pinch-off.

After pinch-off occurs at t = 3.2ns the channel no longer exists and for the interval 3.2ns to 5ns the equivalent circuit is as shown in Figure 4.20(b). Currents i_d' and i_s' flow in the switch's drain and source, where $i_d' \approx i_s' = UC_{OL} = 1.6\mu$A. Without the channel there is no negative feedback. The current i_s discharges C still further and so increases the charge injection error. The total charge injected into C is the integral of the switch's source current during the interval from 1ns to 6ns. The average value of this current for the interval preceding pinch-off (1ns to 3.2ns) is 6.6µA, giving a charge of 14.6fC. The value after pinch-off (3.2ns to 6ns) is a constant 1.6µA, giving a charge of 4.5fC. The total injected charge is 19.06fC which gives an error voltage δV on C (2.08pF) of 9.2mV. The resulting error current in the drain of NMC1 is $\delta V g_m = 2.21$µA (approx. 2.2%). During the next 5ns, i.e. from 6ns to 11ns, both switches NMC2 and NMC3 are open and memory transistor NMC1 holds its drain current at a level 2.21µA below the input current (100µA) which is still flowing in the drain of PMC2. The 2.21µA difference current flows in the diffusion capacitances summing at node 2, ($C_d = 0.3$pF) and causes a slow rise in VN(2). In the interval from 11ns to 16ns, the gate voltage of switch NMC3 is driven high at $U = 1$V/ns.

Input Current JI	V_{gs} (µA)	Injected Charge (V)	Error current q (fC)	Calculated δI (µA)
20	1.202	23.47	1.18	1.12
40	1.352	21.42	1.59	1.52
60	1.467	20.57	1.86	1.80
80	1.565	19.79	2.06	2.01
100	1.652	19.06	2.21	2.19
120	1.730	18.35	2.32	2.34
140	1.803	17.60	2.39	2.46
160	1.871	17.15	2.46	2.57
180	1.936	16.83	2.55	2.59

Table 4.1 Simulated charge injection error in a basic current memory with varying input current. These results are for the test circuit of Figure 4.18. The fall time of the switch clock is 5ns. The calculated error current uses (4.121).

At first the channel is pinched-off and the rising gate voltage causes a constant current of 8µA to flow in the gate-drain overlap capacitance to add to the error current charging C_d. At $t = 13.8$ns VN(2) has charged 123mV above E3 and at that moment the channel of NMC3 has become conductive. This causes C_d to discharge into E3 with a discharge current limited only by the channel resistance of NMC3. Once VN(2) has settled back to 1.652V, the error current of 2.21µA flows in E3.

    The simulation was repeated for input currents ranging from 20µA to 180µA to find the dependence of the charge injection error on input current and these are shown in Table 4.1. At low input currents the V_{gs} of the memory transistor (NMC1) is low and pinch-off of the memory switch's channel occurs later in the turn-off transient. This gives proportionately more charge injected through the channel capacitance and less injected through the overlap capacitance giving an overall increase in total injected charge. The error voltage δV on C is translated by a low value of transconductance of NMC1 to give the output error current, δI. At higher input currents the total charge is lower but the transconductance is higher, giving a net increase in error current with increasing input current. This error current varies from 5.9% at $J_1 = 20\mu A$ to 1.4% at $180\mu A$, resulting in both a gain error and non-linearity.

Table 4.2 shows simulation results of charge injection errors as a function of the fall time of the switch's (NMC2 of Figure 18) clock edge, t_{fall}, with constant input current $J_1 = 100\mu A$. At $t_{fall} = 1ns$, the charge from switch S divides nearly equally between the memory transistor's drain and gate capacitances, as the channel is conducting for such a short time that no significant charge redistribution occurs. As t_{fall} increases, the switch channel stays conducting for longer, giving time for the charge initially injected into the memory transistor's drain capacitance, C_d, to redistribute onto its gate capacitance C and

Error current	Fall Time	Injected charge
t_{fall} (ns)	q (fC)	δI (μA)
1	17.37	2.01
5	19.06	2.21
10	19.52	2.26
20	18.27	2.12
50	13.81	1.62
100	10.11	1.19
200	7.66	0.91

Table 4.2 Simulated charge injection error in a basic current memory with varying clock fall time t_{fall}. These results are for the test circuit of Figure 4.18. The input current (J_1) is 100μA.

thereby increase the injection error. At values of t_{fall} above 10ns the negative feedback loop has time to act and this decreases the charge injection error. At $t_{fall} = 200ns$, $q = 7.66fC$ made up from charge injected into C via the gate-channel capacitance (3.16fC) and via the gate overlap capacitance (4.5fC). This compares with 15.02fC and 4.5fC, respectively, at $t_{fall} = 10ns$, giving nearly a fivefold decrease in the gate-channel contribution due to the feedback loop.

The following observations may be made from the foregoing simulation results. The proportion of the switch charge injected into the storage capacitor C is dependent on many factors. As with the reported results for the voltage sample and hold circuit, unequal charging of capacitance at either end of the switch tends to be equalised while the channel is still present. The significance of this effect depends on the ratio of the time constant $r_s C_d$ and the clock fall time t_{fall}. Additionally, while the channel is still formed, negative feedback reduces the charge remaining on the capacitor C, an effect which is not present in the voltage sample and hold circuit. Its significance depends on the ratio of the loop delay (usually dominated by the time constant C/g_m) and the clock fall-time. The error current also depends on the g_m of the memory transistor. Both the charge and the transconductance depend on the input current.

4.5.3 Analysis of charge injection errors

The charge sharing dynamics are more complicated than in the voltage sample and hold circuit for which no closed-form solution could be found and so no exact analysis for the current memory is attempted here. However, the following approximate analysis may be helpful.

The total charge injected through the gate-channel capacitance of the switch is given by

$$q_{CH} = (V_H - V_{gs} - V_{TS})C_{CH} \qquad (4.107)$$

where V_H is the high excursion of the switch gate voltage, V_{gs} is the memory transistor's gate-source voltage and $C_{CH} = W_S L_S C_{OX}$ is the switch channel capacitance where W_S and L_S are the switch width and length. V_{TS} is the threshold voltage of the switch when biased at the gate voltage of the memory transistor, V_{gs}, and is given by

$$V_{TS} = V_T + \gamma \left(\sqrt{2\,/\phi_F/ + Vgs} - \sqrt{2/\phi_F/} \right) \approx V_T + \frac{\gamma}{3}V_{gs} \qquad (4.108)$$

where the approximation is adequate for the practical range $1V < V_{gs} < 3.5V$. Substituting (4.108) in (4.107) gives

$$q_{CH} \approx \left[V_H - \left(1 + \frac{\gamma}{3} \right)V_{gs} - V_T \right]C_{CH} \qquad (4.109)$$

A fraction, α, of this charge flows in the memory capacitor C where, for most practical situations, $0.5 < \alpha < 1$, depending on the aforementioned factors. The charge injected into the memory capacitor C from the switch gate-diffusion overlap capacitance, C_{OL}, is given by

$$q_{OL} = C_{OL}(V_H - V_L) \qquad (4.110)$$

where V_L is the low excursion of the switch gate voltage. The total charge injected into C is given by

$$q = \alpha q_{CH} + q_{OL} \qquad (4.111)$$

and the resulting error current in the memory transistor is δI where

$$\delta I = g_m \frac{q}{C} \qquad (4.112)$$

which gives a relative charge injection error given by

$$\frac{\delta I}{I} = \frac{g_m}{I}\frac{q}{C} = \omega_{CO}\frac{q}{I} \qquad (4.113)$$

where ω_{CO} is the memory transistor's cut-off frequency given by

$$\omega_{CO} = \frac{g_m}{C} = \frac{3\mu(V_{gs} - V_T)}{2L^2} \qquad (4.114)$$

It can be seen from (4.113) that charge injection errors are kept low by minimising ω_{CO} and q and maximising I. The memory transistor's cut-off frequency ω_{CO} should be no greater than needed for adequate settling performance. The charge q is minimised by using the minimum value for L_S and keeping g_s as small as possible consistent with the required Q-factor (4.58).

High frequency circuit operation is possible as can be seen from (4.114). With $V_{gs} - V_T = 1V$ and $L = 1\mu m$ then $\omega_{CO} \approx 30.10^9$ rads^{-1}. However, it can be seen from (4.113) that the cost of using very high values of ω_{CO} is to incur correspondingly higher charge injection errors unless proportionately higher values of current I are used.

Returning to the analysis, the error current can be found by substituting (4.109), (4.110) and (4.111) in (4.112) giving

$$\delta I = \frac{g_m}{C}\left[\alpha C_{CH}\left(V_H - (1 + \frac{\gamma}{3})V_{gs} - V_T\right) + (V_H - V_L)C_{0L}\right] \qquad (4.115)$$

Substituting $V_{gs} = V_T + 2I/g_m$ gives

$$\delta I = \left[\alpha\frac{C_{CH}}{C}\left(V_H - (2 + \frac{\gamma}{3})V_{gs} - V_T\right) + (V_H - V_L)\frac{C_{0L}}{C}\right]g_m - \left[2\alpha\left(1 + \frac{\gamma}{3}\right)\frac{C_{CH}}{C}\right]I \qquad (4.116)$$

$$= k_A g_m - k_B I \qquad (4.117)$$

and

$$\frac{\delta I}{I} = \frac{2k_A}{V_{gs} - V_T} - k_B \qquad (4.118)$$

where k_A and k_B are constants defined by

$$k_A = \alpha\frac{C_{CH}}{C}\left(V_H - (2 + \frac{\gamma}{3})V_{gs} - V_T\right) + (V_H - V_L)\frac{C_{0L}}{C}$$

$$k_B = 2\alpha\left(1 + \frac{\gamma}{3}\right)\frac{C_{CH}}{C} \tag{4.119}$$

Equation (4.118) was evaluated for I ranging from 20µA to 180µA for $\alpha = 0.7$ in the test circuit shown in Figure 4.18 and the results are included in Table 4.1. The small signal error, ε_q, can be found by differentiating δI with respect to I and putting $I = J$. From (4.117)

$$\delta I = k_A\sqrt{2\beta I} - k_B I$$

$$\varepsilon_q = \frac{d}{dI}(\delta I)_{(I=J)} = \frac{k_A}{V_{gs} - V_T} - k_B \tag{4.120}$$

Clearly, ε_q can be made vanishingly small if

$$V_{gs} - V_T = \frac{k_A}{k_B} = \frac{\alpha\frac{C_{CH}}{C}\left(V_{H}-(2+\frac{\gamma}{3})V_T\right) + (V_H-V_L)\frac{C_{OL}}{C}}{2\alpha\left(1+\frac{\gamma}{3}\right)\frac{C_{CH}}{C}}$$

$$= \frac{\left(V_{H}-(2+\frac{\gamma}{3})V_T\right) + (V_H-V_L)\frac{C_{OL}}{\alpha C_{CH}}}{2\left(1+\frac{\gamma}{3}\right)} \tag{4.121}$$

For the simulated memory cell (Figure 4.18) this condition occurs at $V_{gs} - V_T \approx 2V$. The simulation of charge injection was extended to higher currents to increase $V_{gs} - V_T$ and the results of the simulation and calculation are shown in Figure 4.21. At low currents the clock voltage at which the switch opens is also low resulting in a high injected charge. However, the cell's transconductance is low and this gives a low error current. At high currents the clock voltage at which the switch opens is higher and the injected charge is lower. Even though the cell's transconductance is higher, this gives a lower current error. The maximum occurs when the effect of decreasing injected charge just cancels the effect of increasing transconductance. If memory cells, designed with this optimal value of $V_{gs} - V_T$, are used in cascaded pairs (as occurs in integrator loops, differentiators, and unit delays in FIR filters and sample and hold), then the errors of the two cells are very nearly equal and so almost cancel. For the simulated cell, the error for such a pair is less than 0.05µA for a swing of 500µA (0.01%). Where still greater precision is required, enhanced circuit techniques must be employed and these are discussed in Chapter 6.

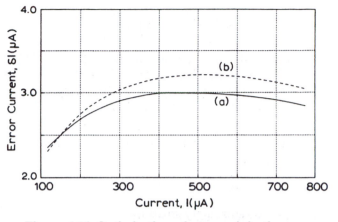

**Figure 4.21 Optimisation of charge injection error
(a) simulated (b) calculated.**

4.5.4   Memory cell response with charge injection errors

Having considered the factors determining the charge injection error, we now analyse the effect of this error on the memory cell's frequency response. The analysis assumes that the memory cell is ideal except for the charge injection error. The approach is to consider transmission through two cascaded memory cells, since some components of the errors cancel, and then to deduce the average response of each memory cell. Consider the cascaded pair of memory cells shown in Figure 4.22. During clock phase ϕ_1 of period *(n-1)* (not shown in Figure 4.22), the drain current in T_1 is

$$I_1(n\text{-}1) = J + i(n\text{-}1) \qquad (4.122)$$

At the end of phase ϕ_1, switch S_1 opens and its charge q_1 causes an error $-\delta I_1$ in the current stored during the next clock phase ϕ_2. The stored current in T_1

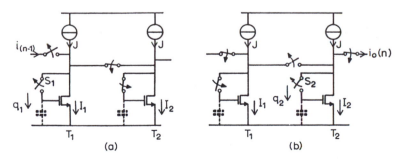

**Figure 4.22 Transmission through cascaded memory cells
with charge injection (a) phase ϕ_1 (b) phase ϕ_2.**

is $I_1(n-1/2)$ where

$$I_1(n-1/2) = I_1(n-1) - \delta I_1 = J + i(n-1) - \delta I_1 \qquad (4.123)$$

and the current in T_2 is $I_2(n-1/2)$ where

$$I_2(n-1/2) = 2J - I_1(n-1/2) = J - i(n-1) + \delta I_1 \qquad (4.124)$$

At the end of clock phase ϕ_2, switch S_2 opens and charge q_2 causes an error $-\delta I_2$ in the current stored during the next clock phase ϕ_1. The stored current in T_2 is $I_2(n)$ where

$$I_2(n) = I_2(n-1/2) - \delta I_2 = J - i(n-1) + \delta I_1 - \delta I_2 \qquad (4.125)$$

and the output current from T_2 is $i_0(n)$ where

$$i_0(n) = J - I_2(n) = i(n-1) - (\delta I_1 - \delta I_2) \qquad (4.126)$$

Using (4.117), the error currents are given by

$$\delta I_1 = k_A g_{m1} - k_B I_1(n-1) \qquad (4.127)$$

$$\delta I_2 = k_A g_{m2} - k_B I_2(n-1/2) \qquad (4.128)$$

$$\delta I_1 - \delta I_2 = k_A(g_{m1} - g_{m2}) - k_B(I_1(n-1) - I_2(n-1/2)) \qquad (4.129)$$

Equation (4.129) may be simplified as follows. From (4.122) and (4.124)

$$I_1(n-1) - I_2(n-1/2) \approx 2i(n-1) \qquad (4.130)$$

and

$$g_{m1} - g_{m2} = \sqrt{2\beta I_1(n-1)} - \sqrt{2\beta I_2(n-1/2)}$$

$$\approx \sqrt{2\beta(J + i(n-1))} - \sqrt{2\beta(J - i(n-1))}$$

$$\approx \frac{g_m}{J} i(n-1) \qquad (4.131)$$

where g_m is the cell transconductance with $I_{ds} = J$. Substituting (4.130) and (4.131) in (4.129) gives

$$\delta I_1 - \delta I_2 = k_A \frac{g_m}{J} i(n-1) - 2k_B i(n-1) = 2i(n-1)\left[\frac{k_A}{V_{gs} - V_T} - k_B\right]$$

Substituting (4.120) gives

$$\delta I_1 - \delta I_2 = 2\varepsilon_q i(n-1) \qquad (4.132)$$

Substituting (4.132) in (4.126) gives

$$i_0(n) = (1 - 2\varepsilon_q)i(n-1)$$

Taking z-transforms gives

$$i_0(z) = (1 - 2\varepsilon_q)i(z)z^{-1}$$

The z-domain transfer characteristic for the pair of memory cells is given by

$$[H(z)]^2 = \frac{i_0(z)}{i(z)} = z^{-1}(1 - 2\varepsilon_q)$$

$$\approx \left[\frac{-z^{-1/2}}{1 + \varepsilon_q}\right]^2$$

The memory cell transfer characteristic is

$$H(z) = \frac{-z^{-1/2}}{1 + \varepsilon_q} = \frac{H_i(z)}{1 + \varepsilon_q}$$

For physical frequencies, putting $z = e^{j\omega T}$,

$$H(e^{j\omega T}) = \frac{H_i(e^{j\omega T})}{1 + \varepsilon_q}$$

Using the approximation of (4.39) gives the following magnitude and phase errors:

$$m(\omega) = -\varepsilon_q$$

$$\theta(\omega) = 0 \qquad\qquad (4.133)$$

4.5.5 Integrator frequency response with charge injection errors

We will now analyse the effect of charge injection errors on the integrator's frequency response. Again, the approach is to re-establish the difference equations taking account of the effects of charge injection. Consider a non-inverting lossless integrator which is ideal in all respects except that a charge q is injected into the integrator's memory capacitors C when the memory switches are opened.

Figure 4.23(a) shows the integrator on phase ϕ_2 of period $(n-1)$. The current in T_1 is $I_1(n-1)$ where

$$I_1(n-1) = J + i_0(n-1) + \alpha_1 i(n-1) \qquad\qquad (4.134)$$

Figure 4.23(b) shows the integrator on phase ϕ_1 of period (n). The current in T_1 is $I_1(n-1/2)$ where

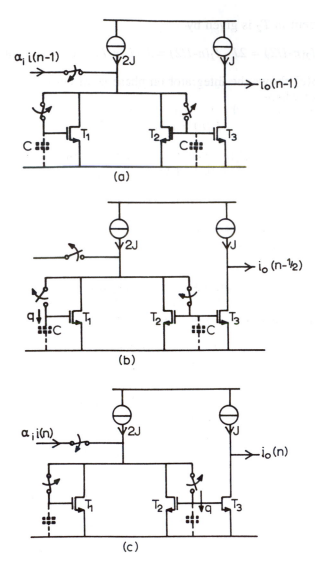

Figure 4.23 Non-inverting lossless integrator with charge injection errors
(a) phase ϕ_2, period $(n-1)$ (b) phase ϕ_1, period (n) (c) phase ϕ_2, period(n).

$$I_1(n-1/2) = I_1(n-1) - \delta I_1 \qquad (4.135)$$

where $\delta I_1 = k_A g_{m1} - k_B I_1$ where k_A and k_B are the constants defined in (4.119) and g_{m1} is the transconductance of T_1 with drain current I_1. Substituting (4.134) in (4.135) gives

$$I_1(n-1/2) = J + i_0(n-1) + \alpha_1 i(n-1) - \delta I_1 \qquad (4.136)$$

and the current in T_2 is given by

$$I_2(n-1/2) = 2J - I_1(n-1/2) = J - i_0(n-1) - \alpha_1 i(n-1) + \delta I_1 \quad (4.137)$$

Figure 4.23(c) shows the integrator on phase ϕ_2 of period *(n)*. The current in T_2 is $I_2(n)$ where

$$I_2(n) = I_2(n-1/2) - \delta I_2$$

where $\delta I_2 = k_A g_{m2} - k_B I_2(n)$ where g_{m2} is the transconductance of transistor T_2 with current I_2. Substituting (4.137) gives

$$I_2(n) = J - i_0(n-1) - \alpha_1 i(n-1) + \delta I_1 - \delta I_2 \quad (4.138)$$

So, the output current $i_0(n)$ is given by

$$i_0(n) = J - I_2(n) = i_0(n-1) + \alpha_1 i(n-1) - (\delta I_1 - \delta I_2) \quad (4.139)$$

This is identical to the corresponding equation for the ideal non-inverting integrator except for the final term $\delta I_1 - \delta I_2$. Comparing (4.136) with (4.138) it can be seen that for δI_1 and $\delta I_2 \ll J$

$$I_1(n-1/2) \approx I_1(n-1) \approx J + i_S$$

$$I_2(n) \approx J - i_S$$

where $i_S = i_0(n-1) + \alpha_1 i(n-1)$. This symmetry may be used to simplify the above error term as before for the current memory ((4.129) to (4.131)).

$$\delta I_1 - \delta I_2 = k_A(g_{m1} - g_{m2}) - 2k_B i_S$$

$$\delta I_1 - \delta I_2 = (k_A \frac{g_m}{J} - 2k_B)i_S = 2\varepsilon_q(i_0(n-1) + \alpha_1 i(n-1)) \quad (4.140)$$

where g_m is the transconductance of T_1 and T_2 at drain current J and ε_q is the average memory charge injection error given by

$$\varepsilon_q = \frac{\alpha \frac{C_{CH}}{C}\left(V_H - (2+\frac{\gamma}{3})V_T\right) + (V_H - V_L)\frac{C_{OL}}{C}}{V_{gs} - V_T} - 2\alpha(1+\frac{\gamma}{3})\frac{C_{OL}}{C}$$

$$= \frac{k_A}{V_{gs} - V_T} - k_B$$

where $V_{gs} - V_T$ is the saturation voltage at drain current J. Note that the average charge injection error ε_q given above is the same as the small signal error of a single memory cell as given in (4.120). Substitution of (4.140) in (4.139) gives

$$i_0(n) = i_0(n-1) + \alpha_1 i(n-1) - 2\varepsilon_q(i_0(n-1) + \alpha_1 i(n-1))$$

Transforming to the z-domain and rearranging gives

$$i_0(z)[1 - (1 - 2\varepsilon_q)z^{-1}] = \alpha_1(1 - 2\varepsilon_q)i(z)z^{-1}$$

$$H(z) = \frac{i_0(z)}{i(z)} = \frac{\alpha_1 z^{-1}}{1 + 2\varepsilon_q - z^{-1}}$$

As before, the response to physical frequencies is found by putting $z = e^{j\omega T}$ and this gives

$$H(e^{j\omega T}) = \frac{H_i(e^{j\omega T})}{1 + \varepsilon_q - j\dfrac{2\varepsilon_q}{\tan\dfrac{\omega T}{2}}}$$

where $H_i(e^{j\omega T})$ is the ideal response ($\varepsilon_q = 0$). Using the approximation again (4.39) this gives the magnitude and phase errors,

$$m(\omega) \approx -\varepsilon_q$$

$$\theta(\omega) \approx \frac{2\varepsilon_q}{\tan\dfrac{\omega T}{2}} \qquad\qquad (4.141)$$

The ideal integrator's unity gain frequency is given by

$$\omega_1 = \frac{2}{T} \sin^{-1}\left(\frac{\alpha_1}{2}\right)$$

and the magnitude and phase errors at the unity gain frequency for $\omega_1 T \ll 1$ are given by

$$m(\omega_1) \approx -\varepsilon_q$$

$$\theta(\omega_1) \approx \frac{4\varepsilon_q}{\alpha_1} \qquad\qquad (4.142)$$

It should be noted that these errors are for small signal currents. With large signal currents, q and $g_{m1,2}$ vary non-linearly and this will result in extra harmonic distortion.

4.6 Noise Errors

During the operation of a switched-current circuit, the input current signal is sampled on one clock phase and then held on the other. Any noise currents introduced with the signal or by the memory cell's own transistors undergo the same sampling process. Noise with frequency components above the Nyquist frequency $(f_{CK}/2)$ are undersampled and therefore create replicas at base-band frequencies. The analysis of this process follows the procedure set out by *Gobet* [17] for switched-capacitor circuits and has been performed by *Daubert* [18] and Chapter 5, and by *Wegmann* and *Vittoz* [19] and Chapter 16. In this section, their analyses are developed for the basic memory cell, augmented by simulation and then extended to the switched-current integrator.

4.6.1 Noise analysis of switched-current memory cells

Of the various types of noise present in analogue circuits [20] two are dominant in switched-current circuits: thermal noise and flicker noise in the MOS transistors of the memory cell. Thermal noise originates from the random motion of electrons and behaves like white noise up to very high frequencies (flat noise power spectrum). The origin of flicker noise is less certain. While some authors propose explanations based on the random trapping of carriers in interface and bulk states, others explain it by mobility variations. It is often referred to as "$1/f$ noise" as this describes approximately the noise power spectrum.

Noise sources occur in the memory cell's transistor, switch and bias current source as shown in Figure 4.24. The switch is an MOS transistor which operates in the linear region when conductive. It carries no DC current and so produces no flicker noise. It has a resistance r_S and an associated thermal noise voltage source $v_{S(th)}$ with a mean squared value given by

$$\overline{v_{S(th)}^2} = 4kT_j r_S \Delta f \qquad 0 < f < \infty \qquad (4.143)$$

where k is Boltzmann's constant and T_j is the absolute temperature.

The memory transistor T operates in the saturated region and has both flicker and thermal noise sources. Flicker noise is modelled by a voltage source in the gate, $v_{T(fl)}$, with a mean squared value given by

$$\overline{v_{T(fl)}^2} = \frac{K_{fl}}{WL|f|} \Delta f \qquad 0 < f < \infty \qquad (4.144)$$

where W and L are the width and length of the channel and K_{fl} is a process-dependent constant. Thermal noise is modelled by a current source between drain and source, $i_{T(th)}$, with a mean square value given by

$$\overline{i^2_{T(th)}} = \frac{8}{3} m_{th} kT_j g_m \Delta f \qquad 0 < f < \infty \qquad (4.145)$$

where g_m is the transconductance of transistor T and m_{th} is a process-dependent constant which in practice ranges from 1 to 2.5. The bias current source, J, operates in saturation and has both flicker noise, $v_{J(fl)}$, and white noise, $i_{J(th)}$, with mean squared values given by

$$\overline{v^2_{T(fl)}} = \frac{K_{fl}}{W_J L_J |f|} \Delta f \qquad 0 < f < \infty$$

$$\overline{i^2_{J(th)}} = \frac{8}{3} m_{th} kT_j g_{m(J)} \Delta f \qquad 0 < f < \infty \qquad (4.146)$$

where W_J and L_J and $g_{m(J)}$ are the bias current transistor's width, length and transconductance, respectively.

The noise model of the memory may be simplified as follows. The switch noise $v_{S(th)}$ will modulate the voltage on capacitor C when the switch is closed

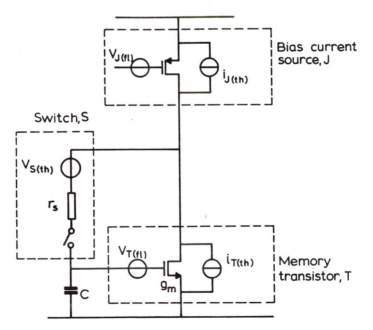

Figure 4.24 Basic switched-current memory cell with noise sources.

only if the impedance at the drain of the memory transistor (due to g_{ds} and C_d) is comparable with that of capacitor C. In practice, this is determined by the ratio C_d/C which is usually a small ratio, and so the switch noise may be neglected in most situations. Next, the flicker noise sources may be 'referred' to their drains to make equivalent drain current noise sources given by

$$\overline{i^2_{J(fl)}} = g^2_{m(J)} \overline{v^2_{J(fl)}}$$

$$\overline{i^2_{T(fl)}} = g^2_m \overline{v^2_{T(fl)}} \tag{4.147}$$

Finally, the four currents summing at the drain of the memory transistor $i_{T(th)}$, $i_{T(fl)}$, $i_{J(th)}$ and $i_{J(fl)}$ are uncorrelated and so may be combined to give the memory cell's mean-squared noise current, given by

$$\overline{i^2_{MN}} = \overline{i^2_{T(th)}} + \overline{i^2_{T(fl)}} + \overline{i^2_{J(th)}} + \overline{i^2_{J(fl)}} \tag{4.148}$$

with a noise power spectral density (PSD) given by

$$S_M(f) = \frac{\overline{i^2_{MN}}}{\Delta f} = \frac{K_{fl}}{|f|}\left(\frac{g^2_m}{WL} + \frac{g^2_{m(J)}}{W_J L_J}\right) + \frac{8}{3}m_{th}kT_j(g_m + g_{m(J)}) \tag{4.149}$$

The simplified noise model for the memory cell is shown in Figure 4.25, and i_{MN} is the memory cell's equivalent noise current defined by (4.148).

The analysis so far has derived the unsampled equivalent noise current. During the sampling process, noise at frequencies above the Nyquist frequency is aliased and this causes the sampled noise power spectral density to change to $S_M^{S/H}(f)$, given by

$$S_M^{S/H}(f) = \left(\frac{\tau}{T}\right)^2 sinc^2(f\tau)\sum_{n=0}^{\infty}S(f-nf_{CK}) \tag{4.150}$$

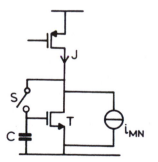

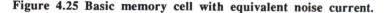

Figure 4.25 Basic memory cell with equivalent noise current.

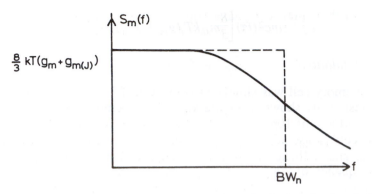

Figure 4.26 Thermal noise spectrum and equivalent noise bandwidth.

where $f_{CK} = 1/T$ is the clock frequency, τ is the hold time and $sinc(x) = sin(\pi x)/(\pi x)$. The summation in (4.150) expresses the addition of aliased replicas of the high-frequency white noise at base-band and can be simplified through the concept of noise bandwidth, BWn, shown in Figure 4.26. The white noise PSD is replaced by a rectangular PSD with constant amplitude, $(8m_{th}/3)kT_j(g_m + g_{M(J)})$, over the frequency band BW_n. For a single-pole lowpass loop, the noise bandwidth is given by

$$BW_n = \frac{\pi}{2} f_{C0} = \frac{g_m}{4C} \qquad (4.151)$$

As the spectral density is constant over BW_n and zero outside this frequency band, the summation of (4.150) is simply evaluated over the range $n = 0$ to $2BW_n/f_{CK}$ with $S_M(f\text{-}nf_{CK}) = (8m_{th}/3)kT_j(g_m + g_{m(J)})$ for all n. This gives

$$S_M^{S/H}(f) = \left(\frac{\tau}{T}\right)^2 sinc^2(f\tau)\left[\frac{8}{3}m_{th}kT_j(g_m + g_{m(J)})\right]\left[\frac{g_m}{2f_{CK}C}\right] \qquad (4.152)$$

The four terms in (4.152) have the following significance. The first term, τ/T, is the fraction of time that the noise is connected to the output; the second term, $sinc(f\tau)$, is the familiar sample and hold transfer function; the third term is the unsampled noise power; and the fourth term is the number of replicas produced by the sampling process. So it may be concluded that the sampling process merely concentrates the unsampled noise at baseband without changing the noise power (the integrals of the unsampled and sampled PSDs are the same). The total noise PSD of the memory cell must include the direct noise which, being uncorrelated, may be simply added to the sampled noise PSD.

$$S_M^{TOT}(f) = \frac{\tau}{T}\left[\frac{8}{3}m_{th}kT_j(g_m + g_{m(J)})\right] +$$

$$\left(\frac{\tau}{T}\right)^2 sinc^2(f\tau)\left[\frac{8}{3}m_{th}kT_j(g_m + g_{m(J)})\right]\left[\frac{g_m}{2f_{CK}C}\right] \qquad (4.153)$$

4.6.1.1 Simulations

A basic memory cell was simulated with SCALP2 to confirm the analysis and the results are shown in Figure 4.27. The memory transistor had a g_m of 10µS and a C of 10pF and with a bias current of 10µA, $(V_{gs} - V_T)$ was 2V. The cell's noise PSD, $S_M(f)$, consisted of a white noise PSD of 2.86E-25 A²/Hz and a flicker noise PSD of 5.72E-21/lflA²/Hz giving a noise corner frequency of 20kHz. The noise bandwidth $(g_m/4C)$ was 0.25MHz, the clock frequency was 100kHz and the sample and hold times were equal $(\tau/T = 0.5)$. It can be seen from Figure 4.27 that the PSD has the characteristic $sinc^2x$ shape with nulls at multiples of 200kHz consistent with $\tau/T = 0.5$. The low-frequency noise PSD, found by evaluating (4.153) at $f = 0$, gives 500E-27 A²/Hz compared with the simulated value of 473E-27 A²/Hz, while the high frequency value is the direct noise PSD which agrees exactly with the first term of (4.153). The discrepancy at low frequency is attributed to the approximation inherent in the use of noise bandwidth to simplify the analysis. Note that flicker noise is not apparent. The memory cell's RMS noise current was simulated with its load band-limited to that of another cell (1Ω in parallel with 1µF) first with both flicker and white noise as before and then with only white noise. The change in simulated RMS noise current was 16.8pA compared with a calculated RMS flicker noise current of 75.6pA, indicating that the flicker noise is considerably diminished by the sampling process.

4.6.1.2 Correlated double sampling

It is noted in the previous simulations that the flicker noise is apparently removed by the sampling process and this is now discussed. During the hold time of the memory, while the sample and hold noise is being output from the memory transistor, the noise source, i_N, also flows directly to the output. This direct noise has the noise PSD expressed in (4.149). Low-frequency components of the noise undergo a correlated double sampling (CDS) process as is now explained. Consider the memory cell shown in Figure 4.28. With switch S closed (Figure 4.28(a)), the current in T is modulated by the noise current i_{MN}. At the moment the switch opens it has the value I where

$$I = J - i_{MN}(t) \qquad (4.154)$$

where $i_{MN}(t)$ is the instantaneous value of the noise current when the switch opens. After one half clock period, the noise has the value $i_{MN}(t+T/2)$ and the current held in the memory transistor is still I. So the output current is $i_0(t +T/2)$ where

$$i_0(t+T/2) = J - I - i_{MN}(t+T/2)$$

$$= i_{MN}(t) - i_{MN}(t+T/2) \qquad (4.155)$$

For low-frequency noise components

$$i_{MN}(t) \approx i_{MN}(t+T/2)$$

and the noise is nearly cancelled.

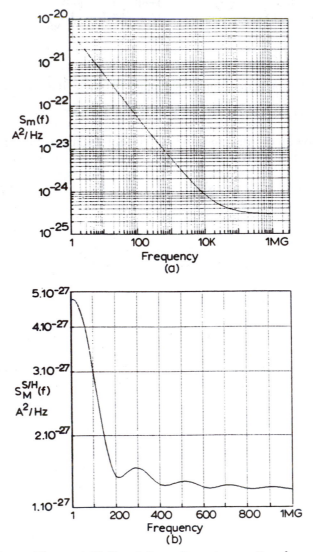

Figure 4.27 Simulation of memory cell noise
(a) PSD of input noise (b) PSD of output noise after sampling.

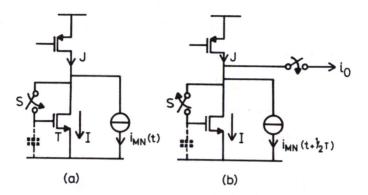

Figure 4.28 Correlated double sampling (CDS) in the memory cell.

4.6.1.3 Dynamic range

The noise power of the memory cell is not changed by the sampling process and is simply given as

$$\overline{i_{MN}^2} = S_M(f)BW_n \qquad (4.156)$$

Assuming that white noise dominates,

$$\overline{i_{MN}^2} \approx \frac{8}{3} m_{th} kT_j(g_m + g_{m(J)}) \frac{g_m}{4C} \qquad (4.157)$$

Putting $g_{m(J)} = k_J g_m$, then

$$\overline{i_{MN}^2} \approx \frac{2m_{th}}{3} \frac{kT_j}{C} (1 + k_J) g_m^2 \qquad (4.158)$$

For a sinusoidal signal with peak amplitude $m_i J$ (where m_i is the signal modulation index) the RMS value is $m_i J/\sqrt{2}$ and the dynamic range of the memory is DR_M where

$$DR_M = 10 \log \left[\frac{m_i^2 J^2/2}{\frac{2m_{th}}{3}(1 + k_J)\left(\frac{kT_j}{C}\right) g_m^2} \right] \qquad (4.159)$$

Substituting $(V_{gs} - V_T)^2 = 4J^2/g_m^2$ gives

$$DR_M = 10 \log \left[\frac{m_i^2 (V_{gs} - V_T)2}{\frac{16 m_{th}}{3} (1 + k_J) \left(\frac{kT_i}{C} \right)} \right] \qquad (4.160)$$

The highest dynamic range occurs when g_m and $g_{m(J)}$ are minimised, or conversely when $(V_{gs} - V_T)$ and $(V_{gs} - V_T)_J$ are maximised, so we need to examine the circuit conditions which limit these voltages. Figure 4.29 shows the basic memory cells connected in cascade.

On phase ϕ_2, transistor T_1 stays saturated provided

$$(V_{gs} - V_T) \leq \frac{V_T}{\sqrt{1 + m_i} - \sqrt{1 - m_i}} \qquad (4.161)$$

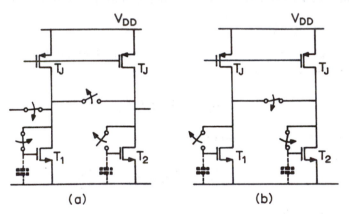

(a)   (b)

Figure 4.29 Cascade-connected basic memory cells
(a) phase ϕ_1 (b) phase ϕ_2.

This equation describes a trade-off between the saturation voltage $(V_{gs} - V_T)$ and the modulation index m_i; values of $(V_{gs} - V_T)$ above $V_T/\sqrt{2}$ can only be accommodated with values of m_i below unity.

On phase ϕ_1, the current source J transistor stays saturated provided

$$(V_{gs} - V_T)_J \leq V_{DD} - V_T - (V_{gs} - V_T)\sqrt{1 + m_i} \qquad (4.162)$$

This equation describes a trade-off between $(V_{gs} - V_T)$ and the saturation voltage of the current source transistor, $(V_{gs} - V_T)_J$ due to the limited supply voltage; a high value of $(V_{gs} - V_T)$ would need a low value of $(V_{gs} - V_T)_J$ with consequent increase in k_J given by

$$k_J = \frac{g_{m(J)}}{g_m} = \frac{(V_{gs} - V_T)}{(V_{gs} - V_T)_J} \qquad (4.163)$$

and potentially worsened dynamic range (see (4.160)). Clearly, choosing the conditions to maximise dynamic range is complex. For any given value of $(V_{gs} - V_T)$, m_i should be maximised using (4.161) and $(V_{gs} - V_T)_J$ should be maximised using (4.162). The dynamic range was computed using this strategy (with $V_{DD} = 5V$, $V_T = 1V$, $k = 1.38E-23$, $T_j = 300K$ and $m_{th} = 2.25$) and the result is shown in Figure 4.30. The dashed line shows the trade-off between $(V_{gs} - V_T)$ and m_i. For $(V_{gs} - V_T) \leq V_T/\sqrt{2} = 0.707V$, m_i is set to unity without violating (4.161) but above this value m_i falls. Dynamic range rises to a maximum at $(V_{gs} - V_T) = 1.14V$, and $m_i = 0.79$, with $(V_{gs} - V_T)_J = 2.48V$. It should be noted that the choice of m_i is also influenced by settling and charge injection errors. These can become excessively large as m_i approaches unity due to signal-dependent variations in the memory transistor's transconductance, g_m. However, the condition for optimum dynamic range ($m_i = 0.79$) should not cause difficulty in this respect. For $C = 10pF$, the maximum dynamic range is 80.5dB and the only way to improve this still further is by increasing C.

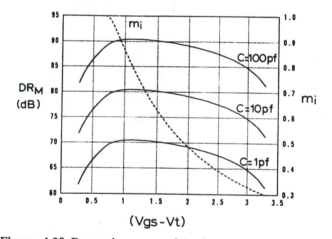

Figure 4.30 Dynamic range of basic current memory cell.

4.6.2 Integrator noise analysis

Consider the damped integrator shown in Figure 4.31. It is assumed that white noise dominates and that each current memory contributes noise. As with the current memory analysis, these may be combined to give a total input referred noise current, i_N, where

$$\overline{i_N^2} = 2\,\overline{i_{MN}^2} \qquad\qquad (4.164)$$

where $\overline{i_{MN}^2}$ is the total mean squared white noise current of each memory cell. The noise current is permanently connected and so is alternately sampled by each memory cell as shown in Figure 4.32.

During the ϕ_2 phase, starting at time $(n-1)T$, T_1 is diode-connected and its current is I_1 where

$$I_1 = 2J + i_N(t) - I_2(n-1)T \tag{4.165}$$

At $t = (n-1/2)T$, its switch opens and its instantaneous value, $I_1(n-1/2)T$, is held in T_1 where

$$I_1(n-1/2)T = 2J + i_N(n-1/2)T - I_2(n-1)T \tag{4.166}$$

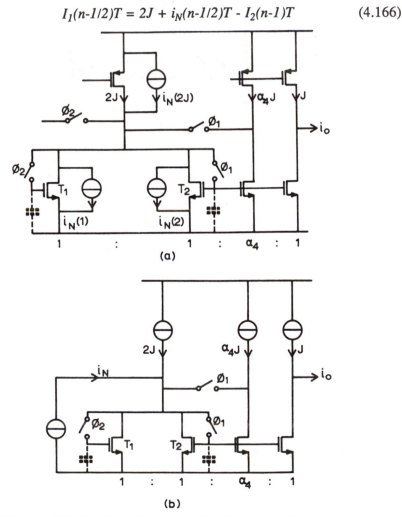

(a)

(b)

Figure 4.31 Non-inverting damped integrator noise sources (a) main noise sources (b) equivalent noise model.

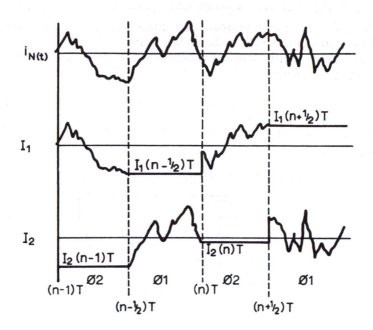

Figure 4.32 Noise waveforms in the integrator.

During the next ϕ_1 phase, starting at time $(n-1/2)T$, T_2 is diode-connected and its current is I_2 where

$$(1 + \alpha_4)I_2 = (2 + \alpha_4)J + i_N(t) - I_1(n-1/2)T \qquad (4.167)$$

At $t = (n)T$, its switch opens and the instantaneous value of its current, $I_2(n)T$, is held in T_2 where

$$(1 + \alpha_4)I_2(n)T = (2 + \alpha_4)J + i_N(n)T - I_1(n-1/2)T \qquad (4.168)$$

Substituting (4.166) in (4.168) gives

$$(1 + \alpha_4)I_2(n)T = (2 + \alpha_4)J + i_N(n)T - (2J + i_N(n-1/2)T - I_2(n-1)T)$$

$$= \alpha_4 J + I_2(n-1)T + i_N(n)T - i_N(n-1/2)T \qquad (4.169)$$

Further, the output signal current is, in general, $i_0 = J - I_2$ and the output current throughout phase ϕ_1 is determined by the value $I_2(n)T$. Substituting $I_2 = J - i_0$ in (4.169) gives, for phase ϕ_1,

$$(1 + \alpha_4)i_0(n)T + i_0(n-1)T = i_N(n)T - i_N(n-1/2)T \qquad (4.170)$$

Now, $i_N(n)T$ and $i_N(n-1/2)T$ are successive samples of a white noise source the noise bandwidth being determined by the bandwidth of the memory cells which for good settling accuracy is necessarily several times higher than the clock frequency. For frequency components of the noise signal below the Nyquist frequency, the samples are correlated and (4.170) becomes, in the z-domain,

$$i_0(z) = -\left[\frac{1 - z^{-1/2}}{1 + \alpha_4 - z^{-1}}\right] i_N(z) \qquad (4.171)$$

The numerator expresses the differentiation which gave correlated double sampling in the memory cell.

Returning to (4.170) for noise frequency components above the Nyquist frequency (the major part of the noise spectrum), the samples $i_N(n)T$ and $i_N(n-1/2)T$ are uncorrelated and (4.170) may be written in terms of the noise PSDs and this gives, in the z-domain,

$$S_I(f) = \frac{\overline{i_0^2}}{\Delta f} = \left[\frac{1}{1 + \alpha_4 - z^{-1}}\right]^2 [2S_F] \qquad (4.172)$$

where S_F is the total folded input noise PSD for the integrator and, assuming that thermal noise dominates, this is given by

$$S_F = 2S_M(f)\frac{2BW_n}{f_{CK}}$$

$$\approx \frac{8m_{th}}{3}\left(\frac{kT_j}{C}\right)\left(\frac{1 + k_J}{f_{CK}}\right)g_m^2 \qquad (4.173)$$

For physical frequencies, substituting $z = e^{j\omega T}$,

$$S_I(f) = |H(e^{j\omega T})|^2 [2S_F] \qquad (4.174)$$

where

$$H(e^{j\omega T}) = \frac{\left(\frac{1}{\alpha_4}\right)e^{-j\omega T/2}}{1 + j\left[\frac{\omega}{2}\frac{\alpha_4}{T}\frac{\alpha_4}{2+\alpha_4}\right]} \qquad (4.175)$$

Equation (4.174) indicates that the integrator noise PSD results from doubling the input-referred folded noise PSD and then multiplying by the square of the integrator's frequency response. At first sight this is a surprising result since the noise source is connected directly to the integrator summing node, a connection which gives amplification rather than integration of signals. This can be explained by referring to the ideal integrator shown in Figure 4.33. The circuit is changed in Figure 4.33(b) by inserting a pair of paralleled anti-phased switches which clearly does not change the circuit function. Next, since the samples of i_N taken on phases ϕ_1 and ϕ_2 are uncorrelated, they could equally come from independent noise sources, as shown in Figure 4.33(c). Now each noise source is processed via an integrator function and summed to give the doubling of the noise PSD.

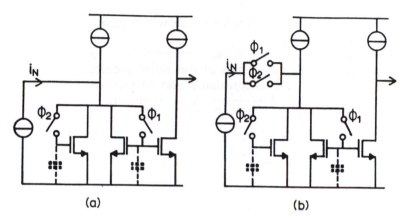

(a)   (b)

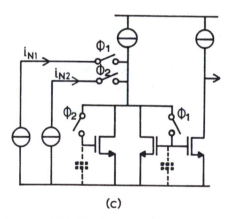

(c)

Figure 4.33 Equivalent integrator noise sources.

4.6.2.1 Simulations

The analysis of the integrator noise performance was confirmed by simulation using SCALP2. Figure 4.34 shows the simulated noise PSDs of a damped integrator made with the same memory cells which were simulated in the last section.

The half-power frequencies agree closely with the integrator cut-off frequencies, $f_0 = f_{CK}\alpha_4/(2 + \alpha_4)\pi$, and the low-frequency values of $S(f)$ are consistent with the square of the integrator low-frequency gain, $a_0 = 1/\alpha_4$. The simulated values of $S(f)$ at 1Hz were found to be about 2% lower than computed from (4.173) and this is attributed to the approximation involved in the use noise bandwidth.

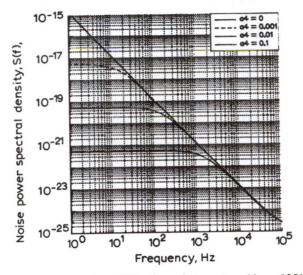

Figure 4.34 Simulated noise PSD of an integrator (f_{CK} =100kHz, α_1 =1).

4.6.2.2 Dynamic range

Returning to the noise analysis, (4.174) describes the integrator noise PSD as a function of frequency. It has a low-frequency value of $S(f)$ given by

$$S_I(0) = 2/|H(0)|^2 S_F$$

$$= \frac{16m_{th}}{3}\left(\frac{1}{\alpha_4}\right)^2\left(\frac{kT_j}{C}\right)\left(\frac{1 + k_J}{f_{CK}}\right)^2 g_m \qquad (4.176)$$

and a first-order roll-off with a half-power frequency f_0 given by

$$f_0 = \frac{f_{CK}}{\pi}\frac{\alpha_4}{2 + \alpha_4} \qquad (4.177)$$

The output noise power, $\overline{i_{NI}^2}$, is the area under this frequency response curve. This area may be estimated using the noise bandwidth concept which for the integrator output noise is given by

$$BW_{nI} = \frac{\pi}{2}f_0 = \frac{f_{CK}}{2}\frac{\alpha_4}{2 + \alpha_4} \qquad (4.178)$$

The noise power is

$$\overline{i_{NI}^2} = S_I(0)BW_{nI}$$

$$\approx \frac{8m_{th}}{3}\left(\frac{kT_j}{C}\right)\left(\frac{1 + k_J}{\alpha_4(2 + \alpha_4)}\right)g_m^2 \qquad (4.179)$$

For a sinusoidal signal with peak amplitude m_iJ the RMS value is $m_iJ/\sqrt{2}$ and the dynamic range is

$$DR_I = 10\ log\left[\frac{m_i^2J^2/2}{\overline{i_N^2}}\right] = 10\ log\left[\frac{m_i^2(V_{gs} - V_T)^2}{\dfrac{32m_{th}}{3}\left(\dfrac{kT_j}{C}\right)\left(\dfrac{1+k_J}{\alpha_4(2+\alpha_4)}\right)}\right] \qquad (4.180)$$

Comparing (4.180) and (4.160), the integrator dynamic range can be related to that of the memory cell as follows:

$$DR_I = DR_M - 10\ log\left(\frac{4}{\alpha_4(2 + \alpha_4)}\right) \qquad (4.181)$$

Equation (4.181) indicates that the integrator carries a noise penalty the significance of which depends on the low-frequency gain determined here by the damping coefficient α_4 and is related to the ratio of clock to integrator cut-off frequency. As an example, it was shown before that for $V_{DD} = 5V$, $(V_{gs} - V_T) = 1V$, and $C = 10pF$, and with optimised saturation voltages, the basic memory cell has a dynamic range $DR_M = 80.5dB$. When an integrator is made from these memory cells with $f_{CK}/f_0 = 50$ (corresponding to a low-frequency gain $a_0 = 7.46$ produced by $\alpha_1 = 1$ and $\alpha_4 = 0.134$) then its dynamic range DR_I is 12dB lower. This aspect of the switched-current integrator could be a serious drawback in designs with a large low-frequency gain since the dynamic range can only be restored by an increase in the value of C with a corresponding increase in area. So, care must be taken in the optimisation of dynamic range of a full filter design to avoid large variations in integrator gain.

4.7 Summary

Non-ideal MOS transistor performance, particularly mismatch errors, non-zero output conductance, finite bandwidth and switch charge injection, produces non-ideal behaviour in switched-current circuits. In this chapter, the effect of these errors has been analysed, first, for the basic current memory and then for the lossless integrator. Table 4.3 gives the principal formulae which are useful in two ways: they give insight into the nature of the errors and into design strategies for their minimisation; they may also be included in macro-models of switched-current sub-circuits (e.g. memory cells and integrators) to provide estimates of the effects of these errors on the total filter frequency response using only an analogue CAD package.

Error Type	Device Error	Memory Error		Integrator Error	
		$m(\omega)$	$\theta(\omega)$	$m(\omega)$	$\theta(\omega)$
Mismatch	ε_M	-	-	ε_M	0
Output/Input Conductance	ε_G	ε_G	0	$\left(1 + \dfrac{\alpha_1}{4}\right)\varepsilon_G$	$\left(1 + \dfrac{\alpha_1}{4}\right)\varepsilon_G \tan\dfrac{\omega T}{2}$
Settling	ε_S	$\dfrac{-\varepsilon_S \sin^2\omega T}{2(1 - \varepsilon_S)}$	$\dfrac{-\varepsilon_S \sin\omega T}{1 - \varepsilon_S}$	non-inverting $-2\varepsilon_S$; inverting $-\varepsilon_S(1+\cos\omega T)$	non-inverting $\dfrac{\sin\omega T}{(1 - \varepsilon_S)^2}$; inverting $\varepsilon_S \sin\omega T$
Charge Injection	ε_q	$-\varepsilon_q$	0	$-\varepsilon_q$	$\dfrac{2\varepsilon_q}{\tan\dfrac{\omega T}{2}}$

Table 4.3 Summary of non-ideal small signal behaviour in the basic switched-current memory and lossless integrator. The formulae assume that the device errors are small.

Mismatch of MOS transistors' threshold voltages (V_T) and gain factors (β) produce no errors in the gain of the basic current memory. However, where mirroring is employed (e.g. integrator output stage and for damping) gain errors occur with variances inversely proportional to transistor area. These errors produce magnitude (but not phase) errors in the frequency response of the lossless integrator. Switched-current filters will have random errors in their coefficients in much the same way that switched-capacitor filters suffer errors resulting from capacitor mismatches. Switched-current gain error variance resulting from MOS transistor mismatch is comparable with the ratio error variance of switched-capacitors with a similar area.

A non-zero output-input conductance ratio results from channel-shortening and from capacitive feedback in the memory transistor and produces a systematic gain error of a few percent, but no phase error. An ideal lossless integrator has a pole at zero frequency and this error moves the pole to a positive frequency and limits the low-frequency gain. It produces a magnitude and phase error at the integrator's unity gain frequency in much the same way that switched-capacitor integrators have magnitude and phase errors resulting from finite op-amp gain. However, the basic switched-current integrator made from memory cells with typical conductance ratio errors has a performance which is only as good as a switched-capacitor integrator using an op-amp with 30dB gain. Clearly, circuit enhancements are needed to improve the ratio of input to output conductance by about 40dB.

The current memory transistor's cut-off frequency $(\omega_{C0} = g_m/C)$ produces settling time errors. With drain capacitance and switch resistance, the step response is second-order and careful design of these parameters is needed to avoid ringing. Settling time errors give the memory cell both magnitude and phase errors while in the lossless integrator they produce an effect identical to that of cascading with a first-order lowpass section. This gives the non-inverting integrator a magnitude error and the inverting integrator both magnitude and phase errors. This is analogous with the effect of op-amp gain-bandwidth product in switched-capacitor filters.

Charge injection effects in switched-current circuits have many similarities with those in switched-capacitor circuits. However, one notable difference results from the connection of the switch in the current memory between the memory transistor's gate and drain. During the switch turn-off, while the channel is still conductive, there is a negative feedback path which helps to reduce charge injection errors, especially effective with long clock fall-times. This extra complexity in the switching dynamics gives switched-current circuits scant chance of an exact analysis of charge injection errors yielding a closed-form solution. Simplified analysis, however, is given and this shows a trade-off between charge injection error and memory transistor cut-off frequency. So, while an extremely high cut-off frequency (several GHz) is possible giving short settling time suitable for high clock frequency systems, this is necessarily only possible with increased charge injection errors. The effect of charge injection errors on the memory cell's and the integrator's frequency response is similar to that introduced by non-zero output conductance; they introduce only magnitude errors in the memory cell while in the integrator they move the pole from zero frequency to a positive value and produce magnitude and phase errors at the integrator's unity gain frequency. Pairs of memory cells (e.g. in the integrator loop) have reduced average error due to the cancellation of some components of the injected charge.

The noise analysis was developed for the basic memory cell and extended to the switched-current integrator. Flicker noise present in the memory cell's transistors was found to undergo a correlated double sampling process and in most cases will be removed from the base band. White noise, on the other hand, is sub-sampled and creates many replicas in the base band. However, the sampling process does not increase the noise power but merely reshapes the noise spectrum. The analysis showed good agreement with simulations. Expressions derived for the memory cell's dynamic range showed that the saturation voltages of the memory and current bias transistors should be maximised for best dynamic range which then depends only on the size of the memory cell's storage capacitor. Values in excess of 90dB are feasible. The noise analysis of the switched-current integrator showed good agreement with simulation. However, it showed that the thermal noise sources within the integrator's memory cells produce a noise power spectral density which is shaped by the integrator's frequency response giving a noise penalty the significance of which depends on the low-frequency gain of the integrator.

Much of the analysis presented in this chapter assumes signals which are small compared with the bias current. The expressions for the various errors (Table 4.3) all indicate a dependence on signal amplitude, implying that harmonic distortion and offsets are also present. Management of the errors through judicious design and enhanced circuit topologies will be a necessary pre-requisite to successful switched-current circuits. Chapter 6 gives a variety of ways for enhancing switched-current circuits to improve their analogue performance.

References

[1] J.-B. Shyu, G. C. Temes and F. Krummenacher, "Random error effects in matched MOS capacitors and current sources," *IEEE J. Solid State Circuits*, vol. SC-19, pp. 948-955, Dec. 1984.

[2] K. R. Lakshmikumar, R. A. Hadaway and M. A. Copeland, "Characterisation and modelling of mismatch in MOS transistors for precision analog design," *IEEE J. Solid-State Circuits*, vol. SC-21, pp. 1057-1066, Dec. 1986.

[3] M. J. M. Pelgrom, A. C. J. Duinmaijer and A. P. G. Welbers, "Matching properties of MOS transistors," *IEEE J. Solid-State Circuits*, vol. SC-24, pp. 1433-1439, Oct. 1989

[4] A. S. Sedra and P. O. Brackett, *Filter Theory and Design, Active and Passive*, Pitman, pp. 401-403, 1979.

[5] K. Martin and A. S. Sedra, "Effects of op amp finite gain and bandwidth on the performance of switched-capacitor filters," *IEEE Trans. Circuits Syst.*, vol. CAS-28, pp. 822-829, Aug. 1981.

[6] DiStefano 111, Stubberud and Williams, *Feedback and Control Systems*, McGraw-Hill, 1976.

[7] K. R. Stafford, P. R. Gray and R. A. Blanchard, "A complete monolithic sample and hold amplifier," *IEEE J. Solid-State Circuits*, vol. SC-9, pp. 381-387, Dec. 1974.

[8] R. E. Suarez, P. R. Gray and D. A. Hodges, "All-MOS charge redistribution analog to digital conversion techniques - part 2," *IEEE J. Solid-State Circuits*, vol. SC-10, pp. 379-385, Dec. 1975.

[9]   D. A. Hodges, "Analog switches and passive elements in MOS LSI," pp. 14-18 in, *Analog MOS Integrated Circuits*, New York: IEEE Press, 1980.

[10]   D. MacQuigg, "Residual charge on a switched-capacitor," *IEEE J. Solid-State Circuits*, vol. SC-18, pp. 811-813, Dec. 1983.

[11]   B. J. Sheu and G. Hu, "Switch-induced error voltage on a switched-capacitor," *IEEE J. Solid-State Circuits,* vol. SC-19, pp. 519-525, Aug. 1984.

[12]   W. B. Wilson, H. Z. Massoud, E. Swanson, R. . George and R. B. Fair, "Measurement and modelling of charge feedthrough in n-channel MOS analog switches," *IEEE J. Solid-State Circuits*, vol. SC-20, pp. 1206-1212, Dec. 1985.

[13]   J.-H Shieh, M. Patel and B. J. Sheu, "Measurement and analysis of charge injection in MOS analog switches," *IEEE J. Solid-State Circuits*, vol. SC-22, pp. 277-281, April 1987.

[14]   G. Wegmann, E. A. Vittoz and F. Rahali, "Charge injection in analog MOS switches," *IEEE J. Solid-State Circuits*, vol. SC-22, Dec. 1987.

[15]   C. Eichenberger and W. Guggenbühl, "Dummy transistor compensation of analog MOS switches," *IEEE J. Solid-State Circuits*, vol. SC-24, pp. 1143-1146, Aug. 1989.

[16]   C. Eichenberger and W. Guggenbühl, "On charge injection in analog MOS switches and dummy switch compensation techniques," *IEEE Trans Circuits Syst.*, vol. CAS-37, pp. 256-2610, Feb. 1990.

[17]   C. A. Gobet and A. Knob, "Noise analysis of switched-capacitor networks," *IEEE Trans Circuits Syst.*, vol. CAS-30, pp. 37-43, Jan. 1983.

[18]   S. J. Daubert and D. Vallancourt, "Operation and analysis of current copier circuits," *IEE Proceedings*, Part G, April 1990.

[19]   E. A. Vittoz and G. Wegmann, "Dynamic current mirrors", Chapter 7 in C. Toumazou, F. J. Lidgey and D. G. Haigh, Eds., *Analogue IC design: the current-mode approach*, London: Peter Perigrinus Ltd, pp. 314-317, 1990.

[20]   P. R. Gray, R. G. Meyer, *Analysis and Design of Analog Integrated Circuits*, John Wiley and Sons, 1977.

Chapter 5

Noise in Switched-Current Circuits

Steven Daubert

5.1 Overview

This chapter will examine the effects of electrical noise on a switched-current circuit. Here electrical noise means the small fluctuations of current in an MOS transistor and is inherent to the operation of the device. Often the term electrical noise is used to describe any unwanted signal, including power supply disturbances and parasitic coupling of extraneous digital signals into the analogue signal path. While disturbances by these unwanted signals may also corrupt the operation of a switched-current circuit, they can be greatly reduced and practically eliminated by the use of careful layout techniques and by using balanced circuits.

Electrical noise, on the other hand, cannot be eliminated by careful layout techniques, and will determine the smallest signal that will be delivered from the circuit. This is true for current-mode and voltage-mode circuits and for switched circuits or continuous time circuits. One important difference between switched circuits and continuous circuits is the aliasing of high frequency noise caused by the switching operation. This often leads to a higher level of inband noise for a switched circuit and creates a more complicated analysis.

The analysis of electrical noise in an analogue circuit is handled by describing the current fluctuations as a random signal. This random signal has a mean-square value, $\overline{i^2}$, and can be measured in a bandwidth, Δf, however the instantaneous value, $i(t)$, cannot be predicted.

Other sources of errors that limit the accuracy of current transfer in a switched-current circuit are deterministic. These types of errors include mismatch of devices and charge feedthrough effects. While these deterministic errors limit the accuracy of transferring large and small current signals, the thermal and $1/f$ noise of the MOS transistors places a practical lower level of current that may be delivered by a switched-current circuit. Accurate prediction and understanding of the noise level is important, for circuit noise determines the dynamic range of a switched-current filter circuit or the obtainable resolution of an A/D or D/A converter.

For comparison purposes, this chapter will begin with a noise analysis of the familiar current mirror and continue with the noise analysis of one type of switched-current circuit, a current copier [1]. To make this a valid comparison, the current copier transconductor will be a single n-channel transistor of the same dimensions as the n-channel transistor in the current mirror. In practice, many switched-current circuits will replace the single n-channel transistor with a transconductor having several transistors. In this case the same analysis techniques may be used, but with the input-referred noise voltage of the transconductor replacing the noise voltage of a single transistor.

As with the deterministic errors, techniques for reducing the random errors due to electrical noise exist. Some of these techniques will be described here, though the number of circuit approaches for reducing noise is more limited.

5.2   Transistor Noise

The noise analysis presented here begins by assuming a basic current copier with a single n-channel transistor serving as the transconductor. A current source-based noise model for the MOS transistor [2] is shown in Figure 5.1.

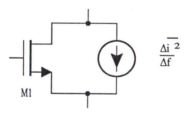

Figure 5.1 Transistor output noise current equivalent circuit.

With this model, a noise current flows from drain to source in addition to the drain current caused by the applied signals to the transistor. The value of the noise current at a given time is not defined - it is a random signal. However, the mean square value of the current $\overline{i^2}$ is a known quantity. Also, within a narrow frequency range, the current noise components are known. This ratio of the mean square noise current to frequency interval, $\Delta \overline{i^2} / \Delta f$, known as the power spectral density (PSD), is a common term used in the analysis of circuits as well as random processes.

For an MOS transistor, the noise current consists of two components originating from two different physical mechanisms. One of these components, thermal noise, is due to random thermal motion of carriers in the channel. This noise increases with increasing temperature, but is independent of frequency. For switched-current applications, the analysis is more straightforward if the power spectral densities are expressed as double-sided quantities with $-\infty < f < \infty$. This notation simplifies the calculations of the current copier output PSD when aliasing effects are considered. For a transistor in saturation with transconductance, g_m, the thermal noise current PSD is

$$\frac{\overline{\Delta i_t^2}}{\Delta f} = 2kT\left(\frac{2}{3}g_m\right) \qquad -\infty < f < \infty \tag{5.1}$$

where k is Boltzmann's constant, T is absolute temperature, and the subscript t is used to represent thermal noise.

The second noise component, $1/f$ noise, has a PSD that decreases with increasing frequency. Although the exact mechanism for $1/f$ noise is not agreed upon, the power spectral density has been characterised and is given by

$$\frac{\overline{\Delta i_t^2}}{\Delta f} = \frac{K_{fn}g_m^2}{C_{ox}'WL|f|} \tag{5.2}$$

where K_{fn} is a process dependent constant, C_{ox}' is the gate capacitance per unit area and W and L are the transistor's width and length respectively.

Both (5.1) and (5.2) give expressions for the noise current PSD of an MOS transistor. For current copier applications, it is more convenient to express the PSD as a voltage source in series with the transistor gate, as shown in Figure 5.2.

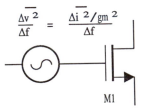

$$\frac{\overline{\Delta v}^2}{\Delta f} = \frac{\overline{\Delta i}^2/\text{gm}^2}{\Delta f}$$

M1

Figure 5.2 Transistor input noise voltage equivalent circuit.

The main advantage of this approach is its ease of extension of analysis results to copiers constructed with more complex transconductors. With a more complex transconductor, input-referred noise can be defined as a voltage source in a manner similar to the circuit in Figure 5.2. This input-referred noise source will have a PSD differing from that of a single transistor, but the transfer function from noise voltage to current output will still be determined by g_m. Although a more generalised transconductor will have additional poles, leading to g_m changing as a function of frequency, by design these poles should be high enough in frequency to have minimal effect on a noise analysis.

For the MOS transistor, the input-referred voltage source is related to the output current noise model by

$$\frac{\overline{\Delta v^2}}{\Delta f} = \frac{\overline{\Delta i^2}}{g_m^2 \Delta f} \tag{5.3}$$

Thus the noise voltage PSD, $S(f)$, is

$$S(f) = \frac{\overline{\Delta v_f^2}}{\Delta f} + \frac{\overline{\Delta v_t^2}}{\Delta f}$$

$$= \frac{K_{fn}}{C_{ox}' WL |f|} + 2kT \left(\frac{2}{3 g_m} \right) \tag{5.4}$$

An example of the noise voltage PSD for an n-channel transistor with W/L = 100/4, a bias current of 10μA and noise parameters for a 1.75μm CMOS process is shown in Figure 5.3.

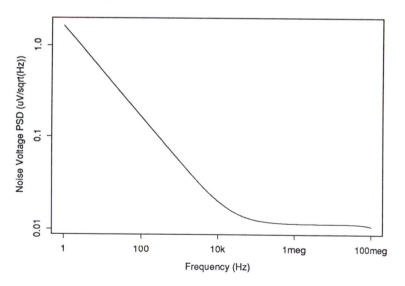

Figure 5.3 Transistor noise voltage PSD.

5.3   Current Mirror Noise

Before considering the noise from a current copier circuit, an analysis of a simple current mirror provides a good basis for comparison. Figure 5.4

shows a two transistor current mirror, biased by an input current I_o and supplying a current to a load. Explicitly shown are the input-referred noise sources of the two transistors, labelled $\overline{v_{M2}^2}$ and $\overline{v_{M1}^2}$.

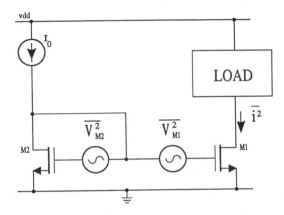

Figure 5.4 Current mirror noise circuit.

Since the noise sources of the two transistors are not correlated, the mean square output noise current components of each source may be added to determine $\overline{i^2}$. With g_{m1} and g_{m2} representing the transconductance of M_1 and M_2 respectively, and g_{o2} representing the output conductance of M_2, the mean square output noise current is

$$\overline{i^2} = g_{m1}^2\, \overline{v_{M1}^2} + g_{m1}^2 \left(\frac{g_{m2}}{g_{m2} + g_{o2}}\right)^2 \overline{v_{M2}^2}$$

$$\approx g_{m1}^2 \left(\overline{v_{M1}^2} + \overline{v_{M2}^2}\right) \qquad\qquad (5.5)$$

Thus the noise of a simple current mirror reflects the noise of the two transistors without significant frequency shaping. Figure 5.5 shows a noise current PSD for a current mirror, again using 100/4 transistors with a 10µA current. This ADVICE [3] simulation shows that the noise of the two transistors appears with equal weight in the output noise current and includes a slight attenuation of the M_2 noise by a high frequency pole at 10MHz.

To lower the circuit noise, it is worthwhile examining the effects of changing the dimensions of the transistors for a simple current mirror. After a review of (5.1) and (5.2), one obvious method of reducing both thermal and 1/f noise components is to lower the transconductance.

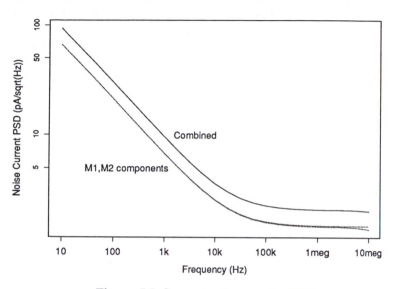

Figure 5.5 Current mirror noise PSD.

With an MOS transistor,

$$g_m = \sqrt{2 \frac{W}{L} \mu C'_{ox} I_o}$$ (5.6)

Thus for a fixed input current I_o, g_m may be lowered by reducing the ratio of W/L. This technique has limitations however. For transistors biased in saturation (needed to create a current mirror), the drain current is given by

$$I_D = \frac{W}{L} \mu C'_{ox} (V_{GS} - V_T)^2 \qquad V_{DS} > V_{GS} - V_T$$ (5.7)

By lowering the ratio of W/L to reduce circuit noise, the gate-to-source voltage of the transistors must be increased to support the desired current. This means the minimum allowable V_{DS} of M_1 must be increased to maintain saturation. Thus there is a trade-off between circuit noise and output compliance; lower noise results in a smaller allowable voltage swing on the output of the mirror. Figure 5.6 shows ADVICE simulations of the noise PSD of a current mirror for several values of W/L, again for transistors biased with $I_o = 10\mu A$.

One additional consideration is possible, and that is to maintain the ratio of W/L while increasing both W and L. As seen from (5.2) this lowers the $1/f$ noise of the transistor. Also, there is a minor change in the transconductance, included in the second order effects of the ADVICE simulation of noise using increased W and L and shown in Figure 5.7. The amount of improvement from this approach depends on the bandwidth needed for the circuit. If the signal frequencies are low, lowering the noise will improve the signal-to-noise ratio (SNR), since the high frequency thermal noise may be removed by filtering. If signal frequencies are high, lowering the *1/f* noise will not improve the SNR, since this approach does not alter the high frequency thermal noise.

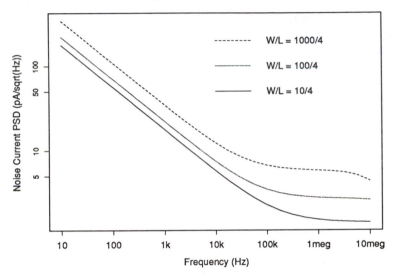

Figure 5.6 Current mirror noise with different transistor widths.

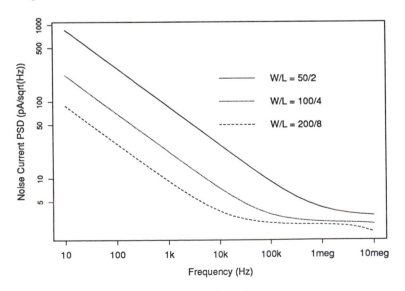

Figure 5.7 Current mirror noise with different gate areas.

5.4 Current Copier Noise

Although current copiers and mirrors perform similar functions they have major differences. First, copiers can be designed to have negligible deterministic errors, while mirrors are dependent on process matching. Secondly, copiers can cancel low frequency random noise while mirrors can not.

As will be shown here, a current copier circuit alters the noise PSD of its transconductor. For a current copier it is not possible to limit the frequency content of the noise source to below the Nyquist frequency, so aliasing and shaping of circuit noise occurs. As shown in Figure 5.5, a current mirror provides only minor shaping of the noise PSD, due to a non-dominant pole lowering the transconductance at high frequencies. The analysis below shows that copier noise-shaping depends on both the sampling clock and the loop bandwidth of the circuit.

For this analysis, it will be assumed the current copier operates such that it "samples a current" for a period τ_1 and supplies this current to a load for a period τ_2, as shown in Figure 5.8. Also assumed for this analysis is a load that receives no current during the sampling period. While most applications will multiplex two current copiers to supply a constant current to the load, this analysis can be easily extended to such an application by summing the output PSDs of two current copiers. The result presented here will be a power spectral density, $S_i(\omega)$, where $\omega = 2\pi f$, representing the random component of the current to the load.

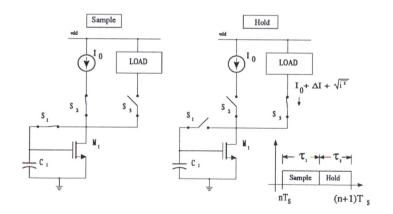

Figure 5.8 Basic current copier operation.

During the sample phase, the circuit of Figure 5.8 can be replaced by a small-signal equivalent, shown in Figure 5.9.

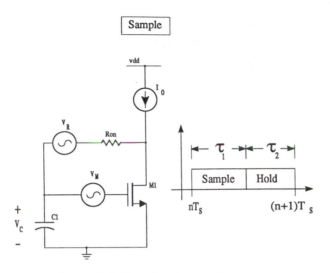

Figure 5.9 Sample phase noise sources.

Here the switching MOS transistor, S_1, is replaced by a resistor representing the on-resistance of the switch. Associated with this resistor is a white noise source v_R with a double-sided PSD,

$$S_R(\omega) = \frac{kTR_{on}}{\pi} \qquad -\infty < \omega < \infty \qquad (5.8)$$

Similarly, the transconductor M_1 is replaced with a "noiseless" transconductor and an input-referred noise voltage, v_M. This noise, containing both white and $1/f$ components, has a double-sided PSD $S_M(\omega)$. For a single transistor,

$$S_M(\omega) = \frac{K_{fn}}{C'_{ox}WL|\omega|} + \frac{kT}{\pi}\left(\frac{2}{3g_m}\right) \qquad (5.9)$$

While current is being sampled by the copier cell, noise from both the switch and the transconductor varies the voltage across C_1. When the switch is turned off, a sample of this noise is stored on C_1.

5.4.1 *Relative importance of switch and transconductor noise in a current copier*

Before considering the sampling operation occurring with current copiers, it is possible to show that, for reasonable choices of g_m and R_{on}, the switch noise will not significantly alter the sample on the capacitor. This is shown with a noise analysis of a copier circuit in the sample phase. Since, for proper settling behaviour $R_{on} \ll 1/g_m$, the loop bandwidth of the copier during sampling is

$$\omega_0 = g_m/C_1 \tag{5.10}$$

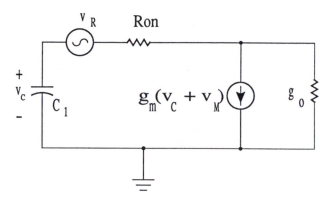

Figure 5.10 Small signal equivalent circuit for sample phase.

Using the small signal equivalent circuit of Figure 5.10, the voltage across the sampling capacitor v_C is related to the switch noise voltage v_R by

$$\frac{v_C}{v_R}(j\omega) = \frac{g_o/(g_m + g_o)}{1 + j\omega C_1(1 + g_oR_{on})/(g_m + g_o)}$$

$$\approx \frac{g_o/g_m}{1 + j\omega/\omega_0} \tag{5.11}$$

Thus, for noise from R_{on}, the power spectral density of noise on the sampling capacitor is

$$S_{C_{Ron}}(\omega) = \left|\frac{v_C}{v_R}(j\omega)\right|^2 S_R(\omega) \tag{5.12}$$

Then

$$S_{C_{Ron}}(\omega) \approx \left[\frac{(g_o/g_m)^2}{1 + (\omega/\omega_0)^2}\right] S_R(\omega) \tag{5.13}$$

Similarly, the voltage across the sampling capacitor, v_C, is related to the transconductor noise voltage, v_M, by

$$\frac{v_C}{v_M}(j\omega) = \frac{g_m/(g_m + g_o)}{1 + j\omega C_1(1 + g_oR_{on})/(g_m + g_o)}$$

$$\approx \frac{1}{1 + j\omega/\omega_0} \tag{5.14}$$

Thus, for noise from M_1, the power spectral density of noise on the sampling capacitor is

$$S_{C_M}(\omega) = \left| \frac{v_C}{v_R}(j\omega) \right|^2 S_M(\omega) \tag{5.15}$$

and

$$S_{C_M}(\omega) \approx \left[\frac{1}{1 + (\omega/\omega_0)^2} \right] S_M(\omega) \tag{5.16}$$

Since the noise sources are uncorrelated, the two PSDs can be added to form the unsampled noise PSD

$$S_C(\omega) = S_{C_{Ron}}(\omega) + S_{C_M}(\omega) \tag{5.17}$$

Examination of (5.13), (5.16) and (5.17) shows that transconductor noise will alter the voltage on the capacitor significantly more than will switch noise.   Since $R_{on} < 1/g_m$, the switch noise will be less than the transconductor noise, and since $g_o/g_m \ll 1$, the switch noise will be attenuated before reaching the sampling capacitor, whereas the transconductor noise is not.   The attenuation of switch noise but not transconductor noise can be explained by examining the location of the two noise sources in the circuit. Noise from the transconductor changes the required gate to source voltage for M_1 to support I_0, thus it is reflected on the voltage across C_1. However, the switch noise acts to change the gate to drain voltage for M_1, having a much smaller effect on the voltage across C_1. For the following analysis, the switch noise will be neglected and (5.17) will be simplified to

$$S_C(\omega) \approx S_{C_M}(\omega) \approx \left[\frac{1}{1 + (\omega/\omega_0)^2} \right] S_M(\omega) \tag{5.18}$$

Figure 5.11 shows an ADVICE simulation result of the capacitor noise PSD for a current copier in its sample configuration. In this simulation, the transistor size is 100/4 with an input current of $I_0 = 10\mu A$, and the switch consists of n- and p-channel transistors, each 2.25/2.25. As predicted by the preceding analysis, the transconductor noise results in a much larger noise on the sampling capacitor than the switch noise. This simulation includes the effects of a zero and a second pole not considered in (5.13) or (5.16), enhancing the value of $S_{C_R}(\omega)$ above 1MHz. This is caused by the output capacitance of M_1, neglected in the small signal model presented here. Even with the second pole included, noise from the transconductor dominates below 100MHz, determining the noise out of the copier.

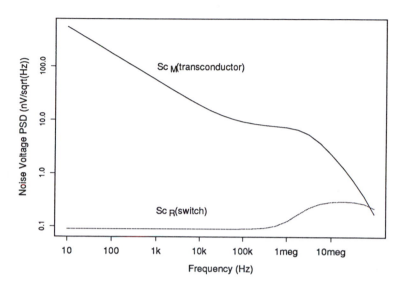

Figure 5.11 Switch and transconductor noise as appearing on a sampling capacitor.

5.4.2 Transfer function approach

The approach taken here to find the output noise current PSD is to write a series of equations describing the state of the current copier circuit [4,5]. The Fourier transform of these time-domain equations is taken to find a frequency-domain transfer function relating the noise source to the output current. Since a current copier is a time-variant system, these transfer functions will differ from sample period to hold period. With time-variant transfer functions, the statistics of the output noise process will change throughout the clock period. The resultant output noise is not a stationary stochastic process, even though the physical noise source is stationary. The average output noise is found by taking the output noise PSD for each clock phase and combining them to find an average PSD.

Writing circuit equations for the sample phase (neglecting switch noise),

$$\frac{dv_{C1}}{dt} = -\frac{g_m}{C} v_{C1}(t) - \frac{g_m}{C} v_M(t) \qquad (5.19a)$$

$$i_1(t) = 0 \qquad (5.19b)$$

Here, v_C is the sampling capacitor voltage, v_M is the transconductor noise source and i_1 is the output current. These expressions apply to only the sampling phase, denoted by the subscript 1. Throughout this analysis we consider g_m to be a constant; not changing from one phase to the next. In most conceivable applications this is a valid assumption, since the transconductor will be designed with a bias current larger than the signal current.

During the hold phase, denoted by subscript 2,

$$v_{C2}(t) = v_{C1}(nT_s + \tau_1) \tag{5.20a}$$

$$i_2(t) = g_m[v_{C1}(nT_s + \tau_1) + v_M(t)] \tag{5.20b}$$

Since $i_2(t)$ represents the current flowing to the load, we begin by taking the Fourier transform of (5.20b). With $i_2(t)$ present only during the hold phase, a gating function $\phi_2(t)$ is needed.

$$\phi_2(t) = \begin{cases} 0, & nT_s < t < nT_s + \tau_1 \\ 1, & nT_s + \tau_1 < t < (n+1)T_s \end{cases} \tag{5.21}$$

Taking the Fourier transform of (5.20b),

$$I_2(\omega) = g_m \int_{-\infty}^{\infty} \sum_n v_{C1}(nT_s + \tau_1)\phi_2(t)e^{-j\omega t}dt + g_m \int_{-\infty}^{\infty} v_M(t)\phi_2(t)e^{-j\omega t}dt \tag{5.22}$$

Simplifying the first integral of (5.22),

$$\int_{-\infty}^{\infty} \sum_n v_{C1}(nT_s + \tau_1)\phi_2(t)e^{-j\omega t}dt$$

$$= \left[\frac{1 - e^{-j\omega\tau_2}}{j\omega}\right] \sum_n v_{C1}(nT_s + \tau_1)e^{-j\omega(nT_s+\tau_1)}$$

$$= \left[\frac{1 - e^{-j\omega\tau_2}}{j\omega T_s}\right] \sum_n V_{C1}(\omega - n\omega_s)e^{-j\omega_s\tau_1} \tag{5.23}$$

where ω_s is equal to $2\pi/T_s$. Although (5.23) contains $V_{C1}(\omega)$, the goal of this analysis is to obtain an expression relating the noise voltage $V_M(\omega)$ to the output current. By using an approximation for $V_{C1}(\omega)$, derived from (5.19a), the relationship between $I(\omega)$ and $V_M(\omega)$ is obtainable. This approximation, based on neglecting the sampling effect on the capacitor voltage is

$$V_{C1}(\omega) \approx \frac{-1}{1 + j\omega/(g_m/C)} V_M(\omega) \tag{5.24}$$

Using (5.24) in (5.23) gives

$$\int_{-\infty}^{\infty} \sum_n v_{C1}(nT_s + \tau_1)\phi_2(t)e^{-j\omega t}dt$$

$$= \left[\frac{1 - e^{-j\omega\tau_2}}{j\omega T_s}\right] \sum_n \left[\frac{-1}{1 + j(\omega - n\omega_s)/(g_m/C)}\right] V_M(\omega - n\omega_s)e^{-j\omega_s\tau_1} \tag{5.25}$$

Likewise, for the second integral in (5.22),

$$\int_{-\infty}^{\infty} v_M(t)\phi_2(t)e^{-j\omega t}dt$$

$$= \sum_n V_M(\omega - n\omega_s)\left[\frac{e^{-jn\omega_s\tau_1} - e^{-jn\omega_s\tau_s}}{jn\omega_s T_s}\right] \tag{5.26}$$

The aim of this analysis is to put this result in the form of a transfer function that is useful even if the input signal is random. This transfer function is denoted by $H_{2,n}$, where the subscript 2 is used to indicate the transfer function is valid for the hold phase only and the subscript n represents the index of the summation. The transfer function $H_{2,n}(\omega)$ is defined by

$$I_2(\omega) = \sum_n H_{2,n}(\omega)V_M(\omega - n\omega_s) \tag{5.27}$$

For a random signal such as noise, $V_M(\omega)$ is not known. Using (5.25) and (5.26) to find $H_{2,n}(\omega)$, and expressing the transfer function in terms of *sinc* functions, where *sinc*(x) is $\sin(\pi x)/\pi x$,

$$H_{2,n}(\omega) = g_m\left(\frac{\tau_2}{T_s}\right)\left[e^{-jn\omega_s(\tau_1+\tau_2/2)}sinc(n\tau_2/T_s) - \frac{e^{-j(n\omega_s\tau_1+\omega\tau_2/2)}sinc(\omega\tau_2/2\pi)}{1 + j(\omega - n\omega_s)/(g_m/C)}\right] \tag{5.28}$$

Similarly, a transfer function relating the output current to the noise source during the sample phase is found by examining (5.19b). Since no current flows to the load during the sample phase,

$$H_{1,n}(\omega) = 0 \tag{5.29}$$

To evaluate the current flowing to the load requires a transfer function that is valid during both the sample phase and the hold phase. This transfer function is simply the sum of the two transfer functions derived in (5.28) and (5.29) [4]:

$$H_n(\omega) = H_{1,n}(\omega) + H_{2,n}(\omega) \qquad (5.30)$$

$H_n(w)$ can be used in (5.27) to find the output current for a deterministic source, v_m. However with a noise source, v_m is not known, but the PSD of v_m is. Thus for noise analysis a relationship between the PSD of the noise source and the output noise current is needed. Using results from linear periodically time-variant (LPTV) systems [6], these PSDs are related by

$$S_i(\omega) = \sum_{n = -\infty}^{\infty} |H_n(\omega)|^2 S_M(\omega - n\omega_s) \qquad (5.31)$$

where $S_i(\omega)$ represents the output noise current PSD. Finally, if the mean square value of the noise is desired, it may be obtained by integrating the PSD. Thus

$$\overline{i^2} = \int_{-\omega_1}^{+\omega_1} S_i(\omega) d\omega \qquad (5.32)$$

where ω_1 represents the bandwidth of interest.

Before proceeding with some examples of current copier noise, it is worthwhile examining the transfer function, $H_n(\omega)$, more closely. The relationship in (5.31) applies to a sampled data system involving current copiers. It should be possible to apply this relationship to two special cases of interest - one is a linear system without sampling and the other is a sampling linear system, but with the input PSD bandlimited to frequencies less than the Nyquist frequency. For a linear system without sampling no aliasing occurs since for $n \neq 0$ the transfer function $H_n(\omega) = 0$. For a sampling linear system with a bandlimited input, again no aliasing occurs, but for a different reason. Here $S_M(\omega - n\omega_s) = 0$ for $n \neq 0$. Both cases result in the simplification of (5.31) to

$$S_i(\omega) = |H_0(\omega)|^2 S_M(\omega) \qquad (no\ aliasing) \qquad (5.33)$$

For this special case, the noise transfer function of the current copier has the desirable property of cancelling low frequency noise. For $n = 0$ the transfer function is

$$H_0(\omega) = g_m \left(\frac{\tau_2}{T_s}\right) \left[1 - \frac{e^{-j\omega\tau_2/2} sinc(\omega\tau_2/2\pi)}{1 + j\omega/(g_m/C)} \right] \qquad (5.34)$$

For small values of ω, both the *sinc* and the $e^{-j\omega\tau_2/2}$ terms approach one, and there is little filtering of the noise during the sample phase. Here H_0 approaches zero. Physically this is explained by realising that the *sinc* term in (5.34) represents a sample of the noise stored on the capacitor, while the unity term represents the noise as it appears at the output during the hold phase. If the sampling rate is fast, compared to the frequency of the noise, the noise sample on the capacitor will correct for the noise of the transconductor. Of course as ω increases, the $e^{-j\omega\tau_2/2}$ creates a phase shift, and the sample and the noise source no longer cancel each other.

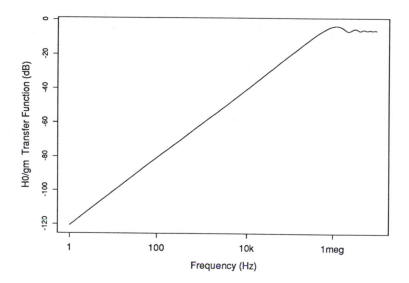

Figure 5.12 H_0/g_m **transfer function** (f_s = 1MHz, f_0 = 3MHz).

Figure 5.12 shows an example of the H_0 transfer function, normalised by dividing by g_m, thus relating the noise current of the transconductor to the output noise of the current copier. In this example, a 100/4 n-channel transistor is biased with a current of 10μA, and the sampling capacitor has a value of 10pF. For such a device in a 1.75μm CMOS technology, g_m = 210μS, resulting in a loop bandwidth $g_m/2\pi C$ of 3MHz. The current copier is clocked at 1MHz, having an equal sample period and hold period. As expected, noise at frequencies much below the sampling frequency is attenuated because the sample of the noise on the capacitor will cancel the noise during the hold period. As the frequency of the noise approaches the sampling frequency, this cancellation no longer occurs, but the transfer function has a minimum attenuation of 6dB, caused by having an output present only half the time. In addition, a slight ripple occurs in the transfer function above 1 MHz, caused by the *sinx/x* sampling effect.

A calculation of the PSD of the current noise involves the evaluation of (5.31) for all values of n, not merely examining H_0 as shown in Figure

5.12. For values of $n \neq 0$, noise from the transconductor will be aliased from high frequencies into the baseband. Thus the cancellation of noise at low frequencies will not be as complete as would be expected if no aliasing had occurred.

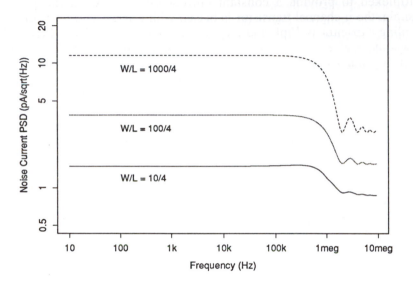

Figure 5.13 Current copier noise with different transistor widths.

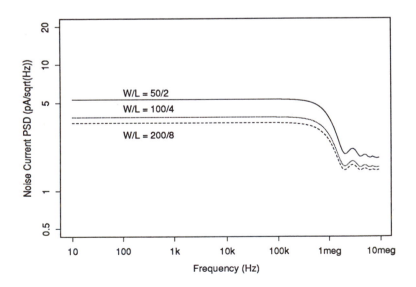

Figure 5.14 Current copier noise with different transistor gate areas.

Figures 5.13 and 5.14 show this evaluation for a single transistor current copier of varying dimensions in much the same way as Figures 5.6 and 5.7 did for a standard current mirror. To make the comparison fair, the simulations for the current copier noise assume that two copiers are multiplexed to provide a constant current output. As in the previous example, the n-channel transistor is biased with a 10μA current source, the sampling capacitor is 10pF and the clock frequency is 1MHz. Figure 5.13 shows the noise current PSD for different transistor widths. With increasing transistor width the transconductance increases, resulting in a wider loop bandwidth and a larger transistor noise current. The output noise PSD of the current copier circuit is still significantly less at low frequencies where the 1/*f* noise dominated the noise PSD of the current mirror. Figure 5.14 shows the noise current PSD for transistors with a constant *W/L* but with different gate areas. With a current mirror circuit, the technique of increasing the gate area to reduce 1/*f* noise is commonly used. With the current copier this technique is not useful, since although the transconductance is slightly decreased with increased gate area, the noise PSD changes only marginally. Since most of the 1/*f* noise is cancelled by the sampling operation, there is much less need to reduce it by increasing the gate area.

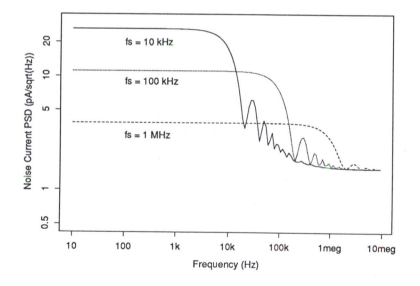

Figure 5.15 Current copier noise with different sampling frequencies.

One important aspect of controlling current copier noise is the design of the loop bandwidth with consideration for the sampling frequency. Figure 5.15 shows examples of the noise PSD for the same current copier circuit, but with three different sampling frequencies. Again the transistor is biased with 10μA, has a *W/L* of 100/4 and the sampling capacitor is 10pF.

The sampling frequency should be adjusted according to the settling accuracy required. For a typical settling accuracy of about 10τ ($\tau = C/g_m$), this copier should be clocked at about 1MHz. This curve is shown, as well as curves for sampling frequencies of 10kHz and 100kHz. If the copier is operated at the lower clock frequencies, aliasing occurs to a greater extent and more noise is folded into the baseband. As an example, the total integrated noise in the voiceband from 300Hz to 3kHz for different clock frequencies is shown in Table 5.1.

Clock frequency	RMS noise current
10 kHz	2.0 nA
100 kHz	0.8 nA
1 MHz	0.3 nA

Table 5.1   RMS noise current (300 to 3000Hz).

5.5   Conclusion

In comparing the noise of one type of switched current circuit, the current copier, to an unswitched current mirror, it becomes clear that several important differences exist. With a current copier, the low frequency transistor $1/f$ noise is sampled and removed from the output current signal, unlike continuous-time current mirrors where the $1/f$ noise is not cancelled. Although this sampling is an advantage for low frequency noise, the sampling operation also creates aliasing of high frequency noise into the signal band. Thus if a current copier is designed with excess high frequency noise due to excess bandwidth in the sampling loop, aliasing will create a high inband noise floor. However, if the loop bandwidth is merely wide enough to provide adequate settling, much less aliasing of noise occurs and the current copier provides a better noise performance.

This approach to noise analysis may also be used if a more complex transconductor is used in place of the single transistor. Since the switch noise may be neglected in most cases, the input-referred noise of the transconductor replaces that of the transistor in determining the output noise current.

References

[1]   S. J. Daubert, D. Vallancourt, and Y. P. Tsividis, "Current copier cells," *Electronics Letters*, vol. 24, pp. 1560-1562, 8 Dec. 1988.

[2]   Y. P. Tsividis, *Operation and modeling of the MOS transistor*, New York: McGraw-Hill, 1987.

[3]   L. W. Nagel, "ADVICE for circuit simulation," in *Proc. IEEE International Symposium on Circuits and Systems*, April 1980.

[4]   M. L. Liou and Y.-L. Kuo, "Exact analysis of switched capacitor circuits with arbitrary inputs," *IEEE Trans. Circuits Syst.*, vol. CAS-26, pp. 213-223, April 1979.

[5]   L. Toth and E. Simonyi, "Explicit formulas for analyzing general switch-capacitor networks," *IEEE Trans. Circuits Syst.*, vol. CAS-34, pp. 1564-1578, Dec. 1987.

[6]   W. A. Gardner, *Introduction to random processes*, New York: Macmillan, 1986.

Chapter 6

Switched-Current Circuit Design Techniques

John B. Hughes, Kenneth W. Moulding and Douglas M. Pattullo

6.1 Introduction

This chapter builds on earlier chapters (Chapters 3 and 4) which described architectures and non-idealities of *basic* switched-current circuits. In Chapter 3, basic cells were developed (delay, integrator and differentiator modules) and their algorithmic behaviour was described in terms of finite difference equations. In Chapter 4, the effects on circuit performance of imperfections in the MOS transistors used to build the circuits were analysed. In this chapter, circuit techniques are described which enhance performance in terms of precision, dynamic range and linearity.

Control of non-ideal behaviour is essential for the successful design of switched-current circuits if specifications of analogue performance are to be met. With the basic circuit topologies described in Chapter 3, the design process involves the choice of MOS transistor dimensions (memory and switch) and operating currents to manage the various errors (mismatch, conductance ratios, settling, charge injection, noise) using the formulae developed in Chapter 4. For undemanding applications this process may be adequate, but to achieve analogue performance that is competitive with state-of-the-art switched-capacitor circuits it will in general be necessary to enhance performance through the use of extra circuit techniques.

This chapter describes circuits developed from the basic current memory cell to give enhanced analogue performance. Two main approaches are adopted:

- The development of circuit structures that reduce conductance ratio errors and maximise dynamic range through the use of negative feedback in ways that allow a large saturation voltage $(V_{gs} - V_T)$.

- The use of fully differential structures to reduce charge injection errors and crosstalk from neighbouring (digital) circuits.

6.2 Feedback Techniques

It was shown in Chapter 4 that a signal transmission error occurs in the basic memory cell through inadequate ratio of input to output conductance.

The input conductance g_i was determined by the diode-connected memory transistor and is given by

$$g_i \approx g_m \qquad (6.1)$$

and the output conductance was determined by channel length modulation and charge feedback and is given by g_0 where

$$g_0 = g_{ds} + g_{ds(J)} + \frac{C_{dg}}{C + C_{dg}} g_m \qquad (6.2)$$

where g_{ds} and $g_{ds(J)}$ are the drain conductances of the memory transistor and the transistor generating the current bias, and C and C_{dg} are the memory transistor's gate-source and drain-gate capacitances. For the transmission error, ε_G, resulting from the ratio of these conductances, to be low, either g_i must be made large or g_0 must be made low, and either can be achieved through the application of negative feedback.

In this section, various ways of applying negative feedback are discussed. In the op-amp and grounded-gate active memory cells, feedback is used to increase the input conductance by creating a 'virtual earth' at the input, while in the various cascode arrangements it is used to decrease the output conductance. Other analogue errors, especially those arising from mismatch, charge injection and noise, are reduced by choosing transistors with large saturation voltage $(V_{gs} - V_T)$. Each feedback arrangement has electrical constraints which restrict this choice, and the impact on dynamic range is examined.

6.2.1 Op-amp active memory cell

The op-amp active memory cell [1,2] is shown in Figure 6.1, slightly modified to allow bipolar signals. The operation of the cell is as follows. On phase ϕ_1, the current $J+i$ differs from that stored in transistor T on the previous period and the net current at the input node has a non-zero value. This disturbs the voltage at the positive input of op-amp OA which drives current into capacitor C causing a change in the drain current of transistor T. Equilibrium is established when the current in T reaches $J+i$. If the gain of amplifier OA is high then the voltage at the summing node is close to V_{bias} and the drain voltage of transistor T is a little lower due to the small voltage drop on its input switch. On phase ϕ_2 the loop is broken and transistor T sustains its current at $J+i$ as long as the output node is held near V_{bias}. This happens quite naturally in cascade connected cells as the output voltage of the driving cell is fixed at the input voltage of the driven cell, i.e. at V_{bias}, as shown in Figure 6.2. The feedback introduced by OA on phase ϕ_1 increases the cell's low frequency input conductance to g_i where

$$g_i \approx g_m A_v \qquad (6.3)$$

where A_v is the voltage gain of *OA*.

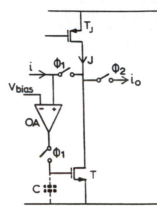

Figure 6.1 Op-amp active memory cell.

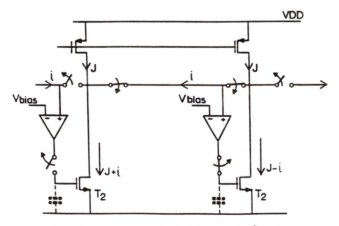

Figure 6.2 Cascade-connected op-amp active memory cells.

The output conductance is the same as that of the basic cell and so the transmission error due to non-zero output to input conductance ratio, ε_G, is also reduced by the factor A_v. Even an op-amp comprising just a simple differential pair can achieve gains of $A_v \approx 100$, making values of $\varepsilon_G = 0.01\%$ quite feasible.

Saturated operation of the memory transistor and the current source transistor is guaranteed as long as

$$V_{DD} - (V_{gs} - V_T)_J \geq V_{bias} \geq (V_{gs} - V_T)\sqrt{1 + m_i} \qquad (6.4)$$

where $(V_{gs} - V_T)$ and $(V_{gs} - V_T)_J$ are the saturation voltages of the memory and current source transistors, respectively, both at $I_{ds} = J$ and m_i is the peak signal modulation index, $i(pk)/J$. This simple condition allows higher values of saturation voltage to be chosen compared with those of the basic memory cell with the prospect of higher dynamic range.

Consider now the noise introduced by amplifier OA. It may be modelled as an input-referred noise voltage which causes slight modulation of the memory transistor's drain voltage. As long as the impedance at the drain node is high, this causes only an insignificant contribution to the cell's noise current. So, neglecting the op-amp's noise, the dynamic range is determined in the same way as for the basic memory cell by using (4.160) which is repeated here for convenience:

$$DR_M = 10 \, log \left[\frac{m_i^2 (V_{gs} - V_T)^2}{\frac{16 m_{th}}{3} (1 + k_J) \left(\frac{kT_j}{C} \right)} \right] \qquad (6.5)$$

where $(V_{gs} - V_T)$ and $(V_{gs} - V_T)_J$ are the saturation voltages of the memory and bias current transistors, respectively, $K_J = (V_{gs} - V_T) / (V_{gs} - V_T)_J$, m_i is the signal modulation index, m_{th} is a process constant for transistor thermal noise, T_j is the junction temperature and C is the memory transistor's gate-source capacitance. For any given values of $(V_{gs} - V_T)$ and m_i, dynamic range is maximised by setting $(V_{gs} - V_T)_J$ to the maximum value allowed by (6.4). Figure 6.3 shows the computed variation of dynamic range, DR_M, with saturation voltage, $(V_{gs} - V_T)$, and modulation index, m_i, (with $(V_{gs} - V_T)_J$ set to its maximum value) for the condition $V_{DD} = 5V$, $V_T = 1V$, $C = 10pF$, $m_{th} = 2.25$, $k = 1.38E-23$, and $T_j = 300K$. It can be seen that for each value of m_i there is an optimum value of $(V_{gs} - V_T)$ giving maximum dynamic range. It can be shown analytically that this optimum occurs when

$$(V_{gs} - V_T) = \frac{1}{\sqrt{1 + m_i}} \left[1 - \frac{3 - \sqrt{1 + 8\sqrt{1 + m_i}}}{4(\sqrt{1 + m_i} - 1)} \right] V_{DD} \approx \frac{0.69 V_{DD}}{\sqrt{1 + m_i}} \quad (6.6)$$

The complete condition for optimum dynamic range may be approximated as

$$V_{bias} \approx 0.69 V_{DD}$$

$$(V_{gs} - V_T)_J \approx 0.31 V_{DD}$$

$$(V_{gs} - V_T) \approx \frac{0.69}{\sqrt{1 + m_i}} V_{DD}$$

$$k_J \approx \frac{2.23}{\sqrt{1 + m_i}} \tag{6.7}$$

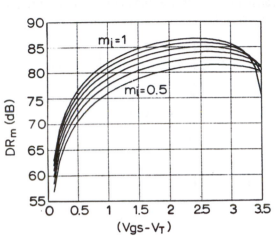

Figure 6.3 Calculated dynamic range of the op-amp active memory cell
$V_{DD} = 5V$, $V_T = 1V$, $C = 10pF$, $m_{th} = 2.25$, $T_j = 300K$.

It can be seen that the op-amp active memory cell allows high saturation voltage with high modulation index and this produces higher dynamic range than the basic memory cell. In practice, the choice of modulation index is also influenced by other factors, particularly by its effect on settling behaviour and charge injection resulting from signal dependent variations in the cell's transconductance, g_m. A practical value is $m_i = 0.8$ which usually gives a reasonable compromise between linearity and dynamic range, and this gives a peak value of dynamic range (for $C = 10pF$) of 85dB at $(V_{gs} - V_T) = 2.57V$. This is 4.5dB higher than that of the basic memory cell.

The major problem for this feedback arrangement is to achieve monotonic settling. Even the basic memory cell behaved as a second-order system because of the switch resistance and the drain capacitance, so the addition of *OA* makes the memory loop potentially a third-order system. Stable systems can result if the parasitic poles of the switch and the op-amp are at a much higher frequency than that of the memory (g_m/C) or if compensation is used. Either way, monotonic settling is hard to achieve, especially when the cell has high bandwidth. Thus, the op-amp active memory cell is most useful in low-frequency systems especially where analogue performance (accuracy, linearity, dynamic range) is demanding.

6.2.2   Grounded-gate active memory cell

An alternative arrangement [3] for creating a memory cell with a 'virtual earth' is shown in Figure 6.4 and has the advantage over the op-amp active memory cell that monotonic settling behaviour is easier to achieve. It uses a grounded-gate voltage amplifier GGA comprising transistors T_P, T_G and cascoded current source T_C, T_N. The bias currents I from transistors T_P and T_N flow in transistor T_G and the bias current J from transistor T_J flows in the memory transistor T.

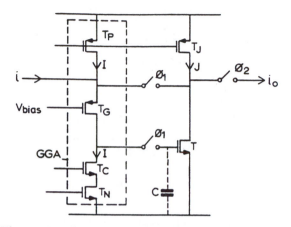

Figure 6.4 Grounded-gate active memory cell.

Operation is as follows. On phase ϕ_1, the signal current i flows into the source of transistor T_G and on into the storage capacitor C. As the voltage on C rises, so does the drain current in transistor T. Equilibrium is established when the current in T reaches $J+i$ and the current in T_G returns to I. If the gain of amplifier GGA is high then the voltage at the summing node is close to a constant value $V_{in} = V_{bias} + V_{gsG}$ where V_{gsG} is the gate-source voltage of T_G at $I_{ds} = I$ and so the arrangement has created a 'virtual earth' at the input node just like the op-amp active memory cell. On phase ϕ_2, the loop is broken and transistor T sustains its current at $J+i$ so long as the output node is held close to V_{in}. This happens quite naturally in cascade connected cells as the output voltage of the driving cell is fixed at the input voltage of the driven cell as before.

The feedback introduced by GGA on phase ϕ_1 increases the cell's low frequency input conductance to g_i where

$$g_i \approx g_m A_{GG} \qquad (6.8)$$

where A_{GG} is the voltage gain of the grounded-gate amplifier. The output conductance is the same as that of the basic cell and so the transmission error due to non- zero output to input conductance ratio, ε_G, is also reduced by the factor A_{GG}. Gains of $A_{GG} \approx 100$ are possible, making values of $\varepsilon_G = 0.01\%$ quite feasible.

Saturated operation the memory transistor and the current source transistors is guaranteed as long as

$$V_{DD} - (V_{gs} - V_T)_J \geq V_{in} \geq (V_{gs} - V_T)_G + V_T + (V_{gs} - V_T)\sqrt{1 + m_i} \quad (6.9)$$

where $(V_{gs} - V_T)$ and $(V_{gs} - V_T)_G$ are the saturation voltages of the memory and grounded-gate transistor at $I_{ds} = J$ and $I_{ds} = I$, respectively, and $(V_{gs} - V_T)_J = (V_{gs} - V_T)_P$ are the saturation voltages of the current source transistors T_J and T_P, respectively. Comparing this condition with (6.4) it can be seen that $(V_{gs} - V_T)$ must be set lower than for the op-amp active memory cell and the bias currents in transistors T_N and T_P contribute extra noise current at the input node, and so the dynamic range will be correspondingly lower. As with the op-amp active memory cell, noise sources in T_G do not cause a significant contribution to the cell's noise current as long as the impedance at the memory transistor's drain is high compared to that at the source. So, dynamic range can be computed from (6.5) again, using the value for k_J given by

$$k_J = \frac{g_{m(N)} + g_{m(P)} + g_{m(J)}}{g_m} \quad (6.10)$$

where $g_{m(N)}$, $g_{m(P)}$ are the transconductances of T_N and T_P at $I_{ds} = I$, and $g_{m(J)}$ and g_m are the transconductances of T_J and T at $I_{ds} = J$. Putting $I = J/2$, and $(V_{gs} - V_T)_N = V_T$ (to ensure saturated operation of T_N) it can be shown that

$$k_J = \frac{3(V_{gs} - V_T)}{2(V_{gs} - V_T)_J} + \frac{(V_{gs} - V_T)}{2V_T} \quad (6.11)$$

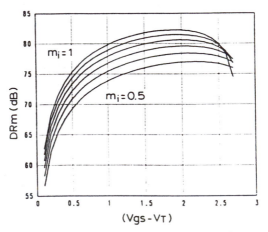

Figure 6.5 Calculated dynamic range of the grounded-gate active memory cell.

Figure 6.5 shows the computed variation of dynamic range, DR_M, with memory transistor saturation voltage, $(V_{gs} - V_T)$, and modulation index,

m_i, (with $(V_{gs} - V_T)_J$ set to its maximum value allowed by (6.9)) for the condition $V_{DD} = 5V$, $(V_{gs} - V_T)_G = 0.1V$, $V_T = 1V$, $C = 10pF$, $m_{th} = 2.25$, $k = 1.38E-23$, and $T_j = 300K$. $V_{DD} = 5V$, $(V_{gs} - V_T)_G = 0.1V$, $V_T = 1V$, $C = 10pF$, $m_{th} = 2.25$, $T_j = 300K$. For the practical value of $m_i = 0.8$, the dynamic range peaks at 80.6dB with $(V_{gs} - V_T) = 2V$ which is close to that of the basic memory cell and 4.4dB lower than that of the op-amp active memory cell. This may be a penalty worth paying in return for the better settling behaviour and reduced complexity. This cell is further developed in Chapter 10.

6.2.3 Simple cascode memory cell

Cascoding is a well-known technique [4] for lowering the output conductance of a MOS transistor. Figure 6.6 shows the cascoded memory cell. The cascode transistors T_C and T_{JC} are biased with fixed voltages such that all transistors are kept in saturation.

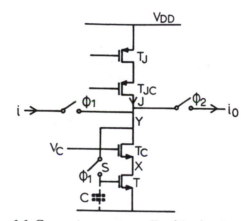

Figure 6.6 Current memory cell with simple cascoding.

Figure 6.7 shows the cells on clock phase ϕ_2. It is assumed for the moment that the current bias arrangement J is ideal. The low-frequency output conductance of the cascoded cell can be shown to be g_{0C} where

$$g_{0C} = g_0 \left[\frac{g_{dsc}}{g_{ds} + g_{dsc} + g_{mc}} \right] \qquad (6.12)$$

where g_0 and g_{dsc} are the output conductance of the open-gate memory transistor and the drain conductance of the cascode transistor, respectively, and g_{mc} is the transconductance of the cascode transistor. Now since g_0, $g_{dsc} \ll g_{mc}$,

$$g_{0C} \approx g_0 \left[\frac{g_{dsc}}{g_{mc}} \right] \qquad (6.13)$$

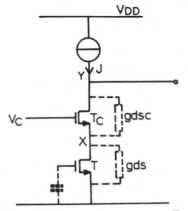

Figure 6.7 Cascoded current memory cell on phase ϕ_2.

Thus, cascoding has decreased the output conductance by a factor approximately equal to g_{mc}/g_{dsc} which is the grounded source voltage gain of the cascode transistor. This is not surprising as the voltage gain of the cascode transistor is the amount by which voltage variations at node Y are reduced at node X. Note that cascoding of the current source transistor would reduce its output conductance by a similar factor. The factor is typically about 100 and gives the cascoded memory cell a transmission error, resulting from non-zero output to input conductance ratios, which is typically 100 times lower than that of the basic memory cell (i.e. below 0.1%).

Next, consider the memory cell on clock phase ϕ_1 as shown in Figure 6.8. The input conductance at node Y is given by

$$g_i(Y) = \frac{g_m\left[1 + \dfrac{g_{mc}}{g_{dsc}}\right] + g_{ds}}{1 + \dfrac{g_{mc}}{g_{dsc}} + \dfrac{g_{ds}}{g_{dsc}}}$$

$$\approx g_m \qquad\qquad (6.14)$$

The input conductance at the intermediate node X is given by

$$g_{i(X)} = g_m\left(1 + \frac{g_{mc}}{g_{dsc}}\right) + g_{ds}$$

$$\approx g_m\left(\frac{g_{mc}}{g_{dsc}}\right) \qquad\qquad (6.15)$$

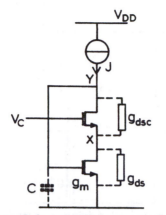

Figure 6.8 Cascoded current memory cells on phase ϕ_1.

So, the diode-connected input conductance at node Y is about the same as that of the basic cell. However, at the intermediate node X the input conductance is increased by the factor g_{mc}/g_{dsc} ; the same factor by which the output conductance was decreased. Unfortunately, node X cannot be used as the cell input if node Y is the output due to the differing bias voltages at these nodes.

Now, consider the problem of cascading current memory cells. Both the memory transistor and the cascode transistor must remain saturated. Two connected cells are shown in Figure 6.9 with current memory T_1 driving the diode connected current memory T_2. The signal current, i, increases the current in T_1 while decreasing the current in T_2, so that T_{1C} has lowest drain voltage when it passes its highest current. When $i = i(pk)$ = m_iJ, the voltage on the drain of T_{1C} is $V_T + (V_{gs} - V_T) \sqrt{1 - m_i}$ and V_C must be chosen low enough to keep T_{1C} saturated. Lowering V_C reduces the drain voltage of T_1 with the risk that it will leave saturation. The condition that both T_1 and T_{1C} remain saturated is (neglecting back-gating effects)

$$2V_T + (V_{gs}\text{-}V_T)\sqrt{1\text{-}m_i} \geq V_c \geq V_T + \sqrt{1+m_i}\,[(V_{gs}\text{-}V_T) + (V_{gs}\text{-}V_T)_c] \quad (6.16)$$

where $(V_{gs} - V_T)$ and $(V_{gs} - V_T)_C$ are approximately the saturation voltages of T_1 and T_{1C} at a drain current of J. Clearly

$$(V_{gs} - V_T) \leq \frac{V_T - (V_{gs} - V_T)_C\sqrt{1 + m_i}}{\sqrt{1 + m_i} - \sqrt{1 - m_i}} \quad (6.17)$$

This condition imposes a trade-off between $(V_{gs} - V_T)$ and m_i similar to that experienced by the basic memory cell. Further, the current source transistors stay saturated provided

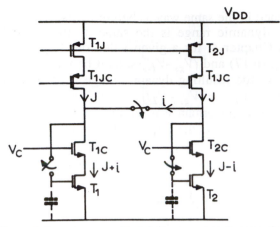

Figure 6.9 Cascoded memory cell driving a second cascoded memory cell.

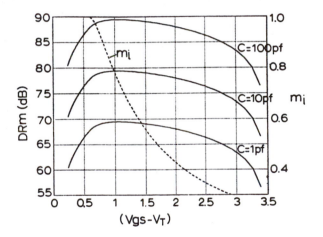

Figure 6.10 Calculated dynamic range of cascoded memory cell V_{DD} = 5V, V_T = 1V, $(V_{gs} - V_T)_C = (V_{gs} - V_T)_{JC}$ = 0.1V, m_{th} = 2.25, T_j = 300K.

$$(V_{gs} - V_T)J \leq V_{DD} - V_T - (V_{gs} - V_T) \sqrt{1 + m_i} - (V_{gs} - V_T)_{JC} \quad (6.18)$$

where $(V_{gs} - V_T)_J$ and $(V_{gs} - V_T)_{JC}$ are the saturation voltages of the current source transistor and its cascode transistor, respectively, at the bias current, J. Again, this condition imposes a trade-off between $(V_{gs} - V_T)_J$ and $(V_{gs} - V_T)$ similar to that experienced in the basic memory cell.

Consider now the noise introduced by the cascode transistor, T_C. It may be modelled as an input-referred noise voltage which causes slight modulation of the memory transistor's drain voltage. So long as the impedance at the drain node is high this causes only an insignificant contribution to the cell's noise current. So, neglecting this, the dynamic range is determined in the same way as before by using (6.5). The strategy for maximising dynamic range is the same as that used for the basic memory cell in Chapter 4, for a given value of $(V_{gs} - V_T)$, m_i should be maximised using (6.17) and $(V_{gs} - V_T)_J$ should be maximised using (6.18). Figure 6.10 shows the resulting dynamic range for the conditions $V_{DD} = 5V$, $V_T = 1V$, $(V_{gs} - V_T)_C = (V_{gs} - V_T)_{JC} = 0.1V$, $m_{th} = 2.25$, $k = 1.38E\text{-}23$ and $T_j = 300K$. The performance is similar to that of the basic memory cell. The dashed line shows the trade-off between $(V_{gs} - V_T)$ and m_i. For $(V_{gs} - V_T) \leq V_T /\sqrt{2} - (V_{gs} - V_T)_C = 0.607V$, m_i is set to unity without violating (6.17) but above this value m_i falls. Dynamic range rises to a maximum at $(V_{gs} - V_T) = 1.07V$, and $m_i = 0.74$, with $(V_{gs} - V_T)_J = 2.48V$. For $C = 10pF$, the maximum dynamic range is 79.5dB which is 1dB lower than that of the basic memory cell.

Monotonic settling is easily achieved with the simple cascode arrangement. It is usually sufficient to ensure that the pole frequency associated with the source of the cascode transistor is at least an order of magnitude higher than that of the memory transistor g_m/C. This can be achieved by giving the cascode transistor a high aspect ratio (W/L) and this is consistent with the requirement of a low cascode saturation voltage.

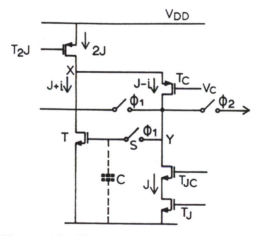

Figure 6.11 Folded-cascode memory cell.

6.2.4   *Folded-cascode memory cell*

The folded-cascode memory cell is shown in Figure 6.11. The cascode transistor, T_C, is of opposite polarity to T (p-channel) and both T and T_C have a bias current, J. The arrangement operates similarly to the simple cascode arrangement already described except that an input current, i, produces a current $J-i$, rather than $J+i$, in the cascode transistor. The cell retains the important property that the drain voltage of transistor T is substantially unchanged from clock phase ϕ_1 to clock phase ϕ_2. The folded-cascode has identical small signal behaviour to the simple cascode circuit and the conductances are the same as calculated before ((6.12) to (6.15)); the conductance at node X with the switch S closed is increased by a factor equal to the grounded source voltage gain of the cascode transistor while the conductance at node Y with switch S open is decreased by the same factor. Whereas, with the simple cascode, the value of $(V_{gs} - V_T)$ was limited by the voltages in the loop formed by closing switch S (6.17), there is no such constraint with the folded cascode. Instead, the limit is set by the voltages occurring in the two connected cells shown in Figure 6.12. Transistors T and T_C are saturated up to the signal $i = i(pk) = m_i J$ as long as

$$(V_{gs} - V_T)_{2J} \leq V_{DD} - V_T - \sqrt{1 + m_i} \, ((V_{gs} - V_T) + (V_{gs} - V_T)_C) \quad (6.19)$$

where $(V_{gs} - V_T)_{2J}$ is the saturation voltage of the *2J* current source. Note that the current source *2J* is already cascoded by T_C and so need only be a single transistor. However, current source *J* must employ cascoding. It stays saturated provided

$$(V_{gs} - V_T)_J \leq V_T + (V_{gs} - V_T)\sqrt{1 - m_i} - (V_{gs} - V_T)_{JC} \quad (6.20)$$

where $(V_{gs} - V_T)_J$ and $(V_{gs} - V_T)_{JC}$ are the saturation voltages of the current source *J* and its cascode transistor, respectively. As with the simple cascode

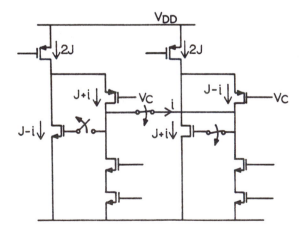

Figure 6.12 Folded-cascode cell driving a second folded-cascode cell.

arrangement, the noise voltage of the cascode transistor will not contribute significantly to the noise current of the cell so long as the impedance at the drain of the memory transistor is high. So, neglecting this noise source, the dynamic range is given again by (6.5) with saturation voltages limited by (6.19) and (6.20). There are two bias current sources, *2J* and *J*, which contribute to the cell's total noise, and the value of K_J for the folded cascode is given by

$$K_J = \frac{g_{m(J)} + g_{m(2J)}}{g_m} = \frac{(V_{gs} - V_T)}{(V_{gs} - V_T)_J} + \frac{2(V_{gs} - V_T)}{(V_{gs} - V_T)_{2J}} \qquad (6.21)$$

The strategy for optimising dynamic range is to make $(V_{gs} - V_T)_J$ and $(V_{gs} - V_T)_{2J}$ as large as possible consistent with (6.19) and (6.20) for any given values of $(V_{gs} - V_T)$ and m_i. Figure 6.13 shows the calculated dynamic range resulting from this strategy for the condition $V_{DD} = 5V$, $V_T = 1V$, $C = 10pF$, $(V_{gs} - V_T)_C = (V_{gs} - V_T)_{JC} = (V_{gs} - V_T)_{2JC} = 0.1V$, $m_{th} = 2.25$, $k = 1.38E-23$, and $T_j = 300K$. It can be seen that, as with the op-amp active memory cell, dynamic range can reach its natural optimum without being constrained by saturation limitations. As with the other arrangements, modulation index will be constrained by settling behaviour and charge injection errors to about $m_i = 0.8$ and this gives an optimum dynamic range at $(V_{gs} - V_T) = 2.0V$, $(V_{gs} - V_T)_J = 1.79V$, $(V_{gs} - V_T)_{2J} = 1.18V$, and reaches a value of 79.7dB which is 0.8dB lower than the basic memory cell. The advantage gained from higher saturation voltages has been nullified by the extra noise in the bias current sources.

As with the simple cascode, monotonic settling is easily achieved with the folded cascode arrangement. It is usually sufficient to ensure that the pole frequency associated with the source of the cascode transistor is at least an order of magnitude higher than that of the memory transistor g_m/C. This can be achieved by giving the cascode transistor a high aspect ratio (W/L) and this is consistent with the requirement of a low cascode saturation voltage.

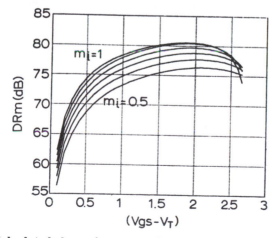

Figure 6.13 Calculated dynamic range of the folded cascode memory cell $V_{DD} = 5V$, $V_T = 1V$, $C = 10pF$, $(V_{gs} - V_T)_C = (V_{gs} - V_T)_{JC} = (V_{gs} - V_T)_{2JC} = 0.1V$, $m_{th} = 2.25$, $T_j = 300K$.

6.2.5 Regulated cascode

An even lower transmission error can be achieved with the regulated cascode memory cell [5] shown in Figure 6.14. It is based on the so-called regulated cascode current mirror [6]. It is similar to the simple cascode cell (Figure 6.6) except that the cascode reference voltage, V_C, is generated by the regulation stage consisting of transistor, T_R, and current source, I. On phase ϕ_1 with the switch S closed, the memory cell is diode-connected and the input conductance at Y is approximately equal to g_m. The V_{gs} of transistor T_R is defined by its threshold voltage, geometry and drain current, I. If the input current is varied, any tendency for the voltage at node X to vary is suppressed by the regulating action of the negative feedback loop formed by transistors T_C and T_R. On phase ϕ_2, with switch S open, the cell delivers output current i_0. Any tendency for the voltage at node X to vary (as a result of voltage variations at node Y) is again suppressed by the regulation loop. It can be shown that the output conductance at node Y is approximately given by

$$g_{0(Y)} \approx g_0 \left(\frac{g_{dsc}}{g_{mc}}\right)\left(\frac{g_{dsR}}{g_{mR}}\right) \tag{6.22}$$

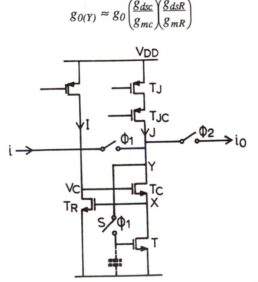

Figure 6.14 Regulated cascode memory cell.

Comparison with (6.13) shows that the output conductance is reduced from that achieved by the simple cascode by a factor equal to the voltage gain of the regulation stage (g_{dsR}/g_{mR}). Since this gain can be typically 100, this gives the regulated cascode memory cell a transmission error about 10,000 times lower than that of the basic memory cell. Of course, cascode transistor T_{JC} should also be regulated (not shown). It should be noted that the negative feedback of the regulated cascode cell also acts to increase the input conductance seen at node X, so

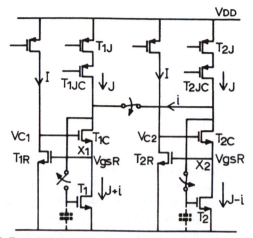

Figure 6.15 Regulated cascode memory cell driving a second cell.

$$g_{i(X)} \approx g_m \left(\frac{g_{mc}}{g_{dsc}} \right) \left(\frac{g_{mR}}{g_{dsR}} \right) \tag{6.23}$$

Now consider the problem of one regulated cascode memory cell driving current into a second similar cell, as shown in Figure 6.15. To achieve the excellent transmission accuracy it is necessary to maintain all devices in saturation and this is ensured provided the following conditions are met.

$$(V_{gs}-V_T)\sqrt{1+m_i} \le V_{gsR} \le V_T +(V_{gs}-V_T)\sqrt{1-m_i} - (V_{gs}-V_T)_C\sqrt{1+m_i} \tag{6.24}$$

which gives

$$(V_{gs} - V_T) \le \frac{V_T - (V_{gs} - V_T)_C\sqrt{1 + m_i}}{\sqrt{1 + m_i} - \sqrt{1 - m_i}} \tag{6.25}$$

Further, current source transistor T_J stays saturated provided

$$(V_{gs} - V_T)_J \le V_{DD} - V_T - (V_{gs} - V_T)\sqrt{1 + m_i} - (V_{gs} - V_T)_{JC} \tag{6.26}$$

It is noted that (6.25) and (6.26) are identical to the corresponding equations for the simple cascode memory cell ((6.17) and (6.18)) and so the circuit constraints which limit dynamic range are the same. This is because, under maximum signal conditions, the regulation loop adjusts V_C to the same value as that used as a fixed reference voltage in the simple cascode arrangement. The noise introduced by the regulation loop is negligible for reasons already discussed in the other arrangements and so the calculated dynamic range is the same as for the simple cascode cell (Figure 6.10).

As with the op-amp active memory cell, the main problem for the regulated cascode arrangement is to achieve adequate settling behaviour. The addition of the regulation amplifier makes the memory loop potentially a third-order system and careful design is needed to produce monotonic settling.

6.2.6 Regulated folded-cascode memory cell

The regulated folded-cascode memory cell is shown in Figure 6.16. It consists of the folded-cascode memory cell (Figure 6.11) with a regulation stage (T_R and current I) to generate the cascode reference voltage V_C. The regulation feedback loop operates in just the same way as that of the regulated cascode cell except that node X is regulated at ($V_{DD} - V_{gsR}$) instead of V_{gsR}. Maximum dynamic range results when, under extreme signals, the regulation loop adjusts V_C to the value used in the optimised (unregulated) folded cascode cell. For this reason, as with the regulated cascode cell, the dynamic range is identical to that of its unregulated counterpart as given in Figure 6.13.

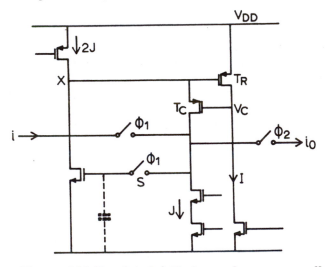

Figure 6.16 Regulated folded-cascode memory cell.

6.2.7 Integrators

Figure 6.17 shows configurations for ideal non-inverting integrators derived from the various current memory cells discussed already. Each is constructed from a pair of oppositely phased memory cells forming an integrator loop (T_1 and T_2) and an output stage (T_3) to mirror the current from T_2. Note that care is taken to use a cascoding style in the output stage which is the same as that used in the integrator loop, to achieve good mirroring of the output signal. Of course, differentiators may also be similarly derived from the various current memory cells.

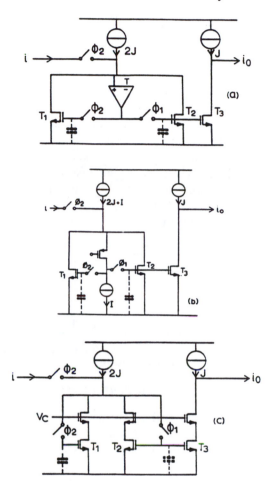

Figure 6.17 Integrator configurations (a) op-amp active (b) grounded-gate active (c) simple cascode.

6.3 Fully Differential Switched-Current Circuits

Fully differential architectures are frequently employed in both continuous-time and switched-capacitor circuits to achieve the highest level of analogue performance. This approach reduces distortion by cancelling even-order harmonics and reduces crosstalk from neighbouring digital circuits through the common-mode and power supply rejection of the amplifiers. In this section, fully differential switched-current circuits are presented with particular emphasis on the additional benefit of reduction in charge injection errors.

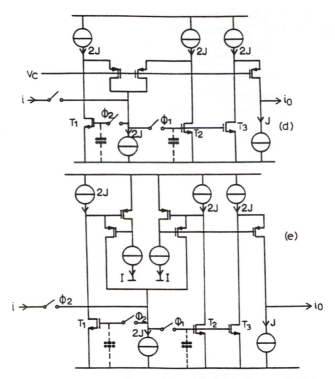

Figure 6.17 Integrator configurations (d) folded-cascode (e) regulated folded-cascode.

6.3.1   *Basic fully-differential current memory cell*

A simplified fully-differential cell is shown in Figure 6.18. The input signal is represented by the balanced pair of input currents, $\pm i$, and the output signal by another balanced pair, $\pm i_0$. On phase ϕ_1, the memory transistors T_a and T_b are diode-connected and their currents are $J-i$ and $J+i$, respectively. On phase ϕ_2, these currents are sustained by the charge held on their gate-source capacitors, and produce the output currents. Ideally, $i_0 = \pm i$ where the inversion may be achieved simply by crossing over the output connections. Filter structures frequently require signal inversion (e.g. state variable filters) and the differential cells' ability to deliver inverted or non-inverted signals can greatly simplify filter architectures. Of course, the simple circuit of Figure 6.18 has the drawback that it requires perfect matching between the upper and lower (J and $2J$) current sources. Figure 6.19 shows an arrangement which achieves this automatically. The tail current, $2J$, is established in transistor T_2 by the external fixed bias V_b. Suppose that, on phase ϕ_1, the voltage at node Z establishes currents in transistors T_{1a} and T_{1b} such that their sum is greater than $2J$. The current entering node X from transistors T_a and T_b will also be greater than $2J$ with the result that the voltage at X is forced to rise. The source follower T_3, biased with current I, translates this voltage to node Z with the result that the currents in T_{1a} and T_{1b} are reduced until their sum is equal to $2J$. Similarly, if the sum of the currents in transistors T_{1a} and T_{1b} is momentarily low, the voltage at X falls, with the result that the currents are increased until they sum to $2J$. Clearly, the loop has negative feedback, with gain determined by the voltage gains of transistors T_{1a} and T_{1b}, which acts to balance the upper and lower current sources. The source follower, T_3, fixes the voltage at node X at a value determined by the sum of the gate-source voltages of T_1 and T_3. It also functions as a common-mode feedback circuit.

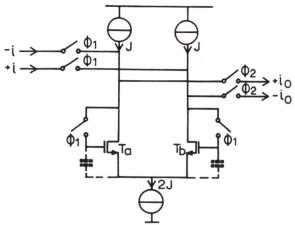

Figure 6.18 Primitive fully-differential memory cell.

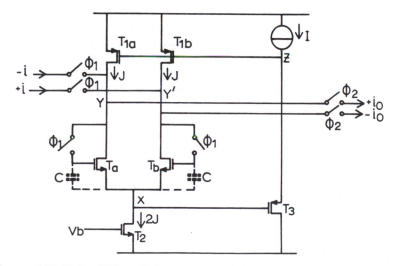

Figure 6.19 Fully-differential current memory cells with automatically balanced current bias.

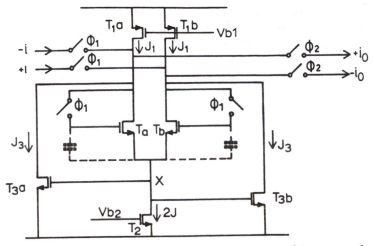

Figure 6.20 Alternative fully-differential memory cell with automatic bias balancing.

If the input currents are $i_{cm} \pm i$, where i_{cm} is a low frequency common-mode signal, then the voltage at node X is disturbed in just the same way as if the bias currents were unbalanced. The negative feedback loop adjusts the currents in T_{1a} and T_{1b} to the value $J - i_{cm}$ leaving the currents in T_a and T_b unaffected at the values $J - i$ and $J + i$. On clock phase ϕ_2, with the memory switches open and the currents fixed in T_a and T_b, the negative feedback loop is broken and ideally the bias currents are left at the values established on phase ϕ_1. It should be noted that the currents in T_{1a} and T_{1b} are held close to $J - i_{cm}$ and the output currents are $-i_{cm} \pm i$. Clearly, the cell transmits rather than rejects common-mode signals.

An alternative common-mode feedback arrangement is shown in Figure 6.20. The upper current source transistors T_{1a} and T_{1b} have a fixed gate-source voltage V_{b1} and produce drain currents J_1, where $J_1 < J$. On phase ϕ_1, with balanced input currents $\pm i$, the memory transistor drain currents are $J_1 - J_3 - i$ and $J_1 - J_3 + i$, where J_3 is the drain current of transistors T_{3a} and T_{3b}. The current entering node X from T_a and T_b is $2(J_1 - J_3)$ and if this is greater than $2J$ (the current leaving node X) then the voltage at node X rises, and increases the currents J_3 until $2(J_1 - J_3) = 2J$. Clearly, the negative feedback loop is adjusting the current J_3 until it equals $J_1 - J$. If the input has a common-mode value i_{cm}, the loop adjusts J_3 to the value $(J_1 - J + i_{cm})$ leaving transistors T_a and T_b conducting $J-i$ and $J+i$, respectively. On phase ϕ_2, the negative feedback loop is broken and the output current is $i_0 = J_1 - (J + i) - J_3$. Since J_3 is ideally left at the value $J_1 - J + i_{cm}$ the output current is $i_0 = -(i + i_{cm})$ and again no common-mode rejection has occurred.

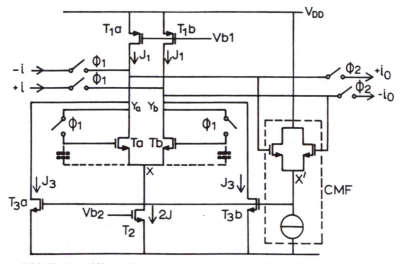

Figure 6.21 Fully differential current memory cell with common-mode feedback.

A solution which achieves common-mode rejection is shown in Figure 6.21. It employs a separate common-mode feedback stage (CMF) to control J_3. On phase ϕ_1, the voltage on node X' matches that on node X and the control loop sets up $J_3 = J_1 - J + i_{cm}$ as above. On phase ϕ_2, T_a and T_b hold currents $J-i$ and $J+i$, respectively, and the currents in both T_{1a} and T_{1b} are J_1. The CMF block detects a common-mode imbalance at the summing nodes Y_a and Y_b and adjusts the voltage at node X' until $J_3 = J_1 - J$ and the output current is $i_0 = -i$. Clearly, the common-mode component of the input signal i_{cm} has been rejected, a consequence of allowing the common-mode feedback circuit to act on both phases of the clock.

6.3.2 *Folded-cascode fully-differential current memory and integrator*

Any of the cascode arrangements discussed earlier for single-ended current memory cells may be used with the fully-differential current memory cell. As an example, the folded-cascode fully-differential memory cell is given in Figure 6.22. It can be easily seen that it combines the single-ended folded-cascode cell shown in Figure 6.11 with the fully-differential cell shown in Figure 6.21, and needs no further explanation. A fully-differential folded-cascode non-inverting lossless integrator is shown in Figure 6.23. It uses two of the fully-differential current memory cells (Figure 6.22) to form the integrator loop and a mirrored output stage in a similar manner to the basic integrator cell. Note that the common-mode feedback circuit is shared by each of the current memories in turn.

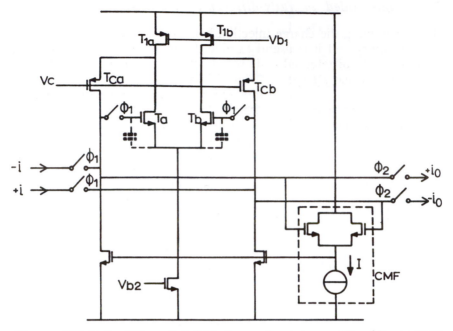

Figure 6.22 Fully-differential folded-cascode memory cell with common-mode feedback.

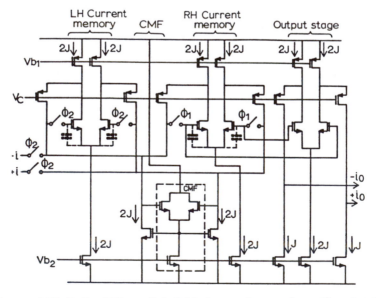

Figure 6.23 Fully-differential folded-cascode non-inverting lossless integrator.

6.3.3 Single-ended to fully-differential equivalence

Before examining the charge injection behaviour of the fully-differential current memory cell it is useful to give the relationship between single-ended and fully-differential cells which gives them equivalent analogue performance in many respects. Figure 6.24 shows this relationship. The fully-differential cell has transistors with half the W/L aspect ratio and half the bias and signal currents of the single-ended cell. Table 6.1 shows a performance comparison of the single-ended and fully-differential memory cells as defined in Figure 6.24. Most of the entries in the table are the result of simple scaling and need no explanation. However, the noise equivalence is less obvious and is explained here. The noise model for the single-ended memory transistor is as shown in Figure 6.25(a). Assuming white noise is dominant, the noise voltage variance, $\overline{v_N^2}$, is given by

$$\overline{v_N^2} = \frac{2}{3} m_{th} \frac{kT_j}{C} \tag{6.27}$$

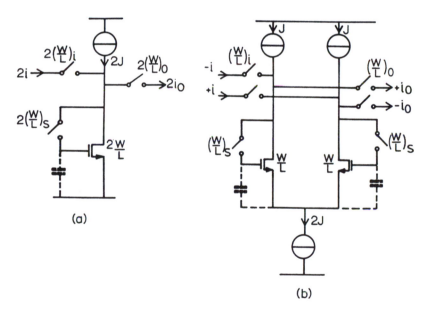

(a)

(b)

**Figure 6.24 Equivalence of current memory cells
(a) basic single-ended (b) basic fully-differential.**

The output current noise variance, $\overline{i_N^2}$, is

$$\overline{i_N^2} = \overline{v_N^2} \, g_m^2 \tag{6.28}$$

where $g_m = \sqrt{2\beta J}$. The maximum RMS signal current is $\sqrt{2}m_i J$ and so the dynamic range is given by

$$DR = 10 \, log \left[\frac{2m_i^2 J^2}{\frac{2}{3} m_{th} \left(\frac{kT_j}{C}\right) g_m^2} \right] \qquad (6.29)$$

The noise model for the fully-differential memory stage is shown in Figure 6.25(b). Again, assuming white noise is dominant, the noise voltage variance of each transistor, $\overline{v_N^2}\,'$, is given by

$$\overline{v_N^2}\,' = \frac{2}{3} m_{th} \left(\frac{kT_j}{C/2}\right) = 2 \, \overline{v_N^2} \qquad (6.30)$$

and the differential noise voltage variance is $\overline{v_{N(D)}^2}$ where,

$$\overline{v_{N(D)}^2} = 2 \, \overline{v_N^2}\,' = 4 \, \overline{v_N^2} \qquad (6.31)$$

The transistor transconductance, g_m', is $\sqrt{2\beta J}$ and the differential transconductance, $g_{m(D)}$, is

$$g_{m(D)} = \frac{g_m'}{2} = \frac{1}{2} \sqrt{2J\beta} = \frac{g_m}{4} \qquad (6.32)$$

Parameter	Single-ended	Fully-differential
Supply current	$I_{DD} = 2J$	$I_{DD(D)} = I_{DD}$
Current swing	$i_{p\text{-}p} = 4J$	$i_{p\text{-}p(D)} = i_{p\text{-}p}/2$
Transconductance	$g_m = 2\sqrt{2J\beta}$	$g_{m(D)} = g_m/4$
Gate capacitance	C	$C_{(D)} = C/4$
Switch resistance	r_S	$r_{S(D)} = 4r_S$
Diffusion capacitance	C_d	$C_{d(D)} = C_d/4$
Bandwidth	$\omega_{C0} = g_m/C$	$\omega_{C0(D)} = \omega_{C0}$
Saturation voltage	$(V_{gs} - V_T) = 4J/g_m$	$(V_{gs} - V_T)_{(D)} = (V_{gs} - V_T)$
Dynamic range	$DR = \dfrac{2m_i^2 J^2}{\frac{2}{3} m_{th} \left(\frac{kT_j}{C}\right) g_m^2}$	$DR_{(D)} = DR$

Table 6.1 Equivalence of single-ended and fully-differential memory cells.

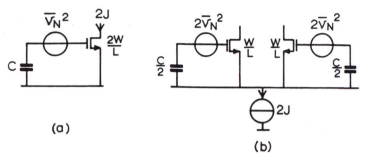

Figure 6.25 Noise models (a) single-ended (b) fully-differential.

The differential output current noise variance is

$$\overline{i_{N(D)}^2} = \overline{v_{N(D)}^2}\, g_{m(D)}^2 = \left(4\;\overline{v_N^2}\right)\left(\frac{g_m}{4}\right)^2 = \frac{\overline{v_N^2}\, g_{m(D)}^2}{4} \qquad (6.33)$$

$$\overline{i_{N(D)}^2} = \frac{\overline{i_N^2}}{4} \qquad (6.34)$$

The maximum differential RMS signal current is $m_i J/\sqrt{2}$ and the dynamic range $DR_{(D)}$ is

$$DR_{(D)} = 10\,log\left[\frac{2m_i^2 J^2}{\frac{2}{3}m_{th}\left(\frac{kT_j}{C}\right)g_m^2}\right] \qquad (6.35)$$

So, the differential current memory with dominant white noise has the same dynamic range as its single-ended counterpart. It can also be seen from Table 6.1 that the single-ended and fully-differential circuits, scaled as shown in Figure 6.24, have equal voltages (neglecting the saturation voltage overhead of the tail current source), equal bandwidth and settling times and nearly equal area. Charge injection behaviour is discussed in the next section.

6.3.4 Charge injection

It was seen in Chapter 4 that charge injection can be a major source of error in the basic memory cell. Even with no signal present, charge injection produces an offset error and the presence of signal modulates the cell's transconductance and the switch's charge non-linearly so that harmonic distortion is produced. Various attempts have been made to reduce charge injection in switched-current circuits. *Yang* [7] proposed extra dummy circuitry to null the offset error and *Daubert* [8] employed a feedback technique used earlier in switched-capacitor sample and hold circuits [9] but this required a multi-phase clock and floating capacitors. *Toumazou* [10] (see also Chapter 22) proposed an algorithmic procedure in which the errors were first stored, inverted and then subtracted from themselves, but this required extra clock phases and increased the complexity about three-fold. Reduction of charge injection errors through the use of fully differential circuits requires no extra clock phases and incurs minimal overhead in extra circuit area.

The dynamics of charge injection in fully-differential current memories is now discussed with the help of simulation results of the test circuit shown in Figure 6.26. It is derived from a single-ended test circuit using the scaling described in the previous section. The clocking sequence is the same as for the single-ended charge injection test described in Chapter 4. The simulation starts with switches NMC3 and NMC4 closed. The drain current in NMC8 is 100µA and in PMC1 and PMC2 the drain currents are 50µA. After 1ns the memory switches' gates (node 4) are driven from +5V to 0V in 5ns to cut off NMC3 and NMC4. After a further 5ns (i.e. at 11ns) the gates of NMC5 and NMC6 are driven from 0V to +5V in 5ns to turn them on and connect nodes 2 and 3 to the voltage generators E5 and E6. These generators were set to the voltages occurring at nodes 2 and 3 when switches NMC3 and NMC4 were closed, so that nodes 2 and 3 settle to the same value on both sample and hold phases thereby eliminating transmission errors arising from the non-zero output conductances of NMC1 and NMC2. The error currents flowing in E5 and E6 are the result of charge injection only.

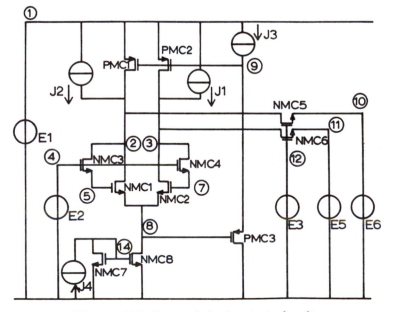

Figure 6.26 Charge injection test circuit.

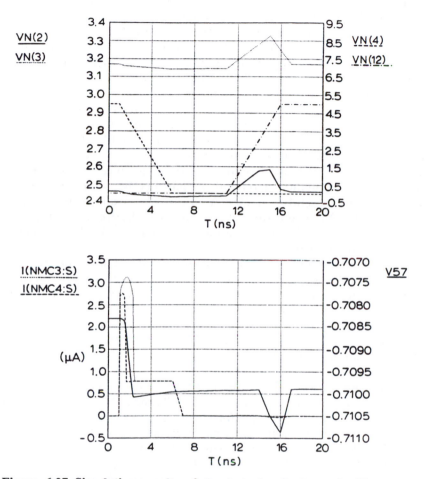

Figure 6.27 Simulation results of the test circuit shown in Figure 6.26.

The results of the simulation are shown in Figure 6.27. During the interval 0 - 1ns, the memory switches are closed and the common-mode feedback circuit sets the voltage at node 8 at 0.939V. The input currents at nodes 2 and 3, J2 and J1, are -40μA and +40μA, respectively, so the drain currents in NMC1 and NMC2 are 10μA and 90μA, resulting in voltages at nodes 2 and 3 (VN(2) and VN(3)) of 2.461V and 3.169V, respectively. During the interval 1ns - 6ns, the memory switch gate voltage (VN(4)) is driven from +5V to 0V. Pinch-off occurs first in NMC4 at time 1.5ns because it is connected to node 3 which is at a higher voltage than node 2. In the interval from 1ns to 1.5ns, channels are present in both switches and they pass similar currents (≈3μA) into the gate capacitances of the memory transistors NMC1 and NMC2. The voltage between nodes 5 and 7 (V57) stays substantially constant indicating equal charging rates of their gate capacitances and negligible charge injection error.

At time 2.3ns, the voltage at node 4 has fallen to 3.7V and the switch NMC3 starts to pinch-off. In the preceding interval, 1.5ns to 2.3ns, switch NMC3 has a channel but switch NMC4 does not. So, the gate of memory transistor NMC1 is charged by the current in the combined gate oxide and overlap capacitance of switch NMC3 ($\approx$3µA), while memory transistor NMC2 is charged only from the overlap capacitance of switch NMC4 ($\approx$0.75µA). Consequently, the voltage at node 5 falls faster than that of node 7 and V57 changes by 1.8mV. This corresponds closely with the product of the charge represented by the area enclosed by the I(NMC3;S) and I(NMC4;S) plots during the interval 1.5ns to 2.3ns ((3µA - 0.75µA)*0.8ns = 1.8fC) and the gate source capacitance of each memory transistor ($\approx$1pF). Note that the current in the source of NMC3 (I(NMC3;S)) has the characteristic rounded peak, observed earlier in the simulation of the single-ended cell, arising from charge distribution from the memory transistors drain capacitance to its gate capacitance via the switch's channel resistance.

During the next interval, from 2.3ns to 6ns, the charge injection error remains fairly constant since both switches are pinched-off and charge is injected equally through the switches' overlap capacitances into the memory transistors' gate capacitances.

The final difference error current flowing into the voltage generators is 120nA for a difference output current of 80µA (an error of 0.15%) which is in fair agreement with twice the product of the error voltage between nodes 5 and 7 (1.8mV) and the differential transconductance (40µS). During the switch turn-off process the common-mode feedback loop operated until the moment that the second switch NMC1 was pinched-off (at 2.3ns). The charge injection errors in NMC1 and NMC2 had a common-mode component of -15nA which caused the voltage at node 8 to fall by 11mV to reduce the current delivered by NMC8. As the common-mode feedback loop is ultimately broken, this error voltage is applied to PMC1 and PMC2 resulting in a common-mode error current flowing into the voltage generators E5 and E6 of 0.7µA. This is an undesirable feature of this common-mode feedback arrangement and it can be reduced by using lower loop gain (higher output conductance for NMC8 or lower transconductance for PMC1 and PMC2) or by using the CMF circuit of Figure 6.22.

The simulation was repeated to find the charge injection errors for input signal currents ranging from +80µA to -80µA. The results are plotted in Figure 6.28 which includes the results for the single-ended circuit. The fully-differential charge injection error has no offset and it varies fairly linearly with input signal (from 0.18% for small input signals to 0.15% for 80% modulation). This compares very favourably with the charge injection error of the single-ended cell which has an offset of 2.21% and varies very non-linearly with signal. The harmonic distortion of the fully-differential memory will be much lower than the single-ended memory. If the offset error of the single-ended cell is discounted, its error for small input signals is 0.7% of the signal amplitude, about 4 times that of the fully-differential memory. This rises to 1.3% for a signal current of -80μA, more than 8 times the error in the fully-differential memory.

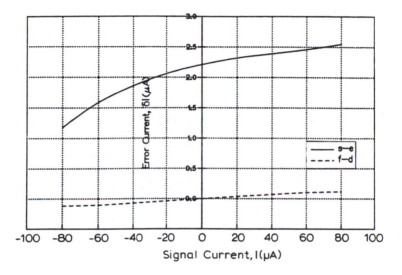

Figure 6.28 Simulated variation of charge injection errors with signal amplitude.

The charge injection errors are now formally analysed with reference to Figure 6.29. Starting with both switches closed, the cell receives an input current, i, so that the memory transistors T_1 and T_2 have drain currents $J+i$ and $J-i$, respectively. The switches are biased at the gate voltages, V_{g1} and V_{g2} ($V_{g1} > V_{g2}$), of the memory transistors.

When the gates of the switches are driven low, the switch biased by V_{g1} starts to pinch-off when the clock voltage reaches V_{CK1}, where

$$V_{CK1} \approx \left(1 + \frac{\gamma}{3}\right)V_{g1} + V_T \qquad (6.36)$$

where γ is the back-gate parameter. A little later, the second switch pinches-off at V_{CK2} where

$$V_{CK2} \approx \left(1 + \frac{\gamma}{3}\right)V_{g2} + V_T \qquad (6.37)$$

The change in clock voltage between these pinch-off points is given by

$$\delta V_{CK} = V_{CK1} - V_{CK2} = (1 + \frac{\gamma}{3})(V_{g1} - V_{g2}) = (1 + \frac{\gamma}{3})(V_{gs1} - V_{gs2})$$

$$= 2(1 + \frac{\gamma}{3})\left(\frac{I_1}{g_{m1}} - \frac{I_2}{g_{m2}}\right) \qquad (6.38)$$

The difference error voltage resulting from charge injection is δV_g where

$$\delta V_g = \delta V_{CK}\, \alpha\, \frac{C_{CH}}{C} = 2\alpha\frac{C_{CH}}{C}(1 + \frac{\gamma}{3})\left(\frac{I_1}{g_{m1}} - \frac{I_2}{g_{m2}}\right) \qquad (6.39)$$

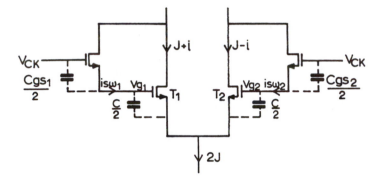

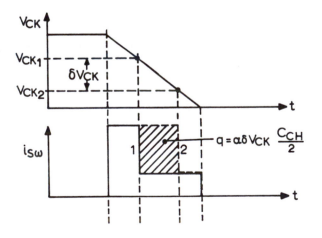

Figure 6.29 Charge injection in a fully differential memory cell.

where α is the factor determined by charge distribution in the switches' channels ($0.5 < \alpha < 1$). The charges injected through the switch overlap capacitances cause only a common-mode error voltage. The resulting difference error current is δi given by

$$\delta i = g_{m(D)} \delta V_g = 2\alpha \frac{C_{CH}}{C}(1 + \frac{\gamma}{3})\left(\frac{I_1}{g_{m1}} - \frac{I_2}{g_{m2}}\right)g_{m(D)} \qquad (6.40)$$

where $g_{m(D)}$ is the differential cell's transconductance given by

$$g_{m(D)} = \frac{g_{m1}g_{m2}}{g_{m1} + g_{m2}} \qquad (6.41)$$

Substituting (6.41) in (6.37) and rearranging gives

$$\delta i = 2\alpha \frac{C_{CH}}{C}(1 + \frac{\gamma}{3})\left(\frac{I_1 g_{m2} - I_2 g_{m1}}{g_{m1} + g_{m2}}\right) \qquad (6.42)$$

Substituting $I_1 = J + i$ and $I_2 = J - i$ in (6.42) gives

$$\delta i = 2\alpha \frac{C_{CH}}{C}(1 + \frac{\gamma}{3})\left[i - J\left(\frac{g_{m1} - g_{m2}}{g_{m1} + g_{m2}}\right)\right] \qquad (6.43)$$

The expression $(g_{m1} - g_{m2})/(g_{m1} + g_{m2})$ can be simplified by substituting

$$g_{m1} = \sqrt{2\beta I_1} = \sqrt{2\beta(J + i)}$$

$$g_{m2} = \sqrt{2\beta I_2} = \sqrt{2\beta(J - i)} \qquad (6.44)$$

This gives, for i ≪ J ,

$$\frac{g_{m1} - g_{m2}}{g_{m1} + g_{m2}} = \frac{i}{2J} \qquad (6.45)$$

Substituting (6.45) in (6.43) gives

$$\delta i = \alpha(1 + \frac{\gamma}{3})\frac{C_{CH}}{C}i \qquad (6.46)$$

and

$$\varepsilon_q = \frac{\delta i}{i} = \alpha(1 + \frac{\gamma}{3})\frac{C_{CH}}{C} \qquad (6.47)$$

This is a much simpler expression than that for the single-ended circuit (4.120). In particular, the error is independent of the clock voltage levels, the switch's gate overlap capacitance and the memory transistor's saturation voltage. Although the analysis assumed small-signal conditions, it will hold to fairly high signal currents because the transconductance of the differential circuit is less signal dependent than that of the single-ended circuit and this will give it better linearity.

6.4 Summary

Switched-current circuits employing only basic structures (Chapter 3) cannot achieve analogue performance which competes with switched-capacitors. Circuit refinements are needed to reduce the errors caused by non-ideal MOS behaviour and this chapter has described two main circuit approaches: in the first, negative feedback techniques have been employed to reduce conductance ratio errors and increase dynamic range; in the second, fully-differential techniques were used to reduce charge injection errors and improve noise immunity.

Conductance ratio errors occur when the voltage on the drain of the memory transistor in the sampling phase changes during the output phase. In the op-amp active memory cell, the feedback stabilises the cell's input voltage against changes in the cell's signal current (i.e. it increases the input conductance) and in so doing stabilises the drain voltage of the cell supplying the signal current. This reduces the cell's conductance ratio error by a factor equal to the op-amp's gain (100-10,000). The memory transistor's saturation voltage may be adjusted to allow the dynamic range to reach an optimum of about 4.5dB above that of the basic memory cell. The main problem of the cell is that the extra poles introduced by the op-amp can make monotonic settling difficult to achieve. The settling behaviour is improved in the grounded-gate memory cell by the use of a grounded-gate voltage amplifier in place of the op-amp. This gives a similar improvement in conductance error and complexity is reduced. However, the dynamic range is little better than that of the basic memory cell.

Several cascode arrangements were described and in each case the feedback stabilises the memory transistor's drain voltage against changes in the cell's output voltage. The conductance ratio error is reduced by a factor equal to the voltage gain of the cascode arrangement which, for the simple and folded cascode, is typically 100. If the cascode's reference voltage is regulated (as in the regulated cascode arrangements), this factor is increased to 10,000 typically. Monotonic settling is achieved in the simple and folded cascodes by ensuring that the extra pole introduced by the cascode transistor occurs at a frequency which is much higher than that of the memory transistor. In the regulated arrangements, compensation may be necessary to ensure monotonic settling. In the simple and regulated cascode circuits, voltage swings are constrained and the dynamic range is limited to about 1dB below that of the basic cell. In the folded-cascode and regulated folded-cascode arrangements, voltage swings are less constrained and may be chosen to optimise dynamic range, but the noise introduced by the extra bias current sources reduces the optimised value to about 0.8dB below that of the basic cell. The dynamic ranges of all the cells are compared in Figure 6.30 for supply voltages of 5V and 3.3V. The curves terminate at the limiting saturation voltages. The only way to enhance further the dynamic range of these cells is to use larger values of memory capacitance, C, (and consequently area) and this increases the dynamic range at 10dB/decade in the value of C.

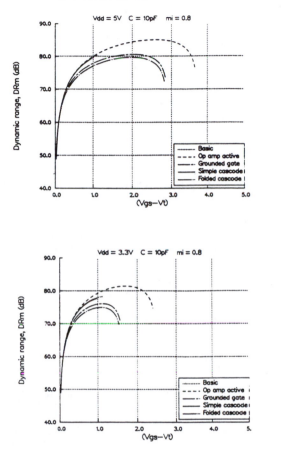

Figure 6.30 Summary of dynamic range for 5V and 3.3V operation.

The second approach to improving analogue performance was to use fully-differential circuit structures. These require a feedback arrangement to balance the drain and tail bias currents and common-mode rejection is possible only with the use of a common-mode feedback circuit which maintains feedback on both input and output phases of the cell's operation. Fully-differential cells designed with half width transistors have similar bandwidth, dynamic range and chip area to single-ended cells with the same supply voltage and current. Further, charge injection errors in the fully-differential cell, were found to be much lower: the offset error is ideally zero and the small signal error is about four times lower. A full analysis of charge injection errors, similar to that given in Chapter 4 for the single-ended cell, has been presented. This shows that the errors depend almost exclusively on the ratio of switch channel capacitance and memory transistor capacitance. This is in contrast with the error in the single-ended cell which depends also on clock voltage levels, switch overlap capacitance, and memory transistor saturation voltage. Linearity will be better in the fully-differential cell both because the charge injection error is smaller and because the transconductance is less signal dependent giving an error which is more linearly related to signal. Although not specifically tested, it may be expected that carefully designed fully-differential circuits will give effective rejection of power supply and substrate noise. A very recently introduced approach, set to reduce all memory cell error sources in a single total error reduction scheme, has been reported in [11] and is discussed briefly in Chapter 22.

References

[1]   D G Nairn and C A Salama, "Current mode analog-to-digital converters," in *Proc. IEEE International Symposium on Circuits and Systems*, pp. 1588-1591, May 1989.

[2]   D Vallancourt, Y P Tsividis and S J Daubert, "Sampled-current circuits," in *Proc. IEEE International Symposium on Circuits and Systems*, pp. 1592-1595, May 1989.

[3]   D. Groeneveld et al " Self-calibrated technique for high resolution D-A converters," *IEEE J. Solid-State Circuits*, vol. SC-24, pp. 1517-1522, Dec. 1989.

[4]   A. A. Abidi, "On the operation of cascode gain stages," *IEEE J. Solid-State Circuits*, vol. SC-23, pp. 1434-1437, Dec. 1988.

[5]   C Toumazou, J B Hughes and D M Pattullo, "A regulated cascode switched-current memory cell," *Electronics Letters*, vol. 26, pp. 303-304, 1 March 1990.

[6]   E. Sackinger and W. Guggenbühl, "A high swing, high-impedance MOS cascode circuit," *IEEE J. Solid-State Circuits*, vol. SC-25, pp. 289-298, Feb. 1990.

[7]   H. C. Yang, T. S. Fiez and D. J. Allstot, "Current-feedthrough effects and cancellation techniques in switched-current circuits," in *Proc. IEEE International Symposium on Circuits and Systems*, pp. 3186-3189, 1990.

[8]   S. J. Daubert, D. Vallancourt and Y. Tsividis, "Current copier cells," *Electronics Letters*, vol. 24, pp. 1560-1562, 8 Dec. 1988.

[9]   H. P. Lie, "Switched-capacitor feedback and hold circuit," *U.S. Patent 4,585,956*, 1986.

[10]   C. Toumazou, N. C. Battersby and C. Maglaras, "High-performance algorithmic switched-current memory cell," *Electronics Letters*, vol. 26, pp. 1593-1595, 13 Sept. 1990.

[11]   J. B. Hughes and K. W. Moulding, "S^2I: A two-step approach to switched-currents, in *Proc. IEEE symposium on Circuits and Systems*, May 1993.

Chapter 7

Class AB Switched-Current Techniques

Nicholas C. Battersby and Chris Toumazou

7.1   Introduction

So far the discussion of switched-current circuits has centred on the design, application and analysis of the second generation memory cell (or current copier) and techniques to enhance the performance of the basic cell (see Chapters 3, 4, 5 and 6).   However, one limitation that has not been considered is that the signal current must always be less than the bias current due to the class A operation of the cell. The consequence of this is that cells have to be operated in a backed-off condition, leading to significant quiescent power consumption in filters, where a large number of bias current sources are frequently needed. Now, whilst it is true that in most mixed-signal applications the analogue portion is small, the overall power budget is often tight and any saving in power consumption is significant.   This is particularly true in portable applications where achieving low power consumption is of critical importance if an acceptable trade-off between product functionality and battery life is to be made.

The extent to which the quiescent power consumption of the conventional class A switched-current memory cell (see Chapter 3) can be reduced is limited by the requirement to maintain bias current levels that are in excess of any incoming signal level. The application of class AB techniques to the design of switched-current memory cells removes this necessity and allows the signal magnitude to exceed that of the bias current. Class AB techniques therefore offer the potential to realise power efficient switched-current filters.

This chapter explores the application of class AB techniques to switched-current filter design. A new class AB memory cell is discussed [1,2] and the application of this cell to the realisation of the fundamental filters building blocks of a delay cell, an integrator and a differentiator described.   The principal limitations of the cell will then be briefly described. Finally, the simulated results of an integrated, although not yet fully tested, lowpass biquadratic filter will be described.

7.2 Memory Cell

The new class AB memory cell is based on the operational amplifier supply current sensing techniques used to realise current conveyors [3,4] and the CMOS implementation of a class AB current conveyor [4,5], which also bears some similarity to a recently proposed class AB current amplifier [6], is shown in Figure 7.1.

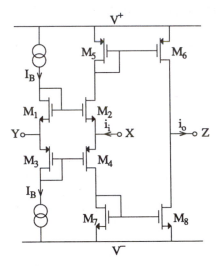

Figure 7.1 Class AB current conveyor.

The circuit (Figure 7.1) exhibits a current following action between nodes X and Z and a voltage following action between Y and X. Assuming that node Y is at a constant potential, the input current at X shares between the sources of M_2 and M_4. This phase split input current then recombines at node Z, where the bias current is cancelled. Thus $i_o = -i_i$ and the circuit realises a CCII+ current-conveyor. Analysis of the current conveyor shows that the input current sharing, assuming matched devices, equal quiescent transconductances and saturated operation, can be described by

$$i_{d4} = \frac{(4I_B + i_i)^2}{16I_B} \quad and \quad i_{d2} = \frac{(4I_B - i_i)^2}{16I_B} \quad for \; |i_i| \le 4I_B \quad (7.1)$$

where i_{d4} and i_{d2} are the drain currents of M_4 and M_2, respectively. When the input current magnitude exceeds $4I_B$, the circuit continues to function correctly, but (7.1) no longer holds since one of the current mirrors now takes virtually all the current, the other one being effectively switched off.

The class AB switched-current memory cell is realised from the class AB conveyor by replacing each of the two output current mirrors with second generation switched-current memory cells, as shown in Figure 7.2.

Note that node Y (of Figure 7.1) has been connected to ground thus holding node X near to ground potential.

The operation of the class AB switched-current memory cell is basically the same as that of the second generation memory cell, from which it is descended, and is as follows. During ϕ_1, M_5 and M_6 are diode-connected and their gate-source capacitances are charged to the voltage required to sink their share of the phase split input current. Then during ϕ_2, the gates of M_5 and M_6 are isolated and they output the currents stored during ϕ_1, these then recombine at the output node to cancel the bias current and form the output current. The switches connecting the drains of M_2 and M_4 to the power rails during the retrieval phase are there to ensure that the input current continues to flow in M_2 and M_4 during ϕ_2.

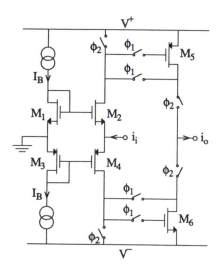

Figure 7.2 Class AB switched-current memory cell.

In the class AB memory cell, the bias chain, consisting of M_1, M_3 and the two current sources, simply provides gate bias voltages to the two input transistors M_2 and M_4. Therefore it is possible to use one bias chain to bias all the memory cells in a filter, resulting in a saving of power and circuitry.

From the above discussion of the operation of the class AB memory cell, it is clear that the input current is not limited by the level of the bias current. In practice, the factor that limits the input current magnitude is the voltage headroom available to the memory transistors, but this limit can be determined at the design stage by the choice of appropriate aspect ratios for M_5 and M_6.

To demonstrate that the magnitude of the input current is not limited by the bias current, the transfer characteristic of the class AB memory cell (Figure 7.2) was simulated using HSPICE and the level 6 parameter set of a typical 1.6μm N-well CMOS process, where the aspect ratios of M_5 and M_6 were both 100μm/20μm *(W/L)*, the power supplies ±2.5V, the bias current 10μA and the clock frequency 2MHz. The results, which are shown in Figure 7.3, show that the maximum input current is around 150μA, which is 15 times greater than the bias current used. To achieve the same signal current levels using class A techniques would require a bias current level of greater than 150μA and consequently much higher power consumption.

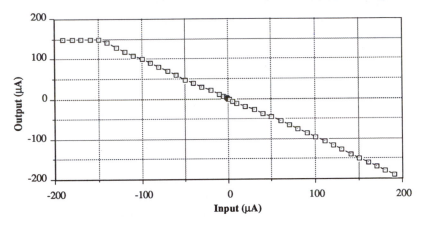

Figure 7.3 Transfer characteristic of the basic class AB memory cell
$(I_B = 10\mu A)$.

7.3   Filter Building Blocks

Having described the principle of operation and advantages of the class AB switched-current memory cell in the previous section, its application to the synthesis of the fundamental filter building blocks of delay cells, integrators and differentiators will now be described.

7.3.1   Delay cells

To generate delays using class AB memory cells it is not necessary to duplicate the entire cell, rather the output of the first stage can be simply cross-coupled to a second stage of memory elements, as illustrated by the two stage delay cell shown in Figure 7.4.

7.3.2   Integrator circuits

The realisation of an integrator using class AB switched-current memory cells can be accomplished by applying the architectures developed originally for second generation memory cells (see Chapter 3). Using the same architecture has the advantage that it allows the class AB circuit to inherit the favourable sensitivity properties of its predecessor.

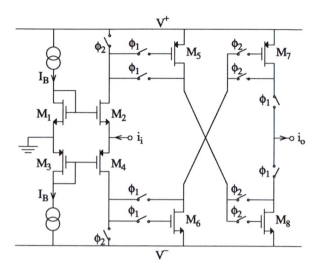

Figure 7.4 Class AB delay cell.

A generalised class AB integrator, which is functionally equivalent to the generalised second generation class A integrator discussed in Chapter 3 (see Figure 3.13), is shown in Figure 7.5.

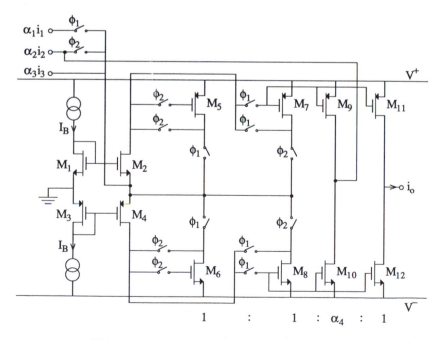

Figure 7.5 Generalised class AB integrator.

The transfer function of this generalised integrator (Figure 7.5) is given by

$$i_o(z) = \frac{A_1 z^{-1}}{1 - B z^{-1}} i_1(z) - \frac{A_2}{1 - B z^{-1}} i_2(z) - \frac{A_3(1 - z^{-1})}{1 - B z^{-1}} i_3(z) \qquad (7.2a)$$

where

$$A_1 = \frac{\alpha_1}{1 + \alpha_4}, \quad A_2 = \frac{\alpha_2}{1 + \alpha_4}, \quad A_3 = \frac{\alpha_3}{1 + \alpha_4}, \quad B = \frac{1}{1 + \alpha_4} \qquad (7.2b)$$

Comparing (7.2) with the equivalent expression from Chapter 3 (3.37) confirms that the transfer function of the generalised class AB integrator is identical to that of its second generation class A counterpart.

Extension of the generalised class AB integrator (Figure 7.5) to realise a lossless bilinear transform integrator can be achieved simply by setting $\alpha_1 = -\alpha_2 = T/2$ and $\alpha_3 = \alpha_4 = 0$.

7.3.3 Differentiator circuits

A generalised class AB inverting differentiator, based on the second generation class A circuit discussed in Chapter 3 (see Figure 3.20), is shown in Figure 7.6.

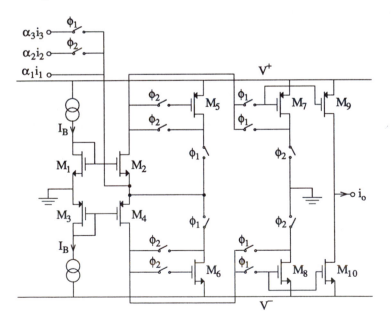

Figure 7.6 Generalised class AB inverting differentiator.

The transfer function of this differentiator (Figure 7.6) is the same as that of its second generation class A counterpart and is given by

$$i_o(z) = -\alpha_1(1 - z^{-1})i_1(z) + \alpha_2 z^{-1}i_2(z) - \alpha_3 i_3(z) \qquad (7.3)$$

Considering Figure 7.6 it will be seen that the output of the second memory (M_7 and M_8) is not used, this is because it is only available during ϕ_2. However, under circumstances where the output is only required during ϕ_2, it can be used and the mirror stage (M_9 and M_{10}) will not be needed.

A generalised class AB non-inverting differentiator can also be derived from its second generation class A counterpart and is shown in Figure 7.7.

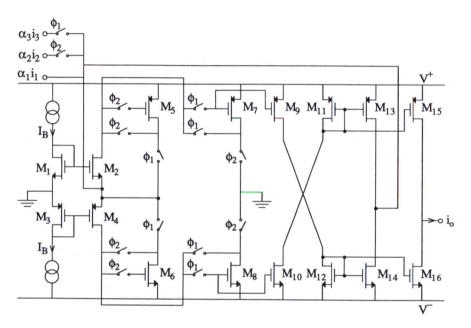

Figure 7.7 Generalised class AB non-inverting differentiator.

This generalised class AB non-inverting differentiator also has the same transfer characteristic as its second generation class A counterpart, which is

$$i_o(z) = \alpha_1 \frac{1 - z^{-1}}{z^{-1}} i_1(z) - \alpha_2 i_2(z) + \frac{\alpha_3}{z^{-1}} i_3(z) \qquad (7.4)$$

7.4 Performance Limitations

The factors that limit the performance of the class AB memory cell are basically the same as those affecting the second generation class A memory cell, which were discussed in Chapters 4 and 5. This section briefly discusses these limitations in the context of the class AB memory cell and centres on how their impact differs from that occurring in the class A cell.

(i) Conductance ratio errors

The impact of having a non-infinite input conductance (g_i) and a non-zero output conductance (g_o) is the same in the class AB memory cell as it was in the class A memory cell (this is described in Section 4.3). For the basic class AB memory cell (Figure 7.2) the small signal input and output conductances are given by

$$g_i = \frac{(g_{m4} + g_{ds4})(g_{m6} + g_{ds6})}{1 + g_{m6} + g_{ds6}} + \frac{(g_{m2} + g_{ds2})(g_{m5} + g_{ds5})}{1 + g_{m5} + g_{ds5}} \qquad (7.5)$$

$$g_o = g_{ds5} + g_{ds6} \qquad (7.6)$$

If (7.5) and (7.6) are simplified, by approximation, it can be seen that the g_i/g_o ratio of the class AB memory cell is of the same order as its class A counterpart and is therefore not adequate for the realisation of high quality filters.

To enhance the g_i/g_o ratio, the techniques described in Chapter 6 to increase output resistance can be applied to each of the two memory elements in the class AB cell.

(ii) Charge injection errors

In the discussion of class AB cells so far, all the switches have been assumed ideal, but in practice they are implemented by MOS transistors in the same way as in the class A memory cells. Thus during their switch turn-off transient the gate isolating switches M_5 and M_6 (of Figure 7.2) will dump an error charge onto the gate capacitances of their respective memory transistors. This will cause each of the two memory transistors to generate an error current, in the same way as described for the class A cell in Chapter 4.

Although the error mechanism by which charge injection errors are generated is the same in both the class A and class AB cells, its impact is different in the class AB cell because of the way in which the errors from the p-channel and n-channel memories combine at the output node. Rather than extending the equations developed in Chapter 4 to cover the class AB cell a more intuitive qualitative approach (backed up by simulation results) will be taken here.

The gate isolating transistor used by the p-channel memory is a p-channel device whilst that used by the n-channel memory is an n-channel device. Consequently, the impact of charge injection is to reduce the output current of each of the two memories and therefore partial charge injection cancellation will occur. As a consequence of the sharing of input current between the two memories, complete cancellation will occur at one particular value of input current. If the memory transistors M_5 and M_6 (of Figure 7.2) have the same aspect ratio and if the aspect ratio of the p-channel gate isolating transistor is approximately 3 times that of the n-channel device this cancellation point will occur when the input current is close to zero. To confirm this a simulated comparison of charge injection in a simple second generation class A memory cell (Figure 3.4) and a simple class AB memory cell (Figure 7.2) was performed. The simulation used the level 6 parameter set of a typical 1.6μm N-well CMOS process, in which all the memory transistors had the aspect ratio 100μm/20μm, the n-channel switches were 2μm/2μm, the p-channel switches were 6μm/2μm, the clock frequency was 2MHz and the power supplies ±2.5V. The simulated errors, for an input current range of ±100μA with a bias current of 10μA in the class AB circuit and 100μA in the class A circuit, are shown in Figure 7.8.

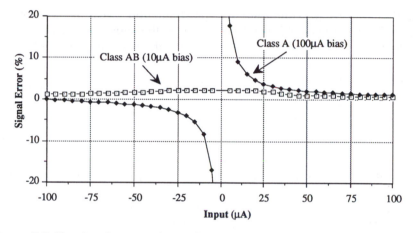

Figure 7.8 Simulated comparison of charge injection errors in class A and AB cells.

From Figure 7.8, the effect of the charge injection cancellation mechanism in the class AB cell can be seen clearly and furthermore the resulting level of charge injection error is much lower in the class AB cell than in the class A cell for low levels of input current.

Cascaded memory cells form the core of all the filter building blocks described in Chapter 3 and it was seen in Chapter 4 that when two class A memory cells are cascaded the signal independent component of the charge injection error is cancelled. This cancellation occurred because as the signal passed from one cell to the next it was inverted and the charge injection error acted alternately to increase and decrease it, thus cancelling the signal independent components. However, this effect does not occur in the class AB memory cell, where the charge injection always acts to attenuate the signal. The consequence of this effect is that integrators constructed from the basic class AB cell would exhibit considerable parasitic loss, resulting in unacceptable filter performance. To circumvent this problem it is possible to use an enlarged dummy switch in one of the memories to invert the polarity of its charge injection and hence introduce partial cancellation into cascaded cells. Other error cancellation techniques are also being investigated.

The absence of a large bias current in the class AB memory cell results in the transconductance of the memory transistors being more sensitive to input current variations than in the class A cell, with the result that harmonic distortion due to charge injection is worse. However, the use of a fully differential class AB structure (a straightforward extension of the basic cell of Figure 7.2) would help to reduce this.

(iii) Settling errors

The settling criteria and the impact of settling errors on the class AB cell is the same as on the class A cell, except that in the class AB cell there are two memories that must be allowed to settle properly.

(iv) Mismatch errors

Since class AB circuits use the same basic structures as second generation class A circuits they also have the same sensitivities to mismatch errors. In class A circuits it was found that distortion components due to mismatch could be reduced by operating with a high quiescent memory transistor saturation voltage $(V_{GS} - V_T)$, but in class AB circuits it is more difficult to achieve this whilst maintaining a low bias current. However, by using careful layout techniques mismatch errors can be reduced to a reasonable level.

(v) Noise errors

The noise generated by the class AB memory cell is similar to that of the class A cell described in Chapters 4 and 5, the principal difference being that the noise at the output now consists of the sampled and direct components from two memories rather than the single memory and current source of the class A cell. The general guidelines for reducing noise in class A cells also apply to the class AB cell, except that, as in the case of mismatch errors, it is more difficult obtain a high quiescent memory transistor saturation voltage in the class AB cell.

Overall, it has been seen that the error mechanisms that apply to the class A memory cell, described in Chapter 4, also apply to the class AB cell. In most cases the impact of these errors is basically the same, the major exception to this being charge injection.

7.5 Biquadratic Filter Example

To demonstrate the feasibility of the class AB switched-current memory cell, a lowpass Butterworth biquadratic filter section, with a -3dB point at one tenth of its clock frequency, was developed [2]. The filter was integrated on a standard 1.2μm digital CMOS process, although since it has yet to be fully tested simulation results will be shown here.

The synthesis procedure and the filter coefficients employed are not described here, since they are identical to those used for the biquadratic filter example presented in Chapter 8.

The core class AB memory cell used in this example is shown in Figure 7.9. This core memory cell is a development of the basic structure (Figure 7.2), in which each of the basic memory elements has been replaced by the regulated-cascode memory cell described earlier in Section 6.2.5 [7,8].

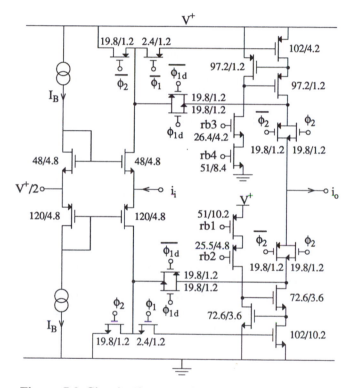

Figure 7.9 Circuit diagram of class AB memory cell.

The bias current for the class AB memory cell is provided by off-chip current sources, to facilitate experimentation. However, the bias voltages *(rb1, rb2, rb3, rb4)* needed to bias the regulated-cascodes are provided by separate integrated bias voltage generators.

To verify the filter's performance, a full transistor level simulation was performed using HSPICE and a circuit netlist extracted directly from the layout file. In the simulation, the filter had an input current magnitude of 100μA (peak), a bias current of 10μA and a clock frequency of 3.33MHz. The resulting frequency response is shown in Figure 7.10.

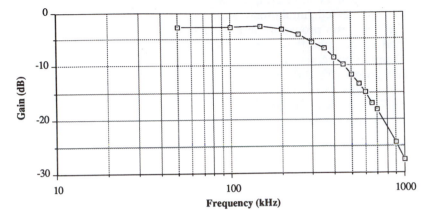

Figure 7.10 Simulated results for class AB biquadratic section (clock frequency = 3.33MHz).

The simulated results show that the class AB biquadratic section does function correctly, although it is subject to a sizeable gain error caused by the accumulation of charge injection errors. However, the quiescent bias currents employed are more than an order of magnitude lower than those needed by an equivalent class A circuit for the same input current magnitude, confirming the technique's potential for low power operation.

7.6 Conclusions

In this chapter, a class AB switched-current technique was proposed to enable switched-current filters with considerably reduced quiescent power consumption to be built. A family of filter building blocks derived from the basic cell was then developed and the principle performance limitations of the technique briefly discussed. Early simulated results for a biquadratic filter showed that charge injection causes significant errors in filter structures and techniques to cancel these errors are currently being investigated.

Acknowledgements

The financial support of the British Science and Engineering Research Council is gratefully acknowledged.

References

[1]   N. C. Battersby and C. Toumazou, "Class AB switched-current memory for analogue sampled data systems, "*Electronics Letters*, vol. 27, pp. 873-875, 9 May 1991.

[2]   N. C. Battersby, *Switched-current techniques for analogue sampled-data signal processing*, Ph.D Thesis, Imperial College of Science, Technology and Medicine (University of London), Feb. 1993.

[3]   A. S. Sedra and K. C. Smith, "A second-generation current conveyor and its applications," *IEEE Trans. Circuit Theory*, vol. CT-17, pp. 132-134, Feb. 1970.

[4]   C. Toumazou, F. J. Lidgey and C. A. Makris, "Extending voltage-mode op amps to current-mode performance," *Proc. IEE Pt. G*, vol. 137, pp. 116-130, April 1990.

[5]   A. S. Sedra and G. W. Roberts, "Current conveyor theory and practice," in C. Toumazou, F. J. Lidgey and D. G. Haigh, Eds., *Analogue IC design: the current-mode approach*, London: Peter Peregrinus Ltd., April 1990.

[6]   Z. Wang, "Wideband class AB (push-pull) current amplifier in CMOS technology," *Electronics Letters*, vol. 26, pp. 543-545, 12 April 1990.

[7]   E. Säckinger and W. Guggenbühl, "A high-swing, high-impedance MOS cascode circuit," *IEEE J. Solid-State Circuits*, vol. 25, pp. 289-298, Feb. 1990.

[8]   C. Toumazou, J. B. Hughes and D. M. Pattullo, "A regulated cascode switched-current memory cell," *IEE Electronics Letters*, vol. 26, pp. 303-305, 1 March 1990.

Chapter 8

Switched-Current Filters

Nicholas C. Battersby and Chris Toumazou

8.1 Introduction

In this chapter, the design of switched-current filters, employing the CMOS building blocks and circuits developed in previous chapters, is considered and the results of two integrated test filters presented.

The chapter starts by considering the synthesis of switched-current filters based on biquadratic sections. Sections based on either integrators or differentiators are described and measured results from an integrated integrator based lowpass section are presented to demonstrate the technique. For the realisation of high-order or high-Q filters, the ladder approach is very popular because of its low sensitivity properties [1,2]. This approach is described, with the aid of an illustrative integrated bilinear fifth order elliptic lowpass filter example, which demonstrates the practical application of the technique. The next three filter synthesis approaches; FIR filters, wave active filters and the transposition of switched-capacitor designs, are only briefly discussed as detailed descriptions can be found in Chapter 10, 11 and 9, respectively. Finally, this chapter proposes a new approach to the realisation of tunable filters on pure digital processes. The proposed switched-transconductance (ST) technique takes advantage of the fact that in many applications switched-current circuits need a voltage interface, and hence also a linear transconductor; by incorporating this transconductor with the switched-current circuits tunable impedances with good inherent accuracy can be realised.

8.2 Biquadratic Filter Sections

Second-order (biquadratic) filter sections are versatile filter building blocks that can easily be cascaded to form high order filters [1]. Consider a high order filter transfer function $H(s)$, where the numerator and denominator are factorised into second-order factors as shown below.

$$H(s) = \left[\frac{a_2 s^2 + a_1 s + a_0}{b_2 s^2 + b_1 s + 1}\right]\left[\frac{c_2 s^2 + c_1 s + c_0}{d_2 s^2 + d_1 s + 1}\right]\cdots\cdots \qquad (8.1)$$

Each of the factors in (8.1) can be realised by a second-order filter section (a biquad), making the realisation of high order filters a simple matter of cascading the appropriate biquads. In the case of an odd order filter it is necessary to include one first-order section in the cascade of biquads.

The modularity and simplicity of the cascade synthesis approach to the realisation of high order filters has led to its wide acceptance in active-*RC*, switched-capacitor and now switched-current filter design. The wide use of the technique has resulted in extensive literature being published in the area [1,2], and widely available design software allows the dynamic range and the pairing of poles and zeros to be optimised [3]. Since these techniques can be directly applied to the synthesis of cascade switched-current filters they will not be discussed further, rather the emphasis is on the implementation of the biquadratic sections themselves.

Returning to the biquadratic filter sections used in (8.1), the transfer function of a single section *H(s)* is commonly written as

$$H(s) = \frac{k_2 s^2 + k_1 s + k_0}{s^2 + \left(\dfrac{\omega_0}{Q}\right)s + \omega_0^2}$$ (8.2)

Applying the bilinear $s \rightarrow z$ transformation to (8.2) gives the *z*-domain transfer function *H(z)* as

$$H(z) = \frac{\left[\dfrac{4k_2 + 2k_1 T + k_0 T^2}{D}\right]z^2 + \left[\dfrac{2k_0 T^2 - 8k_2}{D}\right]z + \left[\dfrac{4k_2 - 2k_1 T + k_0 T^2}{D}\right]}{\left[\dfrac{\omega_0 T^2 + \dfrac{2\omega_0 T}{Q} + 4}{D}\right]z^2 + \left[\dfrac{2\omega_0^2 T^2 - 8}{D}\right]z + 1}$$ (8.3)

where $D = \omega_0^2 T^2 - \dfrac{2\omega_0 T}{Q} + 4$ and *T* is the clock period.

To implement *H(z)* using switched-capacitors, a variety of structures based on a two integrator loop were developed [1,2] and then, more recently, an alternative implementation based on a two differentiator loop was proposed [4]. These switched-capacitor biquadratic sections can easily be translated into the switched-current technique and the next two subsections show how this can be done for both integrator and differentiator based sections [5].

8.2.1 *Integrator based sections*

The realisation of integrator based switched-current biquadratic sections was described by *Hughes* [5], whose approach was to take an existing switched-capacitor structure [1] and implement it using switched-current techniques, and is reviewed here.

The block diagram of an integrator based switched-current

biquadratic section is illustrated in Figure 8.1, and it differs from its switched-capacitor equivalent solely in that the variable used is current rather than voltage.

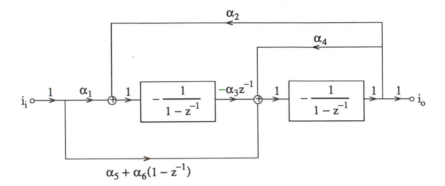

Figure 8.1 Block diagram of an integrator based biquadratic section.

To calculate the filter coefficients, the transfer function of the circuit (Figure 8.1) is first derived and then the coefficients are compared with those of the z-domain biquadratic function (8.3). This data is then used to choose the filter coefficients needed to implement a given transfer function and is given in Table 8.1.

Coefficient	Value
$\alpha_1\alpha_3$	$4k_0T^2/D$
$\alpha_2\alpha_3$	$4\omega_0^2T^2/D$
α_4	$4\omega_0T/QD$
α_5	$4k_1T/D$
α_6	$(4k_2 - 2k_1T + k_0T^2)/D$
D	$\omega_0^2T^2 - 2(\omega_0/Q)T + 4$

Table 8.1 Coefficients for the integrator based biquadratic section.

To implement the block diagram of Figure 8.1, a loop containing two generalised second generation integrators (described in Chapter 3), with appropriate weighted feedback paths, is constructed. The resulting circuit level implementation is shown in Figure 8.2.

In Figure 8.2, basic switched-current cells are shown for clarity, but to obtain good performance it is necessary to employ enhanced cells, such as those described in Chapter 6. Furthermore, the circuit needs to be preceded by a sample-and-hold with multiple scaled output currents.

The implementation of a biquadratic section described above uses a two integrator loop. An alternative design approach using a single integrator loop has been demonstrated recently by *Song* et al [6].

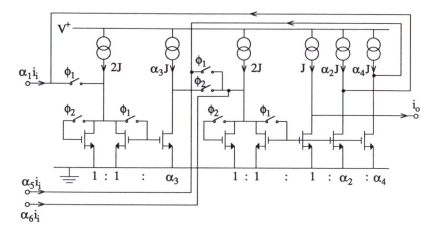

Figure 8.2 Integrator based biquadratic section.

8.2.2 Example: integrated biquadratic filter

As an illustration of the design procedure for the integrator based biquadratic section described above; the design, integration and measured results of a biquadratic filter section are described. The section was designed to realise a second order lowpass Butterworth filter whose -3dB frequency is at one tenth of its clock frequency. The process employed to realise the example filter was a standard 1.2µm digital CMOS technology.

The normalised transfer function of the required biquadratic function in the s-domain is the familiar second-order lowpass Butterworth function i.e.

$$H(s) = \frac{1}{s^2 + s\sqrt{2} + 1} \tag{8.4}$$

The first stage of the design procedure is to prewarp the filter specifications, to take account of the frequency warping caused by the bilinear transformation, according to

$$\omega_d = \frac{2}{T} tan\left[\frac{\omega_a T}{2}\right] \tag{8.5}$$

where ω_a = continuous time frequency and ω_d = discrete time frequency. In this example, the original -3dB frequency of $0.1f_c$ (f_c is the clock frequency) becomes $0.10345f_c$ after being prewarped. The normalised transfer function (8.4) is then denormalised and the resulting values are used to calculate the filter coefficients using (8.2) and Table 8.1. The final filter coefficients, shown in Table 8.2, are then implemented by the choice of appropriate ratios of transistor aspect ratios in the circuit of Figure 8.2.

Coefficient	Value
α_1	0.80848
α_2	0.80848
α_3	0.80848
α_4	1.4228
α_5	0
α_6	0.16341

Table 8.2 Filter coefficients for lowpass Butterworth biquadratic section.

Having chosen the filter architecture and coefficients, the next stage of the design procedure was to design and optimise a high performance memory cell to form the core of the filter, the design of this regulated-cascode cell is shown in Figure 8.3.

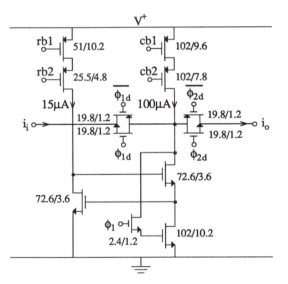

Figure 8.3 Circuit diagram of class A memory cell.

The size of the memory transistor was first optimised, achieving a minimum settling time (to 0.1%) of 19nsec, using a 50μA input current step, a bias current of 100μA, a 2.4μm/1.2μm (minimum geometry) diode-connecting switch transistor, and an aspect ratio of 102μm/10.2μm. The combination of a relatively large memory transistor and a minimum geometry diode-connecting switch transistor was chosen primarily to keep the effect of charge injection errors at a reasonable level. However, the use of a reasonably large memory transistor also has other performance benefits, such as lower output conductance, better matching and improved coefficient resolution, which need to be balanced against the speed advantage of small devices (see Chapter 4). The final choice of transistor dimensions was the result of an optimisation process, which aimed to satisfy both the speed and accuracy requirements of the target filters simultaneously.

Whilst the inclusion of dummy switch compensation in the design, may have reduced the level of charge injection, this was not done. The reason for this was that, the overall design of the test chip did not permit sufficiently tight control of clock edges to ensure that dummy switch compensation would be effective and simulation results indicated that adequate performance could be obtained without them.

The sizes of the memory transistor's cascode device and its associated bias chain were optimised to keep the memory transistor in saturation and maintain critical damping.

The generation of the cell bias (100μA) and the regulated-cascode bias (15μA) currents used cascoded p-channel current sources, whose bias voltages *(cb1, cb2, rb1, rb2)* were provided by separate bias generation circuits.

Stage interconnecting switches were implemented using CMOS transmission gates, which since they do not contribute to charge injection error and must convey the signal current, were made with relatively large transistors having the aspect ratio 19.8μm/1.2μm.

Finally, the coefficients of the filter were implemented using scaled current mirrors attached to the memory cell (Figure 8.3). So that the addition of these mirror transistors did not adversely affect the settling characteristic of the memory cell itself, both the memory and mirror transistors were scaled such that the sum of their areas remained equal to that of the core memory transistor, whilst maintaining the required aspect ratios. Since the value of each coefficient was determined by a ratio of aspect ratios, the resolution attainable was limited by the minimum size increment permitted by the design rules, in this case 0.6μm.

The full circuit diagram of the integrated class A biquadratic filter section, designed according to the coefficients given in Table 8.2, is shown in Figure 8.4. In addition to the two integrator loop discussed earlier, this circuit includes an input sample-and-hold circuit to generate the necessary input signals. Also, extra switches have been included to switch memory cell output currents to a dummy supply of 1.6V 'dump' during phases when they are not required.

The integrated biquadratic filter section was measured at a clock frequency of 1MHz, although operation at higher frequencies was demonstrated.

A comparison of the ideal and measured frequency responses of the biquadratic section is shown in Figure 8.5. The shown ideal response was obtained by simulating an ideal macromodel of the filter using SCALP (a switched-capacitor analysis program from Philips).

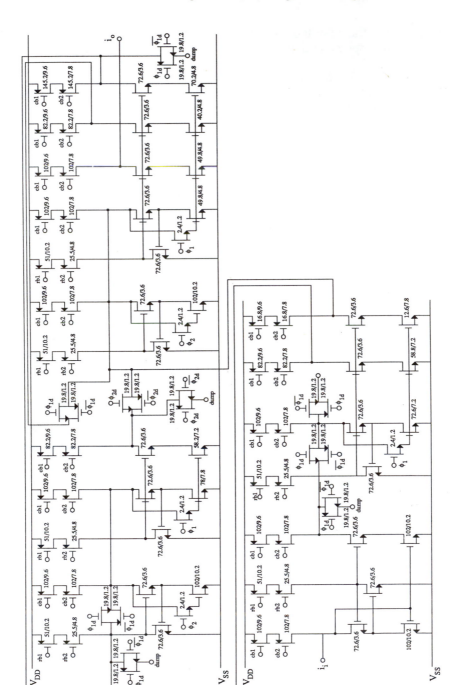

Figure 8.4 Full circuit diagram of class A biquadratic filter section.

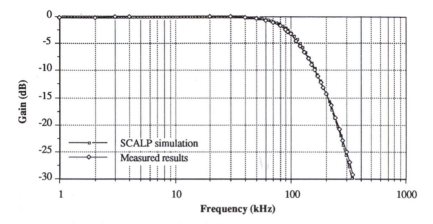

Figure 8.5 Frequency response of class A biquadratic filter section.

The results show that the correspondence between the measured and expected responses is very good. For a 1MHz clock the expected -3dB frequency is 100kHz, the measured gain at this frequency was found to be -3.25dB. Now, since the output of the filter is a sample-and-held current, the results need to be compensated for $sinc(x)$ distortion. This amounts to a droop of 0.14dB at 100kHz (1MHz clock), giving a compensated value of -3.1dB, which is close to the design value. The actual -3dB frequency was found to be 98kHz after compensation for $sinc(x)$ distortion, which is a 2% deviation from the design value. However, because these measured results correspond so closely to the expected values, the reliable measurement limit of ± 0.1dB becomes a significant accuracy limiting factor.

The output spectrum of the biquadratic filter section, for a signal frequency of 10kHz, is shown in Figure 8.6. Measurement of this spectrum showed that the noise floor is 65dB below the signal level for a total harmonic distortion (THD) level of 0.32%.

Performance Characteristic	Value	
	Measured	Expected
Clock frequency	1MHz	-
-3dB frequency	98kHz	100kHz
Gain @100kHz	-3.1dB	-3dB
Noise floor (10kHz signal)	-65dB @ THD=0.32%	-
Power consumption	6.8mW	5.6mW
Filter die area	0.3 mm²	-

Table 8.3 Performance summary of class A biquadratic filter section.

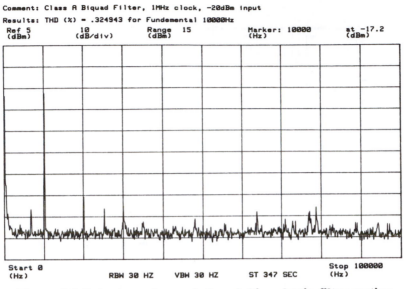

Comment: Class A Biquad Filter, 1MHz clock, -20dBm input
Results: THD (%) = .324943 for Fundemental 10000Hz

Figure 8.6 Output spectrum of class A biquadratic filter section (clock = 1MHz).

The main measured and expected performance characteristics of the biquadratic filter section are summarised in Table 8.3. Overall, the performance of the class A biquadratic filter section is close to what was expected.

The levels of noise and harmonic distortion were not simulated and the measured levels were a little worse than had been hoped for, although they were sufficient for this demonstration. Reduction in the underlying noise level can be achieved by re-optimising the basic memory cell such that the memory transistor operates with as large a saturation voltage as is consistent with the other design constraints. The primary contributor to the observed harmonic distortion is likely to be charge injection. Reduction of charge injection can also be achieved by maximising the saturation voltage of the memory transistors or by employing one of the charge injection cancellation schemes discussed earlier in Chapter 6.

8.2.3 Differentiator based sections

The realisation of switched-capacitor differentiator based biquadratic sections was proposed by *Yu* et al [4]. Subsequently, *Hughes* [5] proposed switched-current differentiator based biquadratic sections and these are briefly described here.

The block diagram of one possible biquadratic section using a loop of two differentiators, one forward Euler and one backward Euler, is shown in Figure 8.7. The table of coefficient values corresponding to this structure is shown as Table 8.4.

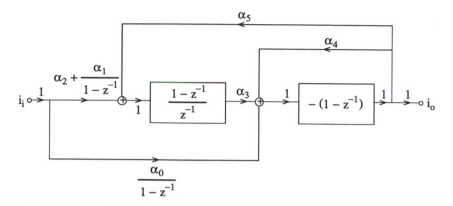

Figure 8.7 Block diagram of a differentiator based biquadratic section.

Coefficient	Value
α_0	k_0/ω_0^2
$\alpha_1\alpha_3$	$k_1/T\omega_0^2$
$\alpha_2\alpha_3$	$(k_0 - 2k_1/T + 4k_2/T^2)/4\omega_0^2$
α_4	$1/QT\omega_0$
$\alpha_5\alpha_3$	$(4/T^2 + 2\omega_0/QT + \omega_0^2)/4\omega_0^2$

Table 8.4 Coefficients for the differentiator based biquadratic section.

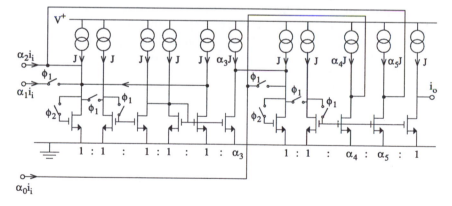

Figure 8.8 Differentiator based biquadratic section.

The circuit level implementation of the biquadratic section (Figure 8.7), using the forward and backward Euler differentiators originally described in Chapter 3, is shown in Figure 8.8. As was the case with the integrator based sections, it is necessary to use enhanced switched-current cells to obtain satisfactory performance and the section should be preceded by a sample-and-hold with multiple scaled output current capability.

In this section, the synthesis of one integrator and one differentiator based biquadratic filter section has been reviewed and the realisation of an integrator based lowpass section described. The synthesis of cascade filters, based on these sections, is relatively straightforward and well documented, since it directly employs those techniques already developed for switched-capacitor synthesis. The range of filters that can be realised using the biquadratic section is very wide, making it an important filter building block.

8.3 Ladder Filters

By cascading single and biquadratic filter sections it is possible to realise a wide range of high order filters. However, in practice, the sensitivity of such cascade structures to component variations can become unacceptably high, especially in high-order or high-Q filters. Therefore, to manufacture such high-Q or high-order filters economically an alternative technique, which is less sensitive to manufacturing tolerances, is needed.

The simulation of doubly terminated LC ladder networks, which produces filters with low component sensitivity, has proved to be particularly well suited to the implementation of these high-order or high-Q filters. The low sensitivity characteristics of these ladder networks has resulted in their wide acceptance as a starting point for switched-capacitor simulation [1,2].

The methods used in the design of switched-capacitor ladder filters can be applied directly to the design of switched-current ladder filters and integrated examples have been produced [7]. In this section, an approach to the synthesis of switched-current ladder filters, based on the bilinear $s \rightarrow z$ transformation is described [8,9]. Early approaches to the design of switched-current ladder filters relied on the use of forward and backward Euler integrators [7,10]. However, these were restricted to applications in which the clock frequency was much higher than the cut-off frequency because of the errors inherent in the transformations. The use of bilinear integrators overcomes these problems and allows the design of filters with low clock frequency to cut-off frequency ratios [1]. The technique proposed here applies the bilinear $s \rightarrow z$ transformation to the synthesis of switched-current filters and results in accurate filters that are well suited to switched-current implementation (see also [11] for a different approach).

8.3.1 Example: integrated elliptic filter

To illustrate the design method, the procedure followed to realise an example ladder filter, designed to meet the design specifications given in Table 8.5, will now be described. This specification was intended as a demonstration for a commercial filtering application in the video arena and represents a significantly more demanding specification than any switched-current ladder filter reported so far [7].

Parameter	Value
Clock frequency	20MHz
Passband edge	3.6MHz
Stopband edge	4.4MHz
Passband ripple	< 0.4dB
Stopband attenuation	> 25dB

Table 8.5 Design specifications for ladder filter example.

The first stage in the design procedure is to prewarp the design specifications to take account of the frequency warping effect of the bilinear transformation (8.5). Doing this gives a prewarped passband edge of 4.04MHz and a stopband edge of 5.27MHz.

Using the prewarped specifications, an appropriate reference passive prototype is now chosen from filter tables [12] and then the component values are denormalised. To satisfy the design specifications (Table 8.5), a lowpass fifth order elliptic filter (C0525) was chosen, giving an ideal passband ripple of 0.28dB, minimum stopband attenuation of 32.3dB and passband to stopband edge ratio of 1.19. The passive prototype filter is shown in Figure 8.9 and its denormalised component values are given in Table 8.6.

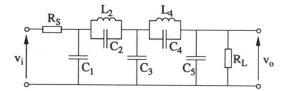

Figure 8.9 Passive fifth order elliptic ladder filter.

Component	Value
C_1	47.3nF
C_2	13.2nF
C_3	62.2nF
C_4	42.7nF
C_5	31.8nF
L_2	40.3nH
L_4	24.2nH
R_S	1Ω
R_L	1Ω

Table 8.6 Denormalised elliptic filter component values.

In order to construct a switched-current simulation of this passive ladder filter (Figure 8.9), the first step is to generate a signal flowgraph description of the structure. This is achieved, using the same techniques as employed for switched-capacitor filters, by writing equations for the state variables of the passive prototype [1]. It is possible to formulate this signal flowgraph in terms of either integrators or differentiators [4], depending on the intended method of implementation. In this case, where bilinear switched-current integrators are to be used, the formulation in terms of integrators is derived and the resulting signal flowgraph is shown in Figure 8.10.

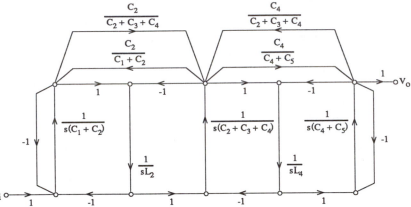

Figure 8.10 Signal flowgraph of fifth order elliptic filter in the s-domain.

The continuous time signal flowgraph (Figure 8.10) can now be converted into the sampled domain by the application of the bilinear transformation $(s \rightarrow (2/T)(1 - z^{-1})/(1 + z^{-1}))$. The 6dB passband attenuation observed in the passive prototype can be removed at the same time, by increasing the weight of the output branch from 1 to 2. The resulting z-domain signal flowgraph is shown in Figure 8.11.

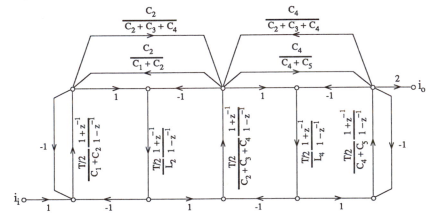

Figure 8.11 SFG of the fifth order switched-current bilinear elliptic filter.

The constant branch weighting coefficients shown in the z-domain signal flowgraph (Figure 8.11) are either constants or design specific parameters, determined by combinations of the clock period (T) and the component values of the passive prototype. Table 8.7 shows these coefficients and their values for the example filter, calculated assuming a clock period of 50ns and the component values given in Table 8.6.

Coefficient	Value
$\dfrac{T/2}{C_1 + C_2}$	0.4129
$\dfrac{T}{2L_2}$	0.6203
$\dfrac{T/2}{C_2 + C_3 + C_4}$	0.2115
$\dfrac{T}{2L_4}$	1.0329
$\dfrac{T/2}{C_4 + C_5}$	0.3355
$\dfrac{C_2}{C_1 + C_2}$	0.2186
$\dfrac{C_2}{C_2 + C_3 + C_4}$	0.1120
$\dfrac{C_4}{C_4 + C_5}$	0.5735
$\dfrac{C_4}{C_2 + C_3 + C_4}$	0.3616

Table 8.7 Elliptic filter coefficients.

The bilinear integration terms shown in the signal flowgraph (Figure 8.11) can now be implemented directly in the switched-current technique using the second generation bilinear integrator structure shown in Figure 8.12, whose transfer function is

$$H(z) = \alpha \frac{1 + z^{-1}}{1 - z^{-1}} \tag{8.6}$$

The multiple scaled output integrators required to implement the signal flowgraph can simply be realised by adding additional output current mirrors to the basic integrator of Figure 8.12. One interesting point to note about the bilinear integrator implementation shown in Figure 8.12, is that additional circuitry is required to generate the necessary complementary outputs, whereas had a fully differential circuit structure been used, these outputs would have been provided naturally.

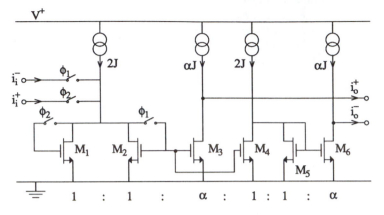

Figure 8.12 Class A bilinear switched-current integrator.

The maximum output current of each class A switched-current integrator in the ladder structure is limited by the applied bias current. Since the maximum output current seen by each integrator, for a given input current, is not necessarily the same, different bias currents are needed in each integrator to maximise efficiency. Alternatively, contour scaling can be applied around the output node of each integrator to ensure that the maximum output current of each of them is equal [1].

Finally, the bilinear switched-current ladder filter should be preceded by a sample-and-hold circuit and enhanced memory cells should be used to obtain satisfactory performance.

The elliptic filter example, described above, was implemented, using the same process and core regulated-cascode memory cell (see Figure 8.3) as the biquadratic filter discussed earlier. The filter used the bilinear integrators shown in Figure 8.12 to implement the signal flowgraph of Figure 8.11, using the coefficient values shown in Table 8.7.

A photograph of the test chip, containing the elliptic filter as well as the class A (see Section 8.2.1) and class AB (see Section 7.5) biquadratic filters and their associated bias circuitry is shown in Figure 8.13. The class AB biquadratic filter section is the long vertical structure on the far left hand side of the die. To the immediate right of this is a similar vertical structure, this is the class A biquadratic filter section. Just above the class AB biquadratic filter are several small structures, these are the bias generators. The elliptic filter consists of the six horizontal sections starting from the top left hand corner of the die. At the bottom of the die are two horizontal structures which are simple test circuits.

The class A bilinear elliptic filter was measured at a clock frequency of 2MHz in order to allow accurate measurements to be made, although higher frequency operation has also been demonstrated [9].

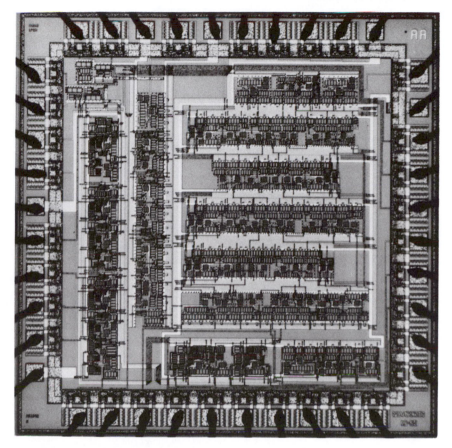

Figure 8.13 Die photograph.

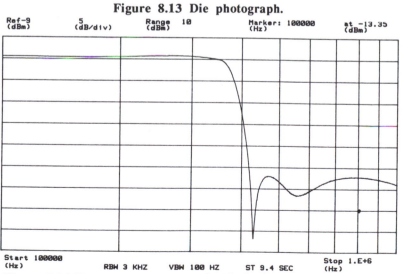

Figure 8.14 Frequency response of elliptic filter (clock = 2MHz).

The frequency response of the elliptic filter operating with a clock frequency of 2MHz is shown in Figure 8.14 and it confirms that the frequency response has the expected form.

To evaluate the accuracy of the frequency response more closely, expanded passband views of the response are shown in Figures 8.15 and 8.16. Figure 8.15 shows the measured gain, the measured gain corrected for $sinc(x)$ distortion and the response of the passive prototype (corrected for its 6dB passband attenuation). In Figure 8.16 the measured response is compared with the simulated response of the transistor level filter circuit (extracted from the layout file).

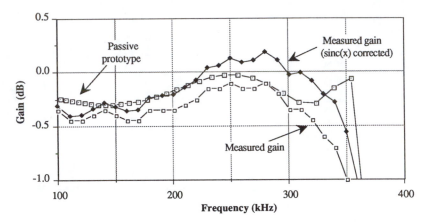

Figure 8.15 Comparison of ideal and measured passband responses of elliptic filter (clock = 2MHz).

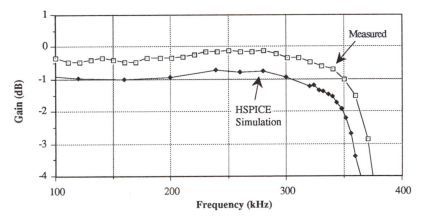

Figure 8.16 Comparison of simulated and measured passband responses of elliptic filter (clock = 2MHz).

The correspondence between the ideal and measured response of the filter (Figure 8.15) is good, except that the measured response exhibits greater passband ripple (0.6dB) and does not roll-off as sharply as the passive prototype. A single cause of these effects cannot be identified, rather it is likely that a combination of error mechanisms is responsible (see Chapter 4). Furthermore, the non-ideal passband response shape was predicted reasonably well by the circuit level simulations shown in Figure 8.16.

The output spectrum of the filter for a 100kHz input signal is shown in Figure 8.17 and from measurements of the spectrum it was found that the noise floor was 60dB below the signal for a THD level of 0.2%. These levels of noise and harmonic distortion are a little worse than had been hoped for, but are consistent with the values obtained for the class A biquadratic filter and the same strategies can be applied to improve performance.

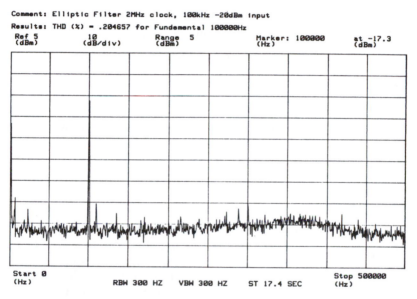

Figure 8.17 Output spectrum of elliptic filter (clock = 2MHz).

The main measured and expected performance characteristics of the elliptic filter are summarised in Table 8.8 (compensation for $sinc(x)$ distortion has been applied). From the performance summary it can be seen that the performance characteristics of the filter fall a little outside their designed values, although they are within the original filter specification in terms of stopband edge, cut-off sharpness and stopband attenuation. Also, because clock generators were not incorporated on the test chip, it was not possible to generate all the required clock phases at very high frequencies and so the maximum clock frequency of the filter could not be determined. In fact in [9] the filter was successfully clocked at 5.75MHz. However, the maximum simulated clock frequency of the filter was 20MHz and we hope to get closer to this with improved measurement procedures.

Performance Characteristic	Value	
	Measured	Expected
Clock frequency	2MHz	-
Passband edge	350kHz	360kHz
Stopband edge	420kHz	412kHz
Passband ripple	0.6dB	0.28dB
Minimum stopband attenuation	26dB	32.3dB
Noise floor (100kHz signal)	-70dB @ THD=0.7%	-
	-60dB @ THD=0.2%	-
Power consumption	28mW	27.4mW
Filter die area	1.5 mm²	-

Table 8.8 Performance summary of elliptic filter.

The measured passband edge is 2.8% lower than its design value of 360kHz, this appears to be mainly due to the lack of sufficient sharpness at the passband edge as mentioned earlier. Another source of error is transistor mismatch, giving rise to small deviations of the filter coefficients from their design values, and hence errors in the filter response. The measured stopband edge is 1.9% higher than its design value of 420kHz, but is still within the 440kHz limit of the original specification (Table 8.5).

8.4 FIR Filters

Finite Impulse Response (FIR) filters are widely used in applications requiring linear phase characteristics or matched filters. The switched-current technique is well suited to the realisation of these filters, since delay elements are simply made by cascading memory cells (see Chapter 3). Practical demonstrations of switched-current FIR filters have been made [13,14] and they confirm both the practicality and simplicity of the approach. In Chapter 10, integrated FIR sections for video frequency applications are described by *Hughes* and *Moulding*.

The basic structure of a switched-current FIR filter, which consists of a series of delay elements whose outputs are weighted and then summed at the output node, is shown in Figure 8.19.

Each z^{-1} delay can be produced by a cascade of two switched-current memory cells and the weighted outputs realised using scaled current mirrors. Overall, the transfer function of a kth order FIR filter is described by

$$H(z) = \sum_{n=0}^{n=k} h_n z^{-n}$$

(8.7)

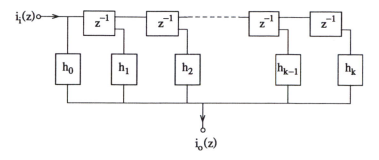

Figure 8.19 FIR filter structure.

8.5 Wave Active Filters

Wave active filter synthesis is an alternative approach to the simulation of passive filters, which is based on the simulation of waves in the filter rather than a set of the node voltages and branch currents as used in conventional ladder synthesis [1]. The approach exploits the use of lossy integrators to realise filter transfer functions.

Wave active filters were originally developed as digital filters [15], but recently a switched-current realisation was proposed by *Rueda*, *Yúfera* and *Huertas* [16,17], and they describe their approach in Chapter 11.

8.6 Transposition of Switched-Capacitor Filters

The synthesis of biquadratic filter sections and ladder filters described previously, has made use of techniques originally developed for the synthesis of switched-capacitor filters. An alternative approach is to transpose the signal flowgraph of an existing switched-capacitor filter, as described by *Roberts* and *Sedra* in Chapter 9 (see also [8]). This approach makes use of the observation that, by transposing the signal flowgraph of a switched-capacitor filter containing multiple-input single-output integrators, a switched-current filter containing single-input multiple-output integrators can be obtained. Furthermore, both the transfer function and sensitivity properties of the two filters are identical. The transposition process itself involves reversing the directions of all the branches and replacing the input/output voltages with output/input currents.

8.7 Switched-Transconductance Filters

A new circuit arrangement, described as a switched-transconductance (ST) circuit, comprising a continuous-time linear tunable transconductor and a discrete time current processor is proposed in this section [18]. This new concept relates to a method of and apparatus for facilitating the accurate and continuously tunable processing of analogue electrical signals in a fully integratable way, without the need for post-manufacture trimming or analogue process options (e.g. double polysilicon). The theoretical basis of the technique is described in this section.

There are two basic ways in which analogue electrical signals can be processed, that is in the continuous or discrete time domains. Continuous time filters, such as the transconductance-capacitor technique [19] offer very high speed potential and are continuously tunable. However, they generally require some form of tuning scheme or trimming to achieve good precision. Complex single or dual loop on-chip tuning schemes are necessary to combat the problem. Furthermore, they cannot generally be realised on a standard digital CMOS process, unless additional analogue process options (such as double polysilicon) are provided.

Discrete time analogue signal processors, such as switched-capacitors or switched-currents, overcome the accuracy difficulties of continuous time processors by relying on the ratio of capacitor values or transistor sizes respectively. However, the maximum signal frequency is much lower than that of their continuous time counterparts for a given technology. Furthermore, it is more difficult to realise continuously tunable filters. Continuous tuning can be achieved by varying the clock frequency, but this approach has the disadvantage of altering the anti-aliasing requirements of the system and requires a relatively complex control loop.

The switched-transconductance technique allows accurate and continuously tunable filters to be realised on a standard digital process, without the requirement for any analogue process options.

8.7.1   Switched-transconductance concept

The basic element of the switched-transconductance (ST) technique is a tunable impedance consisting of a tunable continuous time transconductor with a switched-current processor connected between its output and input as illustrated in Figure 8.20.

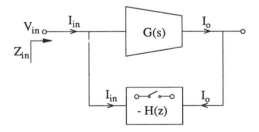

Figure 8.20 Switched-transconductance concept.

If the frequency dependent transconductance gain of the transconductor I_o/V_{in} is $G(s)$ and the transfer function of the switched-current processor I_{in}/I_o is $-H(z)$, then the input impedance of the circuit is

$$Z_{in}(z) = \frac{-1}{G(s)H(z)} \qquad (8.8)$$

Thus the basic switched-transconductance element (Figure 8.20) synthesises a grounded impedance whose value is determined by the value of the transconductor $G(s)$ and the transfer function of the discrete time current processor $H(z)$. By choosing appropriate discrete time functions, $H(z)$, a whole range of impedances can be synthesised, some of which mimic physical components. To illustrate the concept, Table 8.9 shows the switched-current processing functions $H(z)$ that can be used to mimic physical components employing the backward Euler transformation $s \to (1 - z^{-1})$, the forward Euler transformation $s \to (1 - z^{-1})/z^{-1}$ and the bilinear transformation $s \to (2/T)(1 - z^{-1})/(1 + z^{-1})$. In Table 8.9 the transconductance gain G has been taken to be non-inverting, and its cut-off frequency has been assumed to be much higher than the signal frequency, whilst $H(z)$ is inverting; clearly it is possible to reverse this and obtain the same transfer function. Alternatively if $H(z)$ and G have the same sign a negative impedance can be synthesised.

Discrete time current processor	$H(z)$	$Z_{in}(z)$	Equivalent physical component
Scaler	$-A$	$\dfrac{1}{GA}$	Resistor $R = \dfrac{1}{GA}$
Backward Euler integrator	$-\dfrac{A}{1-z^{-1}}$	$\dfrac{1-z^{-1}}{GA}$	Inductor $L = \dfrac{1}{GA}$
Forward Euler integrator	$-\dfrac{Az^{-1}}{1-z^{-1}}$	$\dfrac{1-z^{-1}}{GAz^{-1}}$	Inductor $L = \dfrac{1}{GA}$
Bilinear integrator	$-\dfrac{2A}{T}\dfrac{1+z^{-1}}{1-z^{-1}}$	$\dfrac{T}{2GA}\dfrac{1-z^{-1}}{1+z^{-1}}$	Inductor $L = \dfrac{1}{GA}$
Backward Euler differentiator	$-(1-z^{-1})$	$\dfrac{1}{GA(1-z^{-1})}$	Capacitor $C = GA$
Forward Euler differentiator	$-\dfrac{1-z^{-1}}{z^{-1}}$	$\dfrac{z^{-1}}{GA(1-z^{-1})}$	Capacitor $C = GA$
Bilinear differentiator	$-\dfrac{TA}{2}\dfrac{1-z^{-1}}{1+z^{-1}}$	$\dfrac{2}{GTA}\dfrac{1+z^{-1}}{1-z^{-1}}$	Capacitor $C = GA$

Table 8.9 Impedance synthesis of physical components.

It has been demonstrated above that the switched-transconductance technique can be used to synthesise the grounded impedance of a wide range of components, both physical and imaginary and each component can be tuned by transconductance tuning. The choice of s - z transformation, to realise a physical component, will depend on design specifications.

The use of differential structures allows this technique to be extended to the realisation of identical floating impedances as illustrated in Figure 8.21.

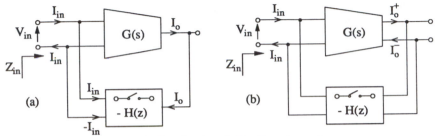

Figure 8.21 Synthesis of floating impedances.

8.7.2 Composite elements

In the previous section the switched-transconductance concept was explained and it was shown that it could be used to synthesise a wide variety of tunable impedances. In this section, the combination of simple switched-transconductance elements to form more complex structures, in which time constants depend on ratios of transconductances and transistor sizes.

The means by which the time constants of composite elements depend on ratios is best explained by considering the simple example of a parallel resistance (R_1) and capacitance (C_2) as shown in Figure 8.22.

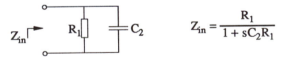

$$Z_{in} = \frac{R_1}{1 + sC_2R_1}$$

Figure 8.22 Parallel RC combination.

From the Figure 8.22 it is obvious that to obtain an accurate time constant requires accurate absolute values of both R_1 and C_2, which is difficult to achieve in practice. However, now consider the case where R_1 and C_2 are replaced by their switched-transconductance equivalents from Table 8.9.

Backward Euler $s \rightarrow z$ mapping

$$Z_{in}(z) = \left[\frac{1}{G_1A_1}\right] \frac{1}{1 + (1 - z^{-1})\left[\frac{G_2A_2}{G_1A_1}\right]} \tag{8.9}$$

Forward Euler $s \rightarrow z$ mapping

$$Z_{in}(z) = \left[\frac{1}{G_1A_1}\right]\frac{z^{-1}}{\dfrac{G_2A_2}{G_1A_1} + z^{-1}\left[1 - \dfrac{G_2A_2}{G_1A_1}\right]} \tag{8.10}$$

Bilinear $s \to z$ mapping

$$Z_{in}(z) = \left[\frac{1}{G_1A_1}\right]\frac{1}{1 + \dfrac{2}{T}\left[\dfrac{1 + z^{-1}}{1 - z^{-1}}\right]\left[\dfrac{G_2A_2}{G_1A_1}\right]} \tag{8.11}$$

Applying their respective $z \to s$ mappings (8.9), (8.10) and (8.11) all yield the following impedance in the s-domain (for signal frequencies much lower than the clock frequency)

$$Z_{in}(s) = \left[\frac{1}{G_1A_1}\right]\frac{1}{1 + s\dfrac{G_2A_2}{G_1A_1}} \tag{8.12}$$

Examining (8.12) it is clear that the pole frequency of the impedance depends on a ratio of transconductances and the ratio of scale factors of the switched-current processor. Since both of these are ratios of similar components they can be accurately controlled, resulting in an inherently accurate time constant without the use of trimming or tuning. The impedance level itself is less well controlled, being the product of a transconductance and a scale factor, but in ladder filter applications this also becomes a ratio [18].

The above discussion has shown how the switched-transconductance technique can be used be used to emulate a parallel resistance and capacitance, with an accurate pole frequency. From this simple example the concept can be extended to encompass a wide range of combinations, to illustrate this the basic series and parallel combinations of a resistor, capacitor and inductor are detailed in Table 8.10.

Probably the most significant advantage of this technique is that a linear floating capacitor can be replaced with a tunable switched-current cell employing pure digital technology.

From Table 8.10, it can be seen that in all the cases shown the pole/zero frequencies of the synthesised impedance depend on the ratio of transconductances and scale factors and can be controlled with a high degree of precision.

8.7.3 *Filter applications*

By combining the basic structures shown in Table 8.10 it is possible to use the switched-transconductance technique to synthesise a wide variety of filter structures and still maintain the pole/zero frequency dependence on ratios.

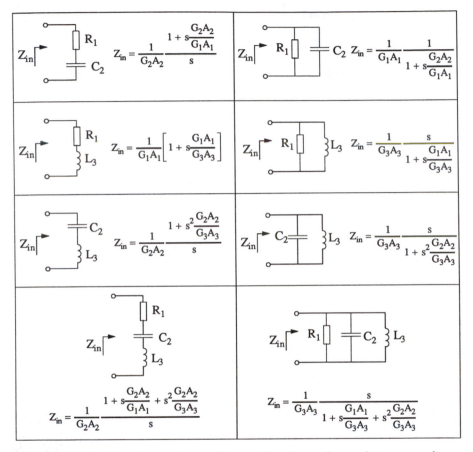

Table 8.10 Basic series and parallel combinations of a resistor, capacitor and inductor.

In addition to the application of switched-transconductance techniques to the synthesis of tunable monolithic passive networks, the technique can usefully be applied to transconductance-capacitor filters [19], active-RC filters [20] and switched-capacitor filters [1]. To demonstrate this, Figure 8.23 illustrates the application of ST techniques in lossless integrators using each of these filter structures.

The advantage of the switched-transconductance technique when applied to transconductance-capacitor and active-RC filters is that precision can be obtained without the need for frequency locked tuning schemes and pure digital process technology can be used. Whilst for the switched-capacitor technique, the advantage is that continuous tuning is made possible and, if the discrete time current processor is a switched-current circuit, the need for linear floating capacitors is eliminated.

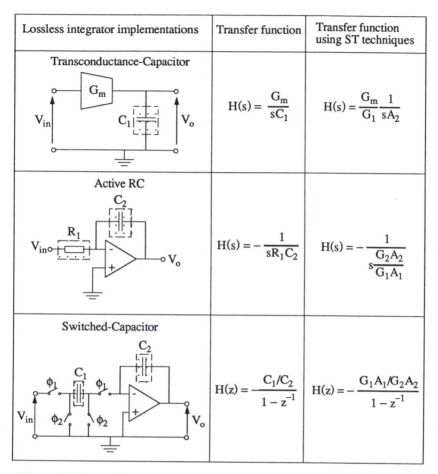

Lossless integrator implementations	Transfer function	Transfer function using ST techniques
Transconductance-Capacitor	$H(s) = \dfrac{G_m}{sC_1}$	$H(s) = \dfrac{G_m}{G_1}\dfrac{1}{sA_2}$
Active RC	$H(s) = -\dfrac{1}{sR_1C_2}$	$H(s) = -\dfrac{1}{s\dfrac{G_2A_2}{G_1A_1}}$
Switched-Capacitor	$H(z) = -\dfrac{C_1/C_2}{1-z^{-1}}$	$H(z) = -\dfrac{G_1A_1/G_2A_2}{1-z^{-1}}$

Figure 8.23 Application of ST techniques to other filter techniques.

In this section we have proposed a new concept, however many practical issues have still to be tackled, in particular the development of tunable current sources for the transconductors.

8.8 Conclusions

This chapter has discussed the synthesis and realisation of switched-current filters based upon the CMOS building blocks and enhanced memory cells described in the previous chapters.

The use of biquadratic filter sections and ladder filter synthesis, as applied to the design of switched-current filters, was discussed with the aid of illustrative examples. As an alternative to the previously proposed switched-current ladder synthesis techniques, a bilinear approach was proposed.

The measured results for the biquadratic section and elliptic filter were very encouraging. The levels of performance obtained exceed, in most respects, those of any previously reported switched-current filters of comparable specification. Performance levels still fall short of those currently attainable by switched-capacitors, although they compare favourably with early switched-capacitor filters. Further development of switched-current circuits is therefore still needed, but the results presented here move switched-current filters a step closer to switched-capacitor performance levels.

The synthesis of FIR filters, wave active filters and the transposition of switched-capacitor designs were briefly described, although these will be discussed in more detail in the next three chapters.

Finally, a new approach to the synthesis of tunable filters on pure digital technology, the switched-transconductance technique, was proposed. This technique combines linear tunable transconductances with switched-current circuit to realise inherently accurate and tunable filter with voltage interfaces. The theoretical basis of this concept has been presented here and work to realise practical filters is currently being pursued.

Acknowledgements

The financial support of the British Science and Engineering Research Council and BNR (Ottawa) Ltd are gratefully acknowledged.

References

[1]   R. Gregorian and G. C. Temes, *Analog MOS integrated circuits for signal processing*, New York: John Wiley & Sons, 1986.

[2]   M. S. Ghausi and K. R. Laker, *Modern Filter Design: Active RC and Switched Capacitor*, Englewood Cliffs, New Jersey: Prentice-Hall Inc., 1981.

[3]   C. Ouslis, M. Snelgrove and A. S. Sedra, "A filter designer's filter design aid: filtorX," in *Proc. IEEE International Symposium on Circuits and Systems*, pp. 376-379, June 1991.

[4]   T.-C. Yu, C.-H. Hsu and C.-Y. Wu, "The bilinear-mapping SC differentiators and their applications in the design of biquad and ladder filters," in *Proc. IEEE International Symposium on Circuits and Systems*, pp. 2189-2192, May 1990.

[5]   J. B. Hughes, *Analogue techniques for large scale integrated circuits*, Ph.D Thesis, University of Southampton, March 1992.

[6]   M. Song, Y. Lee and W. Kim, "A new design methodology of second order switched-current filter," in *Proc. IEEE International Symposium on Circuits and Systems*, pp. 1797-1800, June 1991.

[7]   T. S. Fiez and D. J. Allstot, "CMOS switched-current ladder filters," *IEEE J. Solid-State Circuits*, vol. SC-25, pp. 1360-1367, Dec. 1990.

[8]   N. C. Battersby, *Switched-current techniques for analogue sampled-data signal processing*, Ph.D Thesis, Imperial College of Science, Technology and Medicine (University of London), Feb. 1993.

[9] N. C. Battersby and C. Toumazou, "A 5th order bilinear elliptic filter," in *Proc. IEEE Custom Integrated Circuits Conf.*, San Diego, May 1993.

[10] G. W. Roberts and A. S. Sedra, "Synthesizing switched-current filters by transposing the SFG of switched-capacitor filter circuits," *IEEE Trans. Circuits Syst.*, vol. 38, pp. 338-340, March 1991.

[11] A. C. M. de Queiroz and P. R. M. Pinheiro, "Exact design of switched-current ladder filters," in *Proc. IEEE International Symposium on Circuits and Systems*, pp. 855-858, May 1992.

[12] R. Saal, *Handbook of filter design*, AEG-Telefunken, 1979.

[13] G. Liang and D. J. Allstot, "FIR filtering using CMOS switched-current techniques," in *Proc. IEEE International Symposium on Circuits and Systems*, pp. 2291-2293, May 1990.

[14] J. B. Hughes and K. W. Moulding, "Switched-current video signal processing," *Proc. IEEE Custom Integrated Circuits Conf.*, pp. 24.4.1-24.4.4, May 1992.

[15] A. Fettweis, "Wave digital filters: theory and practice," *Proc. IEEE*, vol. 74, pp. 270-327, Feb. 1986.

[16] A. Rueda, A, Yúfera and J. L. Huertas, "Wave analogue filters using switched-current techniques," *Electronics Letters*, vol. 27, pp. 1482-1483, 1 Aug. 1991.

[17] A. Yúfera, A. Rueda and J. L. Huertas, "Switched-current wave analog filters," in *Proc. IEEE International Symposium on Circuits and Systems*, pp. 859-862, May 1992.

[18] C. Toumazou and N. C. Battersby, "Switched-transconductance techniques: A new approach for tunable, precision analogue sampled-data signal processing," in *Proc. IEEE International Symposium on Circuits and Systems*, Chicago, May 1993.

[19] R. Schaumann and M. A. Tan, "Continuous-time filters," in C. Toumazou, D. G. Haigh and F. J. Lidgey, Eds, *Analogue IC Design: The Current Mode Approach,"* London: Peter Peregrinus Ltd., 1990.

[20] M. Banu and Y. P. Tsividis, "Fully integrated active RC filters in MOS technology," *IEEE J. Solid-State Circuits*, vol. SC-18, pp. 644-651, Dec. 1983.

Chapter 9

A Switched-Capacitor to Switched-Current Conversion Method

Gordon W. Roberts and Adel S. Sedra

9.1 Introduction

During the 1980's the Switched-Capacitor (SC) circuit technique was the dominant analogue sampled-data signal processing technique in any mixed-signal IC. This is still true today; however, one now finds less analogue signal processing on a mixed-signal IC than one did in the past. For the most part, SC circuit techniques are giving way to alternative solutions found using digital signal processing techniques. This has largely been brought on by shrinking transistor dimensions and the advent of area and power efficient A/D and D/A conversion techniques (e.g. oversampled sigma-delta modulators). Although less analogue signal processing is occupying a mixed-signal IC, it has not and cannot be completely eliminated as the real world is largely analogue and requires analogue signals to interface with it. Nevertheless, the manufacturing costs associated with mixed-signal ICs are now dominated by the cost associated with the creation of the analogue portion of the IC. To a large extent, this cost consists of the cost associated with adding extra processing steps to the otherwise standard CMOS process to create the linear floating capacitors required by the SC circuits.

As a means to reduce the dependency of mixed-signal ICs on a non-standard CMOS process and its overall cost, the Switched-Current (SI) technique was recently proposed by *Hughes* et al to perform analogue sampled-data filtering [1] (see also Chapters 3, 8, 10, 11). In contrast to the switched-capacitor (SC) approach, the SI technique can perform accurate signal processing functions in a standard digital CMOS process without the direct use of any capacitor. Moreover, the SI technique does not utilise CMOS op-amps but rather performs all its analogue signal processing with much simpler current mirrors. It is therefore expected that SI circuits will operate over much wider signal bandwidths than present day SC circuits and with less power. Also, the current mode nature of SI circuits should make them less adversely affected by the imminent reduction in power supply voltages. In the end, the cost of manufacturing mixed-signal ICs will be reduced.

A wide array of synthesis methods are currently available for the design of SC filters. The technique proposed in this chapter makes it possible to apply the SC synthesis methods directly to the realisation of SI filters, thus obviating the need for developing new design methods for this emerging filter technology. The basic connection between SC and SI filter circuits is *inter-reciprocity* or simply that one method realises the transposed signal-flow-graph (SFG) of the other [2]. Hence, it is a simple matter to generate an SI filter circuit from an SC filter circuit. Furthermore, since this SFG transposition process does not change sensitivities, all low-sensitivity SC filter circuits can be transformed this way and result in low-sensitivity SI filter circuits. Although we recognise that *Hughes* et al [3] and *Fiez* et al [4] have presented methods for synthesising SI circuits from SC filters, we believe that the approach we propose is more systematic and straightforward.

9.2 The Switched-Current Technique

Precision SC filters are realised using a pair of integrator circuits: one inverting and one non-inverting [5]. In SI technology, corresponding integrator circuits can be constructed with similar transfer functions except the signals involved are currents, not voltages. The basic principle behind these circuits is the current track-and-hold circuit [6] shown in Figure 9.1. Briefly, an input current I_{in} is tracked and delivered to the output for as long as the switch ϕ is closed. However, when the switch is opened, a constant current is delivered to the output at a value equal to the input current at the moment the switch was opened.

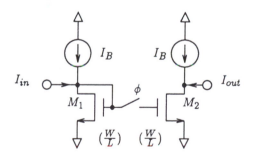

Figure 9.1 A current track-and-hold circuit.

Cascading two track-and-hold stages in a master-slave arrangement, together with an external feedback path, enables present output currents to be combined with past or present inputs to implement the backward or forward Euler approximation to the integration operation [1]. For instance, in Figure 9.2(a) we show an SI circuit that implements the forward Euler integration operation. Assuming that the input is held over a full-period beginning on ϕ_2, it is straightforward to write a set of difference equations in terms of the circuit currents to show that the transfer function to the first output I_{ol} is

$$\frac{I_{ol}}{I_{in}}(z) = K_1 \frac{1}{z-1} \tag{9.1}$$

which can be written in the alternate form as

$$\frac{I_{ol}}{I_{in}}(z) = K_1 \frac{z^{-1/2}}{z^{1/2} - z^{-1/2}} \tag{9.2}$$

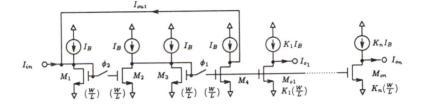

(a)

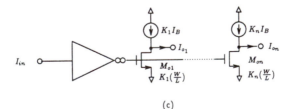

(b)

(c)

Figure 9.2 (a) A multiple-output SI non-inverting integrator circuit, (b) An SFG representation of the circuit shown in (a), (c) Short-hand notation,

The above transfer function is identical to the transfer function for the non-inverting SC integrator and will therefore be referred to in this chapter as the non-inverting SI integrator. In the SC literature, it is common to find the transfer function in (9.2) expressed in terms of the Lossless Digital Integrator (LDI) complex-frequency variable γ. With γ defined as follows

$$\gamma = \frac{1}{2}\left(z^{1/2} - z^{-1/2}\right) \tag{9.3}$$

we can rewrite the transfer function of the non-inverting integrator as

$$\frac{I_{o1}}{I_{in}}(z) = K_1 \frac{z^{-1/2}}{2\gamma} \tag{9.4}$$

Now if we ignore the excess phase factor $(z^{-1/2})$ we see that the above transfer function takes on a form in the γ-plane that is similar to that of an ideal integrator in the s-plane. We choose this notation here as it slightly simplifies our presentation later on. The transfer function from the input to the other outputs associated with the integrator circuit of Figure 9.2(a) have an identical form as that shown in (9.2) or (9.4), except that the scale constant (K_1) will be different depending on the ratio of the area of the output transistor to the area of transistor M_4. Graphically, we can represent the flow of signals from the input to the multiple outputs of this circuit with the SFG shown in Figure 9.2(b). To simplify our diagrams we shall denote this integrator circuit with the symbol or short-hand notation shown in Figure 9.2(c). The triangle with two circles represents the sample-and-hold portion of the integrator (i.e. transistors M_1, M_2, M_3 and M_4). We attach two small circles at the output of this triangle to denote that it is a non-inverting integrator.

An inverting SI integrator circuit is realised with exactly the same circuit structure as the non-inverting integrator; however, the output is taken from the other track-and-hold circuit as is depicted in Figure 9.3(a). The transfer function from the input to one output, for instance I_{o1}, is

$$\frac{I_{o1}}{I_{in}}(z) = -K_1 \frac{z}{z - 1} \tag{9.5}$$

or

$$\frac{I_{o1}}{I_{in}}(z) = -K_1 \frac{z^{1/2}}{2\gamma} \tag{9.6}$$

The SFG representation for the inverting integrator is shown in Figure 9.3(b) and its short-hand notation is provided in Figure 9.3(c). A single circle is attached to the triangular symbol to denote that it is an inverting integrator.

9.3 Comparing the SFGs of SI and SC Integrator Circuits

An important relationship exists between a multiple-output SI integrator and a multiple-input SC integrator of the same polarity. To see this, we show in Figure 9.4(a) a multiple-input SC non-inverting integrator circuit, and in Figure 9.4(b) its SFG. Now, if we reverse the direction of all the

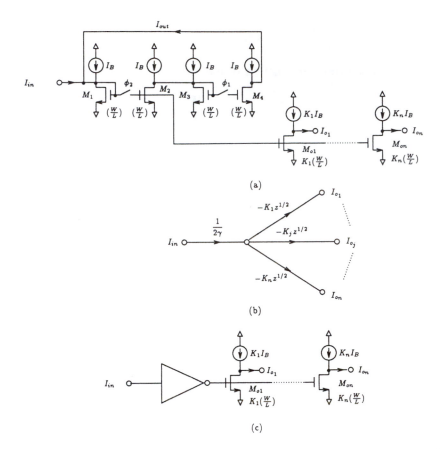

Figure 9.3 (a) A mulitple-output SI inverting integrator circuit, (b) An SFG representation of the inverting SI integrator, (c) Short-hand notation.

branches in this SFG while maintaining the same branch transmittances and convert the summing nodes into branch nodes, and visa-versa, then the SFG shown in Figure 9.4(c) results. Observe that in this transposed SFG the transfer function V_{i1}/V_o is the same as the corresponding transfer function V_o/V_{i1} in the original SFG; V_{i2}/V_o is the same as V_o/V_{i2} and so on. Also observe that the SFG of Figure 9.4(c) is identical to that in Figure 9.2(b) provided that

$$K_1 = C_1/C, \quad K_2 = C_2/C, \ etc.,$$

and that the following signal correspondences are made:

$$V_o <=> I_{in}, \quad V_{i1} <=> I_{o1}, \ etc.$$

Thus, from the theory of SFGs [7], we can state that the SFG for a multiple-input SC non-inverting integrator is equivalent to the transpose of

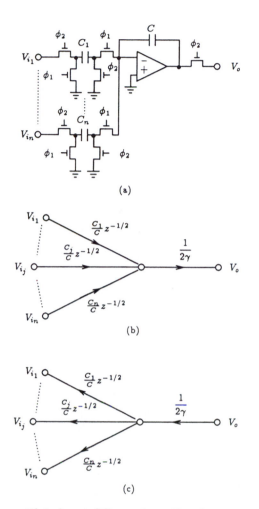

Figure 9.4 (a) A multiple-input SC non-inverting integrator circuit, (b) An SFG representation of the circuit in (a), (c) A transposed version of the SFG shown in (b)

the SFG representation of a multiple-output SI non-inverting integrator. Clearly, the same holds true for the inverting integrator case.

The procedure just described is known as *SFG transposition*. Although, this procedure can be applied to any SFG, we are particularly interested in its application to the SFG of SC filters constructed of multiple-input single-output integrators of the type shown in Figure 9.4(a) (and of course the inverting type). We have shown in the above that the transpose of the SFG of each SC integrator is an SFG of a single-input multiple-output SI integrator. Now, since the transposed SFG will have the same input to output transfer function as that of the original SFG, it constitutes a realisation of the original filter. It follows that by realising the individual integrator SFGs with SI integrator circuits we obtain a SI realisation of the filter. Thus we have a method for converting any integrator-based SC filter realisation into an SI filter realisation.

In the network theory literature, two networks that are topologically equivalent and that realise the same transfer function when the excitation and response of one network are interchanged with the response and excitation of the other are said to be *inter-reciprocal* [8]. We can therefore conclude that SI circuits generated from SC circuits using SFG transposition form inter-reciprocal pairs. As a consequence of their inter-reciprocity, they will also possess identical component sensitivities [9]. This is important because it suggests that the transformation of low sensitivity SC filter circuits will lead to low sensitivity SI filter circuits.

In order to illustrate our design procedure we show in Figure 9.5(a) a third-order SC filter circuit constructed from three single-ended lossless digital integrators, or which have come to be known as LDI integrators. The input is applied to the first integrator consisting of op-amp A_1 and the output is taken as the voltage that appears at the output of the third op-amp A_3. For the time being we shall not assign any particular value to the capacitors, instead we shall keep the discussion as general as possible by using symbolic representations. The corresponding SFG for this circuit is shown in Figure 9.5(b) with the input and output points clearly indicated. Based on the procedure just outlined, it is a simple matter to transform the SFG shown in Figure 9.5(b) to one that can easily be realised with SI integrator circuits. To accomplish this we first transpose the SFG of the SC circuit shown in Figure 9.5(b) into that shown in Figure 9.5(c). Notice from this SFG that the integrators pointing downwards on the page together with the branches that span outwards from these integrators, with transmittances proportional to $z^{-1/2}$, can be replaced with the single-input multiple-output SI non-inverting integrators such as that of Figure 9.2(a). Conversely, the integrators pointing upwards, together with fan-out branches with transmittances proportional to $z^{1/2}$ can be replaced with single-input multiple-output SI inverting integrators such as that of Figure 9.3(a). On doing so, the circuit shown in Figure 9.5(d) results. We have simplified the circuit schematic by representing the sample-and-hold portion of each SI integrator by its short-hand notation (see Figure 9.2(c) and 9.3(c)).

Now at this stage of the development we would like to verify the result of the synthesis procedure. To do so, let us consider the capacitor values (in terms of a unit-sized capacitor) shown in Table 3.1 for a SC circuit implementing a third-order Chebyshev lowpass filter function having a 1 dB passband ripple between dc and 1kHz, and a stopband having no less than 25dB of attenuation beginning at 2.5kHz. The clocking rate is assumed to be equal to 100kHz. If the synthesis is correct then indeed the SI filter circuit should meet these specifications. To check this, we make use of the SI circuit simulation program ASIZ recently developed by de

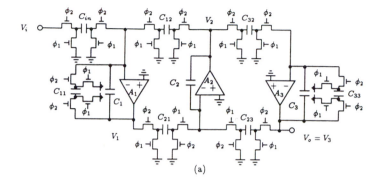

(a)

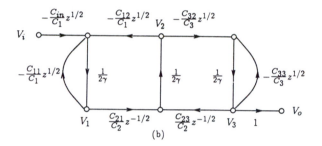

(b)

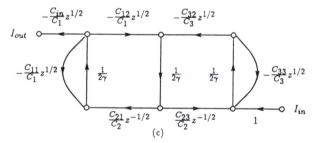

(c)

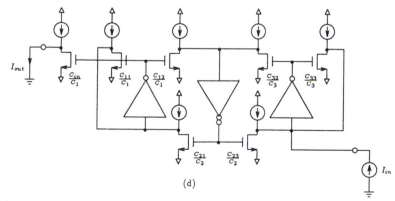

(d)

Figure 9.5 (a) A 3rd-order Chebyshev lowpass SC filter circuit, (b) An SFG representation of the circuit in (a), (c) A transposed version of the

SFG shown in (b), (d) An SI circuit implementation of the SFG shown in (c).

Queiroz et al from the Universidada Federal do Rio de Janeiro [10]. A detailed description of ASIZ and its capabilities are described in Chapter 19 of this text. Simulating our SI circuit leads to the frequency response results shown in Figure 9.6. The graph on the left plots the frequency response over the passband region between dc and 1kHz. Here we see that the attenuation varies between 0 and 1dB, precisely as specified. The right-hand side graph gives a view of both the passband and stopband regions of the filter response between 0 and 5kHz. Here we see that for frequencies above 2.5kHz the attenuation exceeds the required attenuation of 25dB. Clearly then, this SI filter circuit has a frequency response that meets the required frequency specifications.

3rd-Order SC Filter Circuit Capacitor Values		
Cin = 1.5609	C21 = 1.2813	C32 = 2.0824
C11 = 1.0	C23 = 1.0	C33 = 1.0
C12 = 1.6252	C2 = 32.952	C3 = 31.712
C1 = 31.712		

Table 9.1 3rd-order SC filter circuit - capacitor values.

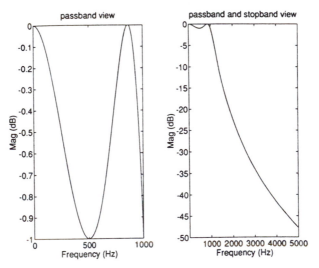

Figure 9.6 Comparing the frequency response of an SI circuit implementation of a 3rd-order Chebyshev filter with its corresponding SC circuit implemtation. Absolutely no difference is evident.

Furthermore, we have superimposed in Figure 9.6 the frequency response of the SC circuit obtained using SWITCAP [11]. Since no differences are evident we can conclude that both the SI and SC circuits behave in exactly the same way. Thus, this example confirms our proposed synthesis procedure.

9.4 SI Filter Circuits with Finite Transmission Zeros

The above example demonstrated the creation of a third-order SI filter circuit from a well-known LDI SC filter circuit composed of several two-integrator-loops. Extending the approach to higher-order SC LDI filter circuits composed of two-integrator-loops should therefore be straightforward. Although important in principle, this class of filter circuit has all of its transmission zeros at infinite frequency and as a result realises filter functions with very limited selectivity (i.e., restricted to a class of all-pole filter functions such as Butterworth or Chebyshev). To realise a more selective frequency response (e.g. an Elliptic filter function) additional cross-coupling branches are incorporated into the basic two-integrator-loop structure, thereby forming a set of finite transmission zeros. One such SC filter realisation used to implement a fifth-order Elliptic filter function is shown in Figure 9.7. As can be seen in the schematic, the SC circuit consists of various two-integrator-loops and several unswitched capacitors. It is the presence of these unswitched capacitors that forms the finite transmission zeros. The additional feed-in branch connecting the sample-and-hold output to the first integrator is necessary to provide an *exact* realisation of (as opposed to an approximation to) the intended transfer function. Interested readers can consult [5] for further details on the particulars of synthesising SC circuits in this way.

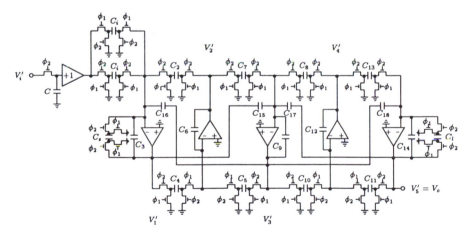

Figure 9.7 A fifth-order SC filter circuit having finite transmission zeros.

To create a SI implementation of the same transfer function implemented by the SC circuit of Figure 9.7, we first must create the SFG of the SC circuit. This is shown in Figure 9.8(a). The basic structure of this SFG is similar to that seen previously in Figure 9.5(b) with several exceptions. The most notable ones are the branches that interconnect the various integrators having transmittances proportional to γ. These branches correspond to the unswitched capacitors in the SC circuit. In addition, one other branch that did not appear in the SFG seen previously in Figure 9.5 is the branch $(C_i/C_3)z^{-1/2}$ that connects the output of the sampled-and-hold to the first integrator. Up to this point we have not yet seen any SI circuits that realise these transmittances; however, as we shall see below, these branches pose no difficulty for SI technology.

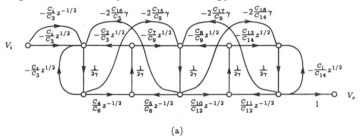

(a)

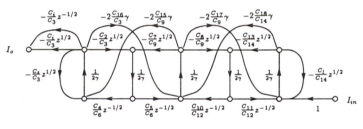

(b)

Figure 9.8 (a) An SFG representation of the 5th-order SC circuit shown in Figure 9.7, (b) A transposed version of the SFG shown in (a).

Let us proceed and transpose the SFG of Figure 9.8(a). The result is shown in Figure 9.8(b). On inspection, we see that each downward pointing integrator branch, together with the $z^{-1/2}$ branches that fan out from its output, can simply be realised with the multiple-output non-inverting SI integrators of the type shown in Figure 9.2(a). In contrast, the integrators that point upwards have various types of transmittances that fan outwards. Only the branches that have transmittances proportional to $z^{1/2}$ can be replaced with the multiple-output inverting SI integrator of Figure 9.3(a). In order to realise the remaining branch transmittances, we extend the structure of the basic SI integrator so that it includes these additional transmittances. This is shown in Figure 9.9(a). It consists of a single current input and three different types of outputs: an inverting integrator, a non-inverting integrator and an inverting amplifier. The transfer function from the input to each output is found by writing a set of difference equations, and transforming the result into the z-domain. The results are:

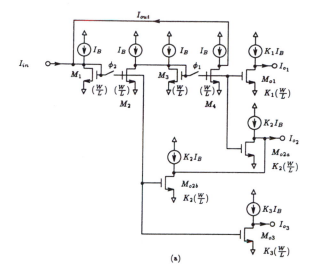

(a)

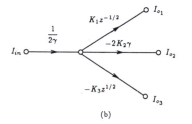

(b)

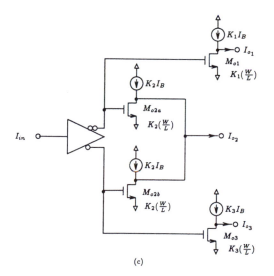

(c)

Figure 9.9 (a) General first-order SI building block, (b) SFG representation, (c) Short-hand notation.

$$\frac{I_{o1}}{I_{in}}(z) = K_1 \frac{1}{z-1} = K_1 \frac{z^{-1/2}}{2\gamma}$$

$$\frac{I_{o2}}{I_{in}}(z) = -K_2$$

$$\frac{I_{o3}}{I_{in}}(z) = -K_3 \frac{z}{z-1} = -K_3 \frac{z^{1/2}}{2\gamma}$$

Extension to multiple outputs of each type is achieved by simply attaching additional current mirror transistors. Inverted versions of any one of these outputs can be obtained by simply placing an additional current mirror in cascade with the output. The SFG representation of this general first-order SI building block is shown in Figure 9.9(b). We denote the sample-and-hold portion of the building block with another triangular symbol having three separate outputs as shown in Figure 9.9(c). Now that we can realise all branches of the SFG in Figure 9.8(b), we can replace each one of them with the appropriate SI circuit and generate the fifth-order SI filter circuit shown in Figure 9.10.

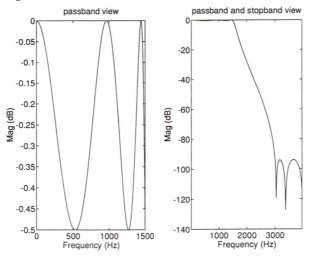

Figure 9.10 A fifth-order SI filter circuit. This circuit was obtained by transposing the SFG of the SC filter circuit shown in Figure 9.7 and replacing the appropriate branches of the transposed SFG with SI circuits.

To verify the design, let us simulate the resulting SI filter circuit shown in Figure 9.10 using the SI simulator ASIZ for the following SC filter case. A fifth-order SC filter circuit is designed using the synthesis method described in [5] to have an Elliptic transfer function with a passband edge located at 1.5kHz and a stopband edge at 3kHz. The passband is designed for an attenuation ripple of 0.5dB, and in addition, the filter is designed for a dc gain of unity. The stopband attenuation is required to exceed 60dB.

Finally, the clock rate is assumed equal to 8kHz. The capacitor values (in terms of a unit-sized capacitor) are provided in Table 3.2. The frequency response of the SI filter circuit is shown in Figure 9.11. As is clearly evident, the frequency response of the SI circuit is Elliptic and satisfies the desired specifications. Moreover, as expected, when we compare the frequency response to that of the SC circuit, as computed using SWITCAP, we get exactly the same response.

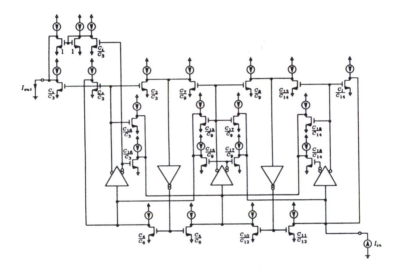

Figure 9.11 Frequency response behaviour of the fifth-order SI circuit of Figure 9.10.

5th-Order Elliptic Filter Circuit Capacitor Values			
$C_i = 1.0$	$C_s = 1.0$	$C_2 = 1.0$	$C_3 = 1.0546$
$C_4 = 1.0$	$C_5 = 1.0$	$C_6 = 0.9062$	$C_7 = 1.0$
$C_8 = 1.0$	$C_9 = 2.4729$	$C_{10} = 1.0$	$C_{11} = 1.0$
$C_{12} = 0.8806$	$C_{13} = 1.0$	$C_{14} = 1.0635$	$C_{15} = 0.2927$
$C_{16} = 0.2927$	$C_{17} = 0.3281$	$C_{18} = 0.3281$	$C_1 = 1.0$

Table 3.2 5th-order SC elliptic filter circuit - capacitor values.

9.5 SI Filter Circuits using Bilinear Integrators

In contrast to the SC filter implementations seen thus far, consisting largely of two-integrator-loops made from LDI integrators, another important class of filter realisations can be formed from two-integrator-loops of bilinear integrators. The major advantage of such an approach is that it enables SC filter circuits to be synthesised *directly* using classical active-RC filter synthesis methods. Based on the method of this chapter, these filter circuits can also be transformed into equivalent SI circuits [12] (see also Chapter 8).

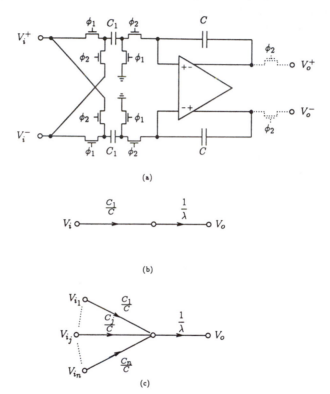

(a)

(b)

(c)

Figure 9.12 (a) A fully-differential SC bilinear integrator, (b) SFG for the circuit of (a), (c) The SFG of a multiple-input bilinear SC integrator circuit.

In Figure 9.12(a) we present the fully-differential SC bilinear integrator with an input-output transfer function $T(z)$ given by

$$T(z) = \frac{V_o{}^+ - V_o{}^-}{V_i{}^+ - V_i{}^-} = \frac{C_1}{C} \frac{z+1}{z-1}$$

or, more conveniently written as

$$T(z) = \frac{C_1}{C} \frac{1}{\lambda}$$

where λ is known as the bilinear transform variable. The corresponding SFG of this circuit is shown in Figure 9.12(b). Extending the circuit structure of Figure 9.12(a) to multiple-inputs is straightforward; additional switched-capacitor feed-in branches are simply connected to the two input summing nodes of the op-amp. The SFG for a bilinear integrator with multiple-inputs is provided in Figure 9.12(c). Based on well-known active-RC filter synthesis methods, high-order filter realisations can be created by interconnecting multiple sets of these two-integrator-loops of bilinear integrators. Figure 9.13 illustrates one such example. In this particular case we present the circuit schematic of a fifth-order fully-differential SC lowpass filter. This circuit is not capable of forming finite transmission zeros, as no unswitched capacitor branches are present. This type of filter circuit would be used to realise an all-pole filter function such as Butterworth or Chebyshev. The SFG corresponding to this circuit is provided in Figure 9.14(a). Transposing this SFG leads to that shown in Figure 9.14(b). To obtain a SI implementation of this SFG requires that we find a single-input multiple-output SI bilinear integrator circuit. Fortunately, this is an easy task.

Consider that the transfer function for a bilinear integrator is given by

$$T(z) = K \frac{z+1}{z-1}$$

where K is an arbitrary constant. Realisation of this transfer function may be facilitated by decomposing $T(z)$ into two separate parts,

$$T(z) = K \frac{z}{z-1} + K \frac{1}{z-1}$$

or, more conveniently written using the LDI variable as

$$T(z) = K \frac{z^{1/2}}{2\gamma} + K \frac{z^{-1/2}}{2\gamma}$$

On inspection, we see that the bilinear transfer function is the sum of the transfer function of a non-inverting LDI integrator and the transfer function of an inverting LDI integrator where both have the same sign. Based on this observation we can form two different SI implementations of the bilinear integrator by simply adding the outputs of the appropriate integrators. These are shown in Figure 9.15.

The circuit of Figure 9.15(a) implements a positive version of the bilinear integrator. This is achieved by simply inverting the output current of the inverting integrator with a simple current mirror circuit and adding it to the output current of the non-inverting integrator. The SFG corresponding to this integrator is shown in Figure 9.15(b). Following along a similar line of thought, the circuit of Figure 9.15(c) implements a negative version of the bilinear integrator. In this case, the output current of the non-inverting integrator is first inverted before combining with the output current of the inverting integrator. Forming additional outputs would be realised by simply adding additional current mirror transistors to the output of the memory circuit in the exact same arrangement as that shown.

Now that the SI bilinear integrator implementation is known we can replace each of the branches of the SFG of Figure 9.14(b) with either the positive or negative version of the SI bilinear integrator and arrive at the SI circuit implementation. On doing so the circuit of Figure 9.16 results.

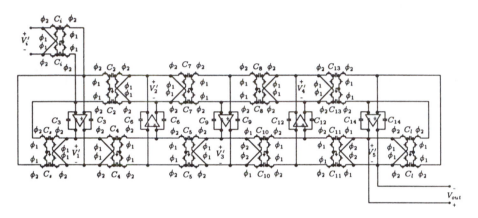

Figure 9.13 A fifth-order SC filter circuit based on bilinear integrators.

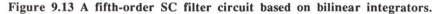

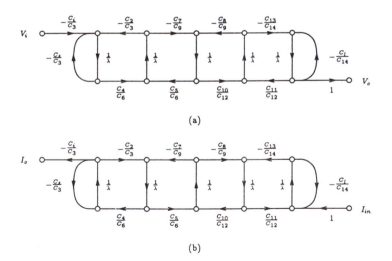

(a)

(b)

Figure 9.14 (a) An SFG representation of the 5th-order SC circuit shown in Figure 9.13, (b)A transposed version of the SFG shown in (a).

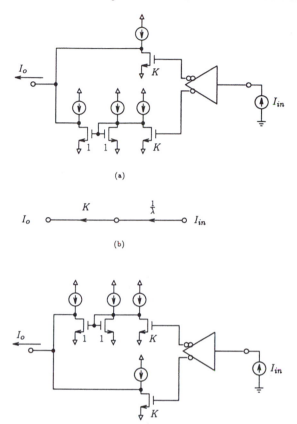

(a)

(b)

(c)

Figure 9.15 (a) A positive SI implementation of a bilinear integrator, (b) SFG of the integrator shown in (a), (c) A negative SI implementation of a bilinear integrator.

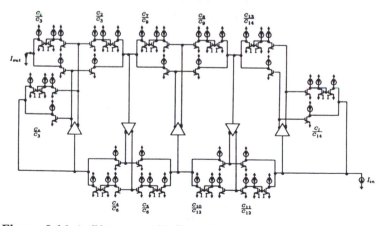

Figure 9.16 A fifth-order SI filter implementation using bilinear integrators.

To verify its frequency response, we simulated its operation using the SI simulation program ASIZ for the case of a fifth-order Chebyshev filter function having a passband attenuation ripple of 1dB, a passband edge located at 1kHz and operated at a clock frequency of 10kHz. The unit-sized capacitor values used in this design are given in Table 3.3. The frequency response of the SI circuit is shown in Figure 9.17. When compared to the frequency response of the corresponding SC circuit, obtained through a SWITCAP simulation, we find identical results.

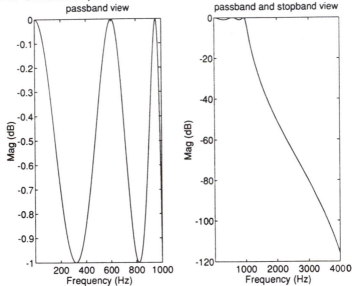

Figure 9.17 Simulated frequency response behaviour of the fifth-order SI filter circuit of Figure 9.16.

5th-Order Bilinear Filter Circuit Capacitor Values			
$C_i = 1.0$	$C_s = 1.0$	$C_2 = 1.0$	$C_3 = 6.5689$
$C_4 = 1.0$	$C_5 = 1.0$	$C_6 = 3.3573$	$C_7 = 1.0$
$C_8 = 1.0$	$C_9 = 9.2336$	$C_{10} = 1.0$	$C_{11} = 1.0$
$C_{12} = 3.3573$	$C_{13} = 1.0$	$C_{14} = 6.5689$	$C_1 = 1.0$

Table 9.3 5th-order bilinear filter circuit - capacitor values.

9.5 Conclusions

This chapter presents a general method for synthesising SI filter circuits using techniques already developed for synthesising SC filter circuits. The method consists of transposing the SFG of a SC filter circuit and realising the single-input multiple-output integrators in the resulting SFG with SI integrators. The resulting SI circuits realise the same transfer function as the SC filter circuit and, in addition, possess identical component sensitivities. As was demonstrated, all types of filter functions are realisable by this method, both all-pole filter functions and those containing finite transmission zeros.

Acknowledgement

The work in this chapter was supported by NSERC and by Micronet, a Canadian federal network of centres of excellence in Microelectronic Devices, Circuits and Systems for Ultra Large Scale Integration.

References

[1] J. B. Hughes, N. C. Bird and I. C. Macbeth, "Switched currents - a new technique for analog sampled-data signal processing," in *Proc. IEEE International Symposium on Circuits and Systems*, pp. 1584-1587, May. 1989.

[2] G. W. Roberts and A. S. Sedra, "Synthesizing switched-current filters by transposing the SFG of switched-capacitor filter circuits," *IEEE Trans. Circuits and Systems*, vol. 38, pp. 337-340, March 1991.

[3] J. B. Hughes, I. C. Macbeth and D. M. Pattullo, "Switched current filters," *IEE Proceedings*, Part G, Special issue on current-mode analogue signal processing circuits, vol. 137, pp. 156-162, April 1990.

[4] T. Fiez and D. Allstot, "A CMOS switched current filter technique," in *IEEE International Solid-State Circuit Conference Digest of Technical Papers*, pp. 206-207, Feb. 1990.

[5] A. S. Sedra, "Switched-capacitor filter synthesis," in Y. Tvisidis and P. Antognetti, Eds, *MOS VLSI Circuits for Telecommunications*, Englewood Cliffs, N.J.: Prentice-Hall, 1985.

[6] E. A. Vittoz, "Dynamic analog techniques," in Y. Tvisidis, and P. Antognetti, Eds, *MOS VLSI Circuits for Telecommunications*, Englewood Cliffs, N.J.: Prentice-Hall, 1985.

[7] S. J. Mason and H. J. Zimmermann, *Electronic Circuits, Signals, and Systems*, New York: John Wiley & Sons, Inc., 1960.

[8] J. L. Bordewijk, "Inter-reciprocity applied to electrical networks," *Applied Science Research*, vol. B6, pp. 1-74, 1956.

[9] A. V. Oppenheim and R. W. Schafer, *Digital Signal Processing*, Englewood Cliffs, New Jersey: Prentice-Hall, 1975.

[10] A. C. M. de Queiroz, P. R. M. Pinheiro and L. P. Calôba, "Systematic nodal analysis of switched-current filters," in *Proc. IEEE International Symposium on Circuits and Systems*, pp. 1801-1804, May. 1991.

[11] S. C. Fang, Y. P. Tsividis and O. Wing, "SWITCAP - A switched-capacitor network analysis program Part 1: Basic features," *IEEE Circuits and Devices Magazine*, pp. 4-10, Sept. 1983.

[12] I. Song and G. W. Roberts, "A 5th order bilinear switched-current Chebyshev filter," in *Proc. IEEE International Symposium on Circuits and Systems*, May 1993.

Chapter 10

Switched-Current Video Signal Processing

John B. Hughes and Kenneth W. Moulding

10.1   Introduction

The switched-current (SI) technique is ideally suited to mixed analogue/digital IC's implemented in standard digital VLSI CMOS processing. This chapter builds on Chapters 3, 4, and 6, developing the technique to enable engineering of video frequency signal processors[1].

The high frequency capability of SI circuits arises from the simplicity of the basic current memory (Figure 10.1). Simplistically, if the switch resistance is assumed to be much lower than $1/g_m$ and the drain capacitance is much lower than C, the cell has a first order settling behaviour and the transistor's drain current rises monotonically with a time constant $\tau = C/g_m$. The settling time is determined by the loop bandwidth, f_{CO} , where

$$f_{CO} = \frac{3\mu(V_{gs} - V_T)}{4\pi L^2} \qquad (10.1)$$

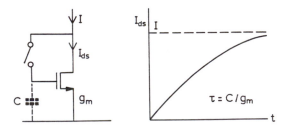

Figure 10.1 Basic current memory and settling behaviour.

in the usual notation. For $(V_{gs} - V_T) = 1$V and $L = 1\mu$m, $f_{CO} = 2.7$GHz and the 0.1% settling time is 0.5ns. This enables filtering at frequencies up to a few 100MHz with clock frequency up to say 1GHz. In practice, such high frequency operation will prove difficult to achieve because of the memory cells' other parasitics and because the memory cells usually include

---

[1] Some of the material in this chapter forms part of a recent journal publication [1]

feedback circuits to improve signal transmission accuracy. Nevertheless, it is the purpose of this chapter to demonstrate that signal processing at video frequency is comfortably within the scope of the SI technique. Although Chapter 21 describes how Gallium Arsenide technology may also be employed to achieve high frequency operation it is the intention of this chapter to consider the speed capabilities of Silicon.

10.2   Basic Signal Processing Cells

In this section, SI cells are developed using only basic circuit structures, since circuit improvements for enhancing analogue performance are already detailed in Chapter 6.

10.2.1   Delay line

The memory cell is a fundamental building block for signal processing. It may be used to generate SI integrators or differentiators for state-variable and leapfrog filters or used directly in delay lines for FIR filter structures [2,3]. Figure 10.2 shows the basic cell.

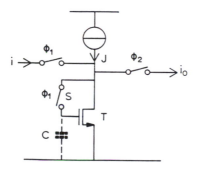

Figure 10.2 Basic SI memory cell.

A delay may be generated by cascading pairs of memory cells, each pair forming a unit delay, as shown in Figure 10.3. In principle, it operates from a bi-phase clock (ϕ_1,ϕ_2) and the phase is reversed on alternate memory cells. *2N* memory cells are required to produce a delay of *N* clock periods *(NT)* and consequently the transmission error resulting from non-ideal behaviour of each memory cell and noise power suffer a *2N*-fold increase. The output is available for only one clock phase (i_{01}) unless a mirrored output is used (i_{02}). With *N*=1, the delay line is a non-inverting unit delay cell or, with a continuous input signal, it is a sample-and-hold circuit. The *z*-domain transfer characteristic is given by,

$$H(z) = z^{-N} \tag{10.2}$$

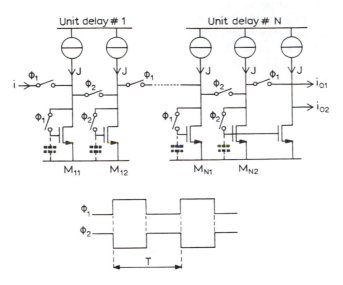

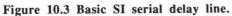

Figure 10.3 Basic SI serial delay line.

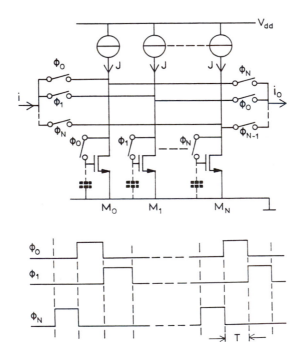

Figure 10.4 Basic SI parallel delay line.

An alternative approach is shown in Figure 10.4. The circuit is an array of $N+1$ paralleled memory cells and is operated from the $N+1$ clock phases shown. On clock phase ϕ_0 , memory cell M_0 receives an input signal while cell M_1 delivers its output. Similarly, on clock phase ϕ_1 , cell M_1 receives input signal while cell M_2 delivers its output. This continues until after cell M_N has received its input signal and cell M_0 has delivered its output signal, and then the cycle repeats. Clearly, each cell delivers its output signal current immediately before its next input and N periods *(NT)* after its previous output phase. With $N=1$, the delay line is an inverting unit delay cell or, with a continuous input signal, it is a sample-and-hold circuit. Note that each memory cell is neither receiving nor delivering signals for $N-1$ clock phases of the cycle. At these times the voltages at the drains of the memory transistors change to produce a match between each bias current (in practice, generated by a PMOS transistor) and the current held in its associated memory transistor. This does not cause an error in a memory cell's output signal so long as the voltage is restored on its output phase. The z-domain transfer characteristic is given by,

$$H(z) = -z^{-N} \tag{10.3}$$

The delay line output is available for the whole clock period without the need for mirroring and the settling interval is double that of the serial delay line. The transmission error and signal-to-noise ratio are in principle the same as those of the memory cell. However, the parallel nature of the structure gives rise to two extra potential sources of error: one resulting from unequal path gains and the other from non-uniform sampling [4]. The path gains will be very close in practice as transmission accuracy is not achieved through component matching. Non-uniform sampling is likely to be of greater importance, especially if the cell is used for the sample-and-hold function.

10.2.2   *Integrator and differentiator cells*

Integrators and differentiators can be derived from cascaded pairs of memory cells and may then be used to synthesise filters from cascaded biquadratic sections or directly by the leapfrog approach. Basic integrator cells have already been described fully in Chapter 3 and will not be repeated here. A practical integrator cell is described in a following section (Figure 10.10).

10.2.3   *FIR cells*

Elemental FIR filter sections are shown in Figure 10.5. Cascading these can produce low-pass, high-pass or band-pass responses. A basic circuit for implementing an FIR cell of the type in Figure 10.5(a) is shown in Figure 10.6. It consists of an input mirror circuit which produces two equal half-amplitude inverted replicas of the input signal current and an inverting parallel delay line $(-z^{-N})$ of the type described. One replica is delayed and inverted by the delay line, and then added to the other replica. The other FIR cell (Figure 10.5(b)) requires an extra inversion unless fully-differential signal processing is used, as will be illustrated later.

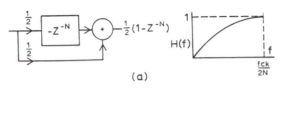

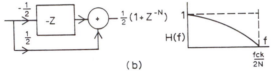

Figure 10.5 Elemental FIR sections.

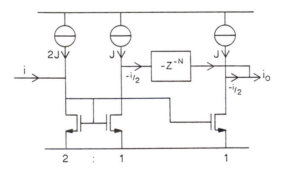

Figure 10.6 Basic SI FIR cell.

10.3 Practical SI Signal Processing Cells

The basic cells described so far do not give adequate analogue performance for two main reasons; channel-shortening effects and charge injection (Chapter 4), and Chapter 6 has already discussed ways to combat these errors.

10.3.1 Active negative feedback

The grounded gate voltage amplifier active memory configuration [5] (Figure 10.7) gives higher speed potential than the op-amp memory configuration (as discussed in Chapter 6).

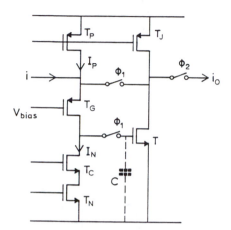

Figure 10.7 SI memory cell with grounded-gate feedback.

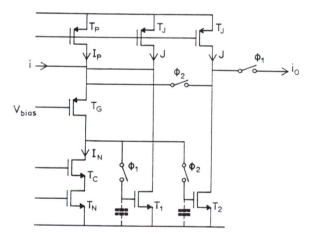

Figure 10.8 Serial unit delay cell ($N=1$) with grounded-gate feedback.

A serial unit delay cell employing the grounded gate feedback configuration is shown in Figure 10.8. It can be seen that the feedback amplifier is multiplexed between memory transistors T_1 and T_2 to maintain the feedback loop, and hence the virtual earth, on both phases. The amplifier bias currents I_P and I_N should be made equal since mismatch causes a small change in the memory transistor's bias current and incurs an offset error.

A parallel delay line with grounded-gate feedback is shown in Figure 10.9. Again the amplifier is multiplexed between the $N+1$ memory cells creating a closed loop, and hence a virtual earth, on all phases. Bias currents I_P and I_N must be closely matched as any difference appears as an offset error at the output.

Figure 10.10 shows a non-inverting lossless integrator with grounded-gate feedback. The ideal integrator loop is formed from the serial unit delay cell (Figure 10.8) by feeding back the output signal to the input. Mismatch between the bias currents I_P and I_N is of no consequence to the integrator loop (T_1, T_2) but causes an offset error at the integrator output.

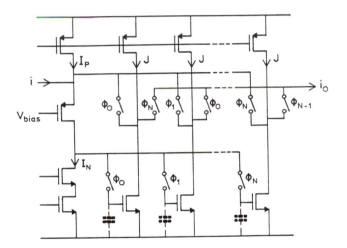

Figure 10.9 Parallel delay line (NT) with grounded-gate feedback.

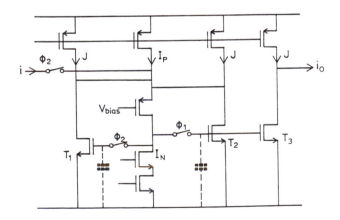

Figure 10.10 Non-inverting lossless integrator with grounded-gate feedback.

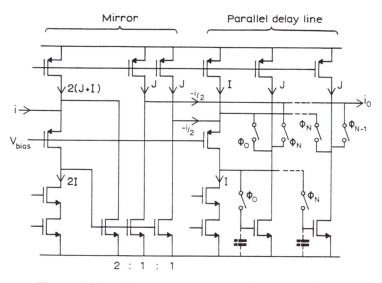

Figure 10.11 FIR cell with grounded-gate feedback.

Figure 10.11 shows the FIR cell with grounded-gate feedback. It is derived directly from the cell shown in Figure 10.6 and the half currents are generated by a mirror which uses the same grounded-gate feedback and so produces the same virtual earth at the cell's input. The delay cell connects each of its $N+1$ memory cells to its amplifier on successive clock phases and so continually generates a virtual earth at its input. The mirror outputs are connected to the virtual earth at either the delay line or at the input of the next cell and so the drains of the three transistors involved in mirroring have equal voltages; a necessary condition for good matching.

The ideal transfer function is given by

$$H(z) = \frac{1}{2}(1 - z^{-N}) \qquad (10.4)$$

and produces a transmission zero at dc. In practice, the degree of cancellation and hence the depth of the zero is limited by the inevitable mismatch of the half signal currents produced by the mirror.

10.3.2 *Fully differential circuits*

As discussed in Chapter 6, fully differential circuits have well-known advantages in signal processing. Where signal inversion is required it can be achieved by simply crossing-over signal pairs and common-mode signals, such as crosstalk from digital circuits on the same IC, are rejected. In SI cells, they bring the added advantage of reducing charge injection errors.

In the basic memory cell (Figure 10.2), the switch S feeds charge from its channel and gate-overlap capacitances into the memory capacitor C every time it opens. This produces an error in the output current even when there is no signal current (i.e. it causes an offset). With signal present, the charge injection produces both transmission gain errors and harmonic distortion. Fully differential circuits and their benefit to charge injection errors have been discussed earlier (Chapter 6) and it will suffice here to describe their use with grounded-gate feedback.

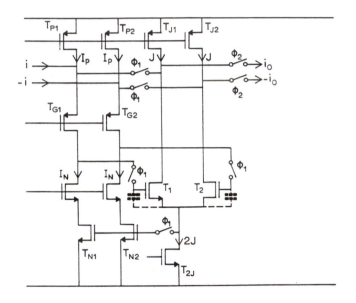

Figure 10.12 Fully differential memory cell.

A SI fully differential memory cell is shown in Figure 10.12 and has been derived directly from the cell shown in Figure 10.7. The only addition is the common-mode feedback connection to transistors T_{N1} and T_{N2}. This ensures that, on phase ϕ_1, the currents flowing at the common source of the differential pair, T_1 and T_2, sum to zero by adjusting the currents I_N in transistors T_{N1} and T_{N2} and the currents in the pair are $J \pm i$. When $i = 0$, the cell is balanced and there is no charge injection error (i.e. no offset error) and when $i \neq 0$, the error is smaller and more linear than that of the single-ended memory cell (Figure 10.7). Common mode input signals are removed by the common-mode feedback loop by the same mechanism and the currents I_N adjust to $I + i_{cm}$ where i_{cm} is the common mode input signal. In complexes of memory cells (integrators, delay lines, FIR cells etc) common mode control is multiplexed as shown for the delay line in Figure 10.13. The direct connection of the multiplexed common sources of the memory cells to the common mode current control sets the voltages at the common source nodes. However, this imposes a larger than necessary drain voltage for the *2J* current sources and may be reduced by a suitable level shift network.

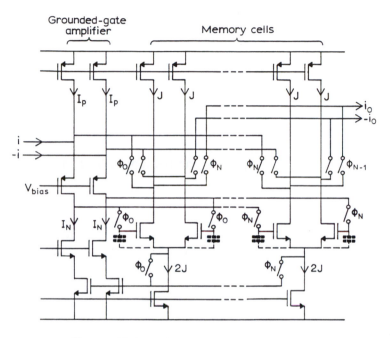

Figure 10.13 Fully differential delay line.

10.4 Memory Cell Design

For the SI circuit to meet system specifications, the memory cell coupled to its grounded-gate feedback amplifier must meet prescribed standards. In this section, the main non-ideal effects of the fully differential memory cell shown in Figure 10.12 (transmission errors, settling behaviour and signal-to-noise ratio) are related to the parameters of its transistors.

10.4.1 Transmission error

Transmission errors occur through conductance ratio errors and charge injection errors.

10.4.1.1 Conductance ratio error

The error is derived in Appendix 10A. This shows that the error is given by

$$\varepsilon_G = -\frac{1}{A_C}\frac{g_{dsN}}{g_m} - \frac{1}{A_{GG}}\left[\frac{g_o}{g_m}\right] \tag{10.5}$$

where $A_{GG}=g_{mG}/g_{dsG}$ and $A_C=g_{mC}/g_{dsC}$ are the open-drain voltage gains of transistors T_G and T_C respectively, where g_{mG} and g_{dsG} are the transconductance and drain conductance of the grounded-gate transistor T_G, and g_{mC} and g_{dsC} are the transconductance and drain conductance of the cascode transistor T_C. Further, g_m and g_0 are the transconductance and output conductance of the memory cell. The first term arises because signal current produces a small change in the bias current from T_N and shows the need for cascoding. The second term results from the large but finite input conductance of the memory cell and shows the benefit of the grounded-gate feedback. The low frequency input conductance is given by

$$g_{in} \approx A_{GG}g_m \tag{10.6}$$

Clearly, the voltage gain of T_G is raising the input conductance to produce a 'virtual earth' at the input.

10.4.1.2 Charge injection errors

Charge injection errors in fully differential SI circuits have been analysed earlier in Chapter 6. This showed that the charge injection error ε_q is given by

$$\varepsilon_q = \alpha\left(1 + \frac{\gamma}{3}\right)\frac{C_{CH}}{C} \tag{10.7}$$

where γ is the back-gate constant of the memory switch, $C_{CH}/2$ is the channel capacitance of each memory switch, $C/2$ is the gate-source capacitance of each memory transistor and α is the charge redistribution constant $(0.5 < \alpha < 1.0)$. So, charge injection [6,7] in fully differential SI circuits does not cause an offset error and potential errors resulting from charge injected through the switches' gate overlap capacitances are rejected as common-mode signals. During the turn-off process of the memory switches, charge is initially injected equally from the switches' drains and sources so that only half is injected into the memory transistors' gate capacitances. Charge redistribution then occurs, increasing this fraction, until the switches' channels are eventually interrupted. For very short clock fall time, negligible charge redistribution takes place and $\alpha \to 0.5$. This leaves close to half of the switches' channel charge injected onto the memory transistors' gate capacitances, and in that case the error can be further decreased by the use of half width dummy switches as shown in Figure 10.14. Note that the switches S, S_i and S_D should be phased to ensure that S opens momentarily before either S_i opens or S_D closes. The resulting charge injection error is ideally zero but in practice the degree of cancellation is limited by mismatch amongst the small-sized switch and dummy switch transistors. This mismatch occurs between the switches on either side of the differential structure and between these switches and their dummies.

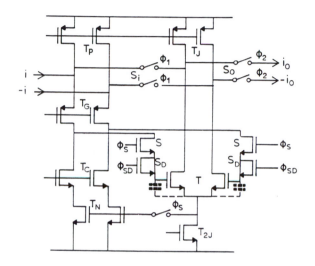

Figure 10.14 Fully differential memory cell with dummy switch transistors.

10.4.2 *Settling behaviour*

10.4.2.1 *Linear settling*

The simplistic discussion of settling behaviour of the basic memory cell in the introduction assumed that the loop formed by the memory transistor and a closed switch gave a first-order system (Figure 10.1). In practice, even the basic cell may have significant switch resistance and memory transistor drain capacitance, and these cause the loop to behave as a second-order system with the possibility of an under-damped response giving non-monotonic settling behaviour. With the inclusion of grounded-gate feedback, still higher order systems may be produced with an attendant risk of poor settling behaviour.

The grounded-gate memory cell's high frequency small signal analysis is given in Appendix 10B. It shows that monotonic settling occurs so long the roots of the following equation are real.

$$s^3 \tau_M \tau_G \tau_S + s^2 \tau_M (\tau_G + \tau_S) + s \tau_M + 1 = 0 \qquad (10.8)$$

where the time constants τ_M, τ_G and τ_S are produced by the memory and grounded-gate transistor transconductances and switch on-conductance respectively, with their associated capacitances (see Appendix 10B for details). It is seen that the grounded-gate memory cell forms a third-order system. Figure 10.15 shows universal curves of normalised settling time

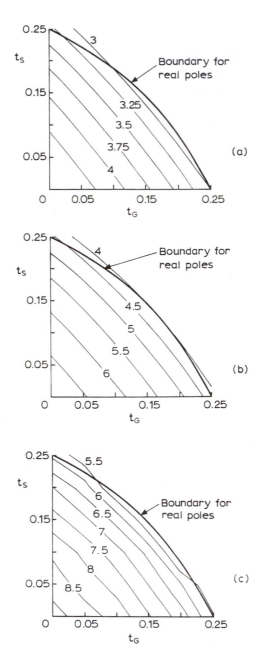

Figure 10.15 Normalised settling time of the grounded-gate memory cell (a) 1% settling accuracy (b) 0.1% settling accuracy (c) 0.01% settling accuracy.

(T_{settle}/τ_M) as a function of normalised time constants of the grounded-gate transistor $(t_G = \tau_G/\tau_M)$ and switch $(t_S = \tau_S/\tau_M)$ for 1%, 0.1% and 0.01% settling accuracy. For monotonic settling, the design must be within the 'quadrant' defined by the axes and the boundary. Designs near the boundary have near critical damping while designs near the origin are heavily overdamped. Designs near to either the t_S or t_G axis are close to a second-order system and designs near the origin approach a first-order system. Clearly, designs which are safely near to the boundary and not too close to either axis benefit from enhanced settling time without requiring extreme values of time constant.

10.4.2.2 Non-linear settling

Consider the memory cell on its input phase as shown in Figure 10.16. When it receives a differential input current step i, the current in T_{G1} increases momentarily to $I+i$ and in T_{G2} it decreases to $I-i$. If $i \ll I$, then the drain currents in T_1 and T_2 settle linearly to $J+i$ and $J-i$ respectively. However, if $i \geq I$, then T_{G2} cuts off, breaking the negative feedback loop, and leaves current sources I to charge the memory capacitors. This state of slewing persists until the drain current of T_2 decreases sufficiently to allow conduction in T_{G2}, and then linear settling is resumed. The slewing process is non-linear and could lead to harmonic distortion so the bias current I should have a sufficiently high value to avoid the problem.

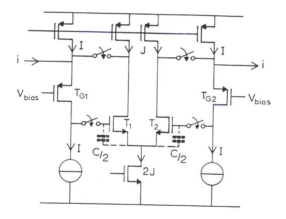

Figure 10.16 Memory cell on its input phase.

10.4.3   Signal-to-noise ratio

Maximum signal-to-noise ratio results when the signal current swings are maximised and the noise current sources are minimised. This condition results from maximising saturation voltages, within the constraints of limited supply headroom, of transistors T_J, T_P, T_N and T. The noise

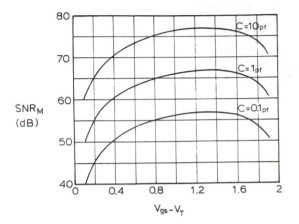

Figure 10.17 Calculated signal-to-noise ratio of the grounded-gate active memory cell V_{DD} = 4.5V, $I=J/2$, $(V_{gs} - V_T)_G$ = 0.1V, V_T = 1V, m_i = 0.75, m_{th} = 2.25, T_j = 300K.

sources in T_G and T_C do not cause a significant contribution to the cell's noise current so long as the impedances at their sources are high and their saturation voltages may be set to low values. As shown in Appendix 10C, signal-to-noise ratio can be computed (Chapter 4) from (10.9)

$$SNR_M = 10log\left[\frac{m_i^2(V_{gs} - V_T)^2}{\frac{16m_{th}}{3}(1 + k_J)\left(\frac{kT_j}{C}\right)}\right] \quad (10.9)$$

where k_J is given by,

$$k_J = \frac{g_{m(P)} + g_{m(N)} + g_{m(J)}}{g_m} \quad (10.10)$$

where $g_{m(P)}$, $g_{m(N)}$, $g_{m(J)}$ and g_m are the quiescent transconductances of T_P, T_N, T_J and T. Figure 10.17 shows the computed variation of signal-to-noise ratio, SNR_M, with memory transistor saturation voltage, $(V_{gs} - V_T)$ and capacitance C for the condition V_{DD} = 4.5V, $I=J/2$, $(V_{gs} - V_T)_G$ = 0.1V, V_T = 1V, m_i = 0.75, m_{th} = 2.25, k = 1.38E-23, and T_j = 300K and with some practical margins to allow for parameter spreads.

10.5 Video IC Test Circuit

In this section, IC test structures for switched-current video applications are described. The test chip comprises the modules shown in Figure 10.18

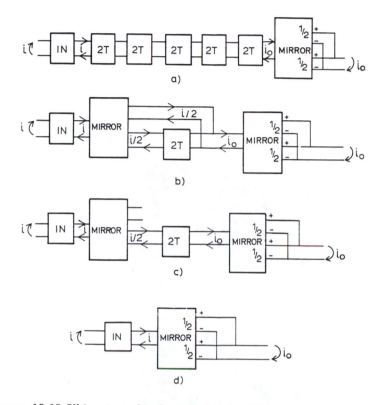

Figure 10.18 Video test circuit modules (a) 10T delay module (b) FIR module (c) 2T delay module (d) in/out module.

[1], designed using fully differential circuits with grounded-gate feedback, and was made in a 1μm N- well CMOS process. The modules were,

- *10T* delay modules made from cascading an input buffer (IN), five *2T* delay cells (each with $(H(z)=-z^{-2})$) and an output buffer (MIRROR).

- FIR modules $(H(z)=(1/2)(1 - z^{-2}))$ made from an input buffer, mirror cell, *2T* delay cell and an output buffer.

- *2T* delay module made by cutting the feed-forward path of the FIR module.

- in/out module made from cascading input and output buffers.

The IC layout photograph is shown in Figure 10.19.

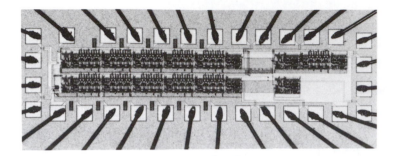

Figure 10.19 Video test circuit microphotograph.

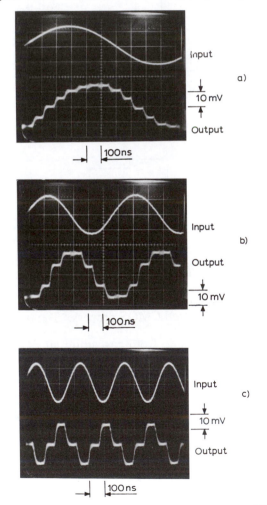

Figure 10.20 Output waveforms (sampling frequency = 13.28MHz) signal frequency (a) 0.83MHz (b) 1.66MHz (c) 3.32MHz.

10.5.1 Output waveforms

The output current waveforms of the *2T* delay module are shown in Figure 10.20. The input signal was a sinusoid and was sampled and held by the *2T* delay module. The signal frequencies chosen (0.83MHz, 1.66MHz, and 3.32MHz) were sub-multiples of the sampling frequency (13.28MHz) to enable synchronisation. The differential output swing is 30μA which corresponds to a modulation index of 93% (the bias current in T_{2J}, Figure 10.12, is 32μA).

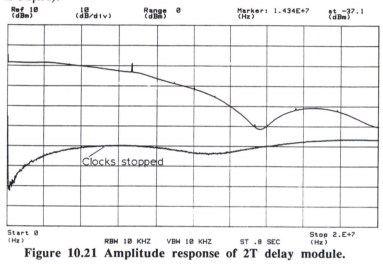

Figure 10.21 shows the spectrum analyser display with the following annotations:

| Ref 10 (dBm) | 10 (dB/div) | Range 0 (dBm) | Marker: 1.434E+7 (Hz) | at −37.1 (dBm) |

Clocks stopped

| Start 0 (Hz) | RBW 10 KHZ | VBW 10 KHZ | ST .8 SEC | Stop 2.E+7 (Hz) |

Figure 10.21 Amplitude response of 2T delay module.

10.5.2 Amplitude responses

The amplitude response of the *2T* delay module is shown in Figure 10.21. As in Figures 10.22-10.25, the response is plotted directly from the spectrum analyser and the scales of the vertical (amplitude) and horizontal (frequency) axes are as shown at the head and foot of the diagram. The general shape is the familiar *sinc(x)* function resulting from the sample and hold action of the *2T* delay module on the continuous input signal. With the clock stopped, there is a residual signal about -40dB below that passing through the module and is the result of the input signal voltage driving current through the parasitic capacitance which exists between the packaged module's input and output pins. With the clock active, the delayed signal interferes with that fed directly through package parasitic capacitance and produces a 'ripple' at a frequency of *1/2T* (6.64MHz) and prevents the trough from being as deep as might be expected.

The amplitude response of the *10T* delay module is shown in Figure 10.22. The *sinc(x)* is produced by the sample and hold action of the first of its *2T* delay cells. The interference of the delayed signal with that fed directly through package parasitic capacitance produces a 'ripple' at a frequency of *1/10T* (1.333MHz).

Figure 10.22 Amplitude response of 10T delay module.

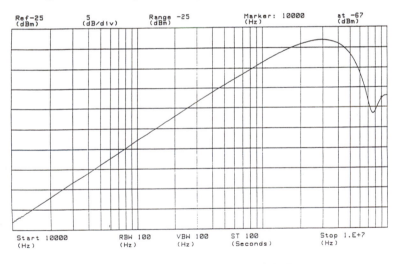

Figure 10.23 Amplitude response of FIR module.

The amplitude response of the FIR module is shown in Figure 10.23. It shows the expected high pass characteristic (Figure 10.5(a)) with its peak close to $1/4T$ and its trough close to $1/2T$. This peak and trough do not occur exactly at these frequencies due to the continuous-time nature of the signal fed forward from the mirror to the output. Though not shown, the response at low frequencies (below 1kHz) levels off at -55dB below the peak response due to the mismatch of mirror output stages and due to the noise floor.

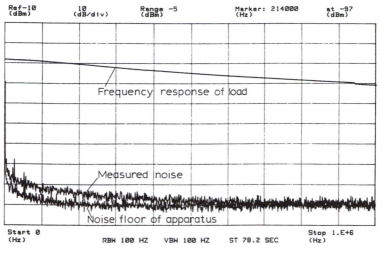

| Ref-10
(dBm) | 10
(dB/div) | Range -5
(dBm) | Marker: 214000
(Hz) | at -97
(dBm) |

Start 0 (Hz)   RBW 100 HZ   VBW 100 HZ   ST 78.2 SEC   Stop 1.E+6 (Hz)

Figure 10.24 Output noise spectrum of the 2T delay module.

10.5.3 Signal-to-noise ratio

The output noise current of the *2T* module, with inputs open circuit and output terminated with a 50kΩ resistor (bandwidth 350kHz), is shown in Figure 10.24. It shows the amplitude response of the load, the output spectrum with the module powered down (to indicate the noise floor of the apparatus), and the noise spectrum with the module in operation. Note that the load bandwidth shapes the noise spectrum. The noise level at 300kHz is -93dBm in a bandwidth of 100Hz and, after correction for the apparatus noise (-0.6dB), the load bandwidth (+2dB) and for a signal bandwidth of 5MHz (+47dB), this increases to -44.6dBm. The signal level for 75% modulation (12μA pk) corresponds to +5.6dBm giving a *SNR* of 50.2dB. This corresponds to a *SNR* for the *2T* module of about 57.2dB which is about 6dB below that calculated directly from Figure 10.17, the difference being due to the noise in the input interface circuit.

10.5.4 Harmonic distortion

The output spectrum of the *2T* delay module for a signal at f_S = 100kHz with amplitude of 12μA (75% modulation) is shown in Figure 10.25. The third harmonic is at -40dB which is somewhat worse than expected due mainly to reactive distortion in the input interface circuit. It drops further with increased signal frequency due mainly to slewing which occurs when the signal has both high frequency and high amplitude, due to insufficient current bias in the grounded-gate stage and could be improved through design optimisation.

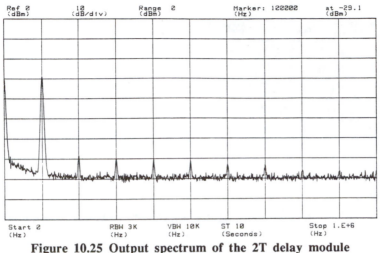

| Ref 0 (dBm) | 10 (dB/div) | Range 0 (dBm) | Marker: 100000 (Hz) | at -29.1 (dBm) |

| Start 0 (Hz) | RBW 3K (Hz) | VBW 10K (Hz) | ST 10 (Seconds) | Stop 1.E+6 (Hz) |

Figure 10.25 Output spectrum of the 2T delay module (signal 75% modulation, 100kHz).

10.5.5 Summary of performance

The performance of the *2T* delay line is summarised in Table 10.1.

Process	1µm, 5V, N-well 'digital' CMOS	
Areas	Memory transistor	11µm * 22µm
	2T delay line	320µm * 160µm
	Mirror	180µm * 160µm
Current Consumption	Memory pair	32µA
	2T delay line	114µA
	Mirror	158µA
Performance of the 2T delay line	Sampling frequency	13.3MHz
	Signal to noise ratio	60dB
	3rd harmonic distortion (100kHz, 75% mod.)	-40dB
	Transmission error	-54dB
	Settling error (T=75ns)	-60dB
	PSRR (1MHz, 1.4V pk-pk on Vdd)	-50dB

Table 10.1 Summary of performance of 2T delay line.

10.6 Discussion and Conclusions

This chapter has demonstrated that in Silicon technology SI cells are capable of video performance. Further improvements in the high frequency capability of the switched-current cell can be obtained through scaling operating currents (λ) and memory transistor width $(\lambda^{1/2})$ and length $(\lambda^{-1/2})$ simultaneously; which scales cut-off frequency (λ) while leaving S/N ratio and charge injection substantially unchanged.

Test circuits made in a 1μm digital CMOS process and described in this chapter have confirmed the potential of the switched-current technique for video frequency performance.

While not explicitly covered in this work, it may be expected that integrators and differentiators constructed with grounded-gate fully-differential memory cells will have correspondingly good high frequency performance enabling video frequency state-variable or leapfrog filtering. It is estimated that SI circuits could be designed to operate at clock frequencies in excess of 100MHz with a 1μm CMOS process and at several 100MHz with future sub-micron low voltage processes [8].

10.7 References

[1] J. B. Hughes and K. W. Moulding, "Switched-current signal processing for video frequencies and beyond," *IEEE J. Solid-State Circuits*, vol. SC-28, March 1993.

[2] J. B. Hughes, I. C. Macbeth and D. M. Pattullo "Switched-Current Filters," *Proc. IEE Part G*, April 1990.

[3] T. S. Fiez, G. Liang and D. J. Allstot, "Switched-Current Design Issues," *IEEE J. Solid State Circuits*, vol. SC-26, pp. 192-202, March 1991.

[4] J. J. F. Rijns and H. Wallinga, "Spectral Analysis of Double-Sampling Switched-Capacitor Filters," *IEEE Trans. Circuits Syst.*, vol. 38, pp. 1269-1279, Nov 1991.

[5] D. Groeneveld et al "Self-calibrated technique for high resolution D-A converters," *IEEE J. Solid-State Circuits*, vol. SC-24, pp. 1517-1522, Dec. 1989.

[6] J.-H. Shieh, M. Patil and B. J. Sheu, "Measurement and analysis of charge Injection in MOS analog switches," *IEEE J. Solid-State Circuits*, April 1987

[7] G. Wegmann, E. A. Vittoz and F. Rahali, "Charge injection in analog MOS switches," *IEEE J. Solid-State Circuits*, Dec. 1987

[8] M. Nagata "Limitations, "Innovations and challenges of circuits and devices into half micrometer and beyond," *IEEE J. Solid-State Circuits*, vol. 27, pp. 465-472, April 1992.

Appendix 10A

Low Frequency Small Signal Analysis of Grounded-Gate Memory Cell

The low frequency small signal equivalent circuit of the grounded-gate memory cell (Figure 10.7) on its input phase is shown in Figure 10.26. Summing the currents at node 1 gives

$$g_{mG}v_{in} + \left(v_{in} - \frac{i_{ds}}{g_m}\right)g_{dsG} = i_{ds}\frac{g_{0C}}{g_m} \qquad (10a.1)$$

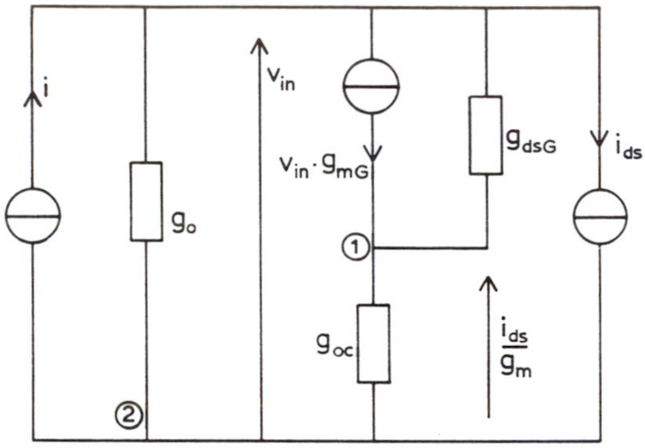

Figure 10.26 Small signal equivalent circuit of grounded-gate memory cell on its input phase.

where v_{in} is the voltage at the input node, i_{ds} is the drain current of the memory transistor, g_m is the transconductance of the memory transistor, g_{mG} is the transconductance of the grounded-gate transistor, g_{dsG} is the drain conductance of the grounded-gate transistor and g_{0C} is the output conductance of the cascoded current source. This reduces to

$$v_{in} = \frac{i_{ds}}{g_m}\left[\frac{g_{0C} + g_{dsG}}{g_{mG} + g_{dsG}}\right] \approx \frac{i_{ds}}{g_m}\left[\frac{g_{dsG}}{g_{mG}}\right] \qquad (10a.2)$$

assuming $g_{0C} \ll g_{dsG} \ll g_{mG}$. Summing currents at node 2 gives

$$i = v_{in}g_0 + i_{ds}\frac{g_{0C}}{g_m} + i_{ds} \qquad (10a.3)$$

where g_{dsJ} is the drain conductance of the bias current transistor T_J and g_{dsP} is the drain conductance of the bias current transistor T_P and g_0 is the memory cell's output conductance given by (Chapter 4)

$$g_0 = g_{ds} + g_{dsJ} + g_{dsP} + \left[\frac{C_{dg}}{C_{dg} + C}\right] g_m \qquad (10a.4)$$

where C_{dg} and C are the memory transistor's gate-drain and gate-source capacitances respectively. This gives,

$$\frac{i_{ds}}{i} \approx 1 - \frac{g_0 C}{g_m} - \left(\frac{g_0}{g_m}\right) \frac{g_{dsG}}{g_{mG}}$$

$$= 1 - \frac{1}{A_C} \frac{g_{dsN}}{g_m} - \frac{1}{A_{GG}} \left[\frac{g_0}{g_m}\right] \qquad (10a.5)$$

where g_{dsN} is the drain conductance of the bias current transistor T_N, and $A_{GG} = g_{mG}/g_{dsG}$ and $A_C = g_{mC}/g_{dsC}$ are the open-drain voltage gains of transistors T_{GG} and T_C respectively. The conductance ratio error ε_G is given by

$$\varepsilon_G = \frac{i_{ds}}{i} - 1 = -\frac{1}{A_C} \frac{g_{dsN}}{g_m} - \frac{1}{A_{GG}} \left[\frac{g_0}{g_m}\right] \qquad (10a.6)$$

The first term arises because signal current produces a small change in the bias current from T_N and shows the benefit of cascoding. The second term results from the large but finite input conductance of the memory cell and shows the benefit of the grounded-gate feedback. The low frequency input conductance is given by

$$g_{in} = \frac{i}{v_{in}} \approx \frac{i_{ds}}{v_{in}} \approx A_{GG} g_m \qquad (10a.7)$$

Clearly, the voltage gain of T_G is raising the input conductance to produce a 'virtual earth' at the input.

Appendix 10B

Settling Behaviour of Grounded-Gate Memory Cell

The high frequency circuit of the memory cell (Figure 10.7) on its input phase is shown with its main parasitics in Figure 10.27(a). With the loop broken at the gate of the memory transistor, this reduces to the equivalent circuit shown in Figure 10.27(b), where

$$C_i = C_{in} + C_{dP} + C_{gsG} + C_{dG} + C_{SW} + C_{dJ} + C_d$$

$$C_a = C_{olC} + C_{dC} + C_{dG} + \frac{1}{2} C_S$$

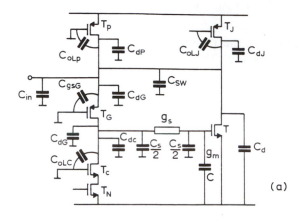

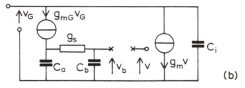

Figure 10.27 High frequency circuit of grounded-gate memory cell
(a) Configuration showing main parasitics
(b) Small signal equivalent circuit with loop broken.

$$C_b = C + \frac{1}{2} C_S \qquad (10b.1)$$

where the various capacitances are diffusion capacitance (C_d), gate-source capacitance (C_{gs}), gate-overlap capacitance (C_{ol}), input capacitance (C_{in}), input switch capacitance (C_{SW}), and memory switch capacitance (C_S). The open loop voltage gain is given by

$$A_{open} = \frac{v_b}{v} = \frac{-g_m g_{mG} g_S}{(g_{mG} + sC_i)(s^2 C_a C_b + sg_s(C_a + C_b))}$$

$$= \frac{-1}{s\tau_M(1 + s\tau_S)(1 + s\tau_G)} \qquad (10b.2)$$

where $\tau_M = (C_a + C_b)/g_m$, $\tau_G = C_i/g_{mG}$, and $\tau_S = C_a C_b/g_S(C_a + C_b)$.

The poles of the closed loop are found by putting $A_{open} = 1$.

$$s^3 \tau_M \tau_G \tau_S + s^2 \tau_M (\tau_G + \tau_S) + s\tau_M + 1 = 0 \qquad (10b.3)$$

Putting $s\tau_M = s'$, $t_G = \tau_G/\tau_M$ and $t_S = \tau_S/\tau_M$ gives

$$s'^3 t_G t_S + s'^2(t_G + t_S) + s' + 1 = 0 \qquad (10b.4)$$

Monotonic settling corresponds to all poles being real. This condition was solved numerically using a root finding routine (C02AGF) and yielded the 'quadrant' shaped region of Figure 10.15. Settling times were then found from a large series of transient simulations and the results are shown as contours in Figure 10.15.

Appendix 10C

Signal-to-noise ratio

Maximum signal-to-noise ratio results when the signal current swings are maximised and the noise current sources are minimised. This condition results from maximising saturation voltages, within the constraints of limited supply headroom, of transistors T_J, T_P, T_N and T. The noise sources in T_G and T_C do not cause a significant contribution to the cell's noise current so long as the impedances at their sources are high and their saturation voltages may be set to low values.

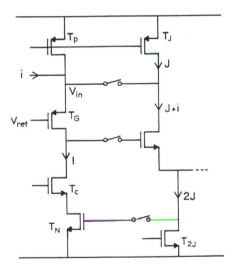

Figure 10.28 Half of memory cell.

Referring to Figure 10.28, saturated operation of transistors is guaranteed so long as

$$V_{DD} - (V_{gs} - V_T)_J \geq V_{in} \geq (V_{gs} - V_T)_G + V_T + (V_{gs} - V_T)\sqrt{1+m_i} + (V_{gs} - V_T)_{2J} \quad (10c.1)$$

and

$$V_{LS} \leq (V_{gs} - V_T)_N + V_{TN} + (V_{gs} - V_T)_{2J} \qquad (10c.2)$$

and

$$(V_{gs} - V_T)_C + (V_{gs} - V_T)_N \leq V_T + (V_{gs} - V_T)\sqrt{1-m_i} + (V_{gs} - V_T)_{2J} \quad (10c.3)$$

where $(V_{gs} - V_T)$, $(V_{gs} - V_T)_G$, $(V_{gs} - V_T)_P$, $(V_{gs} - V_T)_N$, $(V_{gs} - V_T)_C$, $(V_{gs} - V_T)_J$ and $(V_{gs} - V_T)_{2J}$ are the quiescent saturation voltages of T, T_G, T_P, T_N, T_C, T_J, and T_{2J} respectively and m_i is the signal modulation index *(i/J)*. The signal-to-noise ratio can be computed (Chapter 4) from (10c.4)

$$SNR_M = 10\log\left[\frac{m_i^2(V_{gs} - V_T)^2}{\frac{16m_{th}}{3}(1 + k_J)\left(\frac{kT_j}{C}\right)}\right] \quad (10.c4)$$

where k_J is given by,

$$k_J = \frac{g_{m(P)} + g_{m(N)} + g_{m(J)}}{g_m} \quad (10.c5)$$

where $g_{m(P)}$, $g_{m(N)}$, $g_{m(J)}$ and g_m are the quiescent transconductances of T_P, T_N, T_J and T. Putting $I = J/2$, $(V_{gs} - V_T)_N = V_T$ (to conservatively ensure saturated operation of T_N) and $(V_{gs} - V_T)_P = (V_{gs} - V_T)_J$ it can be shown that,

$$k_J = \frac{3(V_{gs} - V_T)}{2(V_{gs} - V_T)_J} + \frac{(V_{gs} - V_T)}{2V_T} \quad (10c.6)$$

Figure 10.17 shows the computed variation of signal-to-noise ratio, SNR_M, with memory transistor saturation voltage, $(V_{gs} - V_T)$ and capacitance C for the condition $V_{DD} = 4.5V$, $(V_{gs} - V_T)G = 0.1V$, $V_T = 1V$, $m_i = 0.75$, $m_{th} = 2.25$, $k = 1.38E-23$, and $T_j = 300K$ and with some practical margins for the inequalities (10c.1-10c.3) to allow for parameter spreads.

Chapter 11

Switched-Current Wave Analogue Filters

Adoración Rueda, Alberto Yúfera and José L. Huertas

11.1 Introduction and Motivations

Applying switched-current (SI) techniques to analogue signal processing has become a cornerstone for its future usefulness. Besides their general advantages in terms of allowing the implementation of mixed analogue-digital systems on standard low-voltage digital VLSI CMOS processes, their feasibility for realising filters which are at least as good as those currently realised by switched-capacitors still has to be proven. Low-sensitivity switched-capacitor (SC) filters are based on three features: (a) the direct emulation of doubly-terminated LC ladder structures, since they exhibit inherently low sensitivity, (b) the transformation between the s and the z domains, which constrains the maximum achievable signal-to-sampling frequency ratio and (c) the actual implementation of the necessary building blocks, essentially SC integrators. In practice, SC filters have been so successful because almost ideal SC integrators can be built, although their complexity is strongly dependent on the particular frequency transformation selected. In general, a crucial issue for discrete-time filters is the relationship between points (b) and (c) above. For instance, the use of the bilinear transformation seems the best choice since the Nyquist limit can be approached; however, bilinear integrators are much more complex than integrators required by other transformations, and thus a trade-off between complexity and maximum frequency must be established. Methods have been developed based on a manipulation of the signal flow graph to allow the use of LDI integrators for synthesising a global bilinear transformation, but again the Q-factor of these LDI integrators has to be high and, strictly speaking, these approaches do not guarantee that the low sensitivity properties are still retained.

Along the same path, quite a few papers have been published during the last three years dealing with SI filter implementations [1-4]. These papers study how the known SC filter synthesis methods can be extended to be efficiently used by SI implementations. Nevertheless, designing high-Q integrators is simpler in SC than in SI because the former are based on high-gain elements (operational amplifiers) whilst the latter rely on less accurate components (current mirrors). Therefore, since the precision attained by SI building blocks is lower, the published results lead us to conclude that there still remains a need for further improvements in SI filters in order to be competitive with their SC counterparts in terms of performance.

A different approach is currently being explored, which is based on the wave concept and is being successfully used in digital contexts because of its low-sensitivity properties. Wave digital filters (WDF) emulate the behaviour of passive lossless filters by transforming passive L and C elements into corresponding one-port digital elements defined by an incident signal, a reflected signal and a port resistance determined by the corresponding analogue element. The interconnection of one-ports with different port resistances requires a multiport element, the so called adaptor (for more details the reader is referred to the now classical papers from Fettweis [5-7]). Wave digital filters present two inherent advantages: they are based on a bilinear transformation between the continuous and discrete frequency domains and maintain the excellent passband sensitivity properties of lossless reference filters. The former means that the only frequency restrictions are those following from the Nyquist limit as well as those imposed by the fabrication process itself; the latter leads to a reduction in the accuracy requirements of the components. An important property of WDFs is that the spread of constant values in any adaptor does not differ significantly from one filter function to another, since they represent reflection coefficients which must always sum to 2.

In spite of both their wide acceptance in digital applications and their theoretical superiority, wave filters have been found to impractical for analogue applications when either continuous-time or switched-capacitor methods were employed. The WDF was extended to sampled analogue filter realisations making use of SC techniques [8,9], but they were of no practical interest as they resulted in more complex circuits than other SC filter structures. However, the basic operations in wave filters are summations and multiplications by a constant, i.e. operations readily available in current-mode.

For this reason the wave concept has been explored for SI filters [10,11] and it has been found that this technique is worth considering as a practical method for implementing filters in low-voltage standard CMOS technologies. A distinguishing feature of SI wave filters is that they do not need integrators, which all the other reported SI realisation schemes do. Besides the difficulty of designing quasi-ideal high-Q integrators using SI techniques, there are other drawbacks associated with the use of these integrators: noise (originating from flicker and thermal effects) and charge injection give rise to errors which are integrated as well as the signal and this imposes a significant limitation on the achievable dynamic range; also, using integrators with multiple outputs raises the capacitive load, thus lowering the bandwidth (although the THD is also reduced) [12]. Finally, the integrator coefficients are fixed by adjusting gains in current mirrors, so the area, power and accuracy properties of each SI integrator depend on the filter transfer function coefficients and hence on the particular filter to be implemented.

Instead, the approach we are about to describe in this chapter has three main advantages: (1) For a given continuous-time reference filter, multiple and very regular wave filter structures are available, all of them based on the exact bilinear mapping; (2) All of these filter structures require SI building blocks which can be simply and accurately implemented; (3) Transistor aspect ratios in adaptor implementations do not have a one-to-one correspondence to the transfer function coefficients, which means that the building blocks can be optimised for many different filter functions.

In this chapter, the present state-of-the-art in SI wave filters is reviewed, starting with an introduction to the main features of the wave filter paradigm. The basic building blocks required are discussed and their implementation using SI techniques is described. Finally, some practical filter implementations are described to illustrate the technique.

11.2 Wave Filters

11.2.1 General principles

Wave filters rely on the simulation of the behaviour of passive filters through the use of wave quantities instead of port voltages or currents. That is, the overall system is only an operational simulator of a scattering description of a given network. Basic components for the simulator are subsystems mimicking the scattering model of the standard passive elements.

Let us consider a general two-port network N connected to resistive terminations, as shown in Figure 11.1(a). In this figure, the wave variables A_k (incident wave) and B_k (reflected wave) are voltages defined as linear combinations of the corresponding port currents, I_k, and voltages, V_k, by

$$A_k = V_k + R_k I_k \qquad (11.1)$$

$$B_k = V_k - R_k I_k \qquad (11.2)$$

where R_k is the port resistance. An equivalent description can be worked out considering currents as wave variables.

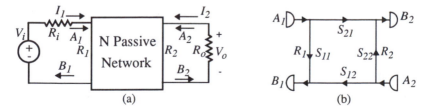

(a)   (b)

Figure 11.1 a) Resistive doubly-terminated two-port, b) Wave description.

The wave model for N is described by the voltage wave scattering matrix defined by

$$\begin{bmatrix} B_1 \\ B_2 \end{bmatrix} = \begin{bmatrix} S_{11} & S_{12} \\ S_{21} & S_{22} \end{bmatrix} \begin{bmatrix} A_1 \\ A_2 \end{bmatrix} \qquad (11.3)$$

The expression above is quite general and hence valid for deriving wave equivalents (henceforth called **wave models**) not only for a complex network but even for the main passive components; thus an actual two-terminal element is made equivalent to a two-port. In practice, our interest is in building a hardware model through which the port relationships of the original two-port are simulated by input-output relationships via the wave formulation.

However, the problem of establishing a link between the network functions of the original network and those of the wave emulator still remains. To solve this we can express the scattering matrix in terms of the chain parameters, which is equivalent to saying that the two-port in Figure 11.1(a) is represented by the signal flow-graph in Figure 11.1(b), where A_1 and A_2 are assumed to be input signals and B_1 and B_2 are output signals. It can be obtained after a simple manipulation that, assuming external port resistances are equal to terminating resistances, $R_i = R_1$ and $R_o = R_2$, the S_{21} parameter will be given by

$$S_{21} = \frac{B_2}{A_1}\bigg|_{A_2 = 0} = 2\frac{V_0}{V_i} = 2H(s) \qquad (11.4)$$

This result is essential for the design of wave filters. Moreover, if N is lossless, the Feldtkeller equation $/S_{11}/^2 + /S_{21}/^2 = 1$ applies, which means that S_{11} also behaves as a transfer function of N, being the complement of S_{21}. Thus, when S_{21} corresponds to a low-pass filter, then S_{11} corresponds to a high-pass filter and vice versa.

Basically the design process for a general *RLC* two-port consists of the implementation of a network by interconnecting wave models for resistors, capacitors and inductors, if the restrictions imposed by the Kirchhoff's laws are still preserved. Hence, instead of the original two-port, we have implemented a two-input two-output network where, since (11.4) applies, a transfer function identical to that of the original network does exist, relating the output variable B_2 to the input variable A_1.

For a discrete-time filter (as is the case for SI filters), the wave model is described in the z-domain and the network N in the ψ-domain, such that the correspondence between domains is ruled by the bilinear transformation,

$$\psi = \frac{2}{T}\frac{1 - z^{-1}}{1 + z^{-1}} \qquad (11.5)$$

where T is the sampling period and $z = e^{sT}$.

Table 11.1 gives a summary of wave models for passive elements as well as for two- and three-port components able to simulate the Kirchhoff's constraints for parallel and series connections. These equivalents have been obtained from (11.1) and (11.2), taking into account (11.5).

Circuit element	Resistance	Wave Model
Resistor	R	$A = 2i$ $B = 0$
Capacitor	$\dfrac{T}{2C}$	$B = z^{-1} \cdot A$
Inductor	$\dfrac{2L}{T}$	$B = -z^{-1} \cdot A$
Unit Element	R_{ue}	$B_0 = z^{-\frac{1}{2}} \cdot A_1$ $B_1 = z^{-\frac{1}{2}} \cdot A_0$
Two-port Adaptor.	R_0, R_1	$\mu = \dfrac{R_0 - R_1}{R_0 + R_1}$ $B_0 = A_1 + \mu \cdot (A_1 - A_0)$ $B_1 = A_0 + \mu \cdot (A_1 - A_0)$
Parallel Three-port Adaptor	R_0, R_1, R_2	$\gamma_j = \dfrac{2G_j}{G_0 + G_1 + G_2}$ $A_N = \gamma_0 \cdot A_0 + \gamma_1 \cdot A_1 + \gamma_2 \cdot A_2$ $B_j = A_N - A_j \ , j = 0, 1, 2$
Series Three-port Adaptor	R_0, R_1, R_2	$\gamma_j = \dfrac{2R_j}{R_0 + R_1 + R_2}$ $A_N = A_0 + A_1 + A_2$ $B_j = A_j - \gamma_j \cdot A_N \ , j = 0, 1, 2$

Table 12.1 Voltage wave models for some circuit elements.

A way to build the wave equivalent circuit is to replace one-by-one every actual passive element by its wave model and connect them through three-port components called adaptors. A general procedure to realise a wave filter consists basically of two steps: first, a reference filter is constructed to meet the required frequency specifications; then, a block diagram with wave models is derived to represent the reference filter. This wave model establishes the relationships between wave signals through the above-introduced modular building blocks. Nevertheless, there is not just one way to derive a wave filter prototype. In practice, there is a variety of ways as a consequence of both the number of reference filters that can be modelled and the different filter structures that can be derived for the same reference filter. The types of wave filters more extensively employed in digital filtering are those obtained either from microwave filters consisting of a chain of unit elements (UEs) or from two-port doubly-terminated *LC* ladder filters. The former can be realised by using wave models of UEs, and the latter are normally implemented in two different forms: the first one is using the method explained above, i.e. passive components are one-to-one replaced by equivalent two-wires connected by adaptors. A second method is based on modelling simultaneously passive components and interconnections; then, every *L* and *C* element in the original passive filter is substituted by a two-port equivalent where the interconnect constraints are included.

Moreover, the number of possibilities is even higher since, when the filter to be designed is not a low-pass one, a wave filter can be built in two ways: either performing first a frequency transformation in the s-domain and then applying the general method we have explained; or making a transformation from LP to the desired characteristics directly in the z-domain and then appropriately altering filter elements.

11.2.2 Wave filter synthesis procedure

Given a particular filter specification and a clock frequency, $1/T$, a prewarping of the cut-off and band edge frequencies is first performed according to the continuous, ω, and discrete, $\acute{\omega}$, frequency transformation

$$\omega = \frac{2}{T} \tan\left(\frac{\acute{\omega}T}{2}\right) \qquad (11.6)$$

Then, the passive reference filter is chosen from continuous-filter tables and the wave filter structure is built as previously explained. For the sake of illustration we show in Figures 11.2 and 11.3 the derivation method for a low-pass filter using the two kinds of reference filters we mentioned above. Figures 11.2(a) and 11.2(b) illustrate the process when a microwave prototype is used as a reference filter. On the other hand, in Figure 11.3(a) an *LC* ladder prototype is shown, and its operational simulation is depicted in Figure 11.3(b). As we can see from Figures 11.2(b) and 11.3(b), each passive component from the reference filter is represented by its incident

(A_k) and reflected (B_k) wave variables and is characterised by an equivalent port resistance whose value depends on the passive component value. Interconnections between components with different port resistances are realised by the corresponding adaptors.

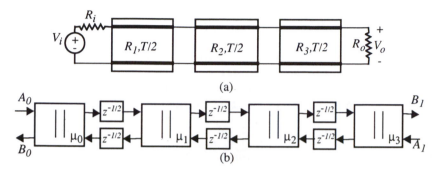

(a)

(b)

Figure 11.2 Generation of wave filters:
a) Microwave reference filter, b) Equivalent wave filter.

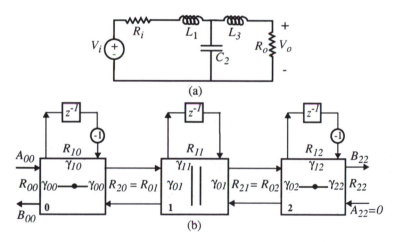

(a)

(b)

Figure 11.3 Generation of wave filters:
a) LC reference filter, b) Equivalent wave filter.

To determine the values of the component parameters, the relationships in Table 11.1 are used. For the case of microwave prototypes, the value of μ for each adaptor is directly derived from the specified R_{UE} values of the reference filter:

$$\mu_0 = \frac{R_i - R_1}{R_i + R_1} \qquad \mu_j = \frac{R_j - R_{j+1}}{R_j + R_{j+1}} \ (j = 1,2) \qquad \mu_3 = \frac{R_3 - R_0}{R_3 + R_0} \quad (11.7)$$

However, in the case of LC ladder structures, only two of the three resistances in each three-port adaptor (except in one of them, usually taken at the middle) are imposed: one by the analogue element and the other for the interconnection between adaptors. Thus,

$$R_{00} = R_i \quad R_{10} = \frac{2L_1}{T} \quad R_{11} = \frac{T}{2C_2} \quad R_{12} = \frac{2L_3}{T} \quad R_{22} = R_0 \qquad (11.8)$$

Therefore, we are free to choose the third resistance which means that **any set of values for the γ_{ij} parameters**, always summing to 2 for the same adaptor, can be used. To obtain the values of these parameters the following relationships, which must be fulfilled for each adaptor, are used

$$\sum_{i=0}^{2} \gamma_{ij} = 2 \qquad (11.9)$$

Series adaptors:

$$\frac{\gamma_{0j}}{R_{0j}} = \frac{\gamma_{1j}}{R_{1j}} = \frac{\gamma_{2j}}{R_{2j}} \qquad (11.10)$$

Parallel adaptors:

$$\frac{\gamma_{0j}}{G_{0j}} = \frac{\gamma_{1j}}{G_{1j}} = \frac{\gamma_{2j}}{G_{2j}} \qquad (11.11)$$

With these equations we can always obtain one port resistance as a function of the other two and the corresponding reflection coefficient. For instance, the R_{2j} of a series adaptor is given by

$$R_{2j} = R_{1j}\left(1 + \frac{R_{0j}}{R_{1j}}\right)\left(\frac{\gamma_{2j}}{2 - \gamma_{2j}}\right) \qquad (11.12)$$

Note that for the particular case where port 2 is reflection-free, $\gamma_{2j} = 1$, with the result that $R_{2j} = R_{0j} + R_{1j}$ (the same applies to the G_{2j} when a parallel adaptor is considered). In WDFs some adaptors must have a reflection-free port to make the interconnection between adaptors digitally realisable. In analogue implementations, as is our case, this restriction is not mandatory. Therefore, we are free to choose any value of R_{2j} and hence any value of γ_{2j}.

A deep knowledge of the influence of the parameter values on the sensitivity, noise and general performance of the filter configuration should allow advantage to be taken of this freedom to optimise practical implementations.

An algorithm based on (11.9)-(11.12) can be written to automatically obtain a set of values for the γ_{ij} coefficients fulfilling custom-imposed constrains, such as for instance, those necessary to improve bandwidth, dynamic range and so on. We have implemented such an algorithm in

WAVER, an in-house program we developed as an aid for the synthesis of wave SI filters.

11.2.3 Scaling techniques

The non-homogeneous distribution of signals inside a filter can produce signal levels larger than the dynamic range of the building blocks. To avoid this important source of distortion, scaling must be applied to equalise the signals throughout the filter. Two basic scaling methods appear in the literature on WDFs: one uses adaptor transformations; and the other is based on the insertion of ideal wave transformers where signal levels must be changed. Figure 11.4 illustrates these two operations.

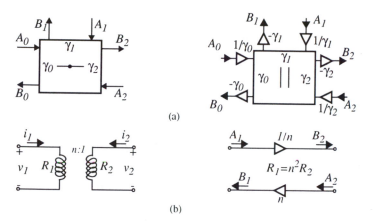

(a)

(b)

Figure 11.4 a) Series-to-parallel adaptor transformation, b) Ideal transformer.

Both scaling methods can be applied in our case. Following the systematic procedure outlined in [13,14], different wave models from the reference wave filter are first determined by applying adaptor transformations. The one involving lower signal levels is chosen, then a second scaling is performed by inserting ideal transformers where needed. A quick simulation facility (such as the one we have in WAVER) is fundamental for carrying out scaling.

11.3 Switched-Current Wave Filter Design

Switched-current wave analogue filters consist of an arrangement of current-mode adaptors (parallel and series), current delay cells and unity-gain inverter amplifiers, establishing a direct correspondence between wave flow diagrams and SI circuits. That is, SI wave filters are implemented using current-mode techniques with currents representing the wave signals a_j and b_j[1]. Although the term applies when either current or

---

[1] Henceforth, lowercase is used to denote time domain wave signals

voltage wave models are used for all the elements in the filter, we will concentrate our attention, without loss of generality, on realisations of voltage wave models.

In practical implementations of wave filters, the elements involved have to be connected in such a way that the incident wave in a port of an element is the reflected wave of the adjacent element. Hence, for SI elements and their interconnections we need a sign criterion for currents; we will take that which follows: currents modelling incident (reflected) waves are considered positive when they enter (leave) the corresponding port.

In this section we will present practical implementations of the required building blocks as well as some design issues for efficient filter realisations.

11.3.1 Basic building blocks

Parallel and series adaptors:

Adaptors are multiport circuits implementing the algebraic relationships between wave variables given in Table 11.1. Therefore, current-mode adaptors only involve the use of summing current amplifiers with different gain factors. There are several possible realisations for the same adaptor depending on how the gain factors are related to the adaptor coefficients γ_j. For instance, a current-mode two-port parallel adaptor can be built as an arrangement of current amplifiers directly implementing the relationships

$$b_0 = (\gamma_0 - 1)a_0 + \gamma_1 a_1$$

$$b_1 = \gamma_0 a_0 + (\gamma_1 - 1)a_1 \qquad (11.13)$$

or alternatively, we can define a new coefficient μ as $\mu = 1 - \gamma_0 = \gamma_1 - 1 = (R_0 - R_1)/(R_0 + R_1)$ and hence implement the following relationships:

$$b_0 = -\mu a_0 + (1 + \mu)a_1$$

$$b_1 = (1 - \mu)a_0 + \mu a_1 \qquad (11.14)$$

The best choice would be one combining the requirements of few transistors and lower power consumption with the accurate realisation of the gain factors. Since these gain factors are in general non-integer and are realised by transistor aspect ratios in current mirrors, an important issue affecting practical accuracy is the quantisation error of the real geometries due to the technological grid. This limitation can be easily considered when adaptor coefficient values are chosen as well as when practical arrangement of amplifiers implementing these coefficients are looked for, i.e. for a particular case of a two-port adaptor, it would be better to use μ coefficients than γ ones. Thus, the freedom in choosing the adaptor implementation can be exploited to obtain the best accuracy. This is an additional advantage of SI wave analogue filters along with the freedom to choose adaptor coefficients.

Figure 11.5 shows a simplified circuit diagram for a current-mode implementation of the two-port parallel adaptor described by (11.14); only n-channel mirrors are used although an alternative arrangement is possible using both n-channel and p-channel mirrors [10]. Also, in practical implementations cascode configurations should be used for both current mirrors and for bias current replications.

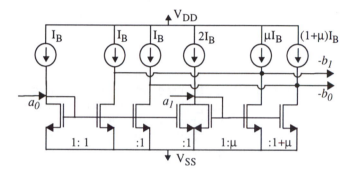

Figure 11.5 Simplified schematic diagram of two-port parallel adaptors.

Likewise, we propose the circuit shown in Figure 11.6 for the realisation of three-port parallel adaptors. It is a direct implementation of the algebraic relationships in Table 11.1; that is, the value $-a_N$ is obtained in node N and is then added to a_j to give b_j at the corresponding j-th output. The proposed realisation is very modular, only the geometries of transistors in the γ_j mirror branches have to be modified for each particular adaptor, and always for ratios less than two.

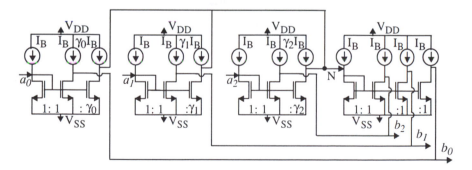

Figure 11.6 Simplified schematic diagram of three-port parallel adaptors.

Concerning series adaptors, they can be transformed into equivalent parallel adaptors as indicated in Figure 11.4(a). Hence, the circuit in Figure 11.6 can be used for series adaptors by modifying the corresponding transistor ratios in such a way that now $-a_N = -(a_0 + a_1 + a_2)$ is obtained at node N and the gains for the output mirror are changed to γ_0, γ_1 and γ_2, respectively. Also, current inverters are necessary at the outputs.

Delay cells:

The output current of the delay element (reflected signal, b) has to be its input current (incident signal, a) delayed by a whole period, i.e. $b(n) = a(n-1)$. But also, since both currents are processed in continuous-time by adaptors, the output must be made available at the same phase that the input is being sampled and hence input and output have to be isolated one from each other.

The cascade of two switched-current memory cells, each built in the conventional form as a current mirror with a switch separating its input and output transistors [1,2], can be used when the respective switches are controlled by two non-overlapping clock phases. Thus, a possible delay cell is that shown in Figure 11.7 in which an improved current mirror is used in order to reduce its output conductance as well as make possible error-free cascading with adaptors realised using cascode current mirrors.

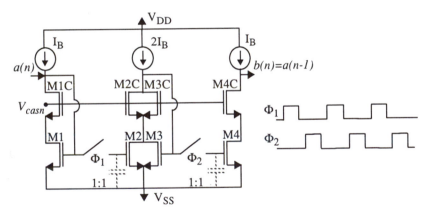

Figure 11.7 Simplified schematic for a whole period delay cell.

An alternative delay cell might be built by resorting to dynamic current mirrors [15,16]. The main advantage of this approach is that it overcomes the matching limitations of standard current mirrors. To provide the required isolation between input and output, two basic dynamic current copiers are needed. Figure 11.8 illustrates two possible arrangements. The circuit in Figure 11.8(a) is a simple cascade connection of two basic current copiers. The first cell samples the input current during the interval Φ_1, the inverse of this current is passed to the second cell during interval Φ_2 and is held as an equivalent voltage in the gate capacitor of transistor M_2. This information is available during the next Φ_1 phase and hence an output which is a whole period delayed version of the input is obtained (this is only a conceptual circuit diagram, for accurate practical implementations more elaborated configurations are required, as can be seen elsewhere in this book). Note that only the mismatch limitation in each basic cell is avoided, the mismatch errors between cells still remains. On the other hand, Figure 11.8(b) illustrates the other alternative which does not suffer from the above limitation because the same transistor is used to sample and to deliver the current. It consists of two cross-coupled basic cells [15], while the input is sampled at the gate capacitor of M_1 during Φ_1, the previous sample, which was held in M_2 during Φ_2, is being delivered. Then, a half-period delayed version of the input is obtained at the output without errors due to mismatching. However, it must be noted that using this half-period delay cell would lead to more complex switching schemes for the overall filter.

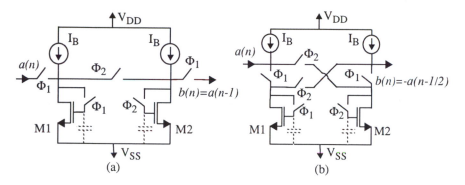

Figure 11.8 Conceptual schematic diagrams of alternative delay cells.

11.3.2   Limitations and non-idealities of the building blocks

In practice, the above circuits suffer from limitations and errors which degrade their actual performance. Here we will focus on studying the speed and accuracy limitations of SI realisations based on standard current mirrors topologies; i.e. for the cells shown in Figures 11.5 to 11.7.

Speed limitation:

The bandwidth of the adaptor building blocks can be approximately studied using the small-signal equivalent circuit for a current mirror shown in Figure 11.9, where R_i and R_o stand for input and output resistances, respectively, R_L and C_L represent the load element, C_1 is the equivalent capacitor of the input transistor and C_2 the equivalent in the output transistor.

Applying this small-signal equivalent to the adaptor in Figure 11.6, the following conclusions can be drawn: (a) Each input signal a_j has two different paths to the output, one through one unity-gain current mirror and the other through two current mirrors; due to the load effect at node N, this latter path imposes the main dynamic limitation; (b) Assuming a

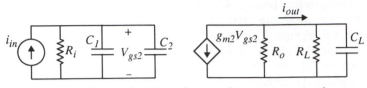

Figure 11.9 Small-signal equivalent for a current mirror.

homogeneous dimensioning in relation to a reference current mirror (that is, all transistors are dimensioned according to the W and L given to transistors of the unity gain current mirror), the transfer function for the most restrictive path can be written as

$$\frac{B_j(s)}{A_j(s)} \approx \frac{\gamma_j g_{m2}^2 R_i R_{ie}}{(1 + 4C_1 R_{ie}s)(1 + (2 + \gamma_j)C_1 R_i s)} \frac{\frac{R_o}{2}(1 + sC_L R_L)}{\left(\frac{R_o}{2} + R_L\right) + sC_L R_L \frac{R_o}{2}} \quad (11.15)$$

Where $R_{ie} = R_i//R_o(\gamma_0)//R_o(\gamma_1)//R_o(\gamma_2) \cong R_i//(R_o/3)$, and where C_1 represents the gate capacitor for the reference transistors; R_i and R_o stand, respectively, for the input and output resistances in the reference mirror and $R_o(\gamma_j)$ represents the output resistance for the output mirror branch corresponding to the γ_j gain; (c) The unloaded parallel adaptor has two poles which can be very close depending on the γ_j values; in the worst case, the speed of the adaptor is limited by a time constant which is four times the time constant of the reference mirror.

On the other hand, provided that switch resistances in the delay cell are negligible compared to the output resistance, the speed of this cell is the same as that of the reference mirror, being limited by a time constant given by $C_1 R_i \cong C_{gs}/g_m$.

In conclusion, adaptors impose the main speed limitation and, as a consequence, appropriate dimensioning for reference mirrors should be carried out taking into account the worst situation in all the filter adaptors in order to allow bandwidths large enough to sufficiently settling of voltage in capacitive nodes.

Accuracy:

There are two main sources of inaccuracy in SI circuits: mismatch (both systematic and random) and clock feedthrough. They cause DC offset, linearity and gain errors. Analytical models for the approximate prediction of these errors have been derived elsewhere for current mirrors and memory cells [3,12]. Practical design issues and layout strategies given in these papers to increase accuracy can be applied directly in our case. Thus, mismatch errors can be significantly reduced by using cascode configurations, appropriate dimensioning and biasing, as well as careful layouts. However, charge injection due to clock feedthrough remains important, becoming the dominant performance limitation unless specific techniques are applied for reducing it. The use of dummy switches or adaptive cancellation circuits are two of these techniques [4]. In practical adaptor implementations, with $I_B = 10\mu A$ and, a reference mirror with transistors of $W/L = 60\mu m/10\mu m$, gain errors in the range 0.01% to 4%, and DC offsets of about $0.2\mu A$ have been obtained. On the other hand, feedthrough produced gain errors of around 19% and DC offsets of $0.2\mu A$ in the whole-period delay cell (measured errors in the half-period delay memory cell were of 11% and $2.14\mu A$, respectively); which confirms the dominant nature of feedthrough induced errors.

The effects of these errors on the performance of each filter component can be analysed by rewriting its wave equations including the new terms. For illustration, let us consider the circuit in Figure 11.10(a) which consists of an equally-terminated capacitor in a parallel branch. The transfer function measured as V_2/V_1 is given by $H(s)=1/(2+ sRC)$.

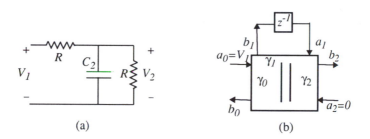

(a) (b)

**Figure 11.10 (a) Capacitor in a parallel branch,
(b) Wave model for circuit in (a).**

The equivalent wave circuit is that shown in Figure 11.10(b), where the adaptor coefficients are given by

$$\gamma_0 = \gamma_2 = \frac{1}{1 + \dfrac{RC}{T}} \qquad \gamma_1 = \frac{\dfrac{2RC}{T}}{1 + \dfrac{RC}{T}} \qquad (11.16)$$

Using the parallel adaptor equations, the following relationships result for the parallel capacitor description

$$\begin{bmatrix} B_0(z) \\ B_2(z) \end{bmatrix} = \begin{bmatrix} \dfrac{(\gamma_0 - 1) + z^{-1}(1 - \gamma_2)}{1 + z^{-1}(1 - \gamma_1)} & \dfrac{\gamma_2(1 + z^{-1})}{1 + z^{-1}(1 - \gamma_1)} \\[3mm] \dfrac{\gamma_0(1 + z^{-1})}{1 + z^{-1}(1 - \gamma_1)} & \dfrac{(\gamma_2 - 1) + z^{-1}(1 - \gamma_0)}{1 + z^{-1}(1 - \gamma_1)} \end{bmatrix} \begin{bmatrix} A_0(z) \\ A_2(z) \end{bmatrix} \qquad (11.17)$$

The transfer function simulating $H(s)$ is that relating B_2 to A_0 when $A_2 = 0$:

$$\frac{B_2(z)}{A_0(z)} = \frac{\gamma_0(1 + z^{-1})}{1 + z^{-1}(1 - \gamma_1)} \tag{11.18}$$

Let us assume now a first-order model for the errors in the current-mode implementation (high-order terms leading to harmonic distortion are neglected); thus, each output current can be described by

$$b_r = \sum_{j=0}^{2} \gamma_j' a_j - a_r + i_{DC,br} \qquad r = 0,1,2 \tag{11.19}$$

where γ_j' are the new coefficients and $i_{DC,br}$ stands for the DC term at the r-th output. Also, clock feedthrough causes the following relationship to apply between the delay cell currents

$$b_1(n) = Ka_1(n-1) + i_{DC,b1} \tag{11.20}$$

where K includes gain errors due to mismatching and feedthrough.

After some manipulation the new relationship between B_2 and A_0 becomes

$$B_2(z) = \frac{\gamma_0'(1 + Kz^{-1})}{1 + z^{-1}K(1 - \gamma_1')} A_0(z) + \frac{(1 + Kz^{-1})i_{DC,b1}}{1 + z^{-1}K(1 - \gamma_1')} +$$

$$\frac{i_{DC,b2}}{1 + z^{-1}K(1 - \gamma_1')} + \frac{\gamma_1' i_{DC,b1}}{1 + z^{-1}K(1 - \gamma_1')} \tag{11.21}$$

Note that in addition to the presence of three new terms due to DC offsets, changes in both the low-frequency gain and the pole location appear as a consequence of inaccuracy in the coefficients γ_0 and γ_1. The same applies for a gain error in the delay cell, but this error also affects the position of the zero.

In evaluating the effects of feedthrough errors, the transfer function (11.21), including only the gain errors due to feedthrough, has been simulated for the particular case of $\gamma_0 = 0.4$, $\gamma_1 = 1.2$ (corresponding to a continuous -3dB normalised frequency of 0.186) and $K = 0.8$. Compared with the ideal response, deviations of 13.7% in the low-frequency gain and of 6.6% in the -3dB frequency have been obtained. In illustrating the same feedthrough effects in the alternative integrator-based SI implementation, numerical simulations have been carried out using a bilinear integrator realised with the same basic memory cell used before. The obtained results were a deviation of 22.7% in the low-frequency gain and of 8% in the -3dB frequency. As a conclusion we can say that similar accuracy is achieved with both SI realisations, and that feedthrough reduction techniques are indispensable in high accuracy applications.

11.4 Programmable Band-pass Filter Structures

11.4.1 Filter synthesis

As mentioned in Section 11.2, one of the methods for realising a band-pass filter is based on an analogue continuous low-pass (LP) reference filter making the LP to BP transformation directly in the z-domain and then appropriately altering filter elements. This method leads to a universal wave filter structure which allows the values of the coefficients to be changed, and hence the filter function to be modified, without altering the global wave structure and not requiring the return to the analogue reference filter [17,18]. We have explored this method, achieving SI programmable filters by providing circuit implementations including variable elements [19].

First, let us illustrate the method in [17]. Consider again the doubly terminated ladder *LC* low-pass filter in Figure 11.3(a) to be the reference filter; its corresponding wave signal flow graph was shown in Figure 11.3(b), where A_{ij} and B_{ij} are the incident and reflected wave signal, respectively, in port i of adaptor j. The following $z^{-1} \rightarrow g(z^{-1})$ transformation is applied to develop a universal WDF structure:

$$g(z^{-1}) = \frac{z^{-2} - \beta(1 + \alpha)z^{-1} + \alpha}{\alpha z^{-2} - \beta(1 + \alpha)z^{-1} + \alpha} \qquad |\alpha| < 1, \ |\beta| \le 1 \qquad (11.22)$$

where parameters α and β characterise the normalised frequency warping $\Omega \rightarrow \underline{\Omega}$ associated with the mapping $z^{-1} \rightarrow g(z^{-1})$. Considering the reference low-pass filter whose bandwidth equals one fourth of the sampling frequency, and calling Ω_u and Ω_l the required upper and lower frequencies, respectively, parameters α and β are given by

$$\alpha = \frac{1 - tan\left(\dfrac{\Omega_u - \Omega_l}{2}\right)}{1 + tan\left(\dfrac{\Omega_u - \Omega_l}{2}\right)} \qquad \beta = \frac{cos\left(\dfrac{\Omega_u + \Omega_l}{2}\right)}{cos\left(\dfrac{\Omega_u - \Omega_l}{2}\right)} \qquad (11.23)$$

It is shown in [17] that the effect of applying the mapping $z^{-1} \rightarrow g(z^{-1})$ is equivalent to making two modifications in the reference wave signal flow graph. One modification affects the port resistances determined by the analogue elements (R_{1j} in Figure 11.3(b)): the new values are obtained by multiplying the resistances of the ports terminated in z^{-1} by a factor $(1-\alpha)/(1 + \alpha)$ and the resistances of ports terminated in $-z^{-1}$ by the factor $(1 + \alpha)/(1 - \alpha)$. The second modification is equivalent to replacing z^{-1} by $-z^1(z^{-1} - \beta)/(1 - \beta z^{-1})$ in the reference filter. This can be realised by using a two-port adaptor with β as the value of its reflection coefficient μ In Figure 11.11 we show the resulting modified wave filter after applying $g(z^{-1})$ to the filter in Figure 11.3(b). In Figure 12.11 R'_{1j} stands for the new values of the R_{1j} which must be obtained as it has previously explained.

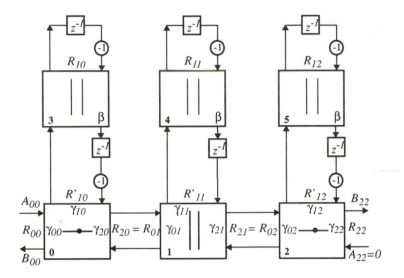

Figure 11.11 Modified wave circuit after applying the $z^{-1} \rightarrow g(z^{-1})$ transformation.

It must be noticed that with this transformation parameter, α controls the bandwidth and parameter β controls the centre frequency. The value of α affects the values of the R_{1j} and then modifies the parameters γ_{ij} in the three-port adaptors. Changing β gives a new value of the reflection coefficient in the two-port parallel adaptors (all of them have the same value). Hence, it can be concluded that we can independently and simultaneously modify the bandwidth and the centre frequency by varying the adaptor's coefficients.

11.4.2 *Programmable filter implementation*

For practical implementations of SI programmable filters we must provide a way to vary the γ_{ij} parameters. Since these parameters are obtained using current mirror gains, we can modify gains by altering the aspect ratio of transistors in the second branch of the mirror. This is easily done with an array of transistors connected in parallel through a set of digitally controlled switches. Figure 11.12 illustrates this approach, where $\gamma_{ij} = (W/L)_{ij}/(W/L)_I$ and the value of $(W/L)_{ij}$ is determined by the binary word $p_0 p_1$ as

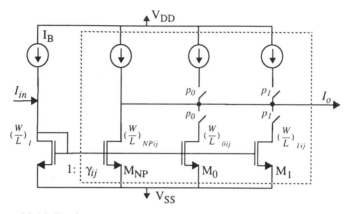

Figure 11.12 Basic programmable current mirror with γ_{ij} gain.

$$\left(\frac{W}{L}\right)_{ij} = \left(\frac{W}{L}\right)_{NPij} + p_o\left(\frac{W}{L}\right)_{0ij} + p_1\left(\frac{W}{L}\right)_{1ij} \qquad (11.24)$$

An important problem to take into account is the quantisation errors due to the limitations imposed by the technological grid. To reduce these errors we will take advantage of the available freedom, choosing parameter values to fulfil two requisites: (a) the quantisation error in each coefficient must be as small as possible and (b) the values of γ_{ij} for each adaptor corresponding to different frequency specifications should be achieved through an array of transistors. In the next section some results obtained for a particular example will be presented.

An alternative way could rely on using mirrors with electrically adjustable gains, but this idea has to be further developed for our case because parameters γ_{ij} have to be changed in such a way that their sum must always remain equal to two.

11.5   Design Examples

To demonstrate the feasibility of the proposed realisation for SI wave filters, we include the following design examples.

Example 1:

A fifth-order Chebyschev low-pass filter with a cut-off frequency one-eighth the clock frequency and a 0.177dB passband ripple has been built using a microwave reference filter. Its wave model consists of six two-port adaptors, interconnected as shown in Figure 11.2, whose respective parameters are $\mu_1 = -\mu_6 = -0.5459$, $\mu_2 = -\mu_5 = 0.8356$ and $\mu_3 = -\mu_4 = -0.8904$. The circuit was designed for a 1.5µm 5V standard CMOS process, the area of the whole chip including pads was 2.4mm². A bias current of 8µA and transistors of the unity-gain mirror having $W/L =$

30μm/6μm were used. The experimental frequency response for a 20kHz clock frequency is shown in Figure 11.13; the corresponding measured data were a 2.6kHz cut-off frequency and a 0.5dB ripple. Also, a dynamic range of 50dB and a THD of 1.5% were obtained in the passband. In Figure 11.14 transfer functions obtained for different clock frequencies are shown.

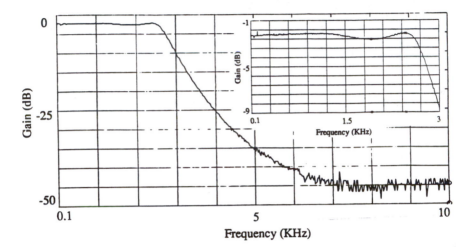

Figure 11.13 Measured frequency response for a 20kHz clock frequency.

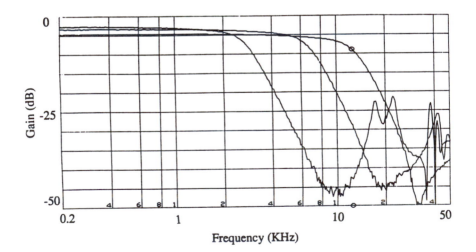

Figure 11.14 Experimental frequency responses for 20kHz, 50kHz and 100kHz clock frequencies.

Example 2:

An *LC* reference filter implementing a third-order Chebyschev low-pass filter was chosen to demonstrate the second method of implementation based on the use of three-port adaptors. The diagram of the filter is given in Figure 11.3(b), $L_1 = L_3 = 1.0315$, $C_2 = 1.1474$ and $R_i = R_o = 1$ are the normalised values of its components. For a $\omega_o T = 1$ and a 0.1dB ripple, the chosen adaptor coefficients were: $\gamma_{00} = 0.3462$, $\gamma_{10} = 0.6537$, $\gamma_{20} = 1$, $\gamma_{01} = \gamma_{21} = 0.2479$, $\gamma_{12} = 1.5040$, $\gamma_{02} = 1$, $\gamma_{11} = 0.5537$ and $\gamma_{22} = 0.3462$. The circuit was designed in a 1.5μm CMOS process, taking a bias current $I_B = 10$μA, and using $(W/L)_I = 60$μm/10μm transistors in the reference mirror. Figure 11.15 illustrates the experimentally obtained frequency response for a 20kHz clock. A 3.22kHz cut-off frequency and a 0.09dB ripple have been measured.

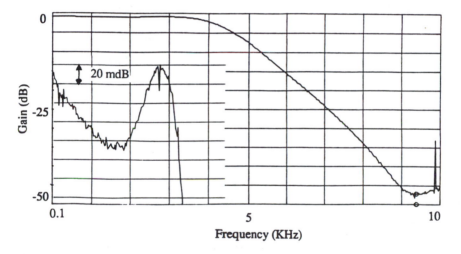

Figure 11.15 Experimental frequency responses for 20kHz clock frequencies.

Example 3:

An *LC* ladder reference filter for a third-order low-pass Chebyschev filter with 0.2dB ripple in the passband and a normalised 0.25 cut-off frequency was chosen to realise a programmable band-pass filter with three different normalised bandwidths ($B_1 = 0.0246$, $B_2 = 0.0563$, $B_3 = 0.0957$) and three normalised centre frequencies ($\Omega_{o1} = 0.184$, $\Omega_{o2} = 0.20$, $\Omega_{o3} = 0.218$), all of them independently controlled.

The values of the analogue elements in the reference filter (which has the same circuit diagram as in Figure 11.3(a)) are $L_1 = L_3 = 1.2275$, $C_2 = 1.1525$ and $R_i = R_o = 1$. Its wave equivalent circuit is equal to that in Figure 11.11. The values of the wave filter parameters are shown in Tables 11.2

and 11.3. Note that the symmetry in the filter adaptors corresponds to the symmetry in the reference filter.

	Ω_{01}	Ω_{02}	Ω_{03}
β	0.40	0.30	0.20

Table 11.2 β coefficients for the three centre frequencies.

	B_1	B_2	B_3
γ_{00}, γ_{22}	0.100	0.200	0.300
γ_{10}, γ_{12}	1.580	1.373	1.187
γ_{20}, γ_{02}	0.320	0.427	0.513
γ_{01}	0.040	0.127	0.239
γ_{11}	1.919	1.746	1.521
γ_{21}	0.040	0.127	0.239

Table 11.3 γ_{ij} values for the three bandwidths.

Simulations were carried out with WAVER to determine the scaling requirements. Since levels greater than 20dB were obtained for some signals, we applied our scaling procedure. The result was a modified wave structure in which all three-port adaptors are of the parallel type and some ideal transformers, with the scaling factor $k_0 = 2.87$dB, $k_1 = -0.5$dB, $k_3 = 2.66$dB, $k_4 = -1.55$dB and $k_5 = -0.97$dB, were needed.

For the switched-current implementation, we used the same building blocks as that in the example 2. We rely on the circuit diagram in Figure 11.12 to implement all γ and β values: a word $p_0 p_1$ controls the γ values while a word $p_2 p_3$ controls the β values. Table 11.4 shows the W of transistors M_{NP}, M_0, and M_1 ($L = 10\mu$m for all of them) corresponding to each γ and β coefficient.

	W_{NPij}	W_{0ij}	W_{1ij}
γ_{00}, γ_{22}	6.00	6.00	6.00
γ_{10}, γ_{12}	71.20	12.40	11.20
γ_{20}, γ_{02}	19.20	6.40	5.20
γ_{01}	2.40	5.20	6.80
γ_{11}	91.20	10.40	13.60
γ_{21}	2.40	5.20	6.80
β	12.00	6.00	6.00

Table 11.4 Transistor Widths (μm) for the three B and Ω.

Electrical simulations of the nine filters have been carried out with HSPICE for a 100kHz clock frequency. Figure 11.16 illustrates the programming of the bandwidth: three responses are shown corresponding to the Ω_{o1} centre frequency. Finally, Figure 11.17 shows the programming of the centre frequencies illustrated for the B_2 bandwidth. The chip is currently under fabrication in a 1.5μm CMOS process.

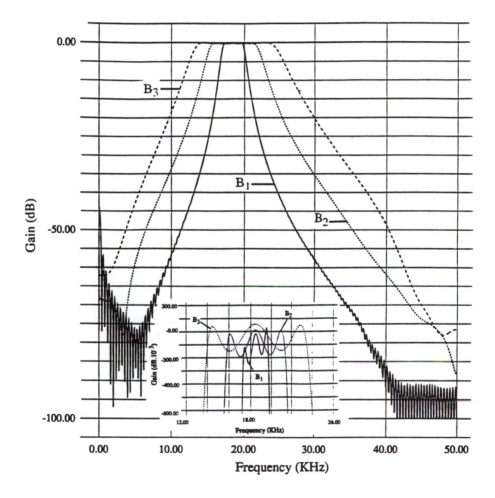

Figure 11.16 Simulated response for the B_1, B_2 and B_3 bandwidths at Ω_{o1} centre frequency.

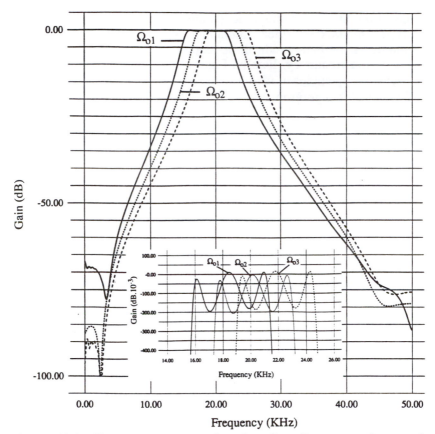

Figure 11.17 Simulated response for Ω_{o1}, Ω_{o2}, Ω_{o3} centre frequencies.

11.6   Conclusions

A new approach for the realisation of switched-current filters has been presented. The simplicity and reliability of the new approach makes it a powerful alternative to other reported realisations based on SI integrators, offering additional advantages such as modularity, easy programmability and simple building blocks, whose performance characteristics are independent of the particular filter. Like other SI filter realisations, the proposed method has to be improved to further reduce the effect of mismatch and feedthrough errors.

References

[1]   J. B. Hughes, I. C. Macbeth and D. M. Pattullo, "Switched currents filters," *Proc. IEE*, Pt. G, vol. 137, pp. 156-162, April 1990.

[2]   J. B. Hughes, "Switched current filters," in C. Toumazou, F. J. Lidgey and D. G. Haigh, Eds, *Analogue IC Design: the current-mode approach*, London: Peter Peregrinus, 1990.

[3]   T.S. Fiez and D. J. Allstot, "CMOS switched-current ladder filters," *IEEE J. Solid State Circuits*, vol. SC-25, pp. 1360-1367, Dec. 1990.

[4]   T. S. Fiez and D. J. Allstot, "Switched-current circuit design issues," *IEEE J. Solid State Circuits*, vol. SC-26, pp. 192-202, March 1991.

[5]   A. Fettweis, "Wave digital filters: Theory and practice," *Proc. IEEE*, vol. 74, pp. 270-327, Feb. 1986.

[6]   A. Fettweis, "Pseudopassivity, sensitivity and stability of wave digital filters," *IEEE Trans. Circuit Theory and Appl.*, vol. CT-19, pp. 668-673, Nov. 1973.

[7]   A. Fettweis and K. Meerkîtter, "Suppression of parasitic oscillations in wave digital filters," *IEEE Trans. Circuits Syst.*, vol. CAS-22, pp. 239-246, March 1975.

[8]   J. Mavor, H. M. Reekie, P. B. Denyer, S. O. Scanlan, T. M. Curran and A. Farrag, "A prototype switched-capacitor voltage-wave filter realized in NMOS technology," *IEEE J. Solid State Circuits*, vol. SC-16, pp. 716-723, Dec. 1981.

[9]   U. Kleine, "Design of wave-SC filters using building blocks," *Int. Journal Circuit Theory and Appl.*, vol. CT-12, pp. 69-87, 1984.

[10]   A. Rueda, A. Yúfera and J. L. Huertas, "Wave analog filters using switched-current techniques," *Electronics Letters*, vol. 27, pp. 1482-1483, August 1991.

[11]   A. Yúfera, A. Rueda, J. L. Huertas, "Switched-current wave analog filters," in *Proc. International Symposium on Circuits and Systems*, pp. 859-862, May 1992.

[12]   R. T. Baird, T.S. Fiez, D. J. Allstot, "Speed and accuracy considerations in switched-current circuits," in *Proc. International Symposium on Circuits and Systems*, pp. 1809-1812, 1991.

[13]   B. Frohlich, A. N. Venetsanopoulos and A. S. Sedra, "On the dynamic range of wave digital filters," *IEEE Trans. Circuits Syst.*, vol. CAS-27, pp. 964-967, Oct. 1980.

[14]   K. S. Thyagarajan and E. Sanchez-Sinencio, "A systematic procedure for scaling wave digital filters," *Proc. IEEE*, vol. 66, pp. 512-513, April 1978.

[15]   E. Wegmann and E. A. Vittoz, "Basic principles of accurate dynamic current mirrors," *Proc. IEE*, vol. 137, Pt. G., pp. 95-100, April 1990.

[16]   E. A. Vittoz and G. Wegmann, "Dynamic current mirrors," in C. Toumazou, F. J. Lidgey and D. G. Haigh, Eds, *Analogue IC Design: the current-mode approach*, London: Peter Peregrinus, 1990.

[17]   H. S. El-Ghoroury and S. C. Gupta, "Wave digital filters structures with variable frequency characteristics," *IEEE Trans. Circuits Syst.*, pp. 624-630, Oct. 1976.

[18]   M. N. Swamy and K. S. Thyagarajan, "Digital bandpass and bandstop filters with variable center frequency and bandwidth," *Proc. IEEE*, pp. 1632-1634, Nov. 1976.

[19]   A. Yúfera, A. Rueda and J. L. Huertas, "A methodology for programmable switched-current filter design," in *Proc. European Conf. Circuit Theory and Design*, 1993.

Chapter 12

Algorithmic and Pipelined A/D Converters

David Nairn

12.1   Introduction

Presently there are two approaches to the implementation of analogue-to-digital converters (ADCs) in VLSI systems. Both approaches seek to minimise the impact of manufacturing variations on the ADC's accuracy but they achieve this goal in radically different ways. In one approach, the signal is sampled significantly faster than the Nyquist rate and then digital signal processing, which is not affected by manufacturing variations, is used to complete the conversion. The second approach uses analogue circuit techniques that are insensitive to manufacturing variations to process the signal at, or near, the Nyquist rate. The first approach, commonly referred to as over-sampling, is described in Chapter 14. The second approach, which may be referred to as Nyquist sampling, is the subject of this chapter.

Nyquist sampling ADCs have been used for many years and have been implemented using many different architectures. Some of the more common architectures include the dual slope, successive approximation and flash ADCs. Each of these architectures offer advantages that make them attractive for specific applications [1]. For these converters, the ADC has usually been implemented as a single-chip or a multi-chip-module. The ADC is then combined with other components to implement a board-level system.

Recent advances in VLSI technology have made it possible to reduce many board-level systems to a single VLSI chip. These single-chip systems are typically smaller with lower power consumption and higher reliability than board-level systems. Consequently, single-chip solutions for many systems are currently being sought.

When the ADC is part of a single-chip mixed analogue/digital system additional demands are placed on the ADC to ensure compatibility with the surrounding system:

- Due to device mismatches and the high cost of post-process trimming, the ADC should not rely on closely matched devices.

- Since most systems consist primarily of digital circuits intended for fabrication using low-cost digital processes, the use of specialised analogue components such as linear capacitors and resistors should be avoided in the ADC.

- Since the ADC is typically a very small part of the final system, the converter's overall size should be minimised.

Switched-current circuit techniques [2-4] can be used to solve the first two problems and the third problem can be solved with an appropriate selection of the ADC's architecture.

In this chapter, the implementation of Nyquist sampling switched-current ADCs is discussed. Since the architecture determines both the converter's performance and the type of subcircuits needed to implement it, the various ADC architectures that have been used to implement switched-current ADCs will first be considered. Then, the necessary switched-current building blocks and their practical implementations will be discussed in the context of the ADC's requirements. Finally, the performance of some switched-current ADCs will be reviewed and their major limitations considered.

12.2 Architectures For Switched-Current ADCs

At the time of this writing, switched-current ADCs tend to be either algorithmic ADCs or pipelined ADCs. The algorithmic type is usually used to achieve a small converter size while the pipelined type is used to achieve high sampling rates. Both of these architectures require relatively little chip area making them attractive for VLSI systems applications.

12.2.1 Algorithmic ADCs

The algorithmic or cyclic ADC technique uses a binary search to perform the analogue-to-digital conversion. The technique is similar to the successive approximation technique but uses much simpler hardware to produce an ADC with a very small size. Based on the available hardware, an algorithmic ADC can be built either with multipliers or dividers.

(a) Multiplying algorithmic ADCs

A multiplying algorithmic ADC performs a conversion by repeatedly adjusting the input signal and comparing it to a fixed reference signal, as shown in Figure 12.1. To perform a conversion, the input signal Y_{in}, which can vary from zero to twice the reference level, Y_{ref}, is first compared to the reference. If Y_{in} is less than Y_{ref}, the digital output is set to zero and the signal is doubled to obtain $2Y_{out}$. If Y_{in} is greater than Y_{ref}, the digital output is set to one and the reference is subtracted from the signal before multiplying it by two. The resulting signal ($2Y_{out}$) is then fed back to the input where it becomes the new input signal, Y_{in}. The procedure is repeated until the desired number of bits have been obtained.

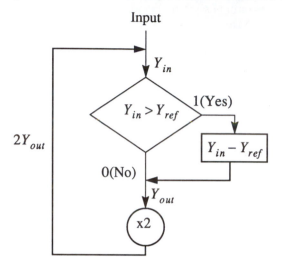

Figure 12.1 Flow chart of a multiplying algorithmic ADC.

(b) Dividing algorithmic ADCs

Unlike a multiplying algorithmic ADC, a dividing algorithmic ADC performs a conversion by adjusting both the input signal and the reference signal, as shown in Figure 12.2.

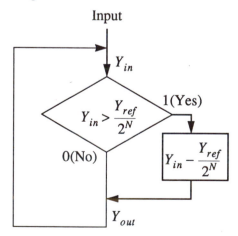

Figure 12.2 Flow chart of a dividing algorithmic ADC.

To perform a conversion, the input signal Y_{in}, which again can vary from zero to twice the reference level, is first compared to the reference, Y_{ref} ($N = 0$). If Y_{in} is less than Y_{ref} the digital output is set to zero and Y_{in} remains unchanged. If Y_{in} is greater than Y_{ref}, the digital output is set to one and the reference is subtracted from the signal. The reference is then divided by two before making the next comparison. Consequently, the reference decreases by half for each successive comparison. Like the multiplying version, the algorithm is repeated until the desired number of bits have been obtained.

Both the multiplying and dividing techniques achieve their small size by making repeated use of some relatively simple hardware. For the multiplying approach, a comparator, a summer and a multiply-by-two are required. For the dividing approach, a comparator, a subtractor and a divide-by-two are required. Consequently, very small converters can be realised using either of these two techniques.

The primary drawback of the algorithmic approach is its relatively slow speed. Since the algorithm is performed sequentially, the conversion time is proportional to the desired resolution. As a result, converters of this type have typically been restricted to telephone and audio applications where sampling rates of 8kHz to 100kHz are typical [5-9].

12.2.2 Pipelined ADCs

To achieve higher sampling rates, the algorithmic technique can be pipelined. As Figure 12.3 shows, a pipelined ADC achieves higher sampling rates by working on more than one sample simultaneously. The complete converter consists of M stages each of which has an ADC, a DAC, a subtractor and a sample-and-hold (S/H). Although not strictly necessary, the resolution of the ADCs in each stage are usually equal, to simplify the design. Hence, the pipelined ADC's resolution will be $M \times N$, where N is the resolution of the ADC used in each stage. The conversion time will be approximately M times that of the N-bit converter but, due to the simultaneous processing of M different samples, the pipelined ADC's sampling rate is determined by the speed of the low resolution N-bit ADCs.

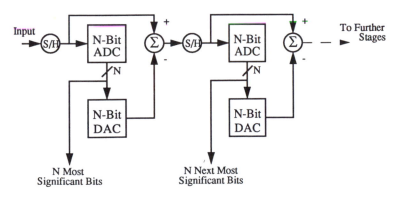

Figure 12.3 Block diagram of a pipelined ADC.

The size and speed of a pipelined ADC is largely determined by the ADCs used in each stage. By simplifying each stage to the 1-bit level, the size of each stage is minimised and the speed is maximised. For this situation, the reference level used in each successive stage will be half that used in the preceding stage and the pipelined ADC will be similar to the dividing algorithmic ADC, except that the signal flows through a cascade of M cells instead of being recirculated through the same cell M times. Alternatively, a gain of two can be added between each stage, thereby allowing identical 1-bit ADCs to be used in each stage. For this situation, the pipelined ADC will be similar to the multiplying algorithmic ADC. Consequently, pipelined ADCs of this type will have similar performance to the algorithmic ADCs except that they will be larger and faster by a factor equal to the converter's resolution [8].

12.2.3   Summary

The component requirements, for both the algorithmic and pipelined ADCs are relatively simple; all converters require a S/H, a summer and a comparator. For the multiplying approach, a multiply-by-two will be required while a divide-by-two, will be required for the dividing approach. As will be shown in the following section, these components are simple and small in size when implemented using switched-current techniques. Consequently switched-current algorithmic and pipelined ADCs can be made very small.

12.3   Building Blocks

The building blocks used in switched-current ADCs consist primarily of summers, current-samplers, current divide-by-twos and current comparators. The first component, the summer, is not really a component at all; a summation is performed simply by connecting the outputs of the desired current sources together. Consequently, to perform addition, subtraction and integer multiplication, all that is necessary is to have the desired signals available as weighted currents. The remaining components, the sampler, divide-by-two and comparator can all be implemented in standard digital CMOS technologies and do not require any special analogue components.

12.3.1   Current-samplers

The basic element in any switched-current circuit is the current sampling circuit [2,7,10,11]. Current-samplers are used both as a sample-and-hold (S/H) and, like current mirrors, to generate multiple copies of a single current. Unlike current mirrors, the accuracy of the copied current in a switched-current circuit is not limited by device matching. Therefore, relatively high accuracy circuits can be fabricated without the need for well matched devices. For more details on current-samplers, their design and their limitations, see Chapters 3-6. In this section the relevant aspects of current-samplers as they apply to ADCs will be considered briefly.

The basic current-sampler is illustrated in Figure 12.4. For this circuit, a copy of the input current is obtained by closing S_1 and S_2 which allows I_{IN} to flow through transistor N_1. When the gate voltage has settled, its value is stored by opening S_1 thereby trapping the gate's charge. Now, by opening S_2 and closing S_3, I_{OUT} is made equal to I_{IN}.

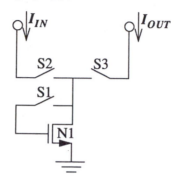

Figure 12.4 A basic current-sampler.

The basic current-sampler is only adequate for unidirectional currents. In many applications bidirectional currents are desirable. Two techniques for generating bidirectional samplers are shown in Figure 12.5.

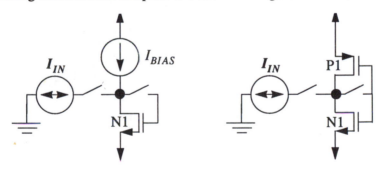

a) **One controlled device**   b) **Two controlled devices**

Figure 12.5 Bidirectional current-samplers.

In the first approach, the basic current-sampler is modified by adding a constant bias current, I_{BIAS} as shown in Figure 12.5a. Consequently, this sampler can source currents up to I_{BIAS}. Typically, I_{BIAS} is made larger than required to ensure that N_1 is not operated with very low current levels which would result in long settling times [12]. The maximum current that can be sunk depends on the size of N_1 and its maximum gate voltage. For the second approach, two controlled transistors are used as shown in Figure 12.5b. For this circuit, the signal swing is determined by the size of N_1

and P_1 along with the allowable voltage swing at their gates. Typically, it is easier to keep the transistors in circuit (a) operating in the desired mode of operation, leading to more consistent performance. On the other hand, circuit (b) can be designed to have a much lower power dissipation for a given current range. Consequently, the choice of which configuration to use will depend on the specific application.

Ideally, for all three current-samplers discussed above, the output current will be identical to the input current. For practical current-samplers though, I_{OUT} deviates from the ideal value due to the finite output resistance, r_o, of the current sampling transistor(s). Although a long channel device could be used to increase r_o, such an approach is not desirable because the use of long channel devices leads to much longer settling times [12]. Two alternatives have been used to reduce this problem. One is to use a cascode configuration to increase the output resistance [13]. The other is to use an active current mirror to reduce the effective input resistance [14].

(a) Cascoded current-samplers

Cascoded current-samplers can be implemented using either the saturation mode as shown in Figure 12.6a or the triode mode as shown in Figure 12.6b.

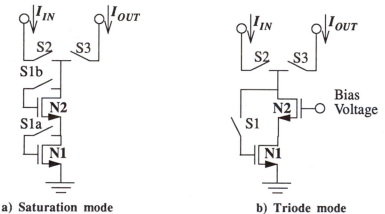

a) Saturation mode   b) Triode mode

Figure 12.6 Structures for cascoded current-samplers.

Although use of the saturation mode leads to a larger output resistance, it also usually leads to a reduced dynamic range and an increased susceptibility to switch-induced charge injection problems. Consequently, the triode mode of operation is usually the preferred choice. To improve the output resistance of the triode mode current-sampler, a regulated cascode structure can be used as shown in Figure 12.7 [15]. Even with the addition of the regulated cascoded circuitry, cascoded current-samplers are relatively small, making them very attractive for use in VLSI systems.

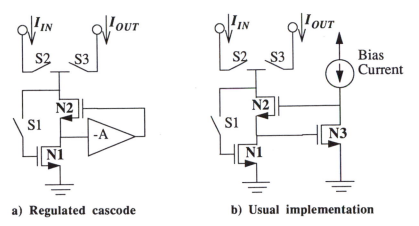

a) Regulated cascode   b) Usual implementation

Figure 12.7 Regulated cascode current-samplers.

(b) Active current-samplers

The alternative to a cascoded current-sampler is the active current-sampler shown in Figure 12.8 [7,14]. The active current-sampler uses an amplifier to allow the gate voltage to be set to the desired level while holding the drain-to-source voltage of the sampling transistor constant. Consequently, for an ideal amplifier, the active current-sampler's input appears as a virtual short to ground and hence, does not load the input current source. In a practical circuit, the current-sampler's effective input impedance is reduced by a factor equal to the amplifier's gain. To ensure that the active current-sampler's output is not loaded by a following sampler's input, the active current-sampler's output should be connected to the input of another active current-sampler with the same bias voltage. Hence, the active current-sampler reduces the effects of the transistor's finite r_o, not by increasing its output resistance but by reducing its input resistance.

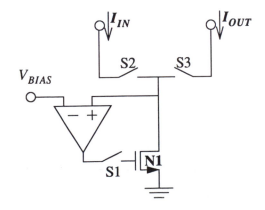

Figure 12.8 An active current-sampling circuit.

The amplifier used in the active current-sampler can be either a voltage amplifier [14] or a transimpedance amplifier [8,10,16]. The use of a transimpedance amplifier has been shown to significantly reduce the current-sampler's settling time, thereby allowing significantly faster sampling rates to be achieved in a given technology [16].

The addition of an amplifier into the current-sampler has both disadvantages and advantages. The primary disadvantages are an increased complexity and power consumption, on the other hand, active current-samplers are typically faster than the cascode current-samplers and provide more flexibility in the circuit design.

12.3.2 Current divide-by-two

A current divide-by-two is used to provide an output current that is exactly half that of the input current. Unlike a current multiply-by-two which can be implemented simply by summing the outputs of two current-samplers that have been used to sample the same input current, an accurate divide-by-two requires a special circuit and an iterative procedure [17,18].

Theoretically, a divide-by-two can be achieved with a current mirror in which the output device is scaled to have half the width of the input device. Due to the inevitable device mismatches though, such a divider will not be very accurate. To achieve better accuracy, the switched-current circuit shown in Figure 12.9 can be used.

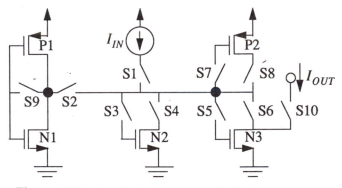

Figure 12.9 A switched-current divide-by-two.

A three-phase clock and a few iterations are required to provide an output current equal to half its input current. The circuit operates as follows. Initially, the bidirectional current-sampler is set to provide zero current[1]. For the first phase, switches S_1 - S_6 are closed thereby causing I_{IN} to flow through N_2 and N_3. The proportion of the current in each device will then depend on their relative sizes. At the end of the first phase, the currents in N_2 and N_3 are sampled by opening S_3 and S_5. In the second phase, only switches S_6, S_7 and S_8 are closed thereby causing the current stored in N_3 to

---

[1] This condition is not strictly necessary for proper operation.

be transferred to P_2. Then in the third phase, switches S_2, S_4, S_8 and S_9 are closed, causing the difference between the currents stored in P_2 (originally stored in N_3) and N_2 to be stored in the bidirectional current-sampler. This operation is then repeated as many times as required to achieve the desired accuracy for the current stored in N_3. Typically, no more than five iterations will be required to achieve good accuracy [17]. Finally, S_{10} is closed to provide an output current equal to $I_{IN}/2$.

Due to their recursive nature and the more complicated circuitry, a divide-by-two is typically both slower and larger than a multiply-by-two. Nevertheless it is an attractive solution to many design problems.

12.3.3 Current comparators

Comparators are an essential part of any ADC. In a current-mode ADC, the signal current is compared to a known reference current and a digital output indicating their relative sizes must be produced both quickly and accurately. Current comparators, like voltage comparators can be implemented in two ways; as open-loop comparators or as closed-loop (positive feedback) comparators.

(a) Open-loop current comparators

A simple open-loop current comparator can be implemented in MOS technology as shown in Figure 12.10. The comparator is a cascade of two CMOS inverters. If I_{IN} is greater than I_{REF}, V_{OUT} will be high, otherwise it will be low. Although the comparator is relatively simple, it is quite small and does not suffer from any DC offsets. However, the comparator's threshold level is affected by manufacturing variations in the n and p channel devices. A better approach is to use a differential input stage such that the comparator's threshold level is determined by a fixed bias voltage. Although the comparator can be accurate, long settling times will occur if $I_{IN} \approx I_{REF}$. In this case, only a very small difference current will be available to charge the input capacitance. Consequently, clamping circuitry is often used to limit the signal swing at the input. In spite of their slow speed, comparators of this type, implemented either with simple inverters or with a differential pair are attractive for use in simple ADCs due to their accuracy and small size.

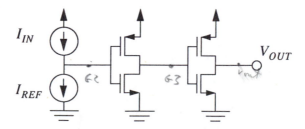

Figure 12.10 A simple current comparator.

(b) Closed-loop current comparators

Closed-loop latching comparators are a better choice if a very fast comparison is desired. Comparators of this type use a two-phase clock and feedback to produce a very fast circuit. The comparator itself is based on a high-gain amplifier to which both negative and positive feedback can be applied. During one clock phase, typically called the setup phase, negative feedback is applied to clamp the input and output at the comparator's threshold level. Then during the second phase, typically called the comparison phase, positive feedback is applied to produce a high-gain, bistable circuit. During the comparison, a small difference between the input signal and the reference is quickly amplified by the regenerative action of the positive feedback until the circuit saturates at either its high or low level depending on the polarity of the original difference at the input. Circuits of this type have typically been implemented using voltage-mode techniques but can also be implemented using current-mode techniques.

A closed-loop current comparator is shown in Figure 12.11 [19] and operates using a two-phase clock. During the setup phase, the Setup control is high thereby clamping the voltage at the two inputs to a value close to V_{REF}. Then, during the comparison phase, Setup is set to zero thereby allowing the comparator to operate as a positive feedback structure that settles quickly to the final value. Circuits of this type are very fast and compact [19,20]. The primary drawback is the inevitable device mismatches that lead to offsets thereby limiting the comparator's resolution. Nevertheless, recent closed-loop comparators have displayed comparison times of under 2ns with a 200nA resolution [19].

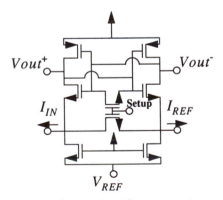

Figure 12.11 A clamped input, latching current comparator [19].

12.3.4   Switches in switched-current circuits

Switches are used for two distinct purposes in switched-current circuits: as current-steering switches and as voltage-sampling switches. The nature of these two tasks is sufficiently different that switches for each purpose should be considered separately.

(a) Current-steering switches

Current-steering switches, such as S_2 and S_3 in Figure 12.4, can lead to two problems. Firstly, the switch's on-resistance leads to a signal dependent voltage drop. Secondly, when the switches are turned off, transient currents are generated that can inject noise into the circuit. Fortunately, both of these problems can be reduced to a reasonable level.

The signal-dependent voltage drop itself has two potential problems. Firstly, if large currents are used in the circuit the switch's voltage drop can limit the maximum current for which the current sourcing or sinking device will remain in saturation. Therefore, short and wide switches should be used if the circuit is to handle large currents. Secondly, due to the finite r_o of the current-samplers, even if the switches are used to switch the current-sampler between nodes with the same potential, differences in the on-resistance of the two switches will lead to differences in the current at the two nodes. Therefore, all current-steering switches connected to a current-sampler should be designed with identical geometries. Consequently, all current-steering switches should be identical and designed for a low on-resistance to avoid problem caused by the switch's on-resistance.

The current-steering switches also generate switching transients that can change the value of the signal stored in a current-sampler. Switching transients arise from three causes:

- If a current-sampler is left open-circuited, the output node will quickly rise or fall to the limits of the power supply.

- If the current-sampler's output is switched between nodes of differing potentials the current-sampler's output node must be charged to the new potential.

- Charge injection from the current-steering switches can be coupled into the signal path.

In all cases, the switching transients are coupled by the current sampling transistor's gate-to-drain capacitance to the gate capacitance, leading to a change in the stored gate voltage and hence the sampled current. To minimise these problems, the current-sampler's output should never be left open-circuited and the current-sampler's output should only be switched between nodes with similar potentials using make-before-break switch timing[2]. If the switching transients are still too large, a circuit such as that shown in Figure 12.12 should be used to buffer the sampled gate voltage. Consequently, with the use of an appropriate switching sequence and by buffering the gate voltage storage capacitor, the effects of the current-steering switching transients can be reduced to an acceptable level.

---

[2] To ensure equal potentials, the use of an active current-sampler is preferred.

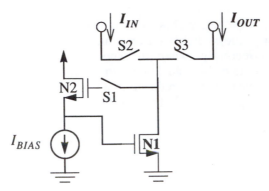

Figure 12.12 A buffered current-sampler.

(b) Voltage-sampling switches

Voltage-sampling switches, such as S_1 in Figure 12.4, are used to sample the required gate voltage. Therefore, charge injection is a potential problem for switches of this type. To reduce the charge injection, the smallest possible switch size should be used. If the charge injection is too large, a charge injection reduction technique such as a charge divider [21] or the use of half-sized dummy switches should be used [2].

12.3.5 Summary

The switched-current circuits discussed above can all be fabricated using standard digital CMOS processes and, through careful attention to their major error sources, can be accurate circuits. Consequently, these circuits are well suited for the implementation of ADCs in a digital process.

12.4 Converter Implementation and Examples

The implementation of ADCs based on switched-current techniques has been restricted to converters based on the algorithmic or the pipelined approaches discussed in Section 12.2. The performance of some published switched-current ADCs are shown in Table 12.1. Typically the algorithmic ADCs are slower than the pipelined ADCs but also consume much less power. Some design issues related to the implementation of the two types of converters will be discussed below.

12.4.1 Algorithmic ADCs

Algorithmic ADCs are usually implemented using the multiplying approach. The resulting ADCs use relatively little hardware and yet, can produce high resolution ADCs with minimal power requirements.

Converter Type	Sampling Rate	Resolution (in Bits)	Power Dissipation	Size and Technology
Algorithmic [7] -uses active samplers	25 ks/s	10	3.5 mW	0.32mm², 3μm CMOS
Algorithmic [8] -uses cascoded samplers	5.7 ks/s	14	2.5 mW	1.0mm², 3μm CMOS
Pipelined [22] (estimate) -uses cascoded samplers	80 ks/s	14	35 mW	14mm², 3μm CMOS
Pipelined [23] -uses active samplers	20 Ms/s	10	1 W	48mm², 2 μm BiCMOS

Table 12.1 Performance of some published switched-current ADCs.

The analogue portion of a multiplying algorithmic ADC and its timing diagram is shown in Figure 12.13 [7]. The circuit consists of three active current-samplers that share a common amplifier, a reference current and some switches. The circuit is relatively simple and hence can be used to perform an analogue-to-digital conversion with relatively little chip area.

The ADC performs a conversion starting with the most significant bit (MSB). Initially S_1, S_2 and S_3 are closed to sample I_{IN} using N_1. Then I_{IN} is also sampled using N_2 by closing S_1, S_4 and S_5. With I_{IN} stored in both N_1 and N_2, $2I_{IN}$ is loaded into P_1 by closing S_2, S_4, S_6 and S_7. Once P_1 has settled, S_6 and S_8 are closed to compare $2I_{IN}$ with I_{REF} by allowing the amplifier to act as a current comparator. After the comparison, the digital output will be "1" if $2I_{IN}$ exceeds I_{REF}, otherwise it will be "0". This completes the conversion of the MSB.

The remaining *N-1* bits are then converted in turn as follows. The signal, now stored in P_1, is loaded into N_1 by closing S_6, S_2 and S_3. Also, if the preceding bit was a "1", S_8 is also opened to subtract I_{REF} from the signal. After N_1 is set, N_2 is set by closing S_6, S_4 and S_5 (and S_8 if the previous bit was a "1"). The signal is then doubled, stored in P_1 and compared with I_{REF} as was done for the MSB. The sequence is repeated until the desired resolution is achieved.

The circuit's resolution is determined primarily by the accuracy of the current-samplers. Problems such as power supply noise, amplifier offsets, charge injection and the finite output resistance of the current sampling transistors can all limit the circuit's resolution. Changes in the stored currents, due to variations in the power supply can be reduced by ensuring that the sampled voltages are held on capacitors that are physically close to their associated transistors. The effects of amplifier input offsets can be virtually eliminated by using only one amplifier, both as an amplifier for all the current-samplers and as the current comparator. The sharing of the amplifier also eliminates the need for a separate current comparator, thereby reducing the circuit's size. The effects of charge injection can be

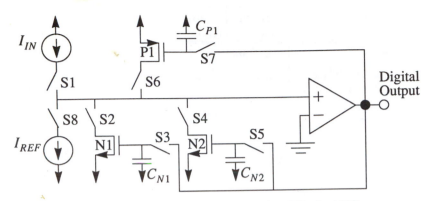

a) Circuit for a switched-current algorithmic ADC.

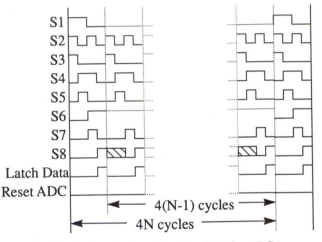

b) Clock signals for the algorithmic ADC.

Figure 12.13 A switched-current algorithmic ADC.

reduced by using the techniques discussed in Section 12.3.4 or by the use of differential circuitry. The final problem is the current sampling transistor's finite output resistance. As discussed previously, either a cascoded or active current sampling should be used. In cases where very high resolutions are desired, a combination of both techniques should be used [8].

The algorithmic converter's conversion time is determined by the clock rate and the desired resolution. Since each bit requires four clock cycles, a complete conversion will require $4N$ clock cycles. Consequently, if short conversion times are desired, a fast settling current-sampler is required. Alternatively, the sampled nature of the circuit will allow it to be used as the basis for a pipelined ADC.

Despite their relatively low speed, current-sampling algorithmic ADCs compete favourably with switched-capacitor (SC) algorithmic ADCs. A comparable resolution switched-capacitor ADC required about 2mm² of chip area and had a power dissipation of 17mW for an 8kHz, 12-bit ADC [5]. Consequently, the switched-current approach to the design of algorithmic ADCs is attractive if minimum chip area and power dissipation are required.

12.4.2 Pipelined ADCs

Pipelined current sampling ADCs can be implemented either by using a circuit similar to that discussed above as a 1-bit ADC in an *N*-stage pipeline [8], or by using multi-bit stages in a shorter pipeline [23]. The first approach relies solely on the accuracy of the current-samplers to achieve a high resolution converter. The second approach uses differential circuitry to reduce the effects of charge injection and uses error correction techniques to relax the accuracy requirements of the analogue circuits. In either case, the use of a pipelined architecture increases the sources of error in the circuit.

In a pipelined circuit two new error sources arise. The first error source is the presence of offsets between the amplifier's used in different stages of the pipeline. Unlike algorithmic ADCs, in which one amplifier can be multiplexed between current-samplers, pipelined ADCs require at least one amplifier or comparator per stage. Therefore if the circuit's accuracy is to be maintained, each current-sampler will have to display a very high output resistance. The second problem that arises in the pipelined architecture is the need for more than one reference signal. Previously, if the algorithmic ADC's reference current was incorrect, a constant gain error occurred in the ADC. In the pipelined ADC, mismatches in the references can lead to non-linearities in the conversion. Therefore, accurately matched current reference sources should be used [10]. Beyond these two error sources, the error sources present in the basic algorithmic ADC will also have to be dealt with if a high resolution pipelined ADC is to be built.

12.5 Limitations

A switched-current ADC's performance is ultimately limited by various systematic and random errors. Ideally, a well designed circuit will have the systematic errors reduced to acceptable levels such that only the random errors remain to limit the performance.

12.5.1 Systematic errors

Systematic errors in switched-current circuits arise from three sources: the settling time of the current-samplers, charge injection from the voltage sampling switches and charge leakage from the sampling capacitors. The sketch shown in Figure 12.14 illustrates that in most circuits, the dominant error source will be determined by the clocking rate. At high speeds, the current-samplers do not have sufficient time to settle, while at low speeds, the leakage currents associated with the switches will discharge the stored voltages on the sampling capacitors. Between these two regions, charge injection places a lower limit on the achievable accuracy. All three factors are inter-related by the size of the sampling capacitor; a small capacitor leads to faster settling times but also to larger errors due to the injected charge and leakage currents. Consequently, for high-speed high-accuracy operation, small sampling capacitors along with charge injection cancellation or reduction techniques will be required. Ideally, the detrimental effects of all three error sources can be reduced to an acceptable level by careful circuit design.

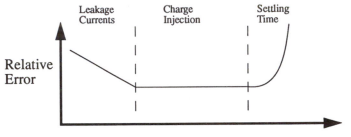

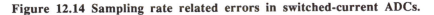

Figure 12.14 Sampling rate related errors in switched-current ADCs.

12.5.2 Random errors

If all the systematic error sources are cancelled or reduced to acceptable levels, random error sources will remain to determine the circuit's fundamental performance limitations. Random errors arise from fundamental noise sources present within the circuit itself such as the thermal noise of the sampling switches and the current sources. For switched-current ADCs, the thermal noise from the switches will give rise to a resolution/area trade-off [24] while the noise from the current sources will give rise to a speed/power trade-off [7]. More details on the limits of switched-current circuits can be found in Chapters 4, 5 and 6.

12.6 Conclusions

The use of sampled current techniques to design algorithmic and pipelined ADCs has been described. Although the ADCs are analogue circuits, they can typically be fabricated in processes with no special enhancements for analogue circuits. The circuits produced to date have been very small in size and had very low power requirements. Consequently, the switched-current approach for ADC design is attractive for mixed analogue-digital system-on-a-chip applications.

References

[1]   D. H. Sheingold, *Analog-Digital Conversion Notes*, Prentice Hall, 1986.

[2]   E. A. Vittoz, "Dynamic analog techniques," Ch. 5 in Y. Tsividis and P. Antognetti, Eds., *Design of MOS VLSI Circuits for Telecommunications*, Prentice-Hall Inc.: Englewood Cliffs, New Jersey, 1985.

[3]   C. Toumazou, F. J. Lidgey and P. Y. K. Cheung, "Current-mode analog signal processing - A review and recent developments," in *Proc. IEEE International Symposium on Circuits and Systems*, pp. 1572-1575, May 1989.

[4]   C. Toumazou, F. J. Lidgey and D. G. Haigh, Eds., *Analogue IC Design: The Current Mode Approach*, Peter Peregrinus Ltd: London (United Kingdom), 1990.

[5]   P. W. Li, M. J. Chin, P. R. Gray and R. Castello, "A ratio-independent algorithmic analog-to-digital conversion technique," *IEEE J. Solid-State Circuits*, vol. SC-19, pp. 828-836, Dec. 1984.

[6]   C. C. Shih and P. R. Gray, "Reference refreshing cyclic analog-to-digital and digital to analog converters," *IEEE J. Solid-State Circuits*, vol. SC-21, pp. 544-554, Aug. 1986.

[7]   D. G. Nairn and C. A. T. Salama, "A ratio-independent algorithmic analog-to-digital converter combining current-mode and dynamic techniques," *IEEE Trans. Circuits Syst.*, vol. CAS-37, pp. 319-325, March 1990.

[8]   P. Deval, J. Robert and M. J. Declercq, "A 14 bit CMOS A/D based on dynamic current memories," in *Proc. IEEE Custom Integrated Circuits Conference*, pp. 24.2.1-24.2.4, 1991.

[9]   S. W. Kim and S. W. Kim, "Current-mode cyclic ADC for low power and high speed applications," *Electronics Letters*, vol. 27, pp. 818-819, 1991.

[10]   W. Groeneveld, H. Schouwenaars and H. Termeer, "A self calibration technique for monolithic high-resolution D/A converters," in *IEEE International Solid-State Circuits Conference Dig. Tech. Papers*, vol XXXII, pp. 22-23, Feb. 1989.

[11]   S. J. Daubert, D. Vallancourt, and Y. P. Tsividis, "Current copier cells," *Electronics Letters*, vol. 24, pp. 1560-1561, 1988.

[12]   D. G. Nairn, "Analytic step response of MOS current mirrors," *IEEE Trans. Circuits Syst.*, to be published.

[13]   G. Wegmann and E. A. Vittoz, "Analysis and improvements of accurate dynamic current mirrors," *IEEE J. Solid-State Circuits*, vol. SC-25, pp. 699-706, June 1990.

[14]   D. G. Nairn and C. A. T. Salama, "High-resolution, current-mode A/D converters using active current mirrors," *Electronics Letters*, vol. 24, pp. 1331-1332, 1988.

[15]   E. Säckinger and W. Guggenbühl, "A high-swing, high-impedance MOS cascode circuit," *IEEE J. Solid-State Circuits*, vol. SC-25, pp. 289-298, Feb. 1990.

[16]   D. G. Nairn, "Amplifiers for high-speed current-mode sample-and-hold circuits," in *Proc. IEEE International Symposium on Circuits and Systems*, pp. 2045-2048, May 1992.

[17]   J. Robert, P. Deval and G. Wegmann, "Very accurate current divider," *Electronics Letters*, vol. 25, pp. 912-913, 1989.

[18]   C. L. Wey and S. Krishnan, "Current-mode divide-by-two circuit," *Electronics Letters*, vol. 28, pp. 820-822, 1992.

[19]   T. N. Blalock and R. C. Jaeger, "A high-speed clamped bit-line current-mode sense amplifier," *IEEE J. Solid-State Circuits*, vol. SC-26, pp. 542-548, April 1991.

[20]   E. Seevinck, P. J. van Beers and H. Ontrop, "Current-mode techniques for high-speed VLSI circuits with application to current sense amplifier for CMOS SRAMs," *IEEE J. Solid-State Circuits*, vol. SC-26, pp. 525-536, April 1991.

[21]   E. J. Swanson, "Echo cancellers: their role and construction," Ch. 16 in Y. Tsividis and P. Antognetti, Eds., *Design of MOS VLSI Circuits for Telecommunications*, Prentice-Hall Inc.: Englewood Cliffs, New Jersey, 1985.

[22]   J. Robert, P. Deval and G. Wegmann, "Novel CMOS pipelined A/D convertor architecture using current mirrors," *Electronics Letters*, vol. 25, pp. 912-913, 1989.

[23]   P. Real, D. H. Robertson, C. W. Mangelsdorf and T. L. Tewksbury, "A wide-band 10-b 20 Ms/s pipelined ADC using current mode signals," *IEEE J. Solid-State Circuits*, vol. SC-26, pp. 1103-1109, Dec. 1991.

[24]   P. R. Gray and R. Castello, "Performance limitations in switched-capacitor filters," Ch. 10 in Y. Tsividis and P. Antognetti, Eds., *Design of MOS VLSI Circuits for Telecommunications*, Prentice-Hall Inc.: Englewood Cliffs, New Jersey, 1985.

Chapter 13

High Resolution Algorithmic A/D Converters based on Dynamic Current Memories

Philippe Deval

13.1   Introduction

Proposed in the mid-1970's, the switched-capacitor (SC) technique quickly proved to be a very attractive approach for making analogue signal processing compatible with MOS technology [1]. This method, however, makes use of high-quality floating capacitors that require additional fabrication steps in a standard CMOS process. Moreover, the linearity, hysteresis and matching characteristics of these capacitors are limiting factors for the SC's ADC performance.

The recently introduced switched-current (SI) technique has become a viable alternative to the traditional SC technique for analogue signal processing in CMOS [2,3,4]. Unlike SC circuits, SI circuits do not require additional fabrication steps in a standard CMOS process. In this approach, a current is stored in a dynamic memory (Figure 13.1) made of a MOS transistor whose gate voltage is stored in a non-critical capacitor. Basic functions such as signal summation, inversion, memorisation or comparison, are very simple to implement and require little chip area. Moreover, the use of dynamic current mirrors eliminates the problem of matching, non-linearity and hysteresis of the components [5-8]. Other limiting factors however appear : data are generally available as voltages, which implies that a voltage-to-current conversion is necessary; it is not possible to distribute a current to several locations at the same time (pipeline converter); a current that is not used cannot be left in an open circuit; and lastly, time constants, charge injection and leakage currents effects are signal dependent.

13.2   CMOS dynamic current memory

One of the key factors in a dynamic memory performance is the effective gate voltage (V_G - V_T) of the memorising transistor Tm (Figure 13.1): this is the actual memorised variable. So a good way to estimate the total memory error is to convert the different disturbance sources in an equivalent gate voltage error ΔV_G. The relative error is given by (13.1)[1]

$$\varepsilon_r = \frac{\Delta I}{I} = 2 \frac{\Delta V_G}{V_{Geff}} \qquad V_{Geff} = (V_G - V_T) \qquad (13.1)$$

The principal error sources that affect a dynamic current memory are: the charge injection and the leakage currents of the sampling switch Tx; the non zero output conductance of Tm; and the noise. Noise considerations alone are sufficient to highlight the importance of Tm gate voltage in the overall memory performance. The memory signal-to-noise ratio is given by (13.2)

$$S/N < \frac{3}{8\,n} \frac{C_G V_{Geff}^2}{K\theta} \qquad (13.2)$$

Since the substrate parameter, n, is between 1 and 2, the theoretical limit of the signal to noise ratio of the memory, given in dB, is as follows:

$$S/N < 10\,log(C_G V_{Geff}^2) + 73\;dB \quad @\;\; \theta = 300\;K \qquad (13.3)$$

where the gate capacitance C_G is expressed in pF.

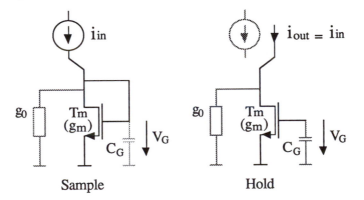

Sample   Hold

Figure 13.1   Basic dynamic current memory.

Example: $C_G = 1pF$, $V_{Geff} = 1V$ => S/N < 73 dB (12 bit)

It must be pointed out that a dynamic current memory reverses the current. But this inversion is not a drawback: combining the ability of memorisation and inversion, a dynamic current memory realises a fundamental function in the current mode signal processing domain. The original value of the current is restored by a second sampling in a complementary type current memory.

---

[1] In weak inversion region, $\Delta I/I = \Delta V_G/nU_T$.

Another important aspect of the dynamic current memory is its sampling time which is inversely proportional to the memorising transistor transconductance: $T \sim C_G/g_m$. As this transconductance is signal dependent, the sampling time is signal dependent as well.

The need for a high effective gate voltage for Tm implies that it is working deeply in strong inversion. Thus the sampling time of the memory is inversely proportional to the square root of the memorised current. The usual method to reduce this dependence is to add a bias current to the signal [4,9]. But this solution adds some noise to the signal and increases the memory error and the power consumption. As a first approximation, noise level, error and power consumption of the memory are increased by a factor $1 + I_{bias}/i_{in}$ [10]. An alternative solution that does not suffer these drawbacks is to restrict algorithmically the dynamic range of the memorised currents [11].

13.3   Conversion algorithm with reduced dynamic range of the memorised currents

The algorithmic conversion principle, as described by many authors such as [12,13], has been and always will be widely used in SC converters. It is also well suited to current mode converters [14,15]. But the basic algorithm proposed in the above references requires biased dynamic current memories (Figure 13.2). A small change in this conversion algorithm allows the use of the basic memory of Figure 13.1, eliminating the drawbacks due to the bias current. The basic algorithmic analogue to digital conversion technique computes $i_{out} = 2\, i_{in} - b_i I_{ref}$. In our case, this equation is replaced by: $i_{out} = 2\, (i_{in} + \overline{b_i} I_{ref}/2) - I_{ref}$. This method (which requires that the bit be determined before proceeding to the multiplication) ensures a value between $I_{ref}/2$ and I_{ref} for the current to be duplicated (Figure 13.3).

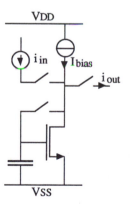

Figure 13.2   Biased dynamic current memory.

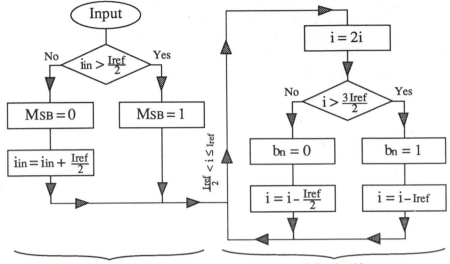

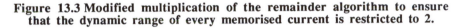

Figure 13.3 Modified multiplication of the remainder algorithm to ensure that the dynamic range of every memorised current is restricted to 2.

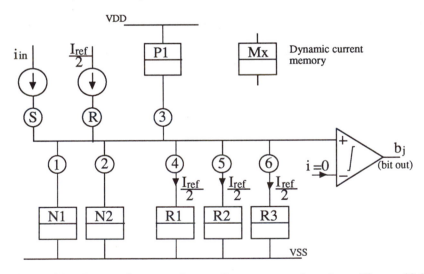

Figure 13.4 Block diagram of a cyclic converter based on Figure 13.3.

Our algorithm takes 4 phases per cycle (Figure 13.4). During the 1st phase of the 1st cycle (which differs from the following n-1 cycles), the input signal i_{in} is compared to $I_{ref}/2$ (S and 4 on) and the MSB, b_1, is extracted. During the following two phases, the current $i_{in} + \overline{b}_i I_{ref}/2$ is successively memorised by the memories N1 (switches S and 1 "on"; switch R "on" if $b_1 = 0$) , and N2 (S, 2: on; R: on if $b_1 = 0$). The current is then transferred to memory P1 where it is both memorised and re-inverted for the second cycle (1, 2 and 3: on). Note that the refresh circuitry of cells R1 to R3 is not shown . The following n-1 cycles proceed as follows (n is the number of bits):

1st phase, determination of bit b_i. The current $i_{in}(i) = i_{out}(i-1)$ calculated in the preceding cycle (in the range of I_{ref} to $2I_{ref}$) is compared to $3I_{ref}/2$ (3,4,5 et 6 : on).

2nd and 3rd phases, doubling of residual. The current $i_{in}(i) - (1 + \overline{b_i})I_{ref}/2$ (in the range of $I_{ref}/2$ to I_{ref}) is successively memorised by the memories N1 (1, 3, 4 : on; 5 : on if b_i =1) and N2 (2, 3, 4 : on; 5 : on if b_i =1) .

4th phase, transfer to P1: same as 1st cycle.

This algorithm requires $I_{ref}/2$, but this is not a problem: highly accurate division by 2 (or by an integer number) is easily achievable with dynamic current memories [16].

13.4 Conversion error

The absolute error of a cyclic converter is equal to the sum of all the errors accumulated during a conversion cycle. Its accuracy, expressed in LSB's is obtained by dividing this absolute error by its full scale range. In practice the absolute value of the error is equal to 2^n times the cycle error[2]. Thus, to achieve n bit within 1/2 LSB precision, the total cycle error, ε_{cy}, must be less than 1/2 LSB. In any dynamic current memory based cyclic converter, the following operations are performed during a cycle: sampling of i_{in} into two N-type memories (doubling i_{in}), temporary transfer of $2 i_{in}$ into a P-type memory (intermediate memorisation and restoring the original sign of the current), bit determination (comparison of $2 i_{in}$ with the reference current) and subsequent reference subtraction (see Figure 13.22). Assuming that the dominant errors that occur during a cycle are those of the dynamic current memories, ε_{cy} is approximated by

$$\varepsilon_{cy} \cong 2 \, \varepsilon_N + \varepsilon_P \cong 4 \, \varepsilon_N \qquad (\varepsilon_P \cong 2 \, \varepsilon_N) \qquad (13.4)$$

where ε_N (ε_P) is the memorisation error of an N-type (P-type) memory.

As ε_P is about twice ε_N for the same settling time (see Figure 13.8), one can see that the memorisation error of a N-type memory must be lower than 1/8 LSB to achieve n bit within 1/2 LSB precision. As an example, ε_N

---

[2] n is the converter resolution.

must be lower than 30 ppm for a 12 bit converter.

13.5 Design of dynamic current memories for use in cyclic ADCs

When a dynamic current memory is designed to be used as a basic building block for cyclic ADC's, both sampling time and memorisation error must be minimised so that a signal can cover the full dynamic range between two successive samples. But generally the sampling time and the memorisation error are inversely proportional (Time∗Error = Constant). So the parameters that influence this Time∗Error product must be highlighted.

13.5.1 Equivalent time constant of the memory

During sampling, the memory may be represented by the simplified equivalent circuit of Figure 13.5. The drain conductance g_{DS} and the drain to gate overlap capacitance C_{DG} of Tm have second order effect on the sampling time and are ignored here.

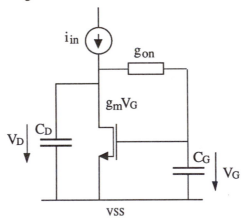

Figure 13.5 Simplified scheme of the memory during sampling phase.

Expressed as a function of the input current, the gate voltage of the memorising transistor is given by the following equation:

$$V_G(p) = \frac{i_{in}(p)/g_m}{p^2 \dfrac{C_G C_D}{g_m g_{on}} + p \dfrac{C_G + C_D}{g_m} + 1} \qquad (13.5)$$

To prevent heavy formulation of the equations, the argument (p) will be omitted for the following calculations. The characteristic and parasitic time-constants of the memory, τ_c and τ_p, are defined by :

$$\tau_C = \frac{C_G + C_D}{g_m} \qquad \tau_p = \frac{C_G}{C_G + C_D} \frac{C_D}{g_{on}} \qquad (13.6)$$

Let **gcrit** be the particular value of the sampling switch "on" conductance for witch the memory is critically damped ($\xi = 1$, $\tau_C = 4 \, \tau_p$).

$$g_{crit} = \frac{4C_GC_D}{(C_G + C_D)^2} g_m = \frac{4C_G}{(C_G + C_D)} \frac{C_D}{\tau_c} \qquad (13.7)$$

A good approximation of the memory equivalent time constant is given by the following equation :

a) $g_{on} \leq g_{crit}$

$$\tau_e = 2 \, \tau_p = \frac{2 \, C_G}{C_G + C_D} \frac{C_D}{g_{on}} = \frac{g_{crit}}{2 \, g_{on}} \tau_c \qquad (13.8\,a)$$

b) $g_{on} > g_{crit}$

$$\tau_e = \frac{1+ \sqrt{1 - g_{crit}/g_{on}}}{2} \tau_c \qquad (13.8\,b)$$

In the next few calculations, the sampling switch "on" conductance will be normalised to gcrit using the factor η :

$$\eta = \frac{g_{on}}{g_{crit}} \qquad (13.9)$$

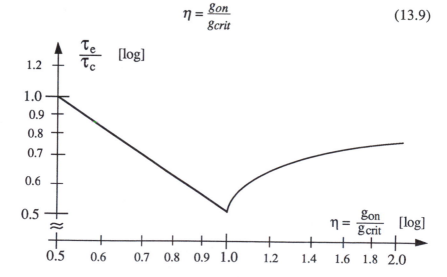

Figure 13.6 Equivalent time constant of the memory as a function of η and τ_c.

Typically the "on" conductance of the sampling switch can be controlled within 10 to 20% of its theoretical value. Taking into account the rapid increase of the equivalent time constant of the memory when the memory enters the over-damped region ($\eta > 1$), a good trade-off to ensure a minimal sampling time is to take η in the order of 0.8 to 0.9.

The input current may be very different between two successive samples and a large signal analysis should be done. In any case, when high accuracy is needed, say 10ppm, the part of the sampling time for which the small signal model is not valid is small compared to the time when it is valid. Therefore, a much simpler small signal analysis is sufficient.

13.5.2 *Charge injection error*

The charge injection of the sampling switch is one of the most important error sources of a dynamic current memory. Due to the intrinsic structure of the current memory, the sampling switch, and thus the charge injection are signal dependent. Consequently, the usual charge injection cancellation techniques (based on signal independent charge injection) cannot be applied here. Among the reported compensation techniques - auxiliary input with low sensitivity [5], differential current memory [6], algorithmic memory [4], current replica [17] - the dummy switches technique seems to be the most suitable: simple to implement, it is very efficient with the compensated switch layout of Figure 13.7.

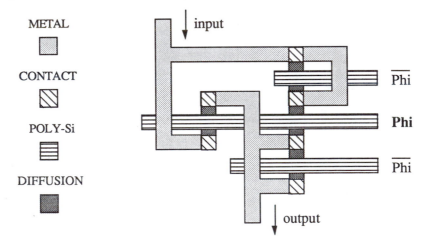

Figure 13.7 Improved layout of a charge injection compensated sampling switch [7, 8].

The total amount of charge that is released on the gate of Tm when the sampling switch Tx is turned off can be divided into two parts: one is due to the channel charges of Tx and the second is due to the charges transferred through its overlap capacitance C_{ol}. References [7,8,18] give a detailed analysis of the first part (channel charge) of the charge injection but only give some qualitative descriptions of the second part. When the error required is in the order of a few hundreds of ppm, this second part of the charge injection can generally be neglected. That is no longer the case when errors lower than 100 ppm are required. Taking into account the overlap capacitances of Tx, the charge injection error may be expressed by the following equation

$$\varepsilon_{inj} = \varepsilon_{ol} \left(1 + \frac{g_{on}}{g_{ol}}\right) \qquad (13.10)$$

with

$$\varepsilon_{ol} = 2k_{ol}\frac{C_{ol}}{C_G}\frac{V_{Tx}}{V_{Geff}} \qquad g_{ol} = \frac{k_{ol}\,C_{ol}V_{Tx}}{k_{ch}\dfrac{L^2}{\mu} + k_{ol}\dfrac{C_{ol}}{\beta}} \qquad (13.11)$$

where

- ε_{ol} is the part of the charge injection error due to the overlap capacitance of Tx.

- g_{ol} is the particular value of g_{on} for which the channel charge contribution to the total injection error is equal to that due to the overlap capacitances (ε_{ol}).

- k_{ch} and k_{ol} are factors relative to the efficiency of the dummy switches compensation.

- V_{Tx} is the threshold voltage of Tx.

V_{Tx} is a quadratic function of the effective gate voltage of Tm, but in practice the following first order approximation is sufficiently accurate

$$V_{Tx} = 1.2\,V_{Geff} + 2\,V_{T0} \qquad (13.12)$$

The k_{ch} and k_{ol} factors are strongly dependent on the compensated sampling switch layout. For the switch of Figure 13.7 their typical value are

$$k_{ch} = 0.05 \qquad k_{ol} = 0.01 \qquad (13.13)$$

The denominator of g_{ol} is proportional to the residual injected charge of the compensated switch and may be taken as a one of its parameters[3]. We will call it k_{on}

---

[3] A detailed description of charge injection can be found in [19, 20, 21].

$$k_{on} = k_{ch}\frac{L^2}{\mu} + k_{ol}\frac{C_{ol}}{\beta} \qquad (13.14)$$

The equivalent time constant of the memory is a function of the sampling switch "on" conductance (13.8). Thus (13.10) can be modified in order to give the charge injection error as a function of the memory sampling time constant:

$$\varepsilon_{inj}(\tau_e) = \varepsilon_{ol}\left(\frac{\tau_{ol}}{\tau_e} + 1\right) \qquad (13.15)$$

where τ_{ol} is the particular value of the memory time constant for which the channel charge contribution to the total injection error is equal to that due to the overlap capacitances. In under-damped mode, τ_{ol} is given by the following equation[4]

$$\tau_{ol} = \frac{C_D}{(C_G + C_D)}\frac{4k_{on}C_G}{k_{ol}C_{ol}\,V_{TX}} \qquad [\eta \le 1] \qquad (13.16)$$

This equation shows that the function $\varepsilon_{inj} = f(\tau_e)$ is independent of the sampled current level. The current level will only fix the memory time constant, thus giving the working point on the curve $\varepsilon_{inj} = f(\tau_e)$.

As we can see in Figure 13.8, injection error as a function of the sampling time constant of the memory is a curve with 2 asymptotes: one parallel to the time axis and one with slope 1/T. A good time/error compromise isn't obtained unless the error can be approximated by the asymptote 1/T. We will therefore use the time*injection error product that includes the working point on the curve $\varepsilon_{inj} = f(\tau_e)$ in the memory evaluation.

$$\tau_e\varepsilon_{inj} = \tau_{ol}\varepsilon_{ol}\left(\frac{I_{ol}}{i_{in}} + 1\right) \qquad (13.17)$$

with

$$\tau_{ol}\varepsilon_{ol} = \frac{8}{C_G/C_D + 1}\frac{k_{on}}{V_{Geff.}} \qquad [\eta \le 1] \qquad (13.18)$$

and

---

[4] For over-damped mode ($\eta > 1$), τ_{ol} is weighted by $\eta\,(1+\sqrt{1-1/\eta}\,) > 1$.

$$I_{ol} = \frac{(C_G + C_D)^2}{C_G C_D} \frac{k_{ol} C_{ol} \, V_{Tx} \, V_{Geff}}{8 \, \eta \, k_{on}}$$ (13.19)

where I_{ol} is the current for which $\tau_e = \tau_{ol}$.

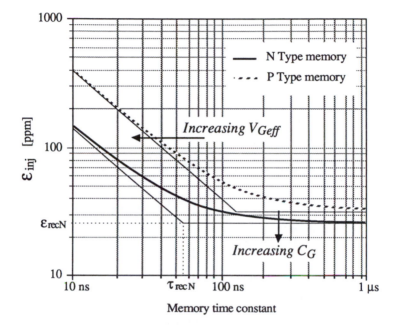

Figure 13.8 Typical relation between the charge injection error and the sampling time constant of a dynamic current memory realised in a standard 2μm CMOS process ($V_{Geff} = 1V$; $C_G = 1$ pF; $\eta=0.8$; minimal size sampling switch with $k_{ch}=5\%$, $k_{ol}=1\%$). The arrows show the main influence of V_{Geff} and C_G on the asymptotes.

The memory performance is improved by minimising the $\tau_{ol}\varepsilon_{ol}$ factor. High effective gate voltage and low drain capacitance are therefore required. But in most applications, the C_G/C_D ratio will be in the range of 2 to 3. The best way to minimise the $\tau_{ol}\varepsilon_{ol}$ factor is thus to maximise the effective gate voltage of Tm.

Note: Because of the difference in carrier mobility between NMOS and PMOS transistors ($\mu_N \approx 2.5\mu_P$), the N-type memory will always have a lower $\tau_{ol}\varepsilon_{ol}$ factor than the P-type. Thus, N-type memories will be preferred, whenever we wish to optimise the trade-off between injection error and sampling time. However, when minimum charge injection error is required, and when the available sampling time allows the memory to function in the region where the injection error is approximated by the horizontal asymptote (refreshing of a constant current, i.e. the reference current), the relative N-type and P-type memories' performances are similar and depend on the technology through the C_{oIN}/C_{oIP} ratio.

13.5.3 Sampled noise

In a high resolution current mode converter, the memory's sampled noise is the second main limitation. The signal-to-noise ratio of the memory is easily derived from the sampled noise voltage V_{Gn}

$$\frac{i_{in}^2}{i_n^2} = \frac{V_{G\,eff}^2}{4\,V_{G\,n}^2} \qquad (13.20)$$

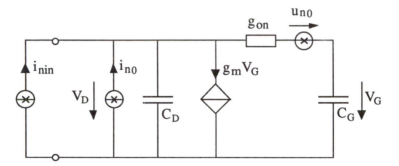

Figure 13.9 Small signal scheme during the sampling phase with the main noise sources.

We calculate from the small signal scheme of Figure 13.9 that

$$V_{Gn}^2 = \frac{1}{g_m^2} \int_0^\infty \frac{\left|\frac{di_{nin}^2}{df}\right| + \left|\frac{di_{no}^2}{df}\right| + \left|\frac{du_{no}^2}{df}\right|\left|p\,\frac{C_D}{g_m}\right|^2}{\left|p^2\,\frac{C_G C_D}{g_m g_{on}} + p\,\frac{C_G + C_D}{g_m} + 1\right|^2}\,df \qquad (13.21)$$

Assuming a much higher working frequency of the memory than its corner frequency, 1/f noise may be neglected. Taking into account only the thermal noise, we have

$$V_{Gn}^2 = (1 + \frac{C_D}{C_G} + \frac{g_{nine}}{g_m})\,\frac{K\theta}{C_G + C_D} \qquad (13.22)$$

Finally the signal-to-noise ratio of the memory is given by (13.23)

$$S_N = \frac{(C_G + C_D)\, V_{Geff}^2}{4\, (1 + \dfrac{C_D}{C_G} + \dfrac{g_{nine}}{g_m})\, K\theta} \qquad (13.23)$$

We notice that increasing the effective gate voltage of Tm always improves the signal-to-noise ratio: the latter is proportional to V_{Geff}^{α}, α ranging from 1 ($g_m \ll (1+C_D/C_G)g_{nine}$) to 2 ($g_m \gg (1+C_D/C_G)g_{nine}$).

In practice, the input current source is another current memory (or an arrangement of several parallel memories). The g_{nine}/g_m ratio in this case becomes independent from the Tm effective gate voltage. The memory (and the converter) signal-to-noise ratio varies together with the square of the effective gate voltage of Tm.

13.5.4 Leakage currents

The leakage currents are the 3rd major limitation to dynamic current memory performance. The main sources of leakage currents in a dynamic current memory are shown in Figure 13.10. The current I_{lm} can be considered as a bias current and thus neglected. The currents I_{lin} and I_{lout} add to the copied current an offset equal to $I_{lout} - I_{lin}$. This offset is negligible compared to the effect of the sampling switch leakage currents I_{lx}. The latter affects the gate voltage of the memorising transistor, resulting in a drift proportional to the hold time T_h. The quadratic relation between current and gate voltage induces a signal-dependent offset which cannot be cancelled.

$$\frac{\Delta i_{out}}{i_{out}}(T_h,\theta) = -2\frac{T_h\, I_{lx}(\theta)}{C_G V_{Geff}} \;\Rightarrow\; \frac{\Delta i_{out}}{i_{out}}(T_h,\theta) \sim -\frac{T_h\, I_{lx}(\theta)}{C_G \sqrt{i_{in}}} \qquad (13.24)$$

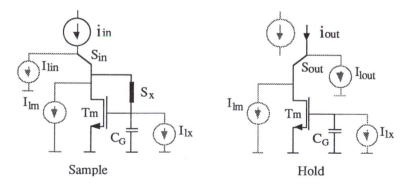

Sample   Hold

Figure 13.10 Different sources of leakage current in a dynamic current memory.

Generally negligible at room temperature, leakage currents rapidly increase with the temperature

$$I_l(\theta) = I_{l,\theta_0} \left(\frac{\theta}{\theta_0}\right)^3 exp\left(\frac{\Delta W}{k_B \theta_0} \frac{\theta - \theta_0}{\theta}\right) \qquad (13.25)$$

where I_{l,θ_0} is the leakage current at the absolute temperature θ_0 and ΔW is the bandgap (1.12 eV for silicon). At room temperature ($\theta_0 = 300K$), I_{lx} typically ranges from 0.1 to 1 pA leading to a drift lower than 1 ppm/μs ($C_G = 2pF$, $V_{Geff} = 1V$). This drift is generally neglected.

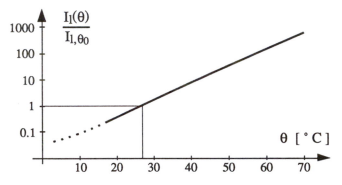

Figure 13.11 Leakage currents as a function of the temperature.

But as shown in Figure 13.11, the leakage currents increase tenfold when the temperature increases by 15 degrees. The leakage currents are therefore 1000 times higher at 70 °C, than they are at room temperature. Thus, an increase in the temperature provokes a rapid deterioration of the memory's performance. The leakage currents effect should be taken into account when the working temperature is over 40 °C.

Conclusion: high accuracy is not possible with dynamic current memories at high temperatures.

13.5.5 Output conductance

For any circuit based on dynamic current memories, the non-zero output conductance of Tm will cause a current error.

Two parasitic effects affect the memory output conductance (Figure 13.12):

- The output conductance g_{DS} of Tm.

- The overlap Drain/gate capacitance C_{DG} of Tm that transfers a fraction of the drain voltage variation to its gate (capacitive divider C_{DG}/C_G).

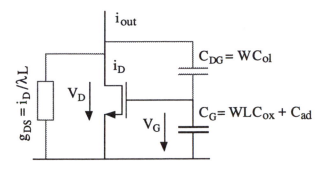

Figure 13.12 Parasitic elements (dashed) cause a non-zero output conductance.

Therefore, the basic memory of Figure 13.1 will have a conduction error, ε_{cond}, in the range of 1 to 3%. This error must be reduced.

A possible solution to reduce the output conductance error is to add a cascode transistor (Figure 13.13). But the output conductance error is only reduced by a factor in the range of 30 to 100 (voltage gain of the cascode transistor) with this solution. Notice that the need of the highest possible gate voltage for Tm eliminates the stacked cascode solution [18].

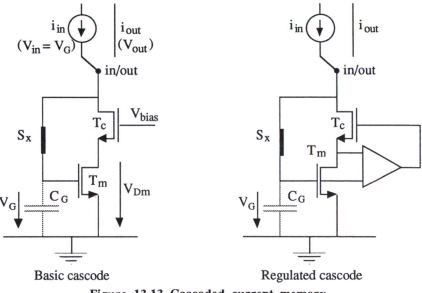

Basic cascode   Regulated cascode

Figure 13.13 Cascoded current memory.

However, this structure is not sufficient to achieve a high precision. An improved version is the regulated cascode where the gate of the cascode transistor is controlled through adaptive circuitry to better regulate the drain voltage of Tm.

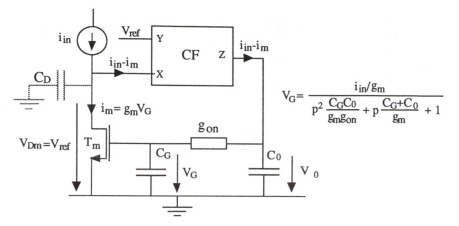

Figure 13.14 Control of the memory input voltage by a current follower.

Another solution is to control the voltage to be the same both at the entrance and at the exit of the memory. In order to achieve the best performance, this control is realised by setting the gate voltage of the sampling transistor through a current follower (Figure 13.14) and not by using an OTA, which makes the control slow or unstable during the sampling phase.

Both the regulated cascode and the current follower controlled memories have an output conductance error about one order of magnitude lower than their injection error. Therefore their output conductance error is negligible. But the current follower controlled memories offers two advantages :

1) Since the output capacitance, C_0, of the current follower is generally smaller than C_D, its time∗injection error product is smaller than that of the base memory[5]

$$\tau_{eCF}\varepsilon_{injCF} = \tau_{olCF}\varepsilon_{ol}\left(1 + \frac{I_{olCF}}{I_n} \right) \qquad (13.26)$$

with

$$\tau_{olCF} = \frac{4k_{on}C_0}{k_{ol}C_{ol}\,V_{Tx}} \quad => \quad \tau_{olCF}\varepsilon_{ol} = \frac{8k_{on}C_0}{C_G\,V_{Geff}} \qquad (13.27)$$

and

$$I_{olCF} = \frac{C_G}{C_0}\,\frac{k_{ol}C_{ol}\,V_{Tx}\,V_{Geff}}{8\,k_{on}} \qquad (13.28)$$

---

[5] The suffix CF is appended to the parameters that are modified when the current follower controlled memory is used.

2) Switching the memories between nodes at the same potential, V_{ref}, eliminates the current spikes that occur when they are switched between nodes at different (signal dependent) potentials.

On the other hand, the current follower adds some noise. Typically the signal-to-noise ratio of the current follower controlled memory is reduced by 2 to 3 dB compared to the basic memory, while it is reduced by less than 1 dB when using the regulated cascode memory.

Figure 13.15 gives an example of a current follower controlled memory. It works as follows. During sampling time, the switches are in position "in". The voltage V_G is established such that T1 sinks $i_{in}+I_{bias}$ with $V_{in} \approx V_{ref}$. T5, used in common-gate configuration, implements the current follower. During hold time, the switches are in position "out". Given that the charge (not shown) fixes $V_{out} = V_{ref}$, T1 sinks a current equal to $i_{in}+I_{bias}$. Transistor T5 is off but transistor T4 conducts I_{biasN} assuring a total current of $i_{out}=i_{in}$. Note that T4 also uses the voltage V_{on}, thus conserving $V_0=V_G$ at the terminals of the source I_{biasN} and avoiding an error due to the variation of this voltage. To improve the performances of the memory, a cascode transistor T2 must be added in order to make the memory insensitive to the finite DC-gain and input offset voltage of the current follower. During sample mode, the memory is easily stable with a phase margin of 60° if the following condition is fulfilled

$$\frac{C_D}{g_{m5}} + \frac{C_0}{g_{on}} < \frac{C_G}{2g_{m1}} \qquad (13.29)$$

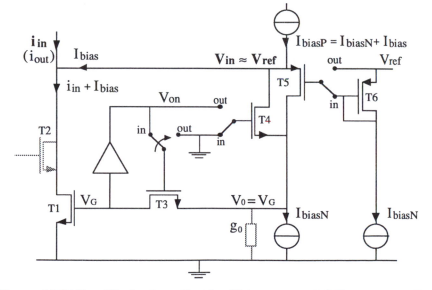

Figure 13.15 Simplified schematic of a N-type current follower controlled memory.

Note that when the memory is used in a converter based on the algorithm of Figure 13.3, the sourcing source, I_{biasP}, is equal to I_{biasN} ($I_{bias} = 0$). Moreover, it can be suppressed when I_{biasN} is significantly lower than $I_{ref}/2$.

13.5.6 *Current comparison*

It is very difficult to make a fast and accurate comparison in the current domain [22]. In practice, a voltage comparison is performed. Its speed depends on: the parasitic capacitance C_p at the comparison node, the LSB current value and the comparator offset voltage V_{off}. The comparison process is divided into 3 parts: *Reset, Initialisation, Comparison.*

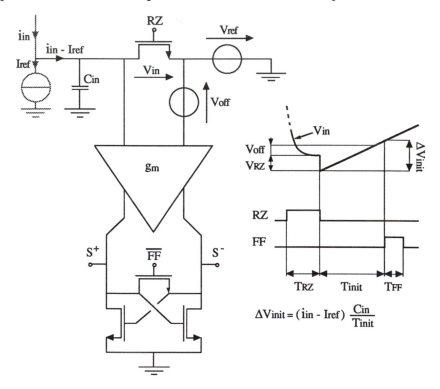

Figure 13.16 Current comparison.

Reset: The input node of the comparator is set to zero via the reset switch (RZ). Time constant is : $\tau_{RZ} = C_p/g_{RZ}$.

Initialisation: The reset switch is turned off thus releasing a charge Q_{RZ} at the input node of the comparator. Therefore the comparison node must ramp a value of $Q_{RZ}/C_p + V_{off}$ before the comparison can start.

Comparison : The comparator compares $Q_{RZ}/C_p + T_I(i_{in} - I_{ref})/C_p$ to its offset voltage V_{off} (T_I is the initialisation phase duration).

The equivalent offset current of such a comparator is given by

$$I_{off} = \frac{C_p(V_{off} + V_{RZ})}{T_I} \qquad V_{RZ} = \frac{Q_{RZ}}{C_{in}} \qquad (13.30)$$

Basically, the offset current of the comparator can be reduced to any desired level by simply increasing the initialisation time. In practice, an auto-zero phase is added in order to compensate the comparator offset and minimise the initialisation phase duration. Thus the total comparison time can be reduced to the duration of a sampling phase (one clock period). This auto-zero phase can be performed at the beginning of each conversion cycle. Therefore it has a negligible effect on the total conversion time of a cyclic converter: However the auto-zero phase increases the conversion time of a pipelined converter by 25%.

    Example: The equivalent offset current of the comparator is equal to 1nA for residual offset voltage (after compensation) of 1mV, a parasitic capacitance of 1pF and an initialisation time of 1µs. This equivalent offset current is equal to 1/3 LSB at 14 bit resolution when the reference current is 50µA.

13.5.7   Comparator offset compensation

Global compensation of all the offsets that affect the comparison can be simply performed without floating capacitors (Figure 13.17).

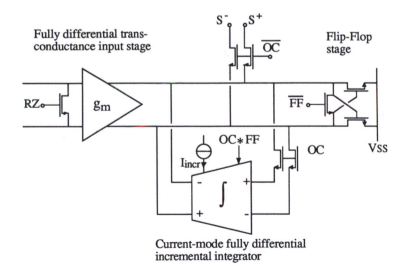

Current-mode fully differential
incremental integrator

Figure 13.17 Block diagram of a comparator with incremental offset compensation, OC is active during the compensation phases.

During each offset compensation phase, the comparator is disconnected from the converter and a complete comparison is performed (reset, initialisation, flip-flop). The comparison result is fed to an incremental current mode integrator that integrates the comparator offset. The output current of the integrator is subtracted from the output current of the input transconductance stage of the comparator. Therefore after a few compensation phases, equilibrium will be reached and the residual offset will oscillate within a range of $\pm I_{incr}/g_m$ (I_{incr} compensation current; g_m input stage transconductance).

$$|V_{off}| = \frac{I_{incr}}{g_m} \quad => \quad |I_{off}| = \frac{T_I}{C_{in}/g_m} I_{incr} \qquad (13.31)$$

A residual offset of less than 200µV (typical 100µV) has been measured on an offset compensated voltage comparator based on this method [23].

13.5.8 Clipping the comparator's input voltage

The aforementioned current comparison method has a drawback: the comparator input node will quickly drift to V_{DD} or V_{SS} (depending on the sign of the input current $i_{in} - I_{ref}$) since the comparison node is floating. This drift may saturate the current memories connected to this node and slow down the subsequent sampling (additional time is required for the memory's de-saturation). Thus a clipping device must be added at the comparator input to prevent memories from saturating. The simplest way to clip the comparator input voltage is to add two diodes in anti-parallel configuration at the input (Figure 13.18).

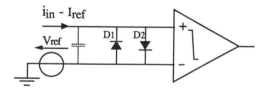

Figure 13.18 Clipping the comparator input voltage.

When floating diodes are not available, one of the diodes should be replaced by a compatible lateral bipolar transistor (CLBT) [24-26] and the other by a MOS connected in common gate configuration, as shown in Figure 13.19. A CLBT is better than a common gate MOS for clipping because it is much smaller and adds less parasitic capacitance for the same efficiency.

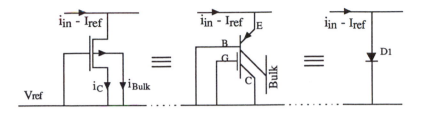

Figure 13.19 Replacing D1 by a CLBT (N-Well process).

13.5.9 Controlling the "on" conductance of the sampling switch

A simple way of controlling the sampling switch "on" conductance to be proportional to the memory transconductance is proposed in Figure 13.20 [18]. Supposing T'_X matched with T_X and T11 matched with T12, we have

$$\frac{g_{on}}{g_m} = n\frac{\sqrt{\beta_{11}\,\beta_x}}{\beta_m} \quad => \quad \frac{\beta_{11}}{\beta_m} = n\left(\frac{g_{on}}{n\,g_m}\right)^2\frac{\beta_m}{\beta_x} \quad (13.32)$$

where **n**, the slope factor, usually around 1.6 for small values of V_G, tends to 1 with increasing V_G.

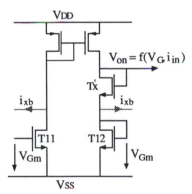

Figure 13.20 Circuit adjusting the "on" conductance (gate voltage) of the sampling switch. Optional bias currents I_{xb} add a constant value to g_{on} :

$$g_{on} = \alpha\ g_m + g_{ono}.$$

13.5.10 Input multiplexer

The input and reference currents are required by the converter only during some specific phases. They must return to the ground during the other phases. Figure 13.21 proposes an improved input multiplexer.

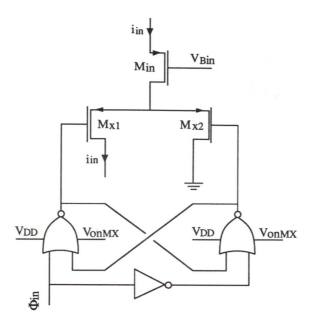

Figure 13.21 Improved input multiplexer. The negative supply voltage for the nor gates is V_{onMX} in order to maintain the "on" transistor M_{X1} (M_{X2}) in the saturation region.

13.6 Technology scaling

The main effect of technology scaling is to reduce the time∗injection error product of the memories. This error is proportional to the square of the sampling switch length (see (13.14), (13.17) and (13.18)). Therefore, technology scaling induces a significant reduction in the sampling time for a given error level (or vice versa). Technology scaling has a smaller effect on the memory's area since the noise characteristics are dependent on the gate capacitance of Tm (the memorising capacitance).

Note that the maximum possible gate voltage of Tm may be reduced by the technology scaling (supply voltage reduction). Therefore, technology scaling may decrease the memory's performance since the effective gate voltage of Tm is a key factor in the memory performance.

13.7 Pipelined converter

A particular feature of current mode operation is that it is not possible to distribute a current to several locations at the same time. Therefore, the reference current needs to be duplicated in each bit cell of a pipelined converter. A highly accurate duplication is achieved with dynamic current memories since there is no speed/accuracy trade-off encountered here (the reference current is constant). Moreover the reference calibration technique described in [6] can be used here. The main difficulty is to refresh the reference memory within a sufficiently short period in order to prevent the action of leakage currents. This problem is solved by pipelining in parallel signal an reference operations.

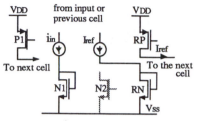

Ø1 Sampling i_{in} in N1 and refreshing of RN

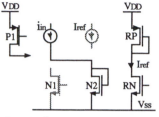

Ø2 Sampling i_{in} in N2 and refreshing of RP

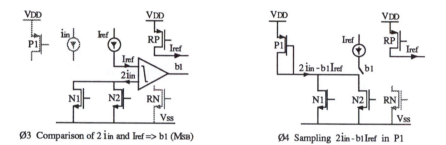

Ø3 Comparison of $2 i_{in}$ and I_{ref} => b1 (MSB)

Ø4 Sampling $2i_{in} - b1 I_{ref}$ in P1

Figure 13.22 Pipeline architecture of a switched current A/D converter. Transistors P1 and RP are signal and reference sources for the next stage.

In a classical remainder multiplication algorithm [14,15], the reference current is unused during the two phases when the signal is doubled (Φ1 and Φ2). Those two phases can be used for the reference refreshing operation (Figure 13.22).

There is no need for additional N-type reference memory cells for the conversion algorithm described in section 13.2. Nevertheless, a P-type current mirror [18, 27] is required (rather than a single P-type memory) since only one phase is available for the reference refreshing operation.

13.8 Experimental results and measurements

A cyclic A/D converter based on the conversion algorithm of Figure 13.3 has been integrated in a standard, single poly, single metal, 3µm CMOS process. Its characteristics are summarised in table 13.1. The measured INL[6] and noise Figure are shown in Figures 13.23 and 13.24. The measured 14 bit linearity proves that each memory cell has a very low error (in the order of 10ppm, see Section 13.3). This proves also that a quasi perfect compensation of charge injection is achieved when using the compensated switch of Figure 13.7.

---

[6] Noise is filtered out by averaging 512 samples for each plotted dot.

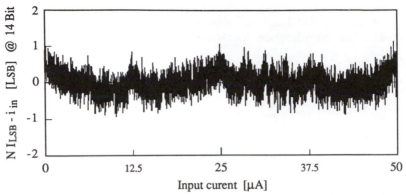

Figure 13.23 Integral linearity of the converter at 14 bit resolution[1].

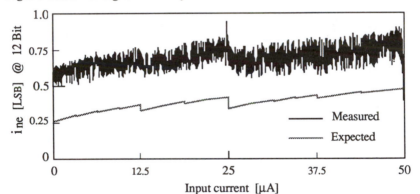

Figure 13.24 Measured and calculated input referred noise of the converter at 12 bit resolution.

The measured input referred current noise is nearly twice its theoretical value. This difference is mainly due to the external logic that adds perturbations onto the input and reference currents.

Specifications	Measured	Expected
Area (without logic and pads)	1 mm²	
Power Supply	5V	5V
Power Dissipation (Average)	2.5mW	2.25mW
Full Scale Current (I_{ref})	50µA	50µA
Conversion Time @ 14 bits	175µs	175µs
Resolution	14 Bits	12 Bits
Integral Non-Linearity	± 0.5 LSB	< 0.5 LSB
Offset & gain error	not measurable	< 0.5 LSB
Input Referred Noise @ 12 bits	< 1 LSB	< 0.5 LSB

Table 13.1 Principal specifications of the converter.

13.9   Summary

This work shows that *the effective gate voltage of the memorising transistor is a key performance factor* of a dynamic current memory to be used in an algorithmic A/D converter. Increasing this voltage reduces both the time*error product and the sampled noise of the memory. It is an intuitive result since the value actually memorised is this voltage. Therefore, the memory's performance will be dramatically reduced in low voltage applications.

It has also been demonstrated that the ratio C_G/C_D must be maximised and has an effect on the memory performance comparable to that of the effective gate voltage. Therefore, the circuit must be arranged in such a way that each memory's *load capacitance* (C_D) is *minimised*.

In any case, the *sampling switch* must be carefully *compensated for charge injection* (Figure 13.7). Moreover its "on" conductance must be controlled in order to maintain the memory in *under-damped operation* during the sampling phase.

Leakage currents *increase tenfold* when the temperature *increases by 15 degrees*. Their effect should be taken into account when the working temperature is over 40 ^{0}C.

The basic memory of Figure 13.1 has a large output conductance. A cascode transistor must be added and/or the memory input/output voltage must be fixed to a constant value. Both solutions are possible in our case., but the second solution offers two advantages:

a) The memory time*injection error product is smaller than that of the base memory.

b) Switching the memories between nodes at the same potential, V_{ref}, eliminates the current spikes that occur when they are switched between nodes at different (signal dependent) potentials. This reduces the sampling time.

On the other hand, controlling the input/output voltage of the memory adds some noise. Typically the signal-to-noise ratio is reduced by 2 to 3 dB compared to the cascode solution.

The most important effect of technology scaling is to reduce the memory's time*injection error product. *This is only valid if the scaling does not imply a reduction in supply voltage.*

Acknowledgements

This chapter is a summary of [28]. The author is very grateful to Professor M. J. Declercq and to Drs J. Robert and G. Wegmann for their help throughout the course of this thesis. The author would also like to thank O. Buset for checking the English used in this chapter.

References

[1]   R. W. Brodsen, P. R. Gray and D. Hodges, "MOS switched-capacitors filters," *Proc IEEE*, vol. 67, pp 61-75, Jan. 1979.

[2]   J. B. Hughes et al, "Switched Currents - A new technique for analog sampled-data signal processing, in *Proc. IEEE International Symposium on Circuits and Systems*, pp. 1584-1587, Portland, May 1989.

[3]   C. Toumazou, J. B. Hughes and D. M. Pattullo, "Regulated cascode switched-current memory cell," *Electronics Letters*, vol. 26, pp. 303-305, 1st March 1990.

[4]   C. Toumazou, N. C. Battersby and C. Maglaras, "High-Performance Switched-Current Memory Cell," *Electronics Letters*, vol. 26, pp. 1593-1595, 13th Sept. 1990.

[5]   J. Daubert, D. Vallancourt and Y. P. Tsividis, "Current copier cells," *Electronics Letters*, vol. 24, pp. 1560-1562, 8th December 1988.

[6]   D. W. J. Groeneveld, H. J. Schouwenaars, H. A. H. Termeer and C. A. A. Bastiaansen, "A self-calibration technique for monolithic high-resolution D/A converters," *IEEE J. Solid-State Circuits*, vol. 24, pp. 1517-1522, Dec. 1989.

[7]   G. Wegmann, *Design and Analysis Techniques for Dynamic Current Mirrors*, Thèse No. 890, EPFL Lausanne, 1990.

[8]   G. Wegmann and E. A. Vittoz. "Analysis and improvements of accurate dynamic current mirrors," *IEEE J. Solid-State Circuits*, vol. 25, pp. 699-706, June 1990.

[9]   J. Robert, P. Deval and G. Wegmann, "Novel CMOS A/D convertor architecture using current mirrors," *Electronics Letters*, vol. 25, pp. 691-692, 25th May 1989.

[10]   P. Deval, G. Wegmann and J. Robert, "CMOS pipelined A/D convertor using current divider," *Electronics Letters*, vol. 25, pp. 1341-1343, 28th Sept. 1989.

[11]   P. Deval, J. Robert and M. J. Declercq, "A 14 bit CMOS A/D converter based on dynamic current memories," In *Proc. IEEE Custom Integrated Circuits Conf.*, pp. 24.2.1-24.2.4, San Diego, May 1991.

[12]   B. D. Smith, "An unusual electronic Analog-to-Digital conversion method," *IRE Trans. on Instrumentation*, vol. PGI-5, pp. 155-160, 1956.

[13]   B. M. Gordon, "Linear electronic A/D conversion architectures, their origin, parameters, limitations and applications," *IEEE Trans. Circuits Syst.*, vol. CAS-25 pp. 391-418, July 1978.

[14]   C. A. T. Salama, D. G. Nairn and H. W. Singor, "Current mode A/D and D/A converters," in C. Toumazou, F. J. Lidgey and D. G. Haigh, Eds, *Analogue IC Design: The Current-mode Approach*, Peter Peregrinus Ltd: London, 1990.

[15]   D. G. Nairn and C. A. T. Salama, "A ratio-independent algorithmic analog-to digital converter combining current mode and dynamic techniques," *IEEE Trans. Circuits Syst.*, vol. 37, pp. 319-325, March 1990.

[16]   J. Robert, Ph. Deval and G. Wegmann, "Very accurate current divider," *Electronics Letters*, vol. 25, pp. 912-913, 6th July 1989.

[17] T. S. Fiez, G. Liang and D. J. Allstot, "Switched-current circuit design issues," *IEEE J. Solid-State Circuits*, vol. 26, pp. 192-201, March 1991.

[18] E. A. Vittoz and G. Wegmann, "Dynamic current mirrors," in C. Toumazou, F. J. Lidgey and D. G. Haigh, Eds, *Analogue IC Design: The Current-mode Approach*, Peter Peregrinus Ltd: London, 1990.

[19] W. B. Willson, H. Z. Massoud, E. J. Swanson, R. T. George and R. B. Fair, "Measurement and modeling of charge feedthrough in N-channel MOS analog switches," *IEEE J. Solid-State Circuits*, vol. 20, pp. 1206-1213, Dec. 1985.

[20] J.-H. Shieh, M. Patil and B. J. Sheu, "Measurement and analysis of charge injection in MOS analog switches," *IEEE J. Solid-State Circuits*, vol. 22, pp. 277-281, April 1987.

[21] G. Wegmann, E. A. Vittoz and F. Rahali, "Charge injection in analog MOS switches," *IEEE J. Solid-State Circuits*, vol. 22, pp. 1091-1097, Dec 1987.

[22] C.-Y. Wu et al, "A 0.5µA offset-free current comparator for high precision current-mode signal processing," in *Proc. IEEE International Symposium on Circuits and Systems*, pp. 1829-1832, Singapore, June 1991.

[23] EPFL, *Rapport scientifique*, Edition 1991, p. 309.

[24] H. C. Lin et al, "Complementary MOS-Bipolar transistor structure," *IEEE Trans. on Electron Devices*, vol. ED-16, pp 945-951, Nov. 1969.

[25] E. A. Vittoz, "MOS Transistors operated in the lateral mode and their applications in CMOS," *IEEE J. Solid-State Circuits*, vol. 18, pp. 273-279, June 1983.

[26] T. W. Pan and A. A. Abidi, "A 50 dB variable gain amplifier using parasitic bipolar transistors in CMOS," *IEEE J. Solid-State Circuits*, vol. 24, pp. 951-961, Aug. 1989.

[27] G. Wegmann and E. A. Vittoz, "Very accurate dynamic current mirrors," *Electronics Letters*, vol. 25, pp. 644-646, 11th May 1989.

[28] P. Deval, *Convertisseurs Analogiques/Numériques cycliques à mémoires de courant CMOS*, Thèse No. 1001, EPFL Lausanne, 1992.

Chapter 14

Building Blocks for Switched-Current Sigma-Delta Converters

Gordon W. Roberts and Philip J. Crawley

14.1 Introduction

Sigma-delta converters have been receiving much attention in the VLSI industry over the past few years [1]. This interest stems from industry's need for high-resolution data converters that can be integrated in fabrication technologies that are optimised for digital circuits and systems. Sigma-delta converters are very well suited to fine-lined VLSI as they make extensive use of digital signal processing and involve very little analogue signal processing. Until recently, the analogue signal processing associated with sigma-delta converters has been implemented using switched-capacitor (SC) techniques. Although this has been more than adequate in terms of achieving high-resolution oversampling A/D converters, SC circuits require non-standard components such as linear floating capacitors to be added to the VLSI process, thereby increasing the cost of the IC. Today, with the analogue portion of the IC chip occupying no more than 10 percent of the total area of the IC, this cost is being called into question.

More recently a new analogue circuit technique called the switched-current (SI) or current copier technique [2,3,4], the subject of earlier chapters, has been proposed and shown to be capable of realising high-performance sampled-data analogue circuits in fully monolithic form using a standard digital CMOS process. This technique is capable of creating discrete-time circuits similar to those realised using the SC technique with the added advantage that only MOS transistors are required. As a result, both the digital and analogue signal processing circuitry associated with an oversampling A/D converter can be implemented directly in a standard digital CMOS process.

One important example that has experimentally demonstrated the high performance achievable with the SI approach is the work of Daubert and Vallancourt [5]. In [5], a Double-Integrator Sigma-Delta Modulator (DISDM) for A/D conversion intended for voice-band telecommunication applications was fabricated using this technique. The resulting circuit presented there consumed 75mW of static power and required a silicon area of 1.3mm^2 in a 0.9μm CMOS process while achieving a peak signal-to-noise ratio (SNR) of approximately 70dB. Although this example clearly illustrates the performance attainable from SI circuits, it was constructed using high-gain transconductance stages much like those used in SC circuits and as a result contributed to an excessive silicon area and power. It is therefore the long term goal of this work to develop a more efficient SI design of a DISDM so that low-voltage and low-power applications can be considered an important use of this technique. It is believed that this can only be achieved if the design of the DISDM is performed at the transistor level. In this way, design trade-offs can be identified and used to select the appropriate transistor sizes such that the area and power requirements are minimised.

In this chapter we shall outline how the SI circuit technique can be used to create two different types of sigma-delta modulator circuits for voice-band (4kHz) telecommunication applications. Together with a description of the basic building blocks used in the DISDM implementation, a step-by-step procedure for selecting the dimensions of the key components of each design is given such that the implementation will meet the desired speed and resolution requirements. Although this work is still in its infancy, experimental results confirm the operation of one of the two designs. More specifically, a SI DISDM has been designed and fabricated in Northern Telecom's 1.2μm CMOS process having an area of 0.5mm^2 and an average power consumption of 2.5mW. Experimental results reveal that the SI DISDM has a peak signal-to-noise ratio of 56dB. Although this particular design example falls somewhat short of the SNR required by the telecommunications industry, this example demonstrates the potential of the SI technique for realising low-power circuits in a standard CMOS process. Of course further research work is necessary to improve the SNR of the DISDM where it is believed that this improvement will come at the expense of using more power and greater silicon area for the design.

14.2 Sigma-Delta A/D Conversion

Let us begin our discussion of SI sigma-delta modulators by first outlining the role that sigma-delta modulators have in the overall operation of an oversampled A/D converter. The basic structure of the oversampled A/D converter is illustrated in Figure 14.1. It has been divided into four separate blocks: the input anti-aliasing analogue filter, the sigma-delta modulator, the decimator, and a digital lowpass filter. The input anti-aliasing filter is usually a low-order passive filter circuit whose 3dB frequency is usually set at some frequency far above the Nyquist interval (i.e. twice the signal bandwidth) of the input signal. This allows for variations in the 3dB frequency of the filter without any signal degradation, as well as minimisation of the effects of group delay distortion created by the low-pass filter. The sigma-delta modulator block performs two important functions: One function is to modulate the bandlimited analogue input signal into a one-bit digital code at a frequency f_s much higher that the Nyquist rate f_N. The other function is to noise-shape the quantisation noise such that the low-frequency input signal is separable from the quantisation noise. Subsequently, the decimator and low-pass filter with additional decimation acts to suppress the quantisation noise, resulting in an n-bit digital output occurring at the Nyquist rate f_N.

f_N : Overall sampling rate of the A/D converter

f_s : Sampling rate of the $\Sigma\Delta$M

$$P \cdot Q = f_s / f_N$$

Figure 14.1 Overview of an oversampling sigma-delta converter in block form.

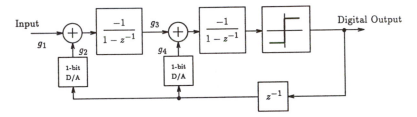

Figure 14.2 A block diagram of a DISDM using a cascade of two first-order integrators.

The overall signal-to-noise behaviour of the sigma-delta converter depends directly on the noise-shaping ability of the sigma-delta modulator, which in turn depends on the structure of the sigma-delta modulator. The most popular structure used is the double-integrator sigma-delta modulator (DISDM) shown in Figure 14.2. It consists of two discrete-time integrators, a comparator, a set of weighted summers and two one-bit D/A converter circuits in the feedback path of the DISDM. The scale factors associated with each of the summers are selected such that the input signal experiences very little attenuation from input to output over the bandwidth of the modulator $f_N/2$, whereas the quantisation noise generated by the comparator circuit is high-pass filtered with its 3dB corner frequency located at a frequency very much higher than $f_N/2$. Provided inband signals to the modulator are not distorted, the performance of the sigma-delta modulator is completely characterised by this high-pass filter function.

   If we assume that the quantisation noise generated by the one-bit A/D converter circuit is uncorrelated with the input, then we can model the DISDM as shown in Figure 14.3 with two separate inputs [6]. The signal

$U(z)$ represents the input to the DISDM in the z-domain, and the other signal, $N(z)$, represents the quantisation noise added by the comparator.

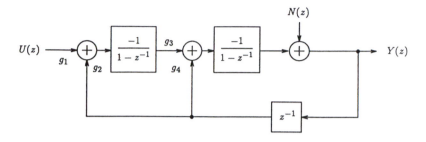

Figure 14.3 Linearised model of the DISDM shown in Figure 14.2. It is assumed that the gain of the D/A circuit is unity.

Computing the transfer functions from the two inputs to the single output $Y(z)$, we obtain

$$\frac{Y(z)}{U(z)} = \frac{g_1 g_3}{1 + (g_4 - g_2 g_3 - 2)z^{-1} - (g_4 - -1)z^{-2}} \qquad (14.1)$$

and

$$\frac{Y(z)}{N(z)} = \frac{(1 - z^{-1})^2}{1 + (g_4 - g_2 g_3 - 2)z^{-1} - (g_4 - 1)z^{-2}} \qquad (14.2)$$

The first term, $Y(z)/U(z)$, denotes the signal transfer function (STF) from the input to the output of the DISDM and the second transfer function, $Y(z)/N(z)$, denotes the noise transfer function (NTF) from the noise source to the output. In practice, the STF is made to be low-pass in nature with a gain very close to unity in the passband of the modulator. In contrast, the NTF is made to have a high-pass behaviour, effectively attenuating the quantisation noise in the modulators passband at the expense of amplifying the quantisation noise at high frequencies.

As an example, with summing coefficients $g_1 = 1$, $g_2 = -1$, $g_3 = 1$, $g_4 = 2$, sampling rate $f_s = 2.048$MHz and the input signal assumed to have a bandwidth of less than 4kHz ($f_N = 8kHz$), the magnitude of STF is held very close to unity throughout the passband. In contrast, the NTF has a minimum attenuation of -88.5dB at the passband edge of the modulator. Borrowing a formula for the in-band SNR of an ideal DISDM from [7], given by

$$SNR \approx 15log_2 M - 13, \, dB \qquad (14.3)$$

where M denotes the oversampling ratio f_s/f_N, the best case SNR is in the neighbourhood of 107dB. Take note that the above DISDM coefficients were not only chosen with behaviour of the STF and NTF in mind but also from a stability perspective. Selecting an incorrect set of coefficients can result in an unstable DISDM.

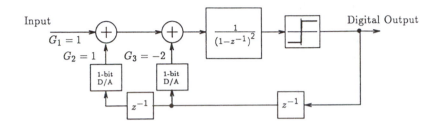

Figure 14.4 An alternative DISDM with a second-order integrator block.

Another possible structure for the DISDM, which we shall find useful in this chapter, is shown in Figure 14.4. It consists of a second-order integrator block, several unit-delay blocks and two single-bit D/A circuits. Following a similar analysis as that used for the previous DISDM, we find that the STFs and NTFs have the same values in the passband region of the modulator. Thus, the overall SNR of this DISDM structure should be quite similar to that of the previous DISDM structure. The usefulness of this DISDM structure is that it allows an SI implementation that is independent of any MOSFET matching requirement. We shall have more to say about this in the next section.

14.3 The SI Circuit Technique

The most common method of implementing DISDM circuits on silicon has been through the use of SC circuit techniques [1]. Using such an approach, fully-monolithic DISDM circuits have been reported with 16-bit resolution accuracy [8]. Although SC circuits have become the main-stay of mixed-signal ICs, especially those employing oversampling A/D techniques, SC circuits are not ideal in these applications. As emphasised throughout this book, SC circuits require high-quality linear capacitors to be added to an otherwise standard digital CMOS process which increases the overall cost of the IC. In contrast, the SI circuit technique has been proposed in order to circumvent the linear capacitor requirement and allow for high-performance analogue circuits to be implemented in a standard CMOS process. In this section we will summarise the SI circuit technique and show how one can use it to create several different types of discrete-time integrator circuits.

The current memory cell

The most fundamental building block of SI circuits is the current copier or current memory cell. The simplest current memory cell is shown in Figure 14.5 where it is composed of two transistors; one acting as the memory transistor (T_1) and the other as a switch (T_2). Depending on the state of the switch, the cell has two modes of operation. With a current I_{in} forced into the node labelled as *In/Out* and the gate voltage of transistor T_2 set high so that the switch is closed, a portion of the input current is temporarily used to charge up the gate-source capacitance C_{gs} of T_1 to a voltage level of V_{gs}. Once the gate voltage reaches a level that allows T_1 to sustain the entire input current I_{in}, the gate charging current decreases to zero and a steady-state is reached. When the switch is opened, the charge stored on T_1 is held constant, thereby causing the drain current of T_1 to remain fixed at a current level of I_{in} for the entire duration that the switch is opened.

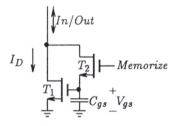

Figure 14.5 A two transistor switched-current memory cell.

The drawback to the simple current memory cell is that it is not very accurate. Currents memorised by the memory cell may differ from those that are retrieved from the cell at a later time owing to the low output resistance of the current memory cell. A simple means by which to increase the output resistance of the current memory cell is to include two additional transistors in the circuit to provide a form of active feedback. Such an arrangement is shown in Figure 14.6 and is referred to as the regulated cascode switched-current memory cell [9]. It is composed of three transistors, a switch and a current source I_{RC}. Transistor T_1 acts as the memory element as in the case of the simple memory cell. Transistors T_2 and T_3, together with the current source, provide a high gain negative feedback loop around the memory transistor which acts to minimise output voltage fluctuations on the drain terminal of T_1. In this way, the drain voltage of T_1 is held nearly constant (at a DC voltage established by transistor T_2) and allows the current of the memory cell to be controlled almost exclusively by the gate voltage on T_1.

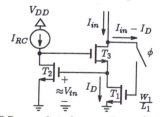

Figure 14.6 An NMOS regulated-cascode switched-current memory cell.

The regulated cascode switched-current memory cell functions as follows: when the switch is closed, i.e. ϕ is high, the gate of transistor T_1 is connected to the input of the circuit. If the drain current of transistor T_1 differs from the input current, then as a result of the connection made by the switch, the difference of the two currents will flow into the gate of transistor T_1. As a result of the negative feedback connection, this will cause the gate voltage of transistor T_1 to increase or decrease until the two currents are equal. When the switch is open, i.e. ϕ is low, the gate of transistor T_1 has no external connection and therefore the voltage on the gate cannot change. Thus, the regulated cascode switched-current memory cell will continue to pull the same current I_{in} as when the switch was closed. For further details of the operation of the regulated-cascode refer to Chapter 6.

The regulated cascode switched-current memory cell has two linear ranges within which to operate. The memory cell can operate with the memory transistor T_1 operating either in the saturation or nonsaturation (triode) region. Although higher output resistance is achievable with the memory transistor operating in saturation, experience has shown us that the regulated cascode switched-current memory cell is less dependent on charge-injection effects, a constant settling time independent of the current level, and more stable when operated with T_1 in the triode region. The reduced output resistance is acceptable for the DISDM designs presented here because the very large loop gain generated by the regulated cascode transistor configuration maintains the output resistance of the memory cell at a very high level.

For the designs of this chapter, we shall design the regulated cascode switched-current memory cell so that the memory transistor operates in its triode region. This is achieved simply by biasing transistor T_2 with current source I_{RC} so that its gate-source voltage is slightly larger than one threshold voltage (V_{tn}). In this way, the voltage at the drain of T_1 is held at about V_{tn} above ground and allows the gate voltage to swing as low as $2V_{tn}$. This provides the greatest signal swing possible on the gate terminal of the memory transistor.

A small-signal analysis reveals that when the memory cell is memorising a current, i.e. when the switch ϕ is closed, the incremental resistance seen looking into the memory cell with T_1 in its triode region is

$$r_{in} = \frac{1}{\mu_n C_{OX}\left(\frac{W_1}{L_1}\right)V_{tn}} \tag{14.4}$$

Note that r_{in} is independent of the memorised current level and is therefore constant over the entire current range of the memory cell. For reasonable engineering values, this resistance is quite low (1kΩ-10kΩ). Alternatively, during the constant-current phase, i.e. when the switch ϕ is open, the incremental output resistance is found to be

$$r_{out} = [g_{m2} r_{o2} g_{m3} r_{o3}] r_{ds1} \qquad (14.5)$$

The first four terms in this expression represent the loop gain of the feedback loop formed by transistors T_2 and T_3. Typical loop gains are in the neighbourhood of 10,000. Thus, with r_{ds1} in the range of 10kΩ, we can expect an output resistance around 100MΩ. For our DISDM application, this provides a reasonable level of accuracy.

Another important aspect of the current memory cell is its current handling capability, i.e. the range of current that it can handle while remaining in its linear region of operation. Consider that the lowest voltage that can appear on the gate of transistor T_1 so that it is still in the triode region is $2V_{tn}$. Therefore, the minimum current that the memory cell can sink is given by

$$I_{min} = \frac{1}{2} \mu_n C_{OX} \frac{W_1}{L_1} V_{tn}^2 \qquad (14.6)$$

If the current being memorised drops below I_{min} non-linear operation can occur. Conversely, the maximum current that this cell can sink depends on the maximum voltage that can appear on the gate of T_1. Unfortunately, this voltage depends on the load attached to the current memory cell and cannot be quantified at this time. However, labelling the maximum gate-source voltage as V_{gsmax} we can write the expression for the maximum current as

$$I_{max} = \mu_n C_{OX} \frac{W_1}{L_1} \left[(V_{gsmax} - V_{tn}) V_{tn} - \frac{1}{2} V_{tn}^2 \right] \qquad (14.7)$$

Once the memory cell is inserted into a particular circuit configuration it is straightforward to determine the maximum cell current.

Finally, to retrieve the signal stored in the memory cell, there are two different approaches: one approach simply samples the output current during a time interval in which the output is held constant and the other approach relies on mirroring the current stored in the memory transistor to the drain terminal of another transistor. The advantage of the latter approach is that the current mirror can provide additional current gain, as well as provide direct access to the input or output current during any time interval. The obvious drawback to the current mirror approach is the requirement for matched transistors. Examples of these two approaches will be illustrated below.

A first-order SI integrator circuit

Fundamental to the DISDM implementation is a realisation of a discrete-time integrator of the form $-K/(1 - z^{-1})$ where K is an arbitrary constant. Using two current memory cells together with a current mirror circuit, a discrete-time integrator with the above transfer function can be realised. Such an integrator was originally proposed by Hughes *et al* [10] and its circuit diagram is shown in Figure 14.7. For the time being, the current memory cells shall be illustrated as single transistor cells in order to keep the circuit diagram simple for the analysis that is to follow.

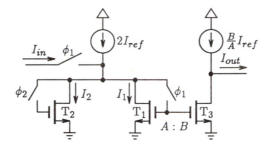

Figure 14.7 A first order SI integrator circuit.

The operation of this integrator can be described by first assuming that clock phase ϕ_1 has just ended on the *(n-1)th* sampling instant and transistor T_1 has memorised a current equal to $I_1(n-1)$. Due to the mirroring action of transistors T_1 and T_3, a scaled version of this current $I_1(n-1)$ is subtracted from the reference current $(B/A)I_{ref}$ to become the output current of the integrator, i.e.

$$I_{out}(n-1) = \frac{B}{A}I_{ref} - \frac{B}{A}I_1(n-1) \tag{14.8}$$

Now on clock phase ϕ_2, the drain current of transistor T_1 equal to $I_1(n-1)$ is subtracted from $2I_{ref}$ and is stored on transistor T_2 as $I_2(n-1/2)$, thus

$$I_2(n-1/2) = 2I_{ref} - I_1(n-1) \tag{14.9}$$

On the next ϕ_1 clock phase, the new input current $I_{in}(n)$ and the reference current $2I_{ref}$ are added together. This sum minus $I_2(n-1/2)$ is stored on transistor T_2 as $I_1(n)$, thus we can write

$$I_1(n) = I_{in}(n) + 2I_{ref} - I_2(n-1/2) \tag{14.10}$$

Substituting for $I_2(n-1/2)$ defined above, we then get

$$I_1(n) = I_{in}(n) + I_1(n-1) \tag{14.11}$$

Due to the mirroring action of transistors T_1 and T_3, a scaled version of the current $I_1(n)$ is subtracted from the reference current $(B/A)I_{ref}$ to become the new output current of the integrator, i.e.

$$I_{out}(n) = \left[\frac{B}{A}I_{ref} - \frac{B}{A}I_1(n-1)\right] - \frac{B}{A}I_{in}(n) \qquad (14.12)$$

Substituting for the terms between the brackets by $I_{out}(n-1)$, we can re-write the above equation and get the final result

$$I_{out}(n) = I_{out}(n-1) - \frac{B}{A}I_{in}(n) \qquad (14.13)$$

Taking the z-transform of the above equation results in the desired integrator transfer function

$$\frac{I_{out}}{I_{in}}(z) = -\frac{B}{A}\frac{1}{(1 - z^{-1})} \qquad (14.14)$$

Thus the SI circuit of Figure 14.7 can be used to implement the two integrators in the block diagram of the DISDM shown in Figure 14.2.

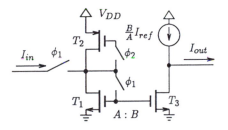

Figure 14.8 A fully-complementary SI integrator circuit using an NMOS and PMOS simple switched-current memory cell.

A very simple innovation which increases the flexibility and performance of the SI integrator is the introduction of a fully-complementary integrator cell as shown in Figure 14.8. This cell allows for both positive and negative input currents to be memorised while maintaining the same transfer function. An important advantage of this transistor arrangement is the increased voltage swing that can appear on the gates of memory transistors T_1 and T_2. A detailed analysis shows that under the assumption of equal p-channel and n-channel threshold voltages, the voltage swing that can appear on the gates of T_1 and T_2 will equal $(V_{DD} - V_t)/2$. This is in contrast to the integrator circuit of Figure 14.7 where a detailed circuit analysis reveals that the voltage on the gates of each memory transistor must not extend outside the range of V_t and $2V_t$ if linear operation is to be maintained. With the simple current memory cells in the fully-complementary integrator cell of Figure 14.8 replaced by regulated cascoded versions, as shown in Figure 14.9, we find that the voltage on the gate of the NMOS memory transistor can swing between $2V_{tn}$ and $V_{DD} - V_{tp}$ while maintaining linear operation and between V_{tn} and $V_{DD} - 2V_{tp}$ for the PMOS memory transistor. With 5V power supplies and MOSFETs with a 0.8V threshold, one can expect a factor of 3 times the improvement in the signal handling capability of this fully-complementary integrator cell over and above that of the simple integrator circuit.

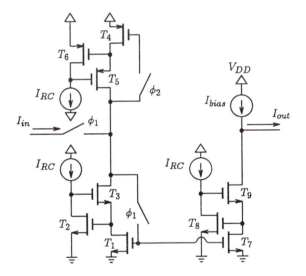

Figure 14.9 An improved regulated-cascode switched-current integrator with a fully-complementary structure.

Based on the above discussion, the minimum and maximum drain current that flows in the NMOS memory transistor of the integrator can be computed from the following two equations:

$$I_{min} = \mu_n C_{OX} \left(\frac{W_n}{L_n}\right) \left[(V_{GSmin} - V_{tn})V_{tn} - \frac{V_{tn}^2}{2}\right] \qquad (14.15)$$

and

$$I_{max} = \mu_n C_{OX} \left(\frac{W_n}{L_n}\right) \left[(V_{GSmax} - V_{tn})V_{tn} - \frac{V_{tn}^2}{2}\right] \qquad (14.16)$$

where $V_{GSmin} = 2V_{tn}$ and $V_{GSmax} = V_{DD} - V_{tp}$. A similar set of equations can also be written for the PMOS current memory cell using the appropriate threshold voltage and aspect ratio. The resulting linear current range of this integrator is then the intersection of the current range of the NMOS and PMOS memory cells. For most CMOS processes, it is reasonable to assume that the threshold voltage for the NMOS and PMOS transistors are quite close in magnitude. Thus, the current limits of the two cells can be made equal, or nearly so, by simply ensuring that $\mu_n(W_n/L_n) = \mu_p(W_p/L_p)$. On doing so, we can quantify the maximum current swing experienced by the memory transistors of this integrator by

$$I_{swing} \triangleq I_{max} - I_{min} = \mu_n C_{OX}\left(\frac{W_n}{L_n}\right)\!\left[V_{DD}V_{tn} - 3V_{tn}^2\right] \qquad (14.17)$$

For symmetrical operation, the bias level I_{bias} of each integrator is selected to be the average value of the current swing at $(I_{max} + I_{min})/2$.

A second-order SI integrator circuit

Another integrator circuit that is useful for DISDM implementation is the second-order SI integrator circuit shown in Figure 14.10. This integrator circuit is composed of a cascade of two first-order integrator circuits. However, this integrator circuit samples the output signal from each integrator during a time interval in which the output is held constant rather than through a transistor mirror circuit. In this way, the behaviour of the integrator does not depend on the matching of different transistors. Moreover, the same transistor partakes in the action of converting a current to a voltage and vice versa during the read and write operations, allowing for very precise operation.

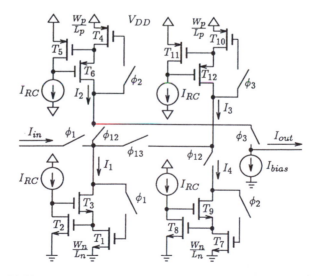

Figure 14.10 A second-order switched-current integrator circuit.

The circuit shown in Figure 14.10 has a 3-phase operation. Clock phases ϕ_1, ϕ_2 and ϕ_3 are assumed to be non-overlapping orthogonal clock signals that follow in the sequence, i.e. ϕ_1, ϕ_2, ϕ_3, ϕ_1, $\cdots$. Clock phases ϕ_{12} and ϕ_{13} represent clock signals that are high for two consecutive clock phases with the subscript denoting the appropriate clock phases. Writing a set of difference equations and solving for the input-output current relationship of the memory cell, we arrive at the following equation:

$$I_{out}(n) = 2I_{out}(n-1) - I_{out}(n-2) + I_{in}(n) \qquad (14.18)$$

Taking the z-transform and expressing the output current as a ratio of the input current results in the following integrator transfer function:

$$\frac{I_{out}}{I_{in}}(z) = \frac{1}{(1 - z^{-1})^2} \qquad (14.19)$$

From the above transfer function we can see that the second-order integrator circuit of Figure 14.10 is non-inverting and has a unity gain constant.

As in the case of the first-order integrator circuit, the current range of the second-order integrator circuit is established as the intersection of the current range of the individual NMOS and PMOS current memory cells. By design, we shall set the minimum and maximum current limits of the NMOS and PMOS cells equal to one another by sizing them appropriately. In this way, symmetrical operation is maintained. Based on our previous analysis of the NMOS cell operation, the minimum cell current would be

$$I_{min} = \frac{1}{2} \mu_n C_{OX} \left(\frac{W_n}{L_n}\right) V_{tn}^2 \qquad (14.20)$$

and the maximum cell current could be determined from

$$I_{max} = \frac{1}{2} \mu_n C_{OX} \left(\frac{W_n}{L_n}\right) \left[(V_{gsmax} - V_{tn})V_{tn} - \frac{1}{2}V_{tn}^2\right] \qquad (14.21)$$

provided V_{gsmax} can be found. By inspection of the second-order integrator circuit of Figure 14.10, we find that the maximum voltage that can appear on the gate of the memory transistor in the NMOS memory cell if linear operation is to be maintained is $V_{gsmax} = V_{DD} - V_{tp}$. Thus, we find the maximum cell current to be

$$I_{max} = \frac{1}{2} \mu_n C_{OX} \left(\frac{W_n}{L_n}\right) \left[(V_{DD} - V_{tp} - V_{tn})V_{tn} - \frac{1}{2}V_{tn}^2\right] \qquad (14.22)$$

The current swing of the second-order integrator would then simply be the difference between I_{max} and I_{min}.

In order to ensure that the integrator is biased midway between I_{min} and I_{max}, the level of the current source I_{bias} is set equal to $(I_{max} + I_{min})/2$.

14.4   Additional Current-Mode Circuits

Other circuits necessary to complete the implementation of the DISDM structures of Figure 14.2 and 14.4 are a current comparator, a one-bit digital-to-analogue feedback circuit with variable gain, a unit delay block with transfer function z^{-1} and a high quality current source. Each of these circuits will now be discussed.

Current comparator

The CMOS inverter circuit shown in Figure 14.11 can act as a simple current comparator. For positive input currents, i.e. current is forced into the input node of the inverter, the voltage at the input of the inverter will ramp up until it hits the upper voltage rail of the circuit (V_{DD}) causing the output of the inverter to go low. On the other hand, for negative input currents, the input is pulled to ground and hence the output is forced high. In this way the inverter distinguishes between positive and negative currents and therefore acts as a current comparator.

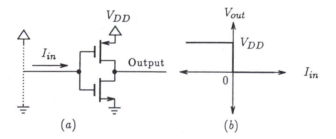

Figure 14.11 (a) A CMOS current comparator circuit, (b) Input-output transfer characteristic.

D/A circuit

Figure 14.12 illustrates two different D/A circuits that we have considered in this work. Both of these circuits are used to convert the one-bit digital output from the DISDM back into a current signal to be applied as an input to the SI integrator. The circuit of Figure 14.12(a) consists of a current source of magnitude $2KI_{ref}$, a current sink of magnitude KI_{ref} and a single MOSFET switch. Depending on the state of the switch, a current of magnitude KI_{ref} is either sourced or sunk at the output terminal of the circuit. The parameter K represents the gain of the D/A circuit and its value is customised to the particulars of the DISDM design. The other D/A circuit of Figure 14.12(b) behaves in much the same way except that it is constructed with a fully complementary structure. There is no particular advantage of one D/A circuit over the other except that when the D/A circuit of Figure 14.12(a) is inserted into the final DISDM circuit a cancellation of several current sources can occur, thereby simplifying the overall DISDM circuit.

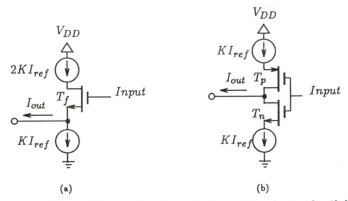

(a) (b)

Figure 14.12 Two different implementations of a single-bit digital-to-analogue feedback circuit with output current $\pm KI_{ref}$.

Delay element

The delay element with transfer function z^{-1} is simply realised with a D-type flip-flop circuit and needs no further explanation.

Current source

Finally, the design of all the current sources was based on the high-swing cascode current source shown in Figure 14.13. A detailed description and analysis of this type of current source is provided in [11]. This type of current source provides high-output resistance with the largest possible linear range of operation.

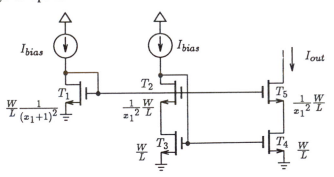

Figure 14.13 An NMOS high-swing cascode current source circuit.

14.5 SI DISDM Design

In this section we shall demonstrate how one can implement the two DISDM structures of Section 14.2 using SI circuit techniques [12,13]. One design will be based on a cascade of two first-order SI integrators with current mirrors and the other will be based on the component-invariant second-order SI integrator circuit. Both designs have been laid out and fabricated in a 1.2μm CMOS process. Of the two designs, the former implementation was successfully fabricated and experimental results will be provided. The component-invariant design experienced latch-up problems associated with the input/output pads and, as a result, experimental results for the component-invariant design are not available at this time.

14.5.1 The current mirror design approach

The first design that we will consider here will be based on the DISDM structure shown in Figure 14.2. Using the first-order SI integrator circuit of Figure 14.9, together with several of the subcircuits of the previous section, we can replace each block in the DISDM of Figure 14.2 and arrive at its circuit implementation.

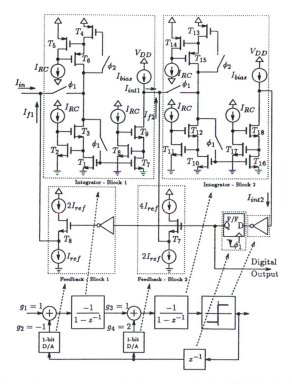

Figure 14.14 Translating the block diagram of the DISDM in Figure 13.2 to an SI design.

Figure 14.14 presents this translation from block to circuit diagram; the broken lines with arrows indicating where each substitution is made. Note that the gains g_1, g_2, g_3, and g_4 associated with the block diagram have been replaced with the values stated in Section 14.2 and that these gains have been incorporated into the circuit implementation by assigning them to either the integrators or the D/A circuits. For instance, with $g_1=1$ the input current signal is applied directly to the input of the first integrator without modification. In the case of $g_2=-1$, and a further assumption that the range of the input current is bounded between $-I_{ref}$ and $+I_{ref}$, the gain of the feedback D/A stage (Feedback-Block 1 in Figure 14.14) is made equal to -1 by inverting the digital control signal that is applied to the switch of the D/A circuit and by switching between the current levels of $-I_{ref}$ and $+I_{ref}$. The third gain term, $g_3=1$, is absorbed into the circuit implementation by setting the gain of the first integrator equal to unity. The gain of the second integrator is by definition assumed equal to unity. Finally, the fourth gain constant, $g_4=2$, establishes the gain of the second feedback D/A circuit. This requires that the current fed back by the second D/A circuit is switched between the current levels of $-2I_{ref}$ and $+2I_{ref}$.

The operation of the DISDM can be described in two phases; during ϕ_1, the input current I_{in}, the feedback current I_{f1}, the bias current of the first integrator $4I_{ref}$ and the drain current of transistor T_1 (denoted as I_1) are added and stored on T_2. The drain current of T_2 is then mirrored by T_3 and subtracted from $2I_{ref}$ to become the output current I_{int1} of the first integrator. This process is repeated for the second integrator, resulting in a current I_{int2} appearing at the output of the second integrator. Finally, this current I_{int2} is passed to the comparator and a comparison is made to see whether the output of the DISDM should be high or low. The digital output is then latched into the flip-flop on clock phase ϕ_2. The digital output from the flip-flop is then fed back to the input of each integrator as two separate currents by the two single-bit D/A circuits.

In order to maximise the dynamic range of the DISDM circuit, two separate issues must be considered if good performance is to be guaranteed. The first issue is that the current swing of the memory transistors of each integrator must be selected such that under peak input conditions the integrator signals occupy the full linear range available without non-linear distortion. The second issue is the selection of the absolute sizes of each memory transistor in each integrator so that the thermal noise voltage is below the expected resolution of the converter. In the following we shall outline a procedure by which to achieve these two objectives. This procedure will consider the effects of: settling time, input and output resistance, thermal noise, parasitic capacitances, clock feedthrough and charge-injection.

Selecting DISDM component values for optimum dynamic range

The design procedure begins with the requirements of the DISDM, which include the bit-resolution n, the clock period T_s and the oversampling ratio M. Although in this design procedure we make use of a first-order model for the MOSFET, more refinement is possible by including higher-order effects through the use of a simulation tool such as SPICE.

(i) Settling time considerations

The first step in the design procedure is to determine the minimum input resistance r_{in} of each current memory cell during the memorising phase so that the necessary settling time requirements are met. Our attention will be focused on the current memory cell containing transistor T_1 in Figure 14.14. The same argument extends to the other memory transistors.

Consider that during a memorising clock phase, a simple RC time constant is formed between the capacitance that appears at the gate of the memory transistor T_1 (denoted C_{g1}) and the resistance seen by this capacitance, which is simply the input resistance r_{in1} of the current memory cell during memorisation (given by (14.4)). Since the incremental input resistance of the cell is constant over the linear range of the memory cell, a simple expression relating the change in the cell current ΔI_{D1} as a function of a change in the input current ΔI_1 is simply

$$\Delta I_{D1}(t) = \Delta I_1 e^{-t/\tau 1} \qquad (14.23)$$

where $\tau_1 = C_{g1} r_{in1}$. Therefore, if the current memory cell is to settle to within the accuracy of the converter's resolution during each phase of memorisation (i.e. $\Delta I_{D1}(T_s/2)/\Delta I_1 < 2^{-(n+1)}$), then the following constraint on the input resistance must be met:

$$r_{in1} < \frac{T_s}{2(n+1)C_{g1}} \qquad (14.24)$$

We should note here that this lower limit was derived under the assumption that the current memory cell must settle within the resolution of the converter if the converters resolution is to be maintained. Recent work suggests that this is probably not necessary in practice [14] and can be relaxed somewhat.

If we denote r_{in1max} as the largest allowable input resistance for the current memory cell, we can write, from (14.24),

$$r_{in1max} = \frac{T_s}{2(n+1)C_{g1}} \qquad (14.25)$$

At this point of the development we cannot assign a value to r_{in1max} until C_{g1} is determined. We shall obtain an estimate of C_{g1} from the thermal noise discussion provided in step (iii) of this procedure. For the mean time, we shall continue under the assumption that r_{in1max} is known.

Now the formula for the input resistance of the memory cell given in (14.4) can be combined with (14.25) to obtain a constraint on the minimum aspect ratio allowable for the memory transistor. On doing so, we find

$$\frac{W_n}{L_n} \geq \frac{1}{\mu_n C_{OX} \, r_{in1max} V_{tn}} \tag{14.26}$$

Similar expressions can be created for the other current memory cells by following the exact same argument. We leave this for the reader to do.

We can conclude from the above discussion that the faster we require the memory cell to settle within a given time interval, the larger the aspect ratio of the memory transistor must be. This, of course, increases the power requirements for the DISDM circuit and it is therefore most desirable to keep the aspect ratio as low as possible. Instead of simply assigning the aspect ratio of the memory cell in each integrator as the minimum of this inequality, we shall utilise one of the two degrees of freedom available and select the aspect ratios so that each integrator is scaled for maximum signal handling capability.

(ii) Scaling for maximum signal handling capability

Let us begin this step of the optimisation procedure by first selecting the aspect ratio of the NMOS memory transistor of the first integrator equal to the smallest aspect ratio allowable under the settling time constraints, i.e.

$$\left(\frac{W_n}{L_n}\right)_1 = \frac{1}{\mu_n C_{OX} \, r_{in1max} V_{tn}} \tag{14.27}$$

Here r_{in1max} denotes the incremental resistance of the NMOS memory cell of the first integrator. Once the aspect ratio is assigned to a particular memory cell, the current swing of that cell is pre-determined. In this particular case, the current swing of this integrator cell would be that given by (14.17) with the appropriate aspect ratio substituted. This, of course, assumes that the aspect ratio of the PMOS memory transistor of the same integrator has been selected so that the current limits are the same or greater than those of the NMOS current memory cell. It is also important to verify that the PMOS memory cell satisfies the settling time requirements. Otherwise, the aspect ratios of the memory transistors in each of the two complementary memory cells making up the integrator should be increased to satisfy the settling requirements.

Assuming that the settling requirements are indeed satisfied, we can refer the current swing of the first integrator back to the input and determine the maximum peak-to-peak value of the input current that can be applied to the DISDM so that the first integrator stays within its linear region. If we denote the maximum signal gain from the input of the DISDM to the memory current signal of the first integrator as $\hat{G}_1$, then we can compute the maximum input peak-to-peak current, which we shall denote as $\hat{I}_{in_{pp}}$, from

$$\hat{I}_{in_{pp}} = \frac{I_{swing1}}{\hat{G}_1} \tag{14.28}$$

If this current range is not sufficient for the application at hand, then the aspect ratios of the memory transistors of the first integrator can be increased, thereby increasing the input current range.

Now to determine the aspect ratio of the memory transistors of the second integrator we make use of the fact that the input current swing given by $\hat{I}_{in_{pp}}$ will cause a maximum current swing in the second integrator given by

$$I_{swing2} = \hat{G}_2 \hat{I}_{in_{pp}} \tag{14.29}$$

where $\hat{G}_2$ denotes the maximum signal gain from the input of the DISDM to the memory current of the second integrator. This expression can be further simplified by substituting for $\hat{I}_{in_{pp}}$ derived in (14.28), leading to

$$I_{swing2} = \frac{\hat{G}_2}{\hat{G}_1} I_{swing1} \tag{14.30}$$

From (14.17) we can see how both I_{swing1} and I_{swing2} relate to the aspect ratio of each NMOS memory transistor of the appropriate integrator. If these two expressions are substituted into (14.30), then we can write the aspect ratio for the NMOS memory transistor of the second integrator in terms of the aspect ratio of the NMOS memory transistor of the first integrator according to

$$\left(\frac{W_n}{L_n}\right)_2 = \frac{\hat{G}_2}{\hat{G}_1}\left(\frac{W_n}{L_n}\right)_1 \tag{14.31}$$

Of course a similar expression can be written for the PMOS memory transistors. Owing to the fact that $\hat{G}_2$ is always larger than $\hat{G}_1$ we can conclude that the aspect ratios of the memory transistors in the second integrator will always be larger than those of the first integrator.

The peak signal gains $\hat{G}_1$ and $\hat{G}_2$ are determined from a non-linear transient analysis of the DISDM with a sinusoidal current signal applied as input. Owing to the highly non-linear behaviour of the DISDM the internal gains will vary with the input level. As a result, a search of the peak signal gains are made by varying both the amplitude of the input signal and its frequency over the bandwidth of the modulator. The amplitude of the input signal is limited to less than $I_{in_{pp}}/2$ because the amplitude of the input signal is never expected to go beyond that level. A transient analysis of the DISDM is then run for a sufficiently long time to capture the worst-case signal maxima at the output of each integrator. Once these peak levels are found, the peak signal gains, $\hat{G}_1$ and $\hat{G}_2$, are determined by dividing the worst-case signal maxima by the amplitude of the input signal.

Once the peak signal gains are found, the aspect ratio of each integrator memory transistor is known. Thus, we can compute the maximum and minimum current of each integrator using (14.15) and (14.16). This in turn allows us to find the current level used to bias each integrator circuit by simply finding the average value of these two current limits, i.e.

$$I_{bias1} = \frac{I_{max1} + I_{min1}}{2} \tag{14.32}$$

and

$$I_{bias2} = \frac{I_{max2} + I_{min2}}{2} \tag{14.33}$$

Finally, through a detailed non-linear analysis of the DISDM, it was found that, for the widest possible range of operation, the maximum input peak-to-peak current $\hat{I}_{in_{pp}}$ should be made 0.85 times smaller than the level of the reference current I_{ref} that is fed back by the D/A circuit. Allowing an input signal swing larger than $0.85 I_{ref}$ will cause excessive distortion and should therefore be avoided.

(iii) Thermal noise consideration

This next step is used to select the actual dimensions of each memory transistor T_1, T_6, T_{10} and T_{15} such that the effect of the thermal noise generated by the circuit is less than the resolution of the sigma-delta converter. This step is simplified by first recognising that the noise of the first integrator dominates the overall thermal noise behaviour of the DISDM. Thus, we only need to consider the thermal noise generated by the first integrator circuit. The conclusions reached for the first integrator will simply be extended to the second integrator circuit.

Due to the sampling nature of the circuit operation, the thermal noise generated by each transistor in the DISDM circuit will collectively contribute a noise power component of magnitude KT/C_g which adds to the sampled voltage signal stored on the gate capacitance of each memory transistor. Since the integrator consists of two memory cells, one constructed from NMOS transistors and the other from PMOS transistors, we can model these noise components as two independent white noise voltage sources with power spectral densities $S_n(f) = (KT/C_{g_n})(1/f_s)$ V²/Hz and $S_p(f) = (KT/C_{g_p})(1/f_s)$ V²/Hz.

Alternatively, if we refer the effect of these two noise sources back to the input in the form of a current signal, we find that the power spectral density of the input-referred noise current signal can be written as

$$S_{Iin}(f) = g_{m_n}^2 S_n(f) + g_{m_p}^2 S_p(f) \qquad (14.34)$$

where g_{m_n} and g_{m_p} are the transconductances of the NMOS and PMOS memory transistors, respectively. To create an integrator with a symmetrical signal swing, the transconductances of the NMOS and PMOS transistors are usually made equal, say to g_m. Thus, we can write (14.34) as

$$S_{Iin}(f) = g_m^2(S_n(f) + S_p(f)) \qquad (14.35)$$

If we further simplify our approach and assume that the gate capacitances of both the NMOS and PMOS transistors are equal (i.e. $C_{gn} = C_{gp} = C_g$) then we can write (14.35) as

$$S_{Iin}(f) = 2g_m^2 \frac{KT}{C_g} \frac{1}{f_s} \qquad (14.36)$$

This assumption is reasonable since it implies that each memory cell generates the same amount of noise power.

Now based on the theory of sampling white noise, the noise power associated with this input-referred current signal is $2g_m^2(KT/C_g)$. But, fortunately, because the baseband signal is oversampled by a factor of M, only $1/M$-th of this noise power appears inside the passband region of the converter. Thus, the effective noise power in the passband of the converter is $2g_m^2(KT/C_g)(1/M)$. If the power associated with the maximum input signal is compared to the noise power of the converter in its passband region then an estimate of the peak SNR of the converter can be found. Since, by design, we limit the input signal to a maximum signal swing given by I_{inpp}, we can assume for a sinusoidal input signal with a peak-to-peak amplitude of I_{inpp} that the power associated with such a signal is $I_{inpp}^2/8$. Thus, the peak SNR can be written as

$$SNR = \frac{\overset{2}{i_{inpp}}/8}{2g_m^2 \dfrac{KT}{C_g} \dfrac{1}{M}} \qquad (14.37)$$

This can be further simplified by substituting (14.17) and (14.28), together with the fact that the transconductance of an NMOS transistor in triode is $g_m = \mu_n C_{OX}(W_n/L_n)V_{tn}$, leading to

$$SNR = \frac{(V_{DD} - 3V_{tn})^2}{16\,\hat{G}_1^2} \frac{C_g}{KT} M \qquad (14.38)$$

Clearly from (14.38) the size of the gate capacitance C_g has a direct impact on the SNR achievable. To ensure that the converter has a resolution of at least n-bits or a peak SNR of approximately $6n$ dB, the following constraint must be met:

$$SNR = \frac{(V_{DD} - 3V_{tn})^2}{16\,\hat{G}_1^2} \frac{C_g}{KT} M > 2^{2(n+1)} \qquad (14.39)$$

Since C_g is the only unknown at this point, we can solve for the minimum gate capacitance necessary for the memory transistors of the first integrator. This same gate capacitance will be assigned to the memory transistors of the second integrator as this will guarantee that the thermal noise behaviour of the first integrator dominates the overall thermal noise behaviour of the DISDM.

Now once C_g is found, the length-width product of each memory transistor is known, i.e.

$$WL = \frac{C_g}{C_{OX}} \qquad (14.40)$$

If this result is combined with the results derived in the previous two steps, then we can determine the actual dimensions of each memory transistor so that all the requirements are met. For instance, if we combine (14.25) and (14.27) with (14.40), we find that the width of the NMOS current memory transistor of the first integrator is found from the following:

$$W_{n1} = \sqrt{\frac{2(n+1)C_g^2}{\mu_n C_{OX}^2 V_{tn} T_s}} \qquad (14.41)$$

Its length would then be calculated directly from either (14.40) or (14.27). Similarly, the width of the PMOS memory transistor of the first integrator would be determined using the relationship based on equal signal swing and equal transistor areas, i.e. $\mu_n(W_n/L_n) = \mu_p(W_p/L_p)$ and $L_{n1}W_{n1} = L_{p1}W_{p1}$. Thus, on combining we find

$$W_{p1} = \sqrt{\frac{\mu_n}{\mu_p}}W_{n1} \tag{14.42}$$

and

$$L_{p1} = \sqrt{\frac{\mu_p}{\mu_n}}L_{n1} \tag{14.43}$$

The width of the NMOS current memory transistor of the second integrator is then computed using (14.25), (14.27), (14.31) and (14.40) resulting in

$$W_{n2} = \sqrt{\frac{\hat{G}_2}{\hat{G}_1}\frac{2(n+1)C_g^2}{\mu_n C_{OX}^2 V_{tn}T_s}} \tag{14.44}$$

and its length would be determined from an expression quite similar to either (14.40) or (14.27). The dimensions of the PMOS memory transistor of the second integrator would then be determined by an expression similar to that provided by (14.42) and (14.43).

(iv) Feedback transistors in regulated-cascode switched-current memory cell

At this stage of the design we are left with determining the size of the feedback transistors of each regulated cascode switched-current memory cell and their appropriate current biasing I_{RC}. For convenience, we shall select the level of I_{RC} equal to the bias level of each integrator circuit (I_{bias}). This choice is somewhat arbitrary and can be altered if one wants to further reduce the power consumed by this integrator circuit. In any event, once the level of biasing is selected, the aspect ratio of the transistor used to hold the source-to-drain voltage of the memory transistor slightly above V_t (say $1.1V_t$) is computed from the following expression:

$$\frac{W}{L} = \frac{200I_{RC}}{\mu C_{OX}V_t^2} \tag{14.45}$$

The dimensions of the cascode transistor in each current memory cell are chosen such that output resistance of the current memory cell is large enough so that the worst-case variation in the output voltage results in a passband current variation that is less than one-half the least-significant bit of the converter. Mathematically, this can be stated as

$$r_{out} > \frac{(V_{DD} - V_{tp} - 2V_{tn})}{2^{-(n+1)}MI_{min}} \qquad (14.46)$$

where r_{out} is given by (14.5). The length of each transistor should be kept reasonably large to minimise channel-length modulation effects and the width of each cascoded transistor in each current memory cell should be kept small in order to minimise the parasitic capacitance that it adds to the output node of the memory cell.

(v) Reducing clock feedthrough and charge-injection effects

Switch charge-injection effects are kept to a minimum by keeping the dimensions of the MOSFETs making up the switches as small as possible. In this way the channel charge released by the MOSFET when turning off is kept as small as possible. A dummy switch scheme can be added to each switch associated with a current memory transistor to minimise the effect of clock feedthrough.

Example

Following the procedure outlined above, we have designed a 14-bit voice-band (4kHz) DISDM clocked at a rate of 2.048MHz for implementation in a 1.2μm CMOS process. Although the circuit structure of the DISDM shown in Figure 14.14 has been designed for single-ended operation, we felt that this would make the overall circuit susceptible to common-mode effects, in particular, the extraneous digital noise generated by the digital logic that accompanies the DISDM design. Thus, to minimise these effects, a differential structure for the SI DISDM was developed. Although the ideal situation would have been to create a DISDM structure using fully-differential SI integrators, one was not apparent at that time so a more global differential structure was used. This differential design is shown in Figure 14.15 and is composed of two single-ended SI DISDMs, several current mirror circuits and additional digital circuits. The current mirrors are used to create both a positive and negative version of the input current. Each of these currents is fed into the input of one of the single-ended SI DISDMs. The result is two output digital bit streams: one for the positive input current and the other for its complement. A digital difference between the two streams results in a three level output code (1, 0, and -1). Digital decimation and filtering is then performed on this digital output to remove the high-frequency quantisation noise.

Using a transistor level simulation (SPICE), a measure of the SNR of the SI DISDM extracted directly from the layout was calculated. With a 1μA sinusoidal signal at a frequency of 4kHz applied as input, a SPICE transient analysis of the DISDM was computed. This required approximately 125 hours of CPU time running on a SPARC II SUN workstation. The spectral content in the resulting digital output was then determined using a Fast Fourier Transform (FFT). From these results, the

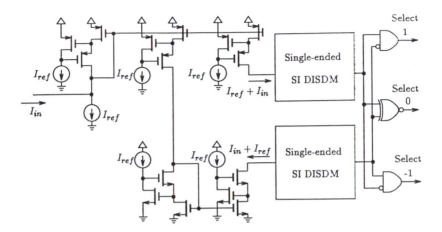

Figure 14.15 Differential SI DISDM design.

SNR in the passband region was calculated to be about 76dB; a result that suggested that the SI DISDM would probably meet the requirements of the Telecom industry.

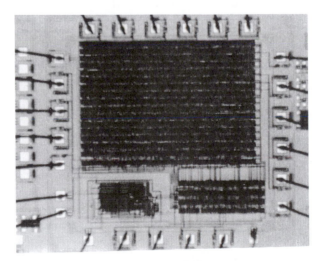

Figure 14.16 A microphotograph of the oversampling A/D converter circuit consisting of an SI sigma-delta modulator using the current mirroring approach and a 16th-order decimator is shown. The sigma-delta is located in the bottom lefthand corner.

The two single-ended SI DISDMs, together with a 16th-order sinc[4] digital decimator [15], was laid out using Northern Telecom's 1.2μm CMOS process. The total area occupied by these circuits is 2400μm by 2560μm. The analogue portion of this IC occupies an area of 550μm by 900μm. A microphotograph of the resulting IC is provided in Figure 14.16. The sigma-delta modulator is located in the bottom lefthand corner and the decimator occupies the rest of the circuitry seen in the microphotograph.

Several ICs were returned from fabrication and all were characterised for their dynamic operation. In particular, a single sinusoidal current tone at 2kHz was applied as input and the digital data from the decimator was collected and stored in a computer. This data was then further decimated and low-pass filtered using a computer software routine. On completion, the frequency content contained in the data was calculated using an FFT. From these results, the SNR ratio was then computed. This was repeated for different amplitudes of the input sinusoidal signal and the results were collected and plotted in Figure 14.17(a) for one specific device. As is evident from this graph, under the best-case input conditions, the SNR peaks at approximately 56dB. From the same spectral information used to create the SNR versus input level curve, the relative change in the gain of the DISDM at 2kHz was also obtained over the wide ranging input current level. The resulting gain tracking curve is shown in Figure 14.17(b). About a 2dB variation in the gain tracking occurs over a 45dB range in the input current level. To provide some sense of how well our SI DISDM design performed, we superimposed the CCITT G.712 system specifications for linear A/Ds used in the telecommunication industry on the two graphs of measured results. As is clearly evident from both graphs, the performance of this particular SI DISDM falls somewhat short of these specifications. The other devices showed similar behaviour.

Finally, the static power dissipated by the sigma-delta modulator with no input applied was indirectly measured to be about 2.5 mW. This is about thirty times less power than that dissipated by the design presented by Daubert and Vallancourt [5]. Unfortunately, the SI design found here had a much lower SNR. We believe that such a low SNR is the result of incorrectly selecting the dimensions of the memory transistors in the DISDM to be too small, and therefore making the design highly susceptible to thermal noise, charge-injection and clock feedthrough effects. This error was caused by an incorrect noise analysis of the DISDM. We have since rectified this situation and are waiting for new silicon to return from fabrication.

14.5.2 The component-invariant design approach

The next design that we shall consider here is based on the DISDM structure shown in Figure 14.4. Following steps similar to those outlined for the previous DISDM circuit, we can translate the DISDM block diagram of Figure 14.4 into an SI circuit implementation according to the procedure shown in Figure 14.18 by using the second-order SI integrator

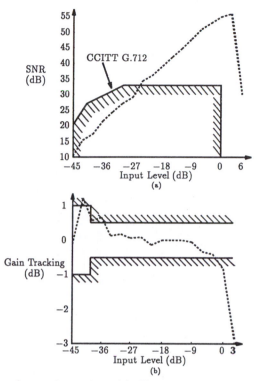

Figure 14.17 Experimental results: (a) Signal-to-noise versus input signal level, (b) Gain tracking versus input signal level.

circuit of Figure 14.10 together with several D/A circuits and delay elements. One D/A circuit is assigned a gain of unity and the other a gain of two. No other scale factors need to be considered.

The component-invariant DISDM of Figure 14.18 operates in much the same way as the previous DISDM design. An input current assumed bounded between $\pm 0.85 I_{ref}$ is added together with the two currents, I_{f1} and I_{f2}, generated by the two D/A circuits in the feedback loop of the DISDM. These currents combine on clock phase ϕ_1 and are then integrated by the second-order SI integrator circuit. The output of the integrator is then read on clock phase ϕ_3 and its sign is then checked by the comparator circuit. The output of the comparator is then latched into one of the flip-flops. The other flip-flop is used to store the previous comparator output. The state of each flip-flop determines the amount of current that is fed back by the D/A circuits to the input of the second-order integrator. Either flip-flop output can be used as the digital output from the DISDM, although we choose to use the flip-flop output closest to the current comparator.

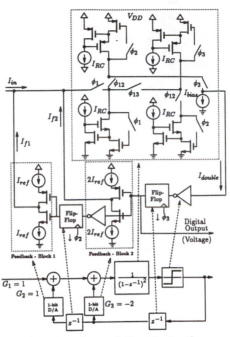

Figure 14.18 Translating the sigma-delta block diagram of the DISDM containing a second-order integrator into an SI circuit implementation.

Example

Following the design procedure outlined in the previous section, another 14-bit $\Sigma\Delta$ voice-band (4kHz) converter operating at a clock rate of 2.048MHz was designed and laid out using a 1.2µm CMOS process. Before committing the design to silicon, a transistor level simulation of the DISDM circuit was run using SPICE where the overall bit-resolution of the converter was found to exceed the 14-bit accuracy. Unfortunately, when the same simulation was run with a circuit extracted directly from the layout it was found that the converters bit resolution decreased to 10-bits. On investigation we found that this design suffers from excessive charge injection and clock feedthrough owing to the many parasitic capacitances found present. We are presently considering increasing the capacitance on the gate of each memory transistor in order to minimise the influence of the charge injection; however, this is on-going work and no results are available at this time.

   Despite the poor resolution expected, we went ahead and had the IC chip fabricated. Only a single-ended version of the DISDM was created. A microphotograph of the SI sigma-delta modulator based on the component-invariant design is shown in Figure 14.19. As is evident from the microphotograph, the DISDM is very small, and can almost be placed in the space between the two bonding pads. Measurements indicate that the

DISDM occupies an area of only 0.2mm^2. Unfortunately, on return of the IC, it was discovered that a problem existed with the protection circuitry associated the input/output bonding pads. As a result, the chip had latched up and we could not extract any useful information about the DISDM circuit.

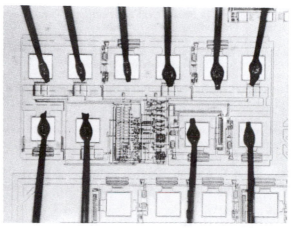

Figure 14.19 A microphotograph of the component-invariant SI DISDM design shown in Figure 14.18.

14.6   Conclusions

This chapter presented the design and implementation of several different building blocks for switched-current double-integrator sigma-delta modulators (DISDMs). Together with a description of these building blocks, a systematic design procedure was given for selecting the dimensions of the key components of each DISDM design such that the implementation meets the desired speed and resolutions requirements. Moreover, with such a concise design procedure, the component of each DISDM can be further selected so that the area and power requirements are minimised. Although this work is still somewhat immature, one DISDM design was successfully fabricated in a 1.2μm CMOS process requiring an area of 0.5mm^2 and had an average power consumption of 2.5mW. The measured peak SNR was 56dB, equivalent to about 9-bits of linear resolution.

Acknowledgements

The work in this chapter was supported by NSERC and by the Micronet, a Canadian federal network of centres of excellence dealing with Microelectronic Devices, Circuits and Systems for Ultra Large Scale Integration. Thanks are also due to the Canadian Microelectronics Corporation for arranging for the designs to be fabricated.

References

[1]   J. C. Candy and G. C. Temes, "Oversampling methods for A/D and D/A conversion," *Oversampling Delta-Sigma Data Converters: Theory, Design, and Simulation*, Eds. J. C. Candy and G. C. Temes, New York: IEEE Press, 1991.

[2]   J. B. Hughes, N. C. Bird and I. C. Macbeth, "A new technique for analog sample data signal processing," in *Proc. IEEE International Symposium on Circuits and Systems*, Portland, Oregon, pp. 1584-1587, May 1989.

[3]   G. Wegmann and E. A. Vittoz, "Very accurate dynamic current mirrors," *Electronics Letters*, vol. 25, pp. 644-646, May 1989.

[4]   S. J. Daubert, D. Vallancourt and Y. P. Tsividis, "Current copier cells," *Electronics Letters*, vol. 24, pp. 1560-1562, Dec. 1988.

[5]   S. J. Daubert and D. Vallancourt, "A transistor-only current-mode $\Sigma\Delta$ modulator," in *Proc. IEEE Custom Integrated Circuits Conf.*, pp. 24.3.1-24.3.4, May 1991.

[6]   B. P. Agrawal and K. Shenoi, "Design Methodology for $\Sigma\Delta M$," *IEEE Trans. Commun.*, vol. COM-31, pp. 360-370, March 1983.

[7]   M. W. Hauser and R. W. Brodersen, "Circuit and technology considerations for MOS delta-sigma A/D converters," in *Proc. IEEE International Symposium on Circuits and Systems*, San Jose, California, pp. 1310-1315, May 1986.

[8]   M. Rebeschini, N. R. Van Bavel, P. Rakers, R. Greene, J. Caldwell and J. R. Haug, "A 16-b 160-kHz CMOS A/D converter using sigma-delta modulation," *IEEE J. Solid-State Circuits*, vol. SC-25, pp. 431-440, April 1990.

[9]   C. Toumazou, J. B. Hughes and D. M. Pattullo, "Regulated cascode switched-current memory cell," *Electronics Letters*, vol. 26, pp. 303-305, March 1990.

[10]   J. B. Hughes, I. C. Macbeth, and D. M. Pattullo, "Second generation switch-current signal processing," in *Proc. IEEE International Symposium on Circuits and Systems*, New Orleans, Louisiana, pp. 2805-2808, August 1990.

[11]   P. J. Crawley and G. W. Roberts, "High-swing MOS current mirror with, arbitrarily high output resistance," Electronics Letters, vol. 28, pp. 361-363, Feb. 1992.

[12]   P. J. Crawley and G. W. Roberts, "Switched-current sigma-delta modulation for A/D conversion," in *Proc. IEEE International Symposium on Circuits and Systems*, San Diego, California, pp. 1320-1323, May 1992.

[13]   P. J. Crawley and G. W. Roberts, "A component invariant second-order switched-current sigma-delta modulator," in *Proc. IEEE International Symposium on Circuits and Systems*, San Diego, California, pp. 1324-1327, May 1992.

[14]   P. J. Crawley and G. W. Roberts, "Predicting harmonic distortion in switched-current memory circuits," accepted for presentation at the *IEEE International Symposium on Circuits and Systems*, Chicago, Illinois, May 1993.

[15]   E. B. Hogenauer, "An economical class of digital filters for decimation and interpolation," *IEEE Trans. on ASSP*, vol. ASSP-29, pp. 155-162, April 1981.

Chapter 15

Continuous Calibration D/A Conversion

Wouter Groeneveld, Hans Schouwenaars,
Corné Bastiaansen and Henk Termeer

15.1   Accuracy of Audio D/A Converters

15.1.1   Introduction

The demands on the linearity of high-resolution D/A and A/D converters for measurement and digital audio equipment are currently so high that the achievable accuracy based on the matching of components in a standard process is not sufficient. Therefore, additional calibration techniques are used to achieve high resolution. A disadvantage of many calibration techniques is the need for a special calibration period [1]. During this period, the converter cannot be used for conversion, which particularly limits the application range. Furthermore, a relatively large amount of chip area is needed to store the error signals. Other calibration techniques, such as laser trimming [2] and external adjustment, take precious time and facilities and are sensitive to ageing and temperature, whilst dynamic element matching [3,4] needs external components. In this chapter, the switched-current technique is used as a self calibration technique. Its use enables the design of D/A converters that need no calibration period, additional trimming or external components, and are insensitive to process variations.

15.1.2   Linearity considerations

Figure 15.1 shows the basic block diagram of an N-bit segmented D/A converter. An array of C coarse current sources is shown, which all deliver the same output current. To increase the resolution, one of the coarse currents, in this case I_c, can be divided into finer current levels by a passive current divider. Depending on the value of the data signal, a number of the currents are switched to the output terminal I_{out} and the remaining currents are dumped to signal ground. It is shown in the literature that a resolution of up to 10 bits can be obtained by a passive current divider stage consisting of equal CMOS transistors [5]. To obtain a higher resolution, the addition of the equal coarse current sources is necessary ($C = 2^{(16-10)} = 64$ for 16 bits). In this case, the linearity of the

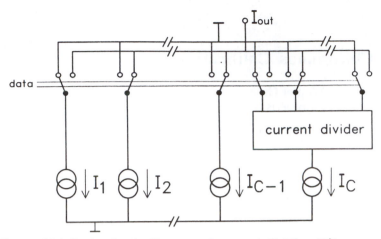

Figure 15.1 Basic circuit diagram of current dividing D/A converter.

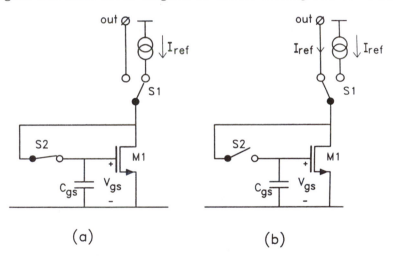

(a)   (b)

Figure 15.2 Calibration principle (a) Calibration (b) Operation.

converter depends mainly on the mutual equality of the coarse currents, which may differ by up to half an LSB-current to achieve maximum linearity. Since it is impossible to obtain a 16-bit accuracy with standard CMOS components [6,7], the switched-current calibration technique is essential to make this type of converter architecture feasible.

15.1.3 Basic calibration principle

The basic circuit of one single current cell is shown in Figure 15.2. The current source transistor M_1 is to be calibrated at the reference current I_{ref}. When the switches S_1 and S_2 are in the depicted state (a), a reference current I_{ref} flows into transistor M_1, since it is connected as an MOS-diode. The voltage V_{gs} on the intrinsic gate-source capacitance C_{gs} of M_1 is then determined by the transistor characteristics. When S_2 is opened and S_1 is switched to the other position (b), the gate-to-source voltage V_{gs} of M_1 is not changed since the charge on C_{gs} is preserved. Provided that the drain voltage is not changed either, the drain current of M_1 will still be equal to I_{ref}. This current is now available at the *out* terminal in Figure 15.2, and the reference current source is no longer needed.

15.1.4 Imperfections

In practice, the switches S_1 and S_2 are MOS transistors. This gives rise to some disturbing effects which cause changes in the gate voltage of M_1 during switching, see Figure 15.3.

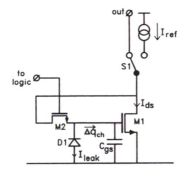

Figure 15.3 Real calibration circuit.

When M_2 is switched off, its channel charge q_{ch} is partly dumped on the gate of M_1, and so the charge on C_{gs} of M_1 is decreased by an amount Δq_{ch}. The charge change implies a sudden decrease of V_{gs} of M_1

$$\Delta V_{gs,q} = \frac{\Delta q_{ch}}{C_{gs}} \tag{15.1}$$

After switching, another effect influences V_{gs}. Although M_2 is switched off, the reverse-biased diode D_1 between its source and the substrate is still present. The DC leakage current I_{leak} of this diode continuously decreases the charge on C_{gs}. Assuming that calibration is carried out until $t = 0$, the gate voltage equals

$$V_{gs,leak}(t) = V_{gs}(0) - \frac{I_{leak}}{C_{gs}} t \tag{15.2}$$

The changes in the gate voltage of M_1 are transformed into changes in the drain current I_{ds} by its transconductance g_m. The voltage drop described by (15.1) causes a drop in the output current just after calibration of

$$I_{ds,q} = I_{ref} - g_m \Delta V_{gs,q} = I_{ref} - g_m \frac{\Delta q_{ch}}{C_{gs}} \tag{15.3}$$

The transconductance of M_1 equals

$$g_m = \sqrt{2\mu C_{ox} \left[\frac{W}{L}\right] I_{ds}} \tag{15.4}$$

in which μ represents the electron mobility and C_{ox} the oxide capacitance per square micron.

The gate-source capacitance in saturation can be rewritten as

$$C_{gs} = \frac{2}{3} W L C_{ox} \tag{15.5}$$

Substituting (15.4) and (15.5) into (15.3) yields

$$I_{ds,q} = I_{ref} - \frac{3}{2} \sqrt{\frac{2\mu}{C_{ox}}} \cdot \frac{\Delta q_{ch}}{L} \sqrt{\frac{I_{ds}}{WL}} \tag{15.6}$$

The leakage effect in the drain current can be calculated in the same way. The time-dependent drain current due to the diode leakage (15.2) is

$$I_{ds,leak}(t) = g_m V_{gs,leak}(t) = I_{ref} - g_m \frac{I_{leak}}{C_{gs}} t \tag{15.7}$$

Substitution of (15.4) and (15.5) into (15.7) yields

$$I_{ds,leak}(t) = I_{ref} - \frac{3}{2} \sqrt{\frac{2\mu}{C_{ox}}} \cdot \frac{1}{L} \sqrt{\frac{I_{ds}}{WL}} I_{leak} t \tag{15.8}$$

Equation (15.8) implies that after a certain time T_r the current cell has to be calibrated again to keep its output current within a specified range. The output current can be represented by the solid line in the graph of Figure 15.4.

The results of (15.6) and (15.8) clearly indicate that the ratio μ/C_{ox} should be small to keep the changes in the drain current small, so an advanced CMOS process is preferable. In addition, both equations contain parameters that can be adjusted in the actual design.

Unfortunately, all adjustable parameters have limits that also depend on other considerations. The current I_{ds} is determined by the output current level of the converter in which the current cell has to be implemented. The

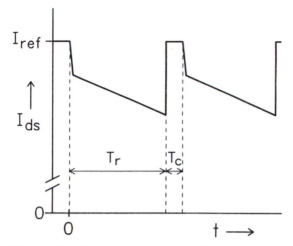

Figure 15.4 Drain current I_{ds} of M_1 against time.

transistor width, and especially its length, should be as large as possible for optimal calibration and *1/f* noise behaviour, but this is limited by layout size considerations that become more important when many cells are used. Furthermore, the *W/L* ratio has a minimum because of limits on V_{gs}.

15.1.5 Numerical example

In the following calculations, a maximum V_{gs} of 1.5V is taken and the RMS value of thermal plus *1/f* noise over the audio band of one current is designed to be 100 dB below its DC value. Note that for *N* currents, the ratio between total DC value and noise will improve by a factor of $\sqrt{N}$, so the mentioned 100dB is sufficient for very high resolution D/A converter design.

	M_1	M_2	Units
V_{th}	1	1	V
C_{ox}	1.7	1.7	nF/mm²
μC_{ox}	70	70	μA/V²
W/L	20/30	1.2/1	
I_{ds}	10		μA
V_{gs}	1.5	3	V
g_m	30		μA/V
C_{gs}	690	1.8	fF
I_{leak}		10	mA/m²
Leakage area		12.5	μm²

Table 15.1 Calibration circuit data.

To evaluate the feasibility of the calibration technique, the values shown in Table 5.1 are used as a starting point. The process used is a standard CMOS NWELL technology with a gate oxide thickness of 20nm. Calculation of $\Delta I_{ds,q}$ is done as follows. The part of the channel charge of M_2 flowing to C_{gs} of M_1 equals

$$\Delta q_{ch,2} = \frac{1}{2}C_{gs2}(V_{gs2} - V_{th2}) = 1.8fC \qquad (15.9)$$

Now (15.6) yields

$$I_{ds,q} = I_{ref} - 78nA \qquad (15.10)$$

Another current cell may have different parameters, resulting in another value of the current step. For a worst-case approach, we can take a mismatch of 10% in switch transistor dimensions, resulting in a deviation of about 8nA from the original current cell.

The leakage effect is calculated using (15.8)

$$I_{ds,leak}(t) = I_{ref} - 5.4*10^{-6}t \qquad (15.11)$$

If a 10-bit current divider stage is used in the basic block diagram of Figure 15.1, the least significant bit current is about $2^{-10}I_{ds} \approx 10nA$. This means that any sum of output currents has to be accurate within 5nA. Equation (15.10) shows that this accuracy is not achieved using the basic calibration technique. Even if only inequalities of output currents are of interest (as in audio applications), the mismatch of 8nA is too large to guarantee the accuracy of any sum of output currents.

Equation (15.11) implies that, for a maximum deviation of 5nA per cell, the calibration repetition time T_r equals 1.0ms for each cell. If in total 100 cells are used ($N = 100$), the calibration time T_c equals 10µs, which is an acceptable value. Settling during calibration is determined by the dominant time constant $C_{gs1}/g_{m1} = 23ns$. For 16-bit accuracy, 10 time constants or 230ns is needed, so leakage is not a problem in this application.

15.1.6 Improved calibration technique

A very simple improvement to the circuit of Figure 15.3 is shown in Figure 15.5. To reduce the absolute value of I_{ds} in the current memory transistor M_1, a main current source I_m is added in parallel to M_1. The current source I_m has a value of about 90% of the reference current. Its value may vary a few percent since calibration is performed on I_m and I_{ds} of M_1 together. This measure decreases the value of M_1's current to about 10%, so its transconductance is decreased by a factor of $\sqrt{10}$. Furthermore, the size of M_1 can now be optimised for the new value of I_{ds}. When V_{gs1} is not changed, a 10 times smaller W/L ratio of M_1 results, giving another transconductance reduction of $\sqrt{10}$. As a result, g_{m1} is reduced by a factor of 10 while keeping the C_{gs} and V_{gs} of M_1 constant.

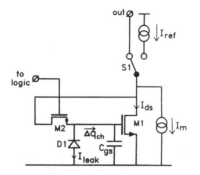

Figure 15.5 Improved calibration circuit.

The main current source I_m is easy to implement; since its value may vary a few percent, it can be derived from the reference current by standard current mirrors. Unfortunately, the addition of a main current source causes a larger mismatch of g_m between several cells since I_m can spread about 2%, and thus $I_{ref} - I_m$ can spread about 20%.

After optimisation of the size of M_1 to $W/L = 6/90$, the calculated results of (15.6) and (15.8) now become for the two outer limits considering the charge effect

$$I_{ds1} = 1.1\mu A: g_{m1} = 3.13\mu A/V, \Delta q_2 = 2.0fC, \Delta I_{ds,q} = 9.0nA \quad (15.12)$$

$$I_{ds1} = 0.9\mu A: g_{m1} = 2.84\mu A/V, \Delta q_2 = 1.8fC, \Delta I_{ds,q} = 7.4nA \quad (15.13)$$

Both the absolute value of the current drop and the difference between two output currents are strongly reduced. For audio applications in which the absolute value is of minor importance, the improved calibration technique is suitable, although the current difference of 1.6nA does not guarantee full 16-bit linearity when more than four cells are used.

The leakage effect becomes

$$I_{ds,leak}(t) \approx I_{ref} - 0.54*10^{-6}t \quad (15.14)$$

The leakage effect is also reduced by a factor of 10 due to the change in transconductance, so for an accuracy of 5nA, the refresh time must be less than 10ms. Although the major time constant is also increased by a factor of 10, settling is still not a problem for this improved calibration technique.

15.1.7 Continuous current calibration

To make the calibration technique suitable for the design of a D/A converter, it must be adapted to an array of current sources as shown in Figure 15.1. Since an audio D/A converter has to operate continuously, the calibration period in which a current cell does not operate normally must be made invisible at the current outputs of the array. This is realised by the continuous current calibration principle which is shown in Figure 15.6. The principle is characterised by the presence of N normal and one spare current cells that generate N equal output currents. The selection of the cell to be calibrated is done by an $N + 1$ stage shift register, shown at the left hand side. Some logic makes sure that only one stage contains a logic 1, while the other outputs are 0. Round coupling ensures that after sequentially calibrating all cells, the first cell is calibrated again, and so on. The switches of all current cells are incorporated in the switching network. This network connects all the output currents of the normally functioning cells to their corresponding outputs. The one cell under calibration is connected to the reference current. Since this cell is now not delivering any current to its output terminal, the output current of the spare cell is switched to this terminal. In this way it is guaranteed that there are always N equal currents available at the output terminals, which can be used in the D/A converter setup of Figure 15.1.

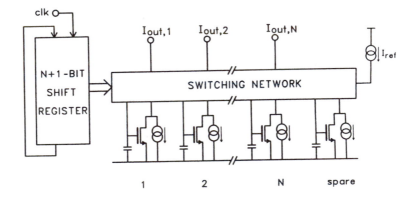

Figure 15.6 Block diagram for the generation of equal continuously flowing currents.

15.2 16-bit Audio D/A Converter Architecture

15.2.1 Block diagram

To investigate the suitability of the continuous calibration technique for high-resolution D/A conversion, a 16-bit D/A converter was designed that meets the demands for digital audio. The basic block diagram is shown in Figure 15.7. The design is based on 64 equal current sources and a spare source. Each current source consists of a complete current cell with the basic architecture of Figure 15.5. A 65-bit shift register selects the cells one by one for calibration. The calibration circuitry consists mainly of the reference source and will be described in detail later on. The current outputs of 63 normally functioning cells are fed to 63 two-way current switches, and one cell is directly connected to a passive 10-bit binary current divider [5]. Depending on the input data, a number of the 63 currents are switched to the output line, and the rest of the 63 currents are dumped to signal ground. In this way, 64 accurate output current levels can be made. The intermediate levels are obtained by routing the 64th current generated by the calibration network to the current divider. The divider output currents are then switched to the output line or to signal ground by two-way current switches, which are directly controlled by the 10 least significant data bits.

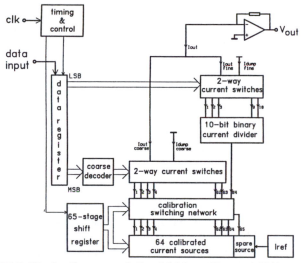

Figure 15.7 Block diagram of the 16-bit calibrated D/A converter.

Finally, the output current is converted into a voltage by means of an external operational amplifier and a resistor. To reduce the supply voltage, all current switches are designed for a small voltage drop (<50mV). Due to the parallel structure the minimum supply voltage is only 3V.

15.2.2 Calibration circuitry and current cell

Figure 15.8a shows the common part of the calibration circuitry. It consists of the reference current source, two PMOS transistors, M_8 and M_9, and two bias current sources. The nodes at the right hand side are connected to all current cells. Figure 15.8b shows one of the 64 cells and part of the switching network. Each cell has a main current source, consisting of one NMOS transistor, M_4. The gate voltage V_g of M_4 is common for all cells.

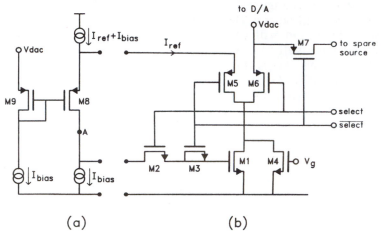

$$(a) \qquad\qquad (b)$$

Figure 15.8 (a) Calibration circuitry (b) Current cell.

In normal operation, the PMOS switch M_6 connects the current sources M_1 and M_4 to the output terminal. When the cell is calibrated, the shift register selects the cell by raising its 'select' signal, and disconnects the cell from the output terminal by routing the current through M_5 instead of M_6. M_5 is also a PMOS switch operating in the linear region. The spare source output current is now routed to the output through M_7. The loop between the drain and gate of M_1 is closed by three transistors. The first is switch M_5, followed by the level shifter M_8, which is biased by a non-critical current I_{bias}. Finally, the compensated switch M_2/M_3 closes the loop. M_3 is a half-sized dummy switch, added to reduce the absolute amount of charge injection of the M_2/M_3 switch.

The level shift stage consisting of M_8 and the bias current sources ensures that the drain voltages of M_1 and M_4 are the same during operation and calibration, instead of depending on the gate voltage of M_1, as is the case in Figure 15.2. During operation, this voltage is determined by the D/A circuit (V_{dac}). During calibration, the drain voltage of the current sources is equal to the source voltage of M_8 within a few millivolts since the voltage drop across switch M_5 is that small. The source voltage of M_8 is copied from V_{dac} by M_9 and M_8.

The clock frequency of the shift register is chosen to be equal to the audio sampling frequency, i.e. 44.1kHz. The resulting calibration period T_c for each of the 65 current cells equals 1.5µs.

15.2.3 Measurement results

The operation of the whole calibration circuit can be clarified by the oscilloscope photograph of Figure 15.9. The top trace shows the output signal of the last shift register stage, which marks the calibration period of the spare source. The lower trace shows the different gate voltages of the M_1 transistors on node A in Figure 15.8a. The DC-amplitude of this signal is about 1.5V, the sawtooth-like ripple has an amplitude of 40mV. It is clear that in this case a linear gradient is present on the main current source array, since the voltages on node A give an impression of the difference between the reference current value and the main current source values. Calculations show that the original mismatch over the current source array equals 0.5%, which would result in a signal to harmonic distortion performance of the uncalibrated D/A converter of only 60dB.

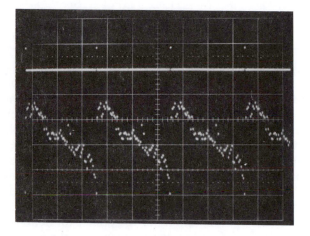

Figure 15.9 Oscilloscope photograph.

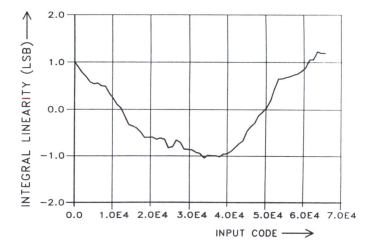

Figure 15.10 Measured integral linearity of the 16-bit DAC.

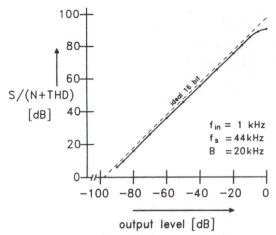

Figure 15.11 Distortion performance of the 16-bit DAC.

The measured integral linearity of the 16-bit D/A converter is shown in Figure 15.10. As can be seen, the integral linearity is within two 16-bit LSBs. In the flat part in the middle of the curve the linearity is much better. This indicates that the converter is very well suited to digital audio, since in this application field a good linearity for small signals is absolutely necessary. This is also visible in Figure 15.11, which shows the measured ratio between the RMS value of the wanted signal against the RMS value of the error signal which includes noise, harmonics and spurious components. Measurement bandwidth is 20Hz to 20kHz. No dependence on the temperature (-10°C to 70°C) or on the frequency of the input sinewave was found. Frequency components related to the refresh cycle of the individual current cells are all below -110dB, thus not affecting the dynamic range of the converter.

Figure 15.12 Die photograph of the 16-bit DAC.

A microphotograph of the test circuit is shown in Figure 15.12. A distinction can easily be made between the 6-bit coarse part and the 10-bit fine part. Most of the bonding pads were added for test purposes. Finally, the most important specifications of the converter are listed in Table 15.2.

Resolution	16 bits
Dynamic range	94dB
S/(N+THD) at 0dB	92dB
at -10dB	84dB
Supply voltage range	3 to 5 V
Max. output current	1 mA at 5 V
Power dissipation	20 mW at 5 V
Temperature range	-10 to 70 °C
Process	1.6 μm CMOS
Active chip area	2.8mm²

Table 15.2   D/A converter specifications.

15.3   Calibrated Noise Shaping D/A Converter

15.3.1   Introduction

In digital audio applications there is a growing interest in highly oversampled noise shaping D/A converters. In these systems, reasonable oversampling ratios are combined with higher order noise-shaping, thus reducing the number of bits and pushing the resulting quantisation noise out of the frequency band of interest. In this way, the performance of the D/A converter as described in the previous section can be maintained or even improved by using a D/A converter of just a few or even 1 bit. In the latter case, the output of the converter is called a bitstream.

The main advantages of 1-bit implementations are obvious. Firstly, the differential linearity is always guaranteed by the bitstream, allowing accurate reconstruction of low level signals. Secondly, there is no need for analogue accuracy as no intermediate analogue levels between the digital "0" and "1" are needed. Thirdly, the 1-bit converter concept mates well with other digital circuitry. Most of the converter itself consists of digital CMOS circuits, while the small analogue part can also be designed in the same conventional CMOS process. Consequently, the integration of the converter together with complex functions, such as digital volume and tone control, becomes possible.

The main disadvantage of 1-bit conversion is the presence of many high frequency components with a relatively large amplitude in the 1-bit data signal. This often causes intermodulation distortion and interference in the 1-bit D/A converter, leading to whizzes and frizzles within the audio band and thus reducing the true dynamic range of the converter. However, the use of a 1-bit switched capacitor D/A converter [8] does result in a 16-

bit performance. Others use multi-stage noise shaping and simple CMOS inverters [9]. To achieve a wider dynamic range, the digital and the analogue converter parts can be separated, eliminating on-chip interference and allowing the best process choice for both parts. This is reported in [10], in which true 18-bit performance has been achieved by a 1-bit D/A converter designed in a BiCMOS process.

The continuous calibration principle enables the design of a multi-bit D/A converter with high linearity and high speed in a standard CMOS process. The use of more bits reduces the high-frequency noise and thus intermodulation problems, resulting in an even wider dynamic range.

In this section, the design of a D/A converter architecture is presented which achieves a dynamic range of 115dB in a conventional CMOS process by using an oversampled 5-bit calibrated D/A converter.

15.3.2 General design considerations

In Figure 15.13 a conversion system which employs an oversampling D/A converter is shown. The input signal is a 20-bit digital sine wave at a sampling frequency of 44kHz. The sampling frequency is increased to 5.6MHz by using an oversampling factor of 128. The 20 bits at 5.6MHz are then applied to a third-order noise-shaper which reduces the 20-bit input data to N-bits. Now, at the output of the noise-shaper the bitstream is available to drive the oversampling D/A converter to produce V_{out}. Figure 15.14 shows the output spectrum of a 1-bit D/A converter with an almost full-scale output sine wave of about 1kHz. The enormous quantisation noise of the 1-bit converter is pushed to higher frequencies by the third-order noise-shaper at the rate of 60dB per decade. By integrating the output spectrum over the audio band of 20Hz to 20kHz, the RMS noise voltage is found and the dynamic range can be determined. A theoretical dynamic range of this conversion system of 120dB can be obtained with an ideal D/A converter at these conditions as is shown by [11] and [8]. In practice, the dynamic range will be determined by intermodulation phenomena in the D/A converter.

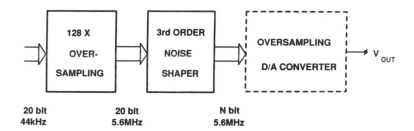

Figure 15.13 General conversion system.

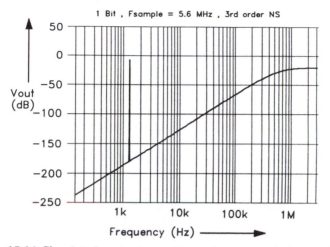

Figure 15.14 Simulated output spectrum of oversampled one-bit D/A converter.

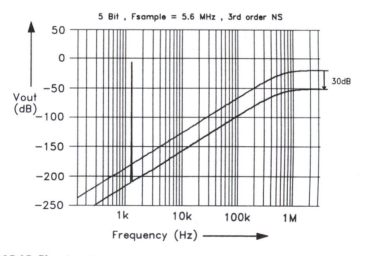

Figure 15.15 Simulated output spectra of oversampled one-bit and 5-bit D/A converters.

The quantisation noise can be reduced using a multi-bit oversampling D/A converter ($N > 1$). The noise reduction is determined by the available number of analogue levels 2^N and the noise transfer characteristic of the noise-shaping loop. This is discussed in [12]. For $N = 5$, the quantisation noise is reduced by 30dB. In Figure 15.15 the output spectra of both a 1-bit and a 5-bit converter are shown. The latter is shifted 30dB down compared to the 1-bit spectrum. This implies that theoretically a 30dB wider dynamic range can be obtained using a 5-bit D/A converter, and at the same time possible intermodulation problems are reduced considerably.

However, the design of a 5-bit D/A converter is much more complicated than the design of a 1-bit converter since a high differential and absolute linearity is required and must be maintained at high sampling frequencies. Therefore, the choice of converter architecture is very important in achieving a wide dynamic range.

15.3.3 Converter architecture considerations

There are several possibilities for the converter architecture. In Figure 15.16 the well-known principle of binary-weighted current conversion is shown for the 3-bit case. The output current of such a converter I_{out} is drawn as a function of time t in Figure 15.16a. In Figure 15.16b the number of matched current sources which contribute to the output current is shown. To generate the top of the sine wave, called full-scale (*F.S.*), all the available current sources are needed. To generate the lowest part of the sine wave, no current sources are needed. Obviously, at the zero crossings of the sine wave, $0.5F.S.$, half of the available current sources are used.

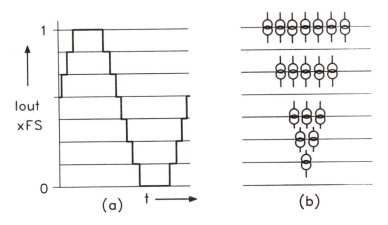

Figure 15.16 Binary-weighted current conversion.

An advantage of this binary weighted current conversion is the easy decoding of the data. The MSB is simply connected to the data switches of half the total number of current sources. The main disadvantage occurs at low signal levels. These signals are often located at half full scale. Many current sources are needed to generate the required DC level and do not contribute to the audio information. The output currents of these sources contain noise and glitches which seriously affect the dynamic range.

A principle which is better suited to obtaining wide dynamic range is called sign-magnitude current conversion and it is shown in Figure 15.17. In this case, the binary input data is first converted into a binary sign-magnitude code before it is applied to the data switches of the current sources. In contrast to the binary weighted conversion, two different types of current sources are needed. The top of the output sine wave in Figure

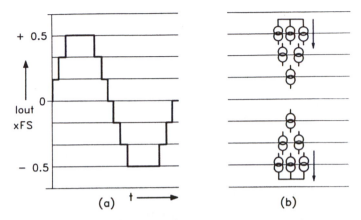

Figure 15.17 Sign-magnitude current conversion.

15.17a is now called +0.5*F.S.* and is generated by all sourcing current sources (Figure 15.17b). The lowest part of the sine wave, now called -0.5*F.S.*, is generated by all sinking current sources. Small signal levels located in the middle of the conversion range are generated by at most one current source. This approach permits a wider dynamic range than the conventional binary solution, at the cost of some encoding circuitry, the need for two types of current sources and accurate matching between the sourcing and the sinking D/A converter.

A further refinement in the encoding circuitry can be made by changing the binary sign-magnitude code into a thermometer sign-magnitude code. This guarantees the differential linearity of the converter since the currents are switched on and off one by one. Furthermore, thermometer encoding minimises the amplitude of the glitches that continuously occur since a noise-shaper generates the input code for the D/A converter.

15.3.4 D/A converter block diagram

Figure 15.18 shows the block diagram of the integrated 5-bit D/A oversampling converter. The 5-bit output data of the noise-shaper at a sampling frequency of 5.6MHz are applied to a binary-to-sign-magnitude encoder which generates two thermometer codes. One controls a 4-bit P-type current D/A, and the other controls a 4-bit N-type current D/A. Each 4-bit D/A consists basically of 15 current sources and current switches. The output currents of both converters are added and converted into voltage by means of an external operational amplifier A_{ex} and its feedback network, which also performs a first order post-filtering action.

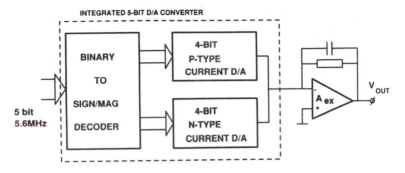

Figure 15.18 Multibit noise-shaping D/A converter block diagram.

15.3.5 Reference current adjustment

Matching between the sourcing (P-type) and the sinking (N-type) D/A converter can be achieved by comparing an N-type and a P-type current cell output, and adjusting the reference current of one of the D/A converters needed for calibration. Figure 15.19 shows the reference current adjustment of the two basic current D/A converters of Figure 15.18. In each D/A, a 16th current cell is added especially for this purpose. The current difference between both 16th cells is integrated by means of an internal operational amplifier A_i and a capacitor C_i. The output voltage of the integrator is used to adjust the reference current source $I_{ref,N}$ of the N-type D/A. In this way, sufficient matching between the P-type and the N-type converter is achieved.

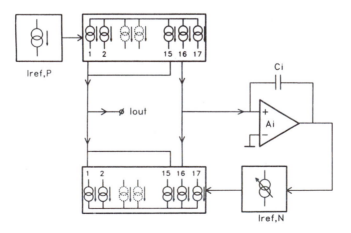

Figure 15.19 Reference current adjustment.

15.3.6 Digital continuous calibration

For operation without special calibration cycles, in both D/A converters an extra 17th current cell is added. These cells replace the output of cells under calibration. In contrast to the converter described in Section 15.2, this replacement is not done by means of current switch transistors, but in the digital domain. Incoming data and information from the shift register, which selects the cell to be calibrated, is combined digitally and, as a result, the needed number of current cells in normal operation is switched to the output terminal.

The major advantage of this digital selection is the reduction of voltage drop in the current path, since only one switch is present between a current cell and the output terminal.

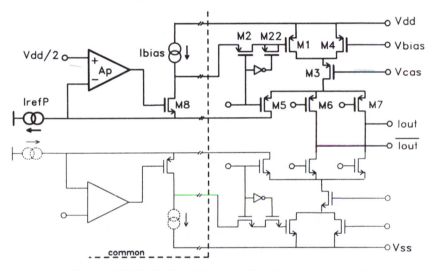

Figure 15.20 Bi-directional calibration current cell.

15.3.7 Bi-directional calibrating current cell

Figure 15.20 shows a so-called bi-directional calibrating current cell which consists of both a P-type and a N-type current cell at the transistor level. For clarity the P-type part is emphasised, and the N-type is shaded as it is fully complementary to the P-part. Only one of the 17 identical cells is drawn. The main current source is transistor M_4, which is biased by V_{bias} at about 90% of the value of $I_{ref,P}$, which equals 70µA. In parallel is transistor M_1 which will supply about 10% current difference between the reference source $I_{ref,P}$ and the main current source M_4. The drains of these transistors are connected to the source of a cascode transistor M_3, biased by V_{cas}. The combined current is routed through one of three transistors M_5-M_7, controlled by digital data. Left of the dashed lines in Figure 15.20 is the circuitry that is needed only once for the 17 cells. It consists of the operational amplifier, A_p, driving the gate of a level-shift transistor, M_8, which is biased by a small current source, I_{bias}. The potential at the source of M_8 is set by A_p at $V_{dd}/2$. The potentials of the two output lines are also set at $V_{dd}/2$ by the external operational amplifier, A_{ex}, of Figure 15.18. The operational amplifiers A_{ex} and A_p have unity gain bandwidths of 10MHz and are Miller compensated.

The most important transistor dimensions of the current cell are given in Table 15.3.

Transistor	M_1	M_2	M_3	M_4	M_5	M_6	M_7	M_8
W/L	100/35	2/1.6	30/5	400/16	10/1.6	10/1.6	10/1.6	20/1.6

Table 15.3 Current cell transistor dimensions.

15.3.8   Measurement results

All presented results are measured unweighted true RMS over a bandwidth of 20kHz at a sampling frequency of 5.6MHz.

Figure 15.21 shows a photograph of the typical output voltage of the oversampling multi-bit D/A converter. A sinewave of 20kHz is generated at about -20dB relative to full-scale. Only four current levels are needed; the signal inbetween two levels is the oversampled and noise-shaped input data. The signal-to-noise ratio including total harmonic distortion (S/(N+THD)) of this small signal is 90dB.

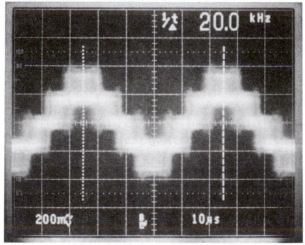

Figure 15.21 Typical output voltage oversampled multi-bit D/A converter.

Figure 15.22 shows the measured S/(N+THD) as a function of the output level of the converter. At small signal levels (up to -30dB) 115dB dynamic range is obtained. At full-scale a S/(N+THD) ratio of 90dB is measured. This is due to distortion caused by the integral linearity of the P-type and the N-type converter, and by the matching between both converters.

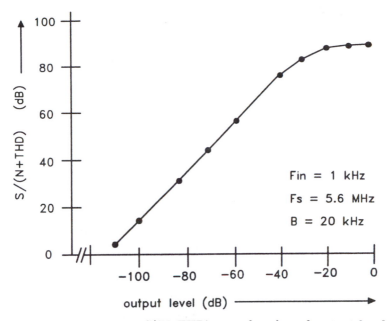

Figure 15.22 Measured S/(N+THD) as a function of output level.

Figure 15.23 Die photograph of the oversampling D/A converter.

A die photograph of the circuit with overlay is shown in Figure 15.23. The most important specifications of the converter are listed in Table 15.4.

Output current	+/- 1mA
S/(N+THD) at 0dB	90dB
at -60dB	55dB
Sampling frequency	5.6MHz
Supply voltage	5 V
Power dissipation	100mW
Temperature range	-10 to 70 °C
Process	1.6 m CMOS
Active chip area	7.5mm²

Table 15.4 D/A converter specifications.

15.4 Conclusions

Using the switched-current copying technique, an array of current sources which are equal to each other within 0.02% can be realised. To reduce the sensitivity for clock feedthrough in switches and for junction diode leakage currents, calibration is carried out using the difference between a main current source and a reference current. The array of calibrated sources is extended with a spare source to ensure that enough calibrated currents are available at any time. The technique can be used in any application that requires equal currents. As a first example, a 16-bit CMOS D/A converter for digital audio was designed in which the technique was incorporated. Measurement results show a linearity of 0.0025% at a power dissipation of 20mW, a minimum supply voltage of 3V and an active chip area of 2.8mm².

A second example shows the application of multi-bit noise-shaping together with sign-magnitude decoding in an oversampled multi-bit current-calibrated D/A converter. The combination of these techniques not only increases the dynamic range of the converter, but also reduces the intermodulation sensitivity. Measurement results show a dynamic range of 115dB and a S/(N+THD) of 90dB at full-scale using a standard CMOS process.

Acknowledgements

The authors thank P. J. A. Naus and P. A. C. Nuyten for their help and advice concerning the third-order noise-shaper.

References

[1] H. S. Lee, D. A. Hodges and P. R. Gray, "A self-calibrating 15b CMOS A/D converter," *IEEE J. Solid-State Circuits*, vol. SC-19, pp. 813-819, Dec. 1984.

[2] J. R. Naylor, "A complete high-speed voltage output 16-bit monolithic DAC," *IEEE J. Solid-State Circuits*, vol. SC-18, pp. 729-735, Dec. 1983.

[3] R. J. van de Plassche, "Dynamic Element Matching for high-accuracy monolithic D/A converters," *IEEE J. Solid-State Circuits*, vol. SC-11, pp. 795-800, Dec. 1976.

[4] H. J. Schouwenaars, E. C. Dijkmans, B. M. J. Kup and E. J. M van Tuijl, "A monolithic dual 16-bit D/A converter," *IEEE J. Solid-State Circuits*, vol. SC-21, pp. 424-429, June 1986.

[5] H. J. Schouwenaars, D. W. J. Groeneveld and H. A. H. Termeer, "A low-power stereo 16-bit CMOS D/A converter for digital audio," *IEEE J. Solid-State Circuits*, vol. SC-23, pp. 1290-1297, Dec. 1988.

[6] M. J. M. Pelgrom, A. C. J. Duinmaijer and A. P. G. Welbers, "Matching properties of MOS transistors," *IEEE J. Solid-State Circuits*, vol. SC-24, pp. 1433-1440, Oct. 1989.

[7] Y. Yamada, M. Kajitani and T. Ohgishi, "A 16-bit CMOS D/A converter for digital audio applications," *IEEE Transactions on Consumer Electronics*, vol. CE-33, pp. 267-273, Aug. 1987.

[8] P. J. A. Naus et al "A CMOS stereo 16-bit D/A converter for digital audio," *IEEE J. Solid-State Circuits*, vol. SC-22, pp. 390-395, June 1987.

[9] Y. Matsuya et al "A 17-bit oversampling D-to-A conversion technology using multistage noise shaping," *IEEE J. Solid-State Circuits*, vol. SC-24, pp. 969-975, Aug. 1989.

[10] B. Kup et al "A bitstream D/A converter with 18-bit resolution," *IEEE J. Solid-State Circuits*, vol. SC-26, pp. 1757-1763, Dec. 1991.

[11] E. F. Stikvoort, "Some remarks on the stability and performance of the noise-shaper or sigma-delta modulator," *IEEE Transactions on Communications*, vol. 36, pp. 1157-1162, Oct. 1988.

[12] P. J. A. Naus and E. C. Dijkmans, "Multibit oversampled sigma-delta A/D converters as front end for CD Players," *IEEE J. Solid-State Circuits*, vol. SC-26, pp. 905-909, July 1991.

Chapter 16

Dynamic Current Mirrors

George Wegmann

16.1 Introduction

A ubiquitous elementary building block in most analogue integrated circuits is the current mirror which is able to multiply and duplicate an imposed input current that contains the information (bias or signal). The reproduced output current is then available for any subsequent processing. Unfortunately, due to random process variations, transistor parameters are affected by a certain variation of the transconductance parameter and of the threshold voltage [1,2]. Hence, the output currents of transistors which have been designed identically are different.

These random variations, the so-called devices mismatch, are a major limitation for most accurate and precise current mode circuit applications. Another main limitation of CMOS circuits is the 1/f flicker noise of the MOS transistors. The standard technique to reduce this 1/f noise and the error due to mismatch is to increase the transistor gate area and to overwhelm the threshold mismatch with a high gate voltage overhead, which simultaneously increases the saturation voltage of the devices.

The performance of a mirror with low saturation voltage can be improved by using lateral bipolar transistors [3]. The resulting current error can be lower than 1%, but the major handicap is that only one type of mirror can be built (source or sink depending on the technology used).

Special circuit techniques allow us to reduce the inherent noise and offset in MOS amplifiers, such as the chopper technique [4,5] and the auto-zero technique [6,7]. The auto-zero technique is also used to ensure adequate biasing of CMOS inverters or analogue-to-digital converters [8]. Dynamic element matching [9] is based on the chopper technique and shifts the error components to higher frequencies. The drawback of this technique is the high residual output ripple, which for most applications must be filtered out by using external components. Furthermore, multiple mirrors are difficult to implement.

Dynamic analogue techniques [10] exploit the absence of gate current to temporarily store some analogue information on the gate capacitance of the MOS device. A reported application of this analogue storage capability is the dynamic comparator [11] which sequentially uses the same transistor as the two devices of a differential pair. With this auto-zero technique the very notion of mismatch disappears. The achievable precision is moved to new limits and depends on the ability to accurately store the signal, mainly limited by charge injection from the MOS transistors used as switches.

16.2 Current Copiers

16.2.1 Principle of current copier and dynamic current mirrors

The basic cell of a dynamic current mirror [12,13], which is also reported as a current copier [14] or sampled current circuit [15], is represented in Figure 16.1. This simple and elegant scheme, which slightly modified is also known as the current matching concept [16] or as the self-calibration technique for D/A converters [17], consists of an elementary sample-and-hold circuit connected to an MOS transistor and a toggle switch (S_y and S_z), which connects the device either to the input or the output.

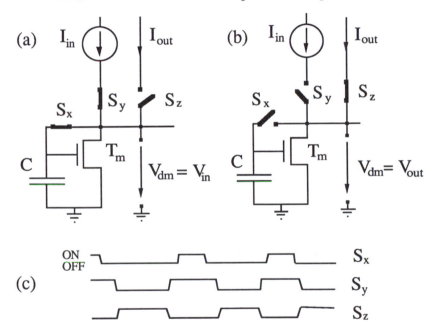

Figure 16.1 Basic principle of a current copier or dynamic current mirror.

The cycle of memorising and reproducing a current is as follows: While the switches S_x and S_y are closed (Figure 16.1(a)), the current source feeds the input current to the diode-connected transistor T_m. During this phase switch S_z is open. After opening sampling switch S_x, capacitor C maintains ("memorises") the gate voltage and thus the drain current of T_m remains equal to I_{in}. When S_y is open and S_z closed (Figure 16.1(b)), the memorised drain current is available at the output.

Figure 16.1(c) shows the corresponding clock phases. Transistor T_m therefore works either as a drain-gate-connected transistor (switch S_x closed) or as an independent current source (switch S_x open).

Note that the value of the output current is independent of the parameters of transistor T_m and of linearity and hysteresis of the capacitor C. On the other hand, the reproduced output current depends on the accuracy of gate voltage storage. Any error voltage ΔV on the storage capacitor C produces an output current error ΔI:

$$\frac{\Delta I}{I_{in}} = \frac{I_{out} - I_{in}}{I_{in}} = \frac{g_{mm} \, \Delta V}{I_{in}} \tag{16.1}$$

where g_{mm} stands for the transconductance parameter of T_m.

(see Appendix 16A for more details on the MOS model, on the symbols and definitions used).

16.2.2 *Principal accuracy limitations*

The performance of the basic cell of a dynamic current mirror is limited by the following major effects:

(a) charge injection of the MOS transistors used as switches :

When switching off an MOS transistor, the mobile charge in the inversion charge layer is shared between drain, source and substrate [18-20]. A part of the charge released by sampling switch S_x is added to the charge already stored on the hold capacitor C and the resulting error voltage ΔV alters the value of the memorised drain current as shown by (16.1).

(b) leakage currents of the reverse biased junctions:

The leakage currents at the gate node of T_m discharge the storage capacitor C. They determine the minimum switching frequency of the current copier for a given accuracy. In the case of very low currents, the leakage currents to ground limit the absolute achievable accuracy.

(c) sampled noise:

White noise is undersampled and the power spectral density of the white noise component in the baseband is increased because of additional foldover terms. 1/f noise is cancelled at low frequencies due to the inherent auto-zero technique, which introduces a double zero at the origin. It will be shown that the undersampled 1/f noise produces an increase of the white noise in the baseband.

(d) drain voltage variations:

Usually V_{in} is different from V_{out}. When switching S_y and S_z, hence when switching from the input to the output, the voltage difference V_{in}-V_{out} is applied to the drain of T_m.

(d.1) transients or spikes:

When S_z is closed the output must produce some additional current to charge the parasitic capacitances at the drain of T_m to V_{out}. The resulting transients or "glitches" at the output have an amplitude which is proportional to the voltage step V_{in}-V_{out}.

(d.2) output conductance :

An important requirement is that the reproduced current should not depend on the output voltage. As in a classical static current mirror, the finite output conductance of T_m has to be minimised to obtain an accurate current mirror.

(d.3) capacitive divider between drain-to-gate capacitance C_{gd} and C:

Gate voltage variations are produced through the capacitive divider formed with the storage capacitance C and the parasitic drain-to-gate capacitance C_{gd} of T_m.

According to the reflections made in Section 16.2(d), V_{dm} should remain as constant as possible in spite of the imposed voltage difference V_{in} - V_{out}.

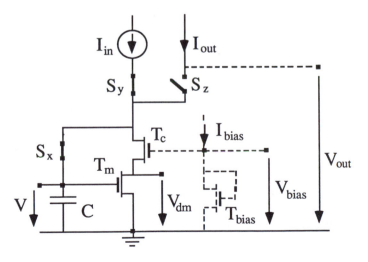

Figure 16.2 Cascoded current copier biased by an external current I_{bias}.

16.2.3 Cascoded structure

An elegant and simple way of achieving a constant voltage V_{dm} is to implement a cascoded structure by means of a common-gate transistor [13,21] in series with the main transistor T_m which leads to the solution represented in Figure 16.2. The cascode combination of T_m and T_c is equivalent to a single transistor T_m with values of Early voltage V_E and of drain-to-gate capacitance C_{gd} increased and decreased, respectively, by a factor equal to the source-to-drain voltage gain of T_c. Because this gain can be as high as several hundred, the errors due to the output conductance g_{ds} and to the drain-to-gate capacitance C_{gd} are drastically reduced.

To increase the output dynamic range of this cell, the bias voltage V_{bias} should be as low as possible. The minimum value is given by the limit needed to keep T_m in saturation, which yields

$$V_{bias} = V_{T0} + nV_{dm} + \sqrt{\frac{2nI_{in}}{\beta_c}} \geq V_{T0} + \sqrt{2nI_{in}}\left(\frac{1}{\sqrt{\beta_c}} + \frac{1}{\sqrt{\beta_m}}\right) \qquad (16.2)$$

with

$$V_{dm} \geq V_{dsat} = \sqrt{\frac{2I_{in}}{n\beta_m}} \qquad (16.3)$$

Equation (16.2) can be expressed as a function of the specific current I_{si} :

$$V_{bias} \geq V_{T0} + 2n\,U_T\left(\sqrt{\frac{I_{in}}{I_{sc}}} + \sqrt{\frac{I_{in}}{I_{sm}}}\right) \qquad (16.4)$$

with

$$I_{si} = 2n\,\beta_i\,U_T^2 \qquad (16.5)$$

The currents I_{si} are determined by the relative dimensions of the corresponding transistor T_i.

The bias voltage V_{bias} can be produced by imposing an external current I_{bias} through a transistor T_{bias} as shown by the dotted line in Figure 16.2. The MOS voltage-current relationship in strong inversion leads to

$$V_{bias} = V_{T0} + 2n\,U_T\sqrt{\frac{I_{bias}}{I_{sb}}} \qquad (16.6)$$

Equations (16.4) and (16.6) yield

$$\frac{I_{bias}}{I_{sb}} \geq \left[\sqrt{\frac{I_{in}}{I_{sc}}} + \sqrt{\frac{I_{in}}{I_{sm}}}\right]^2 \qquad (16.7)$$

With the smallest possible value of V_{bias} given by (16.4), the condition on V_{out} which keeps T_c in saturation is found to be

$$V_{out} \geq 2\, U_T \left(\sqrt{\frac{I_{in}}{I_{sm}}} + \sqrt{\frac{I_{in}}{I_{sc}}} \right) \tag{16.8}$$

During the storage phase (Figure 16.1(a)) T_c and T_m must remain saturated which can be expressed by

$$V_{T0} \geq 2\, U_T \left((1-n) \sqrt{\frac{I_{in}}{I_{sm}}} + \sqrt{\frac{I_{in}}{I_{sc}}} \right) \tag{16.9}$$

The first term of the sum on the right is usually much larger than the second one, because T_c operates as close as possible to weak inversion and T_m to strong inversion. Furthermore the first term is negative, which means that the inequality of (16.9) is satisfied for any positive values of threshold voltage V_{T0}. In the case that inequality (16.9) cannot be satisfied, a source follower T_f introducing a voltage shift can be placed between the storage capacitor C and the gate of T_m [14,22].

16.2.4 Current copier with reduced transconductance $G_{MM\Delta}$

According to (16.1) the achievable current accuracy depends linearly on the error voltage ΔV and on the transconductance g_{mm}. To increase the current precision, the transconductance g_{mm} can be reduced by using a mirror structure with a modified basic cell, which memorises only a current difference $I_{in}-I_0$. This principle, which was reported as the self-calibration technique [17] (see also Chapter 15) or as an improvement for current copiers [23], is shown in Figure 16.3.

The comparison of the transfer parameter β and β_Δ of two current copiers, which are built either with T_m or with $T_{m\Delta}$ and which operate with the same gate-to-source voltage V_{gs}, leads to

$$\beta_\Delta = \frac{I_{in}-I_0}{I_{in}}\, \beta \tag{16.10}$$

which is valid for the transistors operating in strong inversion. Introducing (16.10) in the expression of transconductance $g_{mm\Delta}$ of $T_{m\Delta}$ yields

$$g_{mm\Delta} = \sqrt{\frac{2\beta_\Delta\,(I_{in}-I_0)}{n}} = \frac{I_{in}-I_0}{I_{in}}\, g_{mm} \tag{16.11}$$

But to make this improved cell work correctly, the current difference $I_{in}-I_0$ must be positive to ensure that the diode-connected n-type transistor $T_{m\Delta}$ still is properly biased. Thus an additional current source I_0 whose value is close to, but always smaller than I_{in} must be available. Furthermore V must be higher than the saturation voltage of I_0. Due to the lowered transconductance parameter $g_{mm\Delta}$, a larger voltage swing is needed to compensate for the current mismatch. Therefore, if a dynamic current mirror is built with two improved basic cells, the variations of V_{in} are greater than those of a normal basic cell.

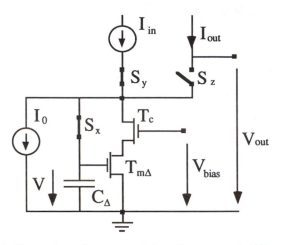

Figure 16.3 Current copier memorising the current difference I_{in}-I_0.

Because the voltage steps V_{in}-V_{out} are still applied to S_y and S_z, the simple cascode cell configurations of Figures 16.2 and 16.3 are only applicable when the output current is used after the transients have faded out. The average current error due to the voltage step V_{in}-V_{out} still reduces the DC accuracy of such a cell.

16.2.5 *Dynamic current mirror structures*

In analogue design a symmetrical structure is generally preferred to an asymmetrical one because parasitic capacitances and leakage currents are balanced, mismatch components reduced and the layout is simplified. The structure of an externally biased dynamic current mirror which is capable of memorising and reproducing an imposed current I_{in}, is represented in Figure 16.4(a). To ensure a continuous output current, two current copiers are needed, which function with complementary clocks (Figure 16.4 (b)). The switching cycle remains identical to that mentioned in Section 16.2.1. Each time S_{xj} is closed (j=0 or 1), V_{in} adapts to the value of gate voltage corresponding to I_{in}. Therefore V_{in} varies stepwise as a function of time with an amplitude proportional to the mismatch of T_{m0} and T_{m1}. The toggle switches S_{yj} and S_{zj} are placed at the sources of the common-gate transistors T_{cj} to reduce the voltage difference which appears at their terminals.

    The drain voltage of T_{mj} must return to the value $V_{dmj} = V_{in}$ during the memorisation phase (switches S_{xj} and S_{yj} closed, S_{zj} open). When the current is restored (switches S_{xj} and S_{yj} open, S_{zj} closed), it must jump to a value V_{out} imposed by the external load of the cell. The voltage step which occurs at the drain of T_{mj} contributes to an important DC error through the output conductances and the parasitic drain-to-gate capacitance C_{gd}.

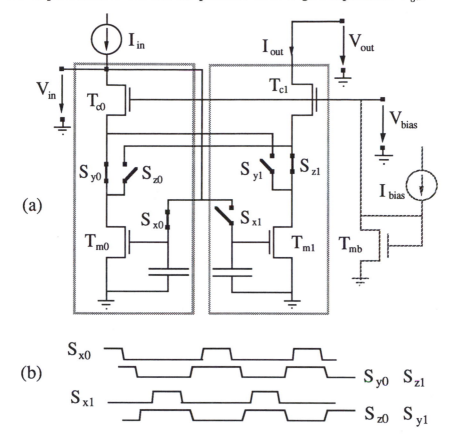

(a)

(b)

Figure 16.4 Dynamic current mirror which is externally biased by I_{bias} (shown by the doted lines).

The condition on V_{bias} is

$$V_{bias} = V_{T0} + nV_{dm} + \sqrt{\frac{2nI_{in}}{\beta_c}} \geq V_{T0} + 2nU_T \left(\sqrt{\frac{I_{in}}{I_{sc}}} + \sqrt{\frac{I_{in}}{I_{sm}}} \right) \quad (16.12)$$

The bias circuit required for the dynamic current mirror is shown dotted. From the externally biased mirror a self-biased current mirror can be deduced by simply connecting V_{in} to V_{bias}. This mirror is much more compact, but the drawback of such a structure is that the two transistors T_{c0} and T_{c1} *must* be located in separate wells, which are connected to their sources. Otherwise the condition, which expresses that V_{dmj} $(j=0,1)$ must be larger than the saturation voltage V_{dsat} of T_{mj}, leads to a relationship which is impossible to satisfy [22].

If separate wells are not available, the connection of S_{xj} of the dynamic current mirror of Figure 16.4 must be modified to obtain the structure of a so-called "stacked" mirror structure [13] represented in Figure 16.5.

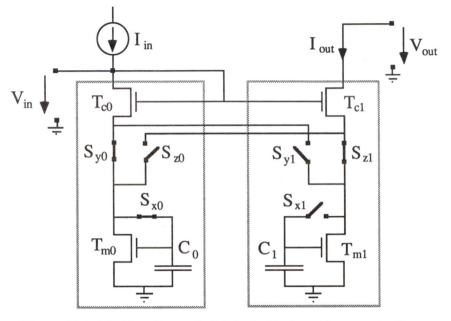

Figure 16.5 Self-biased, so-called "stacked" dynamic current mirror.

The drawback of such a structure is its relatively high saturation voltage which increases V_{out} to

$$V_{out} \geq V_{T0} + 2U_T \left(n \sqrt{\frac{I_{in}}{I_{sm}}} + \sqrt{\frac{I_{in}}{I_{sc}}} \right) \qquad (16.13)$$

16.2.6 *Multiple dynamic current mirrors (1:1: . . :1)*

A multiple dynamic current mirror with n identical current outputs is obtained by simply repeating n times the cascoded current copier, which involves a total of $(n+1)$ copiers. All current copiers, except copier 0, are identical and are connected to their corresponding output. In other words copier 1 is always delivering the current to output 1, copier 2 to output 2 etc. Only copier 0 is sweeping all the outputs and is delivering the output current in place of the cell that is being updated. The corresponding self-biased "stacked" mirror and the switching sequence are shown in Figure 16.6 and 16.7, respectively. The clock phases driving the switches S_{yj} and S_{zj} are complementary and the switches S_{zoj} are driven by the same clocks as S_{yj}. Note that it is also possible to build multiple dynamic mirrors with other mirror structures or with variations of the current copiers as proposed in the preceding sections.

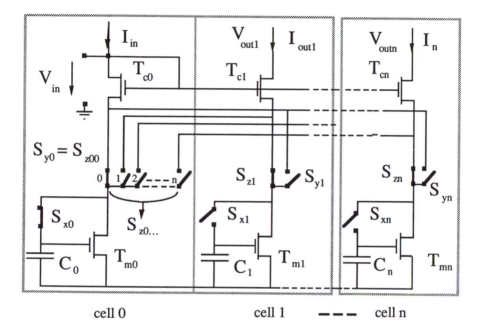

Figure 16.6 Multiple "stacked" dynamic current mirror.

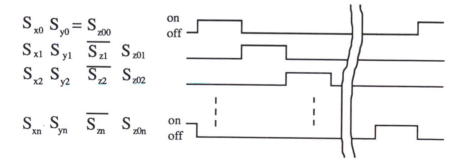

Figure 16.7 Clock phases corresponding to the mirror of Figure 16.6.

16.2.7 *Dividing current mirror: principle of the division by 2* ($I_{out} = \dfrac{I_{in}}{2}$)

Due to the unavoidable transistor mismatch, it is not possible to divide a current I_{in} accurately by simply connecting n current copiers in parallel, since the input current would not be evenly shared among them.

With the help of the calibration scheme shown in Figure 16.8 [24], which performs a division by two, exactly half the input current can be forced to flow through the main transistor T_m, which delivers the output current. The circuit operates in three clock phases and the output current requires a few cycles to reach the correct current values. For the sake of clarity, the three different phases are represented explicitly, and only the transistors which are active during the corresponding phase are represented. Transistor T_p and calibration transistor T_c form a so-called "locked pair" of transistors, since their output current sum equals that of T_m. Each transistor is connected either in a diode configuration (when a new current value is stored) or as an independent current source (when it reproduces the previously memorised current).

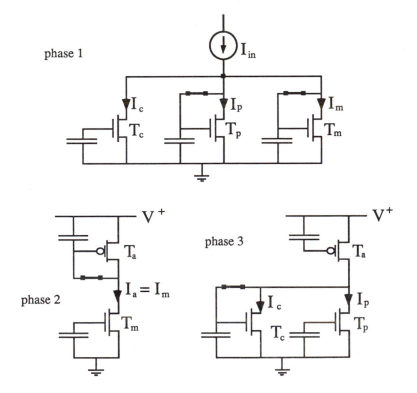

Figure 16.8 Accurate divide-by-two dynamic current mirror: scheme showing the different configurations for each phase.

The principle of accurate current division is the following:

During **phase 1**, transistors T_m and T_p share the input current I_{in} from which the previously established calibration current I_c is subtracted.

Thus, for cycle k

$$I_{mk} = \frac{I_{ink} - I_{c(k-1)}}{2} (1+\varepsilon) \tag{16.14}$$

$$I_{pk} = \frac{I_{ink} - I_{c(k-1)}}{2} (1-\varepsilon) \tag{16.15}$$

where ε represents the mismatch between T_m and T_p. Note that, for the first cycle (k=1), current I_{c1} of T_c is arbitrarily set to zero. Both currents, I_{mk} and I_{pk}, are memorised as voltages on the capacitors C_m and C_p, respectively.

During **phase 2**, the stored current I_{mk} is forced into the p-type transistor T_a, hence

$$I_{mk} = I_{ak} \tag{16.16}$$

During **phase 3**, the difference between current I_{ak} and I_{pk}, which is memorised by T_p, is stored in T_c as the new value of the calibration current I_{ck}:

$$I_{ck} = I_{mk} - I_{pk} = \varepsilon (I_{ink} - I_{c(k-1)}) \tag{16.17}$$

Because in Figure 16.8 T_c is a current sink, I_{ck} can only flow towards ground, meaning that I_{mk} must be larger than I_{pk}. Since I_{ck} is positive, T_m and T_p must be designed to ensure $\varepsilon > 0$. This additional constraint is trivial to implement in a design and reduces the complexity of the circuit dramatically by avoiding the use of T_c either as a current sink or a current source, and therefore simplifies the corresponding control-logic.

The evolution of the calibration current I_{ck} can be obtained by introducing z-transforms into (16.17), which yields

$$I_c(z) = \frac{z\varepsilon}{z+\varepsilon} I_{in}(z) \tag{16.18}$$

Introducing $I_{in}(t) = I_{in}$ while $t \geq 0$ leads to

$$I_c(z) = \frac{z\varepsilon}{z+\varepsilon} \frac{z}{z-1} I_{in} \tag{16.19}$$

which yields the asymptotic or equilibrium value $I_{c\infty}$ of calibration current I_c:

$$I_{c\infty} = \lim_{z\to1} (z-1) I_c(z) = \frac{\varepsilon}{1+\varepsilon} I_{in} \qquad (16.20)$$

The equilibrium value of the output current $I_{m\infty}$ can be found by taking the z-transform of I_{mk}:

$$I_{m\infty} = \lim_{z\to1} (z-1) I_m(z) = \lim_{z\to1} (z-1)\frac{z(1+\varepsilon)}{2(z+\varepsilon)} I_{in}(z) = \frac{I_{in}}{2} \qquad (16.21)$$

and is independent of the mismatch ε. This equilibrium is reached exponentially with a time constant τ [21]:

$$\tau = -\frac{1}{ln(\varepsilon)} \quad [cycles] \qquad (16.22)$$

For example, τ is equal to 0.33 and 0.43 [cycles], while $\varepsilon = 5\%$ and $\varepsilon = 10\%$, respectively. Therefore, after a few cycles the current I_{mk} settles to the desired value of $I_{in}/2$. Obviously the sum of the asymptotic values of the locked-pair corresponds also to half the input current. After the steady state is reached, the three-phase-cycle proceeds to update the stored voltages, which prevents the influence of leakage currents.

Table 16.1 gives numerical values of I_m and I_c during phase 1 for two values of mismatch, $\varepsilon = 5\%$ and $\varepsilon = 10\%$.

	$\varepsilon = 0.05$		$\varepsilon = 0.10$	
$k =$	$I_{m(k)}$	$I_{c(k)}$	$I_{m(k)}$	$I_{c(k)}$
0	.525	0	.55	0
1	.49875	.05	.495	.1
2	.50006	.0475	.5005	.09
3	.49999	.04763	.49995	.091
4	.5	.04762	.50001	.0909
5	.5	.04762	.5	.09091
6	.5	.04762	.5	.09091

Table 16.1 Evolution of I_{mk} and I_{ck} for two mismatch
values ε between T_m and T_p.

16.3   Accuracy Limitations

16.3.1   Influence of drain voltage variations

16.3.1.1 Output conductance G_{DS}

Channel length modulation contributes to an important current error, which can be represented by the drain-to-source conductance g_{ds} (Figure 16.9). The cascoded structures decrease g_{ds} of a current copier by the cascode gain [25] and, thanks to the cascoded structure, the fraction of the output voltage ΔV_{out} which is transmitted to the drain of T_m and which produces drain variations ΔV_{dm}, is reduced to

$$\Delta V_{dm} = \Delta V_{out} \frac{g_{dsc}}{g_{mc}} \tag{16.23}$$

and the resulting current error can be expressed as

$$\frac{\Delta I}{I_D} = \frac{g_{dsm}}{I_D} \Delta V_{dm} = \frac{\Delta V_{dm}}{V_E} = \frac{g_{dsc}}{g_{mc}} \frac{\Delta V_{out}}{V_E} \tag{16.24}$$

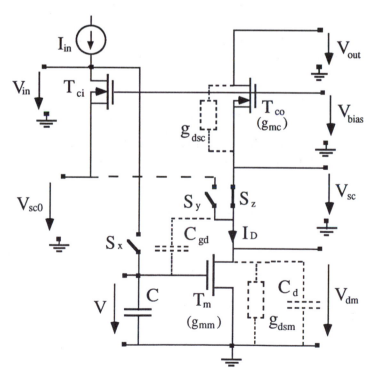

Figure 16.9 Parasitic capacitances and conductances of the current copier.

with the Early voltage V_E given by

$$V_E = \lambda L_{ef} \tag{16.25}$$

V_E is of the order of tens of volts and therefore steps of some hundred millivolts of V_{out} would produce an output current error of about 1% without a cascoded structure.

16.3.1.2 Capacitive divider C_{GD} - C

T_c lowers the effect of output voltage variations on the current accuracy, but there still remains a small voltage difference V_{sc}-V_{sco}, which is minimised if T_{ci} and T_{co} are matched. When switching S_y and S_z a voltage step proportional to V_{sc}-V_{sco} is applied to the drain of T_m. A fraction ΔV is transferred to the capacitance C through the capacitive divider formed by the gate-to-drain capacitance C_{gd} and C, resulting in a current error ΔI :

$$\Delta I = g_{mm} \Delta V = \frac{g_{mm} C_{gd} (V_{sc}\text{-}V_{sco})}{(C + C_{gd})} \tag{16.26}$$

The optimum is calculated for a given value of the storage capacitance C and for a given value of transconductance g_{mm}, hence of a fixed ratio $\frac{W_{ef}}{L_{ef}}$. An increase of the effective channel length L_{ef} reduces the current error ΔI due to g_{dsm} according to (16.24), but increases ΔI due to C_{gd} (16.26), because the effective channel width W_{ef} must be modified to maintain the desired ratio. Hence

$$\Delta I = g_{mm}\Delta V + g_{dsm}\Delta V_{dm} = \Delta V_{dm} I_D \left(\frac{2}{(V\text{-}V_{T0})} \frac{C_{gd}}{(C+C_{gd})} + \frac{1}{\lambda L_{ef}} \right) \tag{16.27}$$

The total capacitance C at the gate node can be split into the gate capacitance C_G and into an additional capacitance C_{ad}.

$$C = C_G + C_{ad} = W_{ef} L_{ef} C'_{ox} (1 + \xi) \tag{16.28}$$

If T_m works in strong inversion, L_{ef} can be extracted and an optimum value ξ_{opt} found, which allows us to express the optimal capacitance ratio as

$$\left(\frac{C_G}{C} \right)_{opt} = \frac{V\text{-}V_{T0}}{2\lambda L_{ov}} = \frac{I_{in}}{g_{mm} \lambda L_{ov}} \tag{16.29}$$

Typical numerical values for the parameters above are (for a standard 3μ CMOS process) :

$$L_{ov} = 0.5\mu m \qquad \lambda = 5V/\mu m \qquad V\text{-}V_{T0} = 0.5V$$

which leads to: $\xi_{opt} = 9$

16.3.1.3 Direct charge flow path

A third effect is due to the additional charge that is supplied from the output to charge the parasitic capacitances C_{gd} and C_d, when switching the current copier from the input to the output. Voltage V_{dm} changes periodically from V_{sc0} to V_{sc} and, when switching back to the input, from V_{sc} to V_{sc0} again. This charge creates a current transient (current "glitch" or spike), which is fairly unimportant if the output current is only used after equilibrium is reached. However, if the current must be available continuously, these glitches produce a DC-current error whose average is given by

$$\Delta I = f_{sw} \left(V_{sc} - V_{sc0} \right) \left(C_d + \frac{C C_{gd}}{C + C_{gd}} \right) \tag{16.30}$$

where f_{sw} is the switching frequency (see also Section 16.5.2.2).

16.3.2 Leakage currents

The input and the output currents are affected by the parasitic leakage currents of several reverse-biased pn-junctions flowing to ground. These leakage currents can be represented by current sources connected between the corresponding node and ground according to Figure 16.10. Note that the leakage currents of the capacitances are negligible in comparison to that of the pn-junctions. Let us first assume that the off conductance g_{xj} of the sampling switch S_{xj} is negligible. During the storage phase the stored drain current I_{Dj} (j=0,1) is the difference between I_{in} and the leakage current sources, hence

$$I_{Dj} = I_{in} - I_{dc0} - I_{s0} - I_{dmj} - I_{xj} \tag{16.31}$$

where I_{xj} corresponds to the leakage currents at the gate node of T_{mj}, I_{dmj} corresponds to the sum of the leakage currents at the drain node of T_{mj}, and I_{dc0} and I_{s0} stand for the leakage currents of T_{c0} at the source and the drain node, respectively.

During the restoring or copying phase t_{hj} the output current is

$$I_{outj}(t) = I_{Dj}(t) + I_{dmj} + I_{s1} + I_{dc1} \tag{16.32}$$

where the stored drain current $I_{Dj}(t)$ has become time dependent because of I_{xj} which discharges C_j. As a consequence, the gate voltage V_j varies linearly as a function of time t as shown in Figure 16.11. The gate voltage variations are assumed to be sufficiently small that the small signal parameter g_{mmj} can be used. Introducing

$$I_{leak} = -I_{dc0} + I_{dc1} - I_{s0} + I_{s1} \tag{16.33}$$

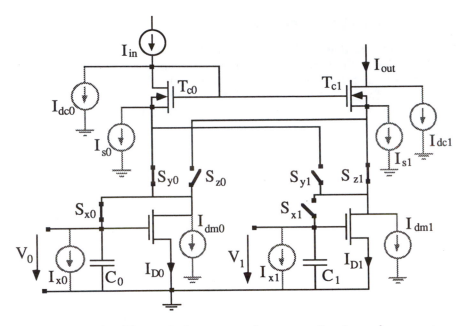

Figure 16.10 Different leakage current sources of a dynamic current mirror.

the output current $I_{outj}(t)$ can be expressed as

$$I_{outj}(t) = I_{in} + I_{leak} - I_{xj} - g_{mmj}\frac{I_{xj}}{C_j} t \qquad (16.34)$$

If both current branches are matched, term I_{leak} is cancelled out because $I_{dc0} \approx I_{dc1}$ and $I_{s0} \approx I_{s1}$. Integrating (16.34) over a time period $t_{h0}+t_{h1}$ for matched branches leads to an average output current ΔI_{out}:

$$\Delta I = \overline{\Delta I_{out}} - I_{in} = -I_x \left(1 + \frac{T_h g_{mm}}{2\,C} \right) \qquad (16.35)$$

which shows clearly the limitation of the achievable accuracy for small currents due to the leakage current I_x of the sampling switch. The last term also sets a limit to the maximum duration T_h of the copying phase.

The voltage drop on the sampling switch conductance g_x due to the leakage current I_x may not be negligible during the storage phase. Assuming $g_x > g_0$, the drain voltage $V_{dm}(t)$ and the gate voltage $V(t)$ at t=0 and t-> ∞ can be expressed as [22]

$$V(0) = V - \frac{T_h I_x}{C} \qquad V(t\text{->}\infty) = V - \frac{I_x}{g_{mm}} \qquad (16.36)$$

$$V_{dm}(0) = V + I_x \frac{g_{mm}\text{-}g_x}{g_x} \frac{T_h}{C} \qquad V_{dm}(t\text{-}>\infty) = V + I_x \frac{g_{mm}\text{-}g_x}{g_x g_{mm}} \qquad (16.37)$$

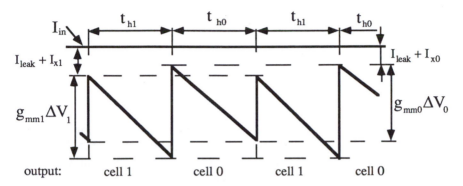

Figure 16.11 Waveform of I_{out} for I_{in} = constant due to leakage currents.

Note that voltage V corresponds to the gate voltage at equilibrium while $g_x > g_{mm}$. If the voltage drop on g_x due to I_x is not negligible, the resulting increase of $V_{dm}(t)$ reduces the output dynamic range. In the case of a self-biased mirror, T_{c1} may leave saturation, whereas for an externally biased mirror T_{c0} may be cut-off.

16.3.3 Charge injection by analogue switches

The scheme of a dynamic current mirror uses several switches, which are all realised with MOS transistors. To close these switches, minority carriers have to be attracted into the channel by the gate voltage. When the switching transistor is turned off, these channel charges are released and alter the voltages on the capacitances at both ends of the transistor [19,26]. The disturbance of the sampled voltage due to such carriers is a major limitation to the accuracy of sampled circuits. The performance of a dynamic current mirror depends on the accurate storage capability of the gate voltage.

16.3.3.1 Interfering parameters

The amount of charge injected into the hold capacitor C depends on the following parameters:

- the normalised switching parameter B, which is given by

$$B = (V_{GON} \text{-} V_{TE}) \sqrt{\frac{\beta}{a \, C}} \qquad (16.38)$$

(where V_{GON} is the gate on voltage, V_{TE} is the effective threshold voltage, β is the transconductance parameter and a is the slope of the gate voltage, which controls the switch)

- the ratio of the hold to source capacitors $\dfrac{C}{C_s}$

- the overlap capacitances C_{gh} and C_{gs} of the switch, which couple the gate to the hold and to the source capacitance, respectively.

16.3.3.2 Strategies

Several strategies are possible which either reduce the amount of charge injected into C or which balance the charges injected into both end capacitors. If the channel charge is shared equally between drain and source, it is possible to compensate the excessive charge ΔQ on the holding capacitor C with "symmetrical dummy switches". This requires an adequate design of the switching parameters and a carefully optimised layout. The residual charge Δq is ultimately limited by the degree of matching between the main switch and the "symmetrical dummy switches".

If B is chosen much larger than 1 and $C_s \gg C$, all charges released flow back into the source capacitor C_s during the decay of V_G and the number of carriers injected into C is reduced. But a high value of B means a small gate slope a, hence a long switch-off time, which is often not acceptable because it limits the operating speed of the circuit.

If $C_s = C$ it is obvious that the channel charges are evenly shared between source and drain, and only a mismatch of the overlap capacitances contributes to a charge difference. It is often not possible to exactly balance the value of a functional capacitor C with that of an estimated, parasitic capacitance C_s.

Another possibility is to choose small values for B, which ensure the equipartition of the total channel charge. The switching parameter B can be reduced while increasing the effective gate voltage V_{GON}, the slope a, and the capacitor C. If "dummy switches" are used, the minimum switch is given by twice the minimum transistor size, which determines the transfer parameter β of the sampling switch. The chosen process determines the minimum possible transistor size, and therefore the minimum β. The slope a depends on the driving capability of the inverters used and on the parasitic capacitances which have to be charged and discharged. The estimation of slope a is therefore very inaccurate, but the larger the slope a, the better the charge equipartition.

The error voltage ΔV depends on the value of C because

$$\Delta V = \frac{\Delta Q}{C} \tag{16.39}$$

The maximum value of C is limited by the occupied area on the chip and by the operating speed (see Section 16.5.2). Reducing $(V_{GON} - V_{TE})$ also reduces the channel conductance g_x, hence the total channel charge Q_{tot}.

16.3.3.3 Reduction of the turn-on voltage of the sampling switch

The total channel charge Q_{tot} of sampling switch S_x is related to its g_x by

$$Q_{tot} = \frac{g_x L_{ef}^2}{\mu}$$

(16.40)

where the conductance is found to be in:

weak inversion:
$$g_x = \frac{I_{Dsat}}{U_T}$$
(16.41)

strong inversion:
$$g_x = \sqrt{2n\beta I_{Dsat}} = \beta(V_{GON} - V_{TE})$$
(16.42)

Note that although $g_x \to 0$, the drain voltage V_{dm} of $T_m \to \infty$ and therefore it is not possible to cancel out Q_{tot} by reducing g_x to zero.

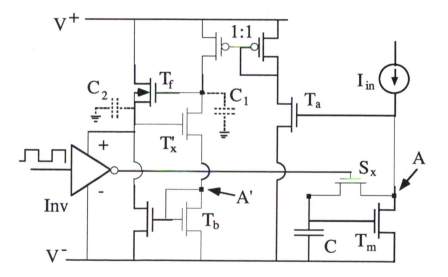

Figure 16.12 Circuit which adjusts the gate turn-on voltage V_{GON} of S_x to track the input voltage V_{in}.

One possible scheme which adjusts V_{GON} of S_x to track V_{in}, hence reduces g_x, is shown in Figure 16.12. The purpose of this circuit is to reproduce the source potential of S_x at the source of T_x', hence to reproduce the potential of node A at node A'. This is implemented with the aid of the two matched transistors T_a and T_b. The sampling switch S_x is turned on by driving its gate by the gate voltage of T'_x , with the aid of voltage follower T_f, and through the inverter *Inv*. The inverter is built either with p-channel and n-channel devices or only with n-channel transistors depending on the value of potential A.

It has to be pointed out that the tuning circuit of Figure 16.12 contains two loops which can introduce instabilities. After calculating the open loop transfer function the phase margin Φ can be approximated as

$$\Phi = \frac{\pi}{2} - arctg \ \frac{g_{mx'} C_2}{g_{mf} C_1} \tag{16.43}$$

where the capacitances C_1 and C_2 are represented in Figure 16.12, and the transconductances are those of T'_x and T_f, respectively. Equation (16.43) can be obtained after cancelling a pole with a zero which reduces the 3rd order equation. As the zero is smaller than the pole, (16.43) is too pessimistic. If Φ approaches zero, hence $\dfrac{g_{mx'} C_2}{g_{mf} C_1}$ is large, instability may occur.

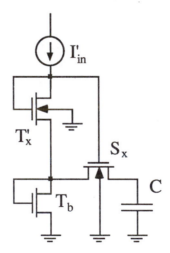

Figure 16.13 Relationship between gate and source voltages of T_x' and S_x.

The relationship between the source and the gate voltages of S_x and T'_x can be calculated with the help of Figure 16.13, where it is assumed that the potential at node A is identical to the potential at node A'. The conductance g_x which is now adapted depends on the operating mode of transistor T'_x and can be expressed as:

T_x' in weak inversion: $\quad g_x = \dfrac{\beta_x}{\beta'_x} g'_x = \dfrac{\beta_x}{\beta'_x} \dfrac{I_{in}}{U_T}$ $\qquad$ (16.44)

T_x' in strong inversion: $g_x = \sqrt{\dfrac{\beta_x}{\beta_x'}} \quad g_x' = \dfrac{\beta_x}{\sqrt{\beta_x'}} \sqrt{2nI_{in}}$ (16.45)

If the turn-off voltage is chosen to fall only a little beyond V_{TE}, the gate voltage swing of S_x is reduced and optimised around V_{in}, hence the amount of charge injected through the coupling of the overlap capacitances is minimised.

16.3.3.4 Two-phase feedback

Another possibility for reducing the effect of charge injection is based on feedback rather than matching [14,21]. The scheme is equivalent to that of offset compensation by a low sensitivity auxiliary input used in the design of operational amplifiers [26]. It needs two steps to store the gate voltage. A capacitive attenuator $\delta C/C$ (with $\delta \ll 1$) reduces the influence of charge injection of the second sample-and-hold, formed by capacitor C' and switch S_x', into the critical gate node G.

A second sample-and-hold is added, and is connected to the critical node through a capacitive attenuator. Two steps are needed to disconnect the gate from the drain. The main switch S_x is opened first, dumping a charge Δq onto C. If the system is allowed to settle to equilibrium, this charge is eliminated by the second loop through δC. The auxiliary switch S_x' is then opened as well, which dumps a charge $\Delta q'$ onto C', but the resulting voltage variation at the critical node G is attenuated by the capacitive divider.

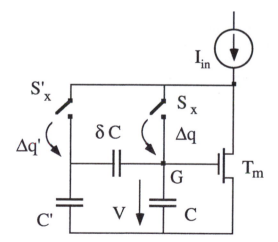

Figure 16.14 Reduction of charge injection with the help of a low sensitivity auxiliary input.

This scheme is equivalent to that of offset compensation by a low sensitivity auxiliary input used in the design of operational amplifiers [26]. The value of d should be as low as possible to attenuate the effect of $\Delta q'$, but large enough to keep a sufficient voltage gain in the auxiliary loop in order to compensate Δq. Thus, there is an optimum value of d that can be calculated with the approach described in [26]. Assuming the worst case with uncorrelated values of $\Delta q/C$ and $\Delta q'/C'$, this calculation yields

$$\delta_{opt} = \sqrt{\frac{|\frac{\Delta q}{C}|_{max}}{A_v |\frac{\Delta q'}{C}|_{max}}} \tag{16.46}$$

where A_v is the gate-to-drain voltage gain limited by the output conductance of the transistor.

This optimum corresponds to a residual voltage step

$$\Delta V = 2 \sqrt{\frac{|\frac{\Delta q}{C}|_{max} \ |\frac{\Delta q'}{C}|_{max}}{A_v}} \tag{16.47}$$

which can be much smaller than the original value given by (16.39)

16.4   Noise Analysis

16.4.1   Noise sources

The noise sources are independent and can be represented in the small signal model of the MOS transistor either by a current generator from drain to source, or by a voltage source in series with the gate. An equivalent input noise resistance R_N can be defined for MOS transistors in saturation [25,27,28]:

$$R_N = \frac{\gamma}{g_m} + \frac{\rho}{f W_{ef} L_{ef}} \tag{16.48}$$

where g_m stands for the transconductance of T_m and W_{ef} and L_{ef} are the effective transistor width and length, respectively. Factor ρ is strongly dependent on the process and its unit is Vm^2/As. The noise factor γ is equal to $n/2$ in weak inversion and to $2n/3$ in strong inversion.

This noise resistance can be associated with the double-sided voltage noise spectral power density $S_V(f)$ and with the double-sided current noise spectral power density $S_I(f)$:

$$S_V(f) = 2k \, \Theta \, R_N \qquad\qquad -\infty > f > +\infty \tag{16.49}$$

$$S_I(f) = g_m^2 \, S_V(f) = 2k\Theta \, \gamma \, g_m \left(1 + \frac{f_k}{|f|}\right) \quad -\infty > f > +\infty \tag{16.50}$$

where f_k is the so-called corner frequency, corresponding to the frequency at which white and 1/f noise spectral densities are equal, and Θ is the absolute temperature. If the MOS transistors operate as switches, hence, in the triode region, their equivalent noise resistance R_N is given by

$$R_N = \frac{1}{\beta_{sw}(V_G - V_{T0} - nV_s)} = \frac{1}{g_{on}} \tag{16.51}$$

with β_{sw} being the transfer parameter of the switch, V_{T0} its threshold voltage and V_s the source voltage. $S_I(f)$ yields

$$S_I(f) = g_{on}^2 S_V(f) = 2k\Theta g_{on} \qquad -\infty > f > +\infty \tag{16.52}$$

Integrating S_V or S_I over the frequency domain leads to total voltage components V_N^2 or current noise components I_N^2, respectively.

16.4.2 *Analysis of direct noise sources*

As for a sampled data circuit, two kinds of noise sources are present:

(i)   direct noise, which occurs when the output current is sunk and which is due to the continuously generated noise of T_m;

(ii)   sampled noise, which is due to the storage of the instantaneous value of the noise voltage on capacitor C.

The total output current noise component I_{Nout}^2 is the superposition of these two noise currents. In Figure 16.15 a self-biased dynamic current mirror [13] is represented with its noise sources.
   The total output current noise spectral power density $S_{Iout}(f)$ is the superposition of these sources

$$S_{Iout}(f) = S_{Ioc}(f) + S_{Iom}(f) + S_{Ioz}(f) \tag{16.53}$$

where $S_{Iom}(f)$ and $S_{Ioz}(f)$ are the contributions to $S_{Iout}(f)$ of T_{mj} and g_{zj} (j=0,1), respectively. $S_{Ioc}(f)$ stands for the noise contribution of T_{c1} and for all the noise sources of the input branch of the circuit. These contributions are:

(a) $$S_{Ioz}(f) = [\frac{(ng_{mc} + g_{dsc})\, g_{dsm}}{g_z\,(ng_{mc} + g_{dsm} + g_{dsc})}]^2 \, S_{Iz}(f) \approx g_{dsm}^2 \, S_{Vz}(f) \tag{16.54}$$

A high switch conductance g_z is desirable, hence $S_{Vz}(f)$ is very small.

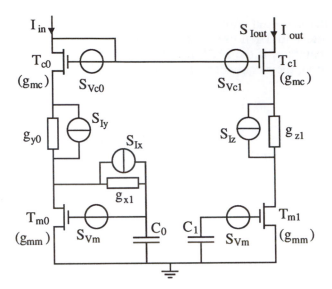

Figure 16.15 Noise sources of a cascoded dynamic current mirror; only the conducting switches are represented.

(b) $S_{Ioc}(f)$, for a transistor T_{c1} located in a common well, is found to be

$$S_{Ioc}(f) \approx \frac{g_{dsm}^2}{n^2} [S_{Vc1}(f) + \frac{S_{Iin}(f)}{g_{mc}^2}] \qquad (16.55)$$

In this expression n is replaced by one if T_{c1} is in a separate well which is connected to the source.

(c) $S_{Iom}(f)$ can be expressed, regardless of whether T_{c1} is located in a common well or in a separate well, as

$$S_{Iom}(f) \approx g_{mm}^2 \, S_{Vm}(f) \qquad (16.56)$$

From (16.54) to (16.56) it can be seen that the noise spectral power densities $S_{Ioz}(f)$ and $S_{Ioc}(f)$ are negligible compared to $S_{Iom}(f)$, hence only this noise spectral power density has to be taken into account. Note that the considerations made in this paragraph remain valid for any kind of cascoded structure.

16.4.3 Description of sampling and autozeroing effects

The small signal representation of a basic cell and the waveforms of the different interfering currents are represented qualitatively in Figure 16.16. The current variations are assumed to be small compared to the steady current, so that a small signal representation is acceptable. The noisy current source $I_N(t)$ stands for all the noise sources of a current copier.

The switch conductance g_x of S_x is assumed to be negligible to simplify the qualitative representation.

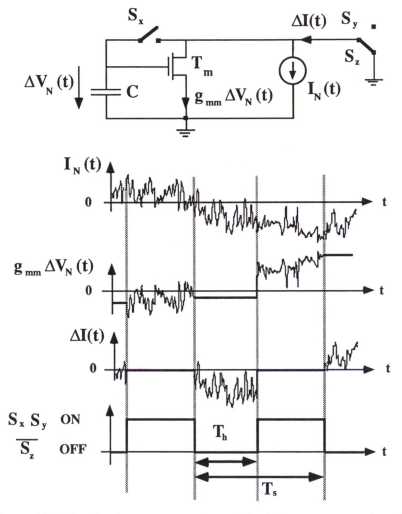

Figure 16.16 Qualitative representation of the inherent autozeroing effect.

When switches S_x and S_y are closed, the instantaneous current error $\Delta I(t)$ is forced to be zero, because the drain node is connected to the input current source I_{in}, hence to a high impedance:

$$\Delta I(t) = I_N(t) + g_{mm}\Delta V_N(t) = 0 \qquad nT_s+T_h<t<(n+1)T_s \qquad (16.57)$$

The DC noise component or offset is cancelled, because the closed feedback loop forces $\Delta V_N(t)$ to be proportional to $I_N(t)$. At time $t=nT_s$, the sampling switch S_x is opened and the instantaneous value of $\Delta V_N(t)=\Delta V_N(nT_s)$ is frozen on C. $\Delta I(t)$ becomes

$$\Delta I(t) = I_N(t) + g_{mm} \Delta V_N(nT_s) \qquad nT_s<t<nT_s+T_h \qquad (16.58)$$

where T_s is the sampling period, n an integer value and T_h the duration of the hold time. Due to the periodical updating of the gate voltage $\Delta V_N(t)$ the current error $\Delta I(t)$ is forced to zero each time S_x and S_y are closed. Because the gate voltage is sampled, the following noise analysis refers to the gate voltage variations $\Delta V_N(t)$ on the storage capacitor C instead of the current variations $\Delta I_N(t)$, but both representations are equivalent.

Figure 16.17 shows the functional block diagram in which the noise source $I_N(t)$ has been split into $I_{NT}(t)$ and $I_{Ng}(t)$, which stand for the noisy transistor T_m and the noisy switch S_x, respectively. *A priori* it is not possible to neglect the noise of S_x, since a low sampling switch conductance g_x might be desired to reduce the charge injection in C.

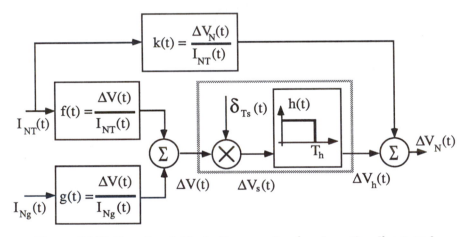

Figure 16.17 Functional block diagram showing how the direct and sampled noise components on C are related to the noise sources $I_{NT}(t)$ and $I_{Ng}(t)$.

During the storage phase the two noise sources $I_{NT}(t)$ and $I_{Ng}(t)$ produce variations of the gate voltage $\Delta V(t)$ through their transfer function $F(f)$ and $G(f)$, respectively. These gate voltage variations enter the sample-and-hold block, which in Figure 16.17 is surrounded by a dotted line, and are frozen on C.

When T_m is switched back to the output to restore the current, the sampled noise component is added to the only direct noise component which is due to T_m. Because the direct noise component influences the output current during the restoring phase, its transfer function $k(t)$ is valid only during the hold time T_h.

Finally the voltage noise component $\Delta V_N(t)$ on C can be expressed as

$$\Delta V_N(t) = \left(I_{NT}(t)*f(t) + I_{Ng}(t)*g(t)\right)\delta_{Ts}*h(t) + I_{NT}(t)*k(t) \qquad (16.59)$$

where the "*" symbol stands for the convolution operator.

16.4.4 Analysis of sampling and autozeroing: transfer functions

Figure 16.18 represents the small signal model of a basic cell during the storage phase. C_{pi} stands for all the parasitic capacitances at the input node, at the drain of T_m, and includes the parasitic switch capacitances, whereas g_0 includes all the parasitic conductances at this node. Assuming that the conductance of the external current source I_{in} is much smaller than g_0, the only remaining noise sources during current storage are $I_{NT}(f)$ and $I_{Ng}(f)$. The Fourier transfer functions $F(f)$ and $G(f)$, during the storage phase, are found to be

$$F(f) = \frac{\Delta V(f)}{I_{NT}(f)}\Big|_{I_{Ng}(f)=0} = -\frac{F'(f)}{(g_{mm}+g_0)} = -\frac{1}{(g_{mm}+g_0)\,D(f)} \qquad (16.60)$$

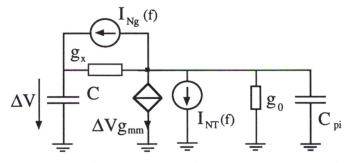

Figure 16.18 Small signal scheme during the storage phase with noise sources $I_{NT}(f)$ and $I_{Ng}(f)$.

and

$$G(f) = \frac{\Delta V(f)}{I_{Ng}(f)}\Big|_{I_{NT}(f)=0} = -\frac{g_0\ G'(f)}{(g_{mm}+g_0)g_x} = -\frac{g_0}{(g_{mm}+g_0)g_x}\,\frac{1 + 2\pi jf\,\dfrac{C_{pi}}{g_0}}{D(f)} \qquad (16.61)$$

Denominator $D(f)$ is given by

$$D(f) = 1 + 2\pi jf\left[\frac{C(g_x+g_0)+C_{pi}g_x}{(g_{mm}+g_0)\,g_x}\right] + (2\pi jf)^2\left[\frac{CC_{pi}}{(g_{mm}+g_0)\,g_x}\right]$$

$$= 1+p\tau_d\,(1+p\tau_p) \qquad (16.62)$$

with time constants

$$\tau_d = \frac{C(g_x+g_0)+C_{pi}g_x}{g_x\,g_{mm}} = \frac{1}{2\pi f_c} \underset{\underset{C \gg C_{pi}}{g_x \gg g_0}}{\approx} \frac{C}{g_{mm}} \tag{16.63}$$

$$\tau_p = \frac{C\,C_{pi}}{C(g_x+g_0)+C_{pi}g_x} = \frac{1}{2\pi f_{c2}} \underset{\underset{C \gg C_{pi}}{g_x \gg g_0}}{\approx} \frac{C_{pi}}{g_x} \tag{16.64}$$

A global settling time constant τ_s may be reasonably approximated by

$$\tau_s < \tau_d + 2\tau_p \approx \frac{g_x(C+C_{pi})^2 + 2\,g_{mm}CC_{pi}}{g_x g_{mm}(C+C_{pi})} \tag{16.65}$$

Comparing the noise components induced on C due to $I_{NT}(f)$ or to $I_{Ng}(f)$ leads to

$$\frac{\Delta V_{Ng}(f)}{\Delta V_{NT}(f)} = \frac{g_0\left(1+\dfrac{2\pi jf\,C_{pi}}{g_0}\right)I_{Ng}(f)}{g_x\,I_{NT}(f)} = \frac{g_0\left(1+\dfrac{2\pi jf\,C_{pi}}{g_0}\right)}{\sqrt{g_x\,g_{mm}}} \approx \frac{2\pi jf\,C_{pi}}{\sqrt{g_x\,g_{mm}}} \tag{16.66}$$

where the approximation is valid if $2\pi f > \dfrac{g_0}{C_{pi}}$. In (16.66) it is apparent that the contribution of the sampling switch noise increases with respect to frequency. For frequencies lower than $\dfrac{\sqrt{g_x\,g_{mm}}}{2\pi\,C_{pi}}$, the noise contribution of S_x is found to be negligible even if $g_x \approx g_0$.

Assuming that the noise contributions of the main transistor and the sampling switch are independent, the Fourier transform $\Delta V(f)$ of the noise error $\Delta V(t)$ on C during storage is the sum of both noise contributions, hence

$$\Delta V(f) = I_{NT}(f)\,F(f) + I_{Ng}(f)\,G(f) \tag{16.67}$$

According to the considerations made in Section 16.3, only the direct noise component of source $I_{NT}(f)$ contributes to the output current noise. The direct transfer function $K(f)$ is given by the relationship between the current noise spectral power density and the voltage noise spectral power density.

$$K(f) = \frac{\Delta V_K(f)}{I_{NT}(f)} = \frac{1}{g_{mm}} \tag{16.68}$$

16.4.5   Calculation of sample-and-hold component $\Delta V_h(F)$

The sample-and-hold operation is based on two operators: an ideal sampler and a hold element (dotted in Figure 16.17). The first operator multiplies the input signal $\Delta V(t)$ with a sequence of Dirac distributions δ_{Ts}. Due to the multiplication of $\Delta V(t)$ with δ_{Ts}, the continuous gate voltage is translated into a sampled one, $\Delta V_s(t)$. The sequence of samples at time instants nT_s is re-transformed by convolution with the second operator, the hold function $h(t)$, and yields

$$\Delta V_h(t) = \Delta V(t)\ \delta_{Ts} * h(t) = h(t) * \sum_{n=-\infty}^{+\infty}\Delta V(nT_s)\ \delta(t\text{-}nT_s) \quad (16.69)$$

The convolution in the time-domain can be replaced by a multiplication in the frequency-domain, and the bilateral Fourier transform of the sampled-and-held noise voltage $\Delta V_h(f)$ can be expressed [5,22,29] as

$$\Delta V_h(f) = \frac{T_h}{T_s}\ sinc(\pi f T_h)\ e^{-j\pi f T_h} \sum_{n=-\infty}^{+\infty}\Delta V(f\text{-}nf_s) \quad (16.70)$$

where the infinite sum is due to the sampling process and corresponds to noise components which are shifted to the integer values of the sampling frequency f_s. The term $sinc(\pi f T_h)$ is due to the hold function $h(t)$, and the exponential term stands for the delay of this function. Note that for a dynamic current mirror of ratio 1:1, the sampling frequency f_s is twice the switching frequency f_{sw}. Or in other words, the output sees only one transistor which is updated instantaneously with f_s and where $T_h = T_s$.

16.4.6   Calculation of the direct component $\Delta V_D(f)$

The direct noise component is modulated by a pulse train as shown in Figure 16.17. Because the noise is undersampled, the voltage component on C can be reasonably approximated by [30]

$$\Delta V_D(f) \approx I_{NT}(f)\ K(f)\ \sqrt{\frac{T_h}{T_s}} \quad (16.71)$$

16.4.7   Autozero transfer function

According to the functional block diagram of Figure 16.17 and with the aid of the transfer functions found in the preceding sections, the total voltage noise $\Delta V_N(f)$ on C in the frequency-domain is found to be [5,22,29]

$$\Delta V_N(f) = \Delta V_H(f) + \Delta V_D(f) = \Delta V_{NT}(f) + \Delta V_{Ng}(f)$$

$$= \frac{T_h}{T_s}\ sinc(\pi f T_h)\ e^{-j\pi f T_h}\sum_{n=-\infty}^{+\infty}I_{NT}(f\text{-}nf_s)F(f\text{-}nf_s) + I_{NT}(f)\ K(f)\ \sqrt{\frac{T_h}{T_s}}$$

$$+ \frac{T_h}{T_s} \, sinc(\pi f T_h) \, e^{-j\pi f T_h} \sum_{n=-\infty}^{+\infty} I_{Ng}(f-nf_s) G(f-nf_s) \qquad (16.72)$$

where $\Delta V_{NT}(f)$ is due to the transistor and $\Delta V_{Ng}(f)$ is due to the sampling switch. The correlation between the direct noise and the sampled noise of T_m leads to the autozeroing effect, because the noise contributions are of opposite sign. The contribution of the noise current $I_{Ng}(f)$ of the sampling switch can be written as

$$\Delta V_{Ng}(f) = - \frac{g_0}{(g_{mm}+g_0)g_x} \frac{T_h}{T_s} \, sinc(\pi f T_h) e^{-j\pi f T_h} \sum_{n=-\infty}^{+\infty} I_{Ng}(f-nf_s) G'(f-nf_s) \qquad (16.73)$$

and the contribution of the noise current $I_{NT}(f)$ of the main transistor T_m as

$$\Delta V_{NT}(f) = \frac{1}{g_{mm}} \sum_{n=-\infty}^{+\infty} X_n(f) \, I_{NT}(f-nf_s) \qquad (16.74)$$

where $X_n(f)$ is defined as

$$\sqrt{\frac{T_h}{T_s}} - \frac{g_{mm} \, g_x \dfrac{T_h}{T_s} \, sinc(\pi f T_h) \, e^{-j\pi f T_h}}{D(f)} \qquad \text{for } n=0 \qquad (16.75)$$

$$X_n(f) = \frac{g_{mm} \, g_x \dfrac{T_h}{T_s} \, sinc(\pi f T_h) \, e^{-j\pi f T_h}}{D(f-nf_s)} \qquad \text{for } n \neq 0 \qquad (16.76)$$

and $D(f)$ is given by (16.62).

The magnitude of $X_n(f)$ for low frequencies, hence in the baseband, is found to be:

for n=0

$$|X_0(f)| = \sqrt{[\sqrt{\frac{T_h}{T_s}} - \frac{g_{mm}}{g_{mm}+g_0} \, \frac{sin(2\pi f T_h)}{2\pi f T_h}]^2 + [\frac{T_h}{T_s} \frac{g_{mm}}{g_{mm}+g_0} \, \frac{1-cos(2\pi f T_h)}{2\pi f T_h}]^2}$$

for n≠0 $\qquad\qquad\qquad\qquad\qquad\qquad\qquad\qquad\qquad\qquad\qquad (16.77)$

$$|X_n(f)| = \frac{T_h}{T_s} \frac{g_{mm}}{(g_{mm}+g_0)} \sqrt{sinc^2(\pi f T_h)} \qquad (16.78)$$

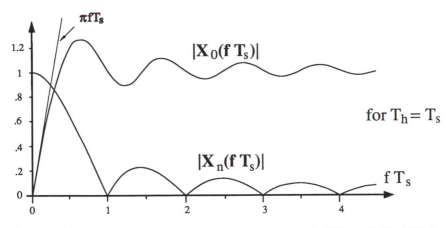

Figure 16.19 Magnitude of the transfer functions $X_0(fT_s)$ and $X_n(fT_s)$ for $T_s = T_h$.

The magnitudes $/X_0(fT_s)/$ and $/X_n(fT_s)/$ assuming $T_h = T_s$, are plotted in Figure 16.19 for small values of fT_s. They are similar to the ideal transfer functions of an autozero amplifier [5,26]. It can easily be seen that the limit of $/X_0(f)/$ while $f \rightarrow 0$ and for $g_{mm} \gg g_0$ is

$$\lim_{f \to 0} /X_0(f)/ = \left(\sqrt{\frac{T_h}{T_s}} - \frac{g_{mm}}{g_{mm}+g_0}\right) \stackrel{T_h=T_s}{=} \left(1 - \frac{g_{mm}}{g_{mm}+g_0}\right) \stackrel{g_{mm} \gg g_0}{\approx} 0 \qquad (16.79)$$

The first term of (16.79) is due to the direct noise, and the second term stands for the sampled noise. For $T_h = T_s$, the function $X_0(f)$ equals zero for $f \rightarrow 0$, which is in accordance with the considerations made on the inherent autozeroing effect of a current copier. The offset of a transistor, which can be defined as the deviation of its threshold voltage from the ideal value, can be represented as a peak at $f=0$. According to (16.79) this peak is not completely cancelled at the origin for $T_h = T_s$, but rather attenuated by the ratio $\dfrac{g_0}{g_0+g_{mm}}$.

16.4.8 Voltage noise spectral power density $S_V(f)$

From (16.72), the noise voltage spectral power density $S_V(f)$ on C is deduced:

$$S_V(f) = S_{Vbaseband}(f) + S_{Vfoldover}(f)$$

$$= \frac{S_{IT}(f)}{(g_{mm}+g_0)^2} /X_0(f)/^2 + \frac{1}{(g_{mm}+g_0)^2} \sum_{\substack{n=-\infty \\ n \neq 0}}^{+\infty} /X_n(f-nf_s)/^2 S_{IT}(f-nf_s)$$

$$+ \; \frac{g_0{}^2}{(g_{mm}+g_0)^2 g_x{}^2} \; sinc^2(\pi f T_h) \sum_{n=-\infty}^{+\infty} |G'(f-nf_s)|^2 \, S_{Ig}(f-nf_s) \qquad (16.80)$$

The infinite sums of the low-pass filtered, translated current noise spectral power densities $S_{IT}(f)$ and $S_{Ig}(f)$, which are due to the undersampling, must be evaluated analytically. These current noise spectral power densities are split into their white noise and 1/f noise foldover components. The following section describes how these sums can be approximated either for the white or the 1/f noise.

16.4.9   Analysis of aliasing effects using the equivalent noise bandwidth technique

The "equivalent bandwidth" Δf is defined as the bandwidth needed to obtain the same voltage power at the output of an ideal low-pass filter as at the output of a low-pass filter with transfer function $|A(f)|$, with both DC gains identical. In the case of unity gain, hence $|A(f=0)| = 1$, the output voltage power can be expressed as

$$V_{Nw}{}^2 = \int_{-\infty}^{+\infty} S(f)|A(f)|^2 \, df = \int_{-\Delta f}^{+\Delta f} S(f) \, df \qquad (16.81)$$

Note that the "equivalent bandwidth" Δf depends on the spectral power density $S(f)$. An undersampling factor N can be defined [5,30] as the closest integer number to

$$N = \frac{2\Delta f}{f_s} = 2 \, \Delta f \, T_s \qquad (16.82)$$

and corresponds to the maximum number of shifted spectra which still increase the noise density $S_V(f)$ in the "equivalent bandwidth" and which have to be taken into account.

16.4.9.1   White noise in first and second-order low-pass filters

For a first-order, unity-gain low-pass filter with the transfer function given by

$$A(f) = \frac{1}{1 + j\dfrac{f}{f_c}} \qquad (16.83)$$

the white noise voltage component $V_{Nw}{}^2$ is found to be

$$V_{Nw}^2 = \int_{-\infty}^{+\infty} \frac{S_{Nw}\, df}{(1+(\frac{f}{f_c})^2)} = S_{Nw}\, \pi f_c = \int_{-\Delta f_w}^{+\Delta f_w} S_w\, df = 2\Delta f_w\, S_{Nw} \quad (16.84)$$

where S_{Nw} is the "white noise spectral power density", which is frequency independent. Δf_w is the "equivalent white noise bandwidth" and f_c is the cut-off frequency of the first-order low-pass filter. From (16.84) Δf_w can be expressed as

$$\Delta f_w = \frac{1}{4\tau_d} = \frac{\pi f_c}{2} \quad (16.85)$$

which is a well known result.

For a second-order, unity-gain low-pass filter the open loop transfer function of the current mirror during the storage phase can be represented by

$$B(f) = \frac{1}{1+2\pi jf\tau_d\,(1+2\pi jf\tau_p)} = \frac{1}{1 + j\frac{f}{f_c}\,(1+j\frac{f}{f_{c2}})} \quad (16.86)$$

with the cut-off frequencies f_c and f_{c2}.

The "equivalent white noise bandwidth" Δf_w is found to be

$$\Delta f_w = \frac{\pi f_c}{2} \quad (16.87)$$

which surprisingly only depends on the first cut-off frequency f_c [22,31].

16.4.9.2 White noise aliasing effect in a current copier

The foldover terms, which are due to the white noise current spectral density $S_{ITw}(f)$ of the main transistor T_m, can now be evaluated with the aid of the techniques mentioned above and yield

$$S_{VTwfold}(f) \approx \frac{T_h^2\, sinc^2(\pi fT_h)}{(g_{mm}+g_0)^2\, T_s^2} \sum_{\substack{n=-\frac{N}{2} \\ n\neq 0}}^{+\frac{N}{2}} S_{ITw}$$

$$\approx 2k\Theta\gamma\, \frac{T_h^2\, sinc^2(\pi fT_h)}{T_s^2\, g_{mm}} \left(\frac{\pi f_c}{f_s}-1\right) \quad (16.88)$$

where $S_{ITw}(f) = S_{ITw} = 2k\Theta\gamma\, g_{mm}$ and $g_{mm} \gg g_0$. The contribution of the white noise $S_{Vgw}(f)$ of the sampling switch S_x is found to be, after using a similar approach to that described in Appendix 16B,

$$S_{Vgw}(f) = \frac{g_0^2}{g_x^2}\, \frac{T_h^2\, sinc^2(\pi f T_h)}{T_s^2\, g_{mm}^2}\, \frac{\pi f_c}{f_s}\left(1 + \frac{f_c f_{c2}}{f_0^2}\right)S_{Igw}(f)$$

$$\begin{array}{l} T_h = T_s \\ \approx \end{array} \quad k\Theta T_s\, \frac{C_{pi}}{C\,(C+C_{pi})}\; sinc^2(\pi f T_s) \qquad (16.89)$$

with $S_{Igw}(f) = 2k\Theta g_x$ and $f_0 = \dfrac{g_0}{2\pi\, C_{pi}}$.

The foldover terms of T_m and S_x can be compared to each other and yield

$$\frac{S_{VTwfold}(f)}{S_{Vgwfold}(f)} = \gamma\, \frac{C}{C_{pi}} \qquad (16.90)$$

16.4.9.3 1/f noise in first and second-order low-pass filter

The 1/f noise voltage component $V_{N1/f}^2$ is given by the integral over the frequency domain of the 1/f noise density according to (16.81). But $V_{N1/f}^2$ is divergent because of the pole at the origin. Nevertheless the 1/f noise power of a low-pass filter can be compared to that of an ideal low-pass filter using the definition given in (16.84), which leads to the "equivalent 1/f noise bandwidth" $\Delta f_{1/f}$.

The first-order low-pass filtered 1/f noise voltage component V_{N1}^2 can be expressed as

$$V_{N1}^2 = K \int_{f_a}^{\infty} \frac{df}{\left(1+\left|\frac{f}{f_c}\right|^2\right)f} = K \lim_{b\to\infty} \int_a^b \frac{dx}{(1+x^2)\, x} \qquad (16.91)$$

which must be compared to the first-order ideal low-pass filtered 1/f noise voltage component:

$$V_N^2 = K \int_{f_a}^{\Delta f_{1/f}} \frac{df}{f} = K \int_a^{\Delta x} \frac{dx}{x} = K\, ln\frac{\Delta x}{a} \qquad (16.92)$$

with

$$\Delta x = \frac{\Delta f_{1/f}}{f_c} \qquad (16.93)$$

where f_c corresponds to the cut-off frequency of the low-pass filter.

The integral of (16.92) is found to be

$$V_{NI}^2 = K \lim_{b\to\infty} \left[ln\sqrt{\frac{b^2(1+a^2)}{a^2(1+b^2)}} \right] = K\frac{1}{2}ln\frac{(1+a^2)}{a^2} \tag{16.94}$$

Identifying (16.92) and (16.94) leads to:

$$ln\frac{(1+a^2)}{a^2} = ln\frac{\Delta x^2}{a^2} \tag{16.95}$$

Hence

$$\Delta x = \frac{\Delta f_{1/f}}{f_c} = \lim_{a\to 0}\ (1+a^2) = 1 \tag{16.96}$$

which means that the equivalent 1/f noise bandwidth $\Delta f_{1/f}$ is equal to the cut-off frequency f_c of the low-pass filter or, in other words, the 1/f noise power of an ideal first-order low-pass filter with cut-off frequency f_c is identical to the power of a low-pass filter with the same cut-off frequency f_c.

In the case of a second-order filter with real poles, it is possible to express the voltage noise component function $V_{NI/f}^2$ as follows:

$$V_{NI/f}^2 = 2K\int_{f_a}^{f_b}\frac{df}{f(1+|\frac{f}{f_c}|^2)(1+|\frac{f}{f_{c2}}|^2)} = 2K\int_{a}^{b}\frac{dx}{x(1+|x|^2)(1+|\rho x|^2)} \tag{16.97}$$

with $x = \frac{f}{f_c}$ and $\rho = \frac{f_c}{f_{c2}}$, $\rho < 1$. After comparing the voltage noise power of (16.97) to the ideal low-pass filtered voltage noise power the following relationship is found:

$$\Delta f_{1/f} \overset{\rho<1}{<} f_c \tag{16.98}$$

meaning that the "equivalent bandwidth" may be approximated by f_c under the above mentioned conditions.

16.4.9.4 1/f aliasing effect in a current copier

The 1/f foldover term can be written as

$$S_{V1/f\text{-fold}} = \frac{T_h^2 sinc^2(\pi fT_h)}{T_s^2}\frac{2k\Theta\rho}{W_{ef}L_{ef}}\sum_{\substack{n=-\infty \\ n\neq 0}}^{+\infty}(\frac{1}{f-nf_s}\frac{1}{|D(f-nf_s)|^2}) \tag{16.99}$$

The discontinuities for $f = nf_s$ are cancelled out by the double-zero of the $sinc^2(f\pi T_h)$-function when $T_h = T_s$ [5]. The infinite sum $S_{V1/f\text{-}fold}(f)$ is approximated by only a limited number $N = \dfrac{2f_c}{f_s}$ of terms, which leads to ($g_0 \ll g_{mm}$):

$$S_{V1/f\text{-}fold} \approx \frac{T_h^2 sinc^2(\pi f T_h)}{T_s^2} \frac{2k\Theta\rho}{W_{ef}L_{ef}} \sum_{\substack{n=-\frac{N}{2} \\ n\neq 0}}^{+\frac{N}{2}} \frac{1}{f-nf_s} \qquad (16.100)$$

In the baseband, hence for $f < f_s$, this sum can be evaluated and yields [5,22]

$$S_{V1/f\text{-}fold} \approx \frac{T_h^2 sinc^2(\pi f T_h)}{T_s} \frac{2k\Theta\rho}{W_{ef}L_{ef}} \left(\frac{2f_s^2}{f_s^2 - f^2} + ln \frac{f_c^2}{(1.5f_s)^2 - f^2} \right) \qquad (16.101)$$

For $f \ll f_s$, the aliasing effect of 1/f noise in the baseband can be considered as an increase of the white noise.

$$S_{V1/f\text{-}fold} \approx \frac{T_h^2 sinc^2(\pi f T_h)}{T_s} \frac{4k\Theta\rho}{W_{ef}L_{ef}} \left(K_{Eul} + ln\frac{f_c}{f_s} \right) \qquad (16.102)$$

where $K_{Eul} = 0.57721566...$ is the so-called *"Euler's constant"*, which is found in finite integrals or summations. The 1/f noise of the sampling switch S_x is negligible, because it corresponds to a modulation of the switch conductance g_x.

16.4.10   *Noise spectral power density* $S_V(f)$

16.4.10.1 White noise due to main transistor T_m

The total double-sided white noise voltage spectral power density $S_{VTw}(f)$ on C due to T_m can be expressed as

$$S_{VTw}(f) = \frac{2k\Theta\gamma}{g_{mm}} |X_0(f)|^2 + S_{VTwfold}(f) \qquad (16.103)$$

If the aliasing effect dominates, $\dfrac{\pi f_c}{f_s} \gg 1$, $S_{VTw}(f)$ can be approximated by

$$S_{VTw}(f) = \frac{2k\Theta\gamma}{g_{mm}} \left(|X_0(f)|^2 + \frac{T_h^2 sinc^2(\pi f T_h)}{2 T_s \tau_d} \right) \qquad (16.104)$$

For $T_h = T_s$ and $f > f_s$, $|X_0(f)|$ is approximately unity, which yields

$$S_{VTw}(f) = k\Theta\gamma \left(\frac{2}{g_{mm}} + \frac{T_s \, sinc^2(\pi f T_s)}{(C+C_{pi})} \right) \qquad (16.105)$$

where the first term corresponds to the direct noise of T_m.

16.4.10.2 Noise spectral power density $S_{VT}(f)$ in the baseband

In the baseband, hence for $f < f_s$, the voltage noise spectral power density $S_{VT}(f)$ on C due to T_m can be summed according to (16.102) and (16.105). In the case of $T_h = T_s$ $S_{VT}(f)$ this reduces to

$$S_{VT}(f) = \frac{2k\Theta}{g_{mm}} \left(\gamma + \frac{g_{mm} \, \rho}{|f|W_{ef}L_{ef}} \right) |X_0(f)|^2 + 2k\Theta\gamma \frac{sinc^2(\pi f T_s)}{g_{mm}} \left(\frac{\pi f_c}{f_s} - 1 \right)$$

$$+ 2k\Theta \, sinc^2(\pi f T_s) \frac{T_s \rho}{W_{ef}L_{ef}} \left(\frac{2f_s^2}{f_s^2 - f^2} + \ln \frac{f_c^2}{(1.5f_s)^2 - f^2} \right) \qquad (16.106)$$

In the baseband $|X_0(f)|$ can be approximated by $\pi f T_s$ which yields

$$S_{VT}(f) \approx \frac{2k\Theta\gamma}{g_{mm}f_s} \left(\frac{\pi^2}{f_s} f^2 + [\pi f_c - f_s] \, sinc^2(\pi f T_s) \right)$$

$$+ \frac{2k\Theta\rho}{f_s W_{ef}L_{ef}} \left(\frac{\pi^2}{f_s} f + [\frac{2f_s^2}{f_s^2 - f^2} + \ln \frac{f_c^2}{(1.5f_s)^2 - f^2}] \, sinc^2(\pi f T_s) \right) \qquad (16.107)$$

For $f \ll f_s$ and $\frac{\pi f_c}{f_s} \gg 1$ (16.104) can be approximated by

$$S_{VT}(f) \approx k\Theta T_s \, sinc^2(\pi f T_s) \left(\frac{\gamma}{C+C_{pi}} + \frac{4\rho}{W_{ef}L_{ef}} [0.6 + \ln\frac{f_c}{f_s}] \right) \qquad (16.108)$$

where the first term is due to the undersampled white noise and the second is due to the undersampled 1/f noise, and K_{Eul} has been approximated by 0.6. From (16.108) it can be seen that the foldover terms of the white noise depend linearly on the ratio $\frac{f_c}{f_s} = \frac{g_{mm}}{2\pi(C+C_p)f_s}$, whereas the foldover terms of the 1/f noise only increase logarithmically with the ratio $\frac{f_c}{f_s}$. In the baseband the effect of 1/f foldover noise can be considered as an increase of the white noise. The voltage noise spectral power density $S_{Vg}(f)$ in the baseband is given by (16.88). Because $S_{Vg}(f)$ and $S_{VT}(f)$ are uncorrelated, the total double-sided voltage noise spectral power density $S_V(f)$ in the baseband is simply the sum of (16.88) and (16.107). The voltage noise spectral power density of S_x on C due to white noise in the case of $T_h = T_s$, $S_{Vgw}(f)$ reduces to

$$S_{Vgw}(f) = kT \frac{g_0^2}{g_x} \frac{sinc^2(\pi f T_s)}{g_{mm}^2} \frac{T_s}{\tau_d} \left(1 + \frac{\tau_0^2}{\tau_d \tau_p}\right)$$

$$\approx kTT_s \frac{C_{pi}}{C(C+C_{pi})} sinc^2(\pi f T_s) \qquad (16.109)$$

Because $S_{Vg}(f)$ and $S_{VT}(f)$ are uncorrelated, the total double-sided voltage noise spectral power density $S_V(f)$ is simply the sum of (16.108) and (16.109).

16.5   Dynamic Behaviour

Dynamic current mirrors are time varying circuits. This means that a different circuit configuration is present at each switching event. The analysis of the dynamic behaviour of dynamic current mirrors providing non-unity ratios can be reduced to one with a 1:1 ratio, because only two basic cells are interfering while switching. To reduce the complexity of the notation the two interfering cells are named 0 and 1, as for a mirror of ratio 1:1. For continuous-time applications the amplitude and the main time constant of the transients (glitches) are very important. Their importance is limited when the currents must only be available in a specific time window and it is possible to wait until the transients have faded.

16.5.1   Critical switching configurations

16.5.1.1 Influence of clock delay

Clearly the sampling switch S_{xj} must only be closed when S_{yj} is closed, otherwise the gate voltage memorised on C_j is not representative of the input current I_{in} and an output ripple is generated. Until now we have assumed that S_{yj} and S_{zk} ($j=0,1$ and $k=1,0$) can be switched simultaneously, which corresponds to an ideal case impossible to perform in practice. The influence of overlapping or non-overlapping clocks on the circuit configuration in the case of a "stacked" current mirror is shown in Figure 16.20 and Figure 16.21 [32]. The configurations corresponding to an externally biased mirror can easily be deduced. If the clocks overlap, S_{yj} and S_{zk} ($j=0,1$ and $k=1,0$) are all closed at the same time (all four switches) and the internal nodes are connected together as shown in Figure 16.19. A new current mirror is formed with T_{c0} and with T_{c1}. The internal parasitic capacitances C_{xx}, represented by broken lines, are charged by currents provided by this new current mirror and are discharged by the currents memorised in the basic cells. Hence the internal node voltage varies proportionally to the mismatch of T_{c0} and T_{c1}.

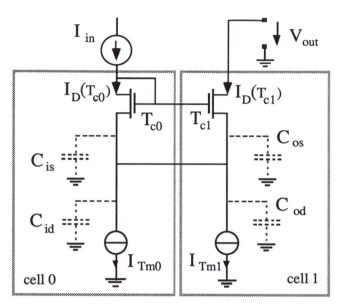

Figure 16.20 Circuit configuration with overlapping clocks for S_{yj} and S_{zk}.

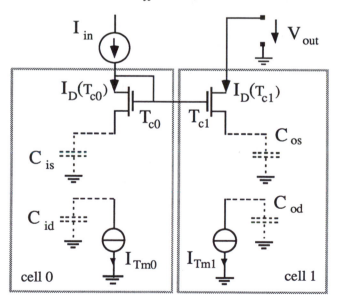

Figure 16.21 Configuration with non-overlapping clocks for S_{yj} and S_{zk}.

In the case of non-overlapping clocks, S_{yj} and S_{zk} are all open at the same time and the different parts of the mirror are disconnected from each other as shown in Figure 16.21. As a consequence, parasitic capacitances C_{is}, C_{os}

are charged by the input currents, whereas C_{id}, C_{od} are discharged by the currents memorised in the basic cells. It can be shown that, even if the time spent in such a configuration is small, large transients and non-linear effects occur and important AC components are produced; the transistors may temporarily come out of saturation, and the DC accuracy of the mirror is completely spoiled. Hence it corresponds to the worst case situation and must absolutely be avoided. [22]

Due to important glitches, the transistors building up the dynamic current mirror may temporarily exit saturation and non-linear effects may be generated. Furthermore, important AC components are added to the DC current and the DC accuracy of the mirror can be completely lost. For well designed clocks, where the time spent in the "all-switches-closed" configuration (Figure 16.20) is minimised, the glitches can be reduced to a small percentage of the input current and depend on: (a) the mismatch between the transistors T_{m0}, T_{m1} and T_{c0}, T_{c1}; (b) the output voltage through the output conductance of T_{c1}; (c) the parasitic capacitances C_{xx}; (d) the switch resistances and (e) the clock overlapping of the switches.

It has to be pointed out that during a cycle the glitches may all be of the same polarity, thus they do not cancel each other out. Therefore the DC accuracy of a dynamic current mirror depends on its switching frequency and on the above mentioned parameters [23]. Special attention has to be paid to the design of the clock phases, as the amplitude of the transients is increased by non-synchronous switching of S_{yj} and S_{zk}.

16.5.1.2 Influence on the output current - AC and DC

Assuming that both branches are matched, hence $C_{ox} = C_{ix}$, and for an externally biased current mirror, the current transients $\Delta I_{out}(t)$ occurring at the output can be approximated by [22]

$$\Delta I_{out}(t) \approx g_{mc} \frac{C_{od}}{C_o} (V_{s1}-V_{s0}) e^{(-t/\tau_{mo})} \tag{16.110}$$

where the main time constant τ_{mo} of the output cell is

$$\tau_{mo} = \frac{C_o}{g_{mc}} \tag{16.111}$$

V_{s0}, V_{s1} are the source voltages of the common gate transistors of the input and the output, respectively, and $C_{px} = C_{xs} + C_{xd}$ $(x = i, o)$. Note that the output time constant due to external loads is not included in τ_{mo}. The average DC current error due to glitches can be found after integrating (16.110), which leads to

$$\overline{\Delta I_{out}} = \frac{1}{T_{sw}} \int_0^{T_{sw}} \Delta I_{out}(t)\, dt = (V_{s1} - V_{s0})\, f_{sw}\, C_{od} \tag{16.112}$$

where $f_{sw} = \dfrac{1}{T_{sw}}$ is the switching frequency. The average output current error $\overline{\Delta I_{out}}$ depends linearly on the switching frequency f_{sw} and is independent of I_{in}, because the current-dependent term g_{mc} of (16.110) is cancelled out by τ_{mo} after integration. If $V_{out} \approx V_{in}$, the voltage difference $V_{s1} - V_{s0}$ may be approximated by the threshold mismatch ΔV_T between T_{c1} and T_{c0}. In the case of a self-biased dynamic current mirror a similar result is found [22]. Furthermore it has to be pointed out that $V_{s1} - V_{s0}$ may remain of the same sign during the whole switching cycle. All the transients then have the same polarity and do not cancel each other out. The observed DC output error is the sum of all these errors due to the glitches.

16.5.2 Speed-accuracy trade-off

16.5.2.1 Settling time constant

A global settling time constant τ_s has been approximated in (16.65) as

$$\tau_s = \tau_d + 2\tau_p = \frac{C^2(1+\frac{g_0}{g_x})^2 + C_{pi}^2 + 2CC_{pi}(1+\frac{g_0}{g_x}+\frac{g_{mm}}{g_x})}{g_{mm}\,(C(1+\frac{g_0}{g_x})+C_{pi})} < t_0 \qquad (16.113)$$

and it must be approximately seven times smaller than the duration of the storage phase t_0 to ensure that equilibrium is reached with an accuracy of better than 1000ppm, and 10 times smaller for an accuracy better than 50ppm. If g_x is high, τ_d is larger than $4\tau_p$ and τ_s is given by

$$\tau_d \approx \frac{C+C_{pi}}{g_{mm}} \qquad (16.114)$$

If g_x is small, an overshoot of the gate voltage appears and τ_s can be approximated by

$$2\tau_p = \frac{C\,C_{pi}}{g_x\,[C(1+\frac{g_0}{g_x})+C_{pi}]} \qquad (16.115)$$

Since the sampling frequency $f_s < \dfrac{1}{t_0}$ (16.114) and (16.115) put an upper limit on f_s and C, and a lower limit on g_x and g_{mm} (see also Section 16.4.4).

16.5.2.2 Speed-accuracy trade-off

The sampling switch conductance g_x can be expressed as

$$g_x = \frac{\mu Q_{tot}}{L_{ef}^2} \qquad (16.116)$$

The residual charge injection after compensation is $\Delta Q_{inj} = \alpha Q_{tot}$ and induces an error voltage ΔV on the storage capacitor C, hence an output current error ΔI_{error}, which can be written as

$$\frac{\Delta I_{error}}{I_{in}} = \frac{g_{mm} \Delta V}{I_{in}} = \frac{\alpha L_{ef}^2}{I_{in} \mu} \frac{g_{mm} g_x}{C} \qquad (16.117)$$

The first term depends on the technology of the device, whereas the second term can be expressed as a function of the storage time constants τ_d and τ_p. For the realistic situation where $g_x \gg g_0$ and $C \gg C_{pi}$, (16.117) simplifies to

$$\frac{\Delta I_{error}}{I_{in}} = \frac{\alpha L_{ef}^2 C_{pi}}{I_{in} \mu} \frac{1}{\tau_d \tau_p} \qquad (16.118)$$

which shows that neither τ_d nor τ_p should be too small.

After introducing the settling time τ_s, the optimum value of τ_d and τ_p which minimises (16.118) is found to be

$$\tau_d = 2\tau_p = \frac{\tau_s}{2} \qquad (16.119)$$

which yields [21]
$$\frac{\Delta I_{error}}{I_{in}} = \frac{8 \, \alpha L_{ef}^2 C_{pi}}{I_{in} \mu} \frac{1}{\tau_s^2} \qquad (16.120)$$

Introducing the numerical values of a standard process shows the significance of (16.120):

$L = 2\mu m$	$C_{pi} = 0.2pF$	$I_{in} = 4\mu A$
$\alpha = 0.5$	(no compensation of charge injection)	
$\mu = 700 \text{ cm}^2/\text{Vs}$	(n-channel)	
$\tau_s = 0.1 \, \mu s$	(compatible with $f_s = 1\text{MHz}$)	

which results in $\dfrac{\Delta I_{error}}{I_{in}} = 1000\text{ppm}$. Furthermore (16.120) shows the direct dependence of the charge injection compensation α on the accuracy.

16.6 Measurements

The measuring system used calibrated resistances (to several ppm) to accurately convert the different currents into voltages, which then could be measured by highly accurate, calibrated voltmeters and by an oscilloscope which could average its input to reduce the magnitude of the observed noise. The accuracy of the measuring system was about 50 ppm at 1μA, decreasing to 100 ppm for 50nA. The technology used was a 3μ p-well CMOS technology with self-aligned contacts (SACMOS3μ) [33]. With the help of built-in auxiliary switches, the different dynamic current mirror configurations were obtained on the same chip, hence were built with the same transistors. So the comparison of the different kinds of mirrors was possible and valid.

16.6.1 AC measurements

16.6.1.1 Variations of input voltage $V_{IN}(t)$ and output current $I_{OUT}(t)$

In Figure 16.22 the measured variations of the input voltage $V_{in}(t)$ and of the output current $I_{out}(t)$ of a self-biased n-type current mirror of ratio 1:4 are shown. The layout of this mirror consisted of an array of five transistors with a ratio $\dfrac{W}{L} = \dfrac{20\mu}{10\mu} = 0.5$.

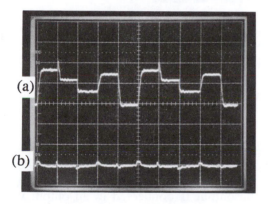

(a)

(b)

time-axis:
$\Delta t = 200\ \mu s/\text{div}$;

y-axis:
(a) $V_{in}(t)$: 5mV/div;
(b) $I_{out}(t)$: 5nA/div

Figure 16.22 Waveforms of $V_{in}(t)$ and $I_{out}(t)$: n-type mirror of ratio 1:4, at $I_{in}=1\mu A$.

Due to the mismatch of the five mirror transistors T_{mj}, the input voltage $V_{in}(t)$ (Figure 16.22a) varies stepwise. In Figure 16.22b the variations of the output current $I_{out}(t)$ are much smaller than one division corresponding to 1250 ppm.

16.6.1.2 Basic cell with a reduced transconductance $g_{mm\Delta}$

Figure 16.23 shows $V_{in}(t)$ and $I_{out}(t)$ for a self-biased n-type current mirror of ratio 1:1, which has a modified basic cell with a reduced transconductance $g_{mm\Delta}$. All the clock phases work correctly, and the ratio of the bypassing current I_0 to the input current I_{in} is equal to 0.85.

According to Section 16.2.5 the voltage variations $V_{in}(t)$ are increased due to the reduced transconductance parameter. Nevertheless the current variations $I_{out}(t)$ are much smaller then one division, which corresponds to 1250 ppm.

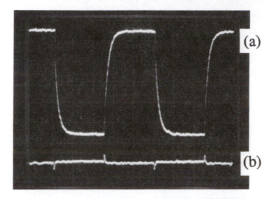

time-axis:
$\Delta t = 100 \, \mu s/div$;

y-axis:
(a) $V_{in}(t)$: 5mV/div;
(b) $I_{out}(t)$: 5nA/div

Figure 16.23 Waveforms of (a) $V_{in}(t)$ and (b) $I_{out}(t)$; n-type mirror of ratio 1:1, @ $I_{in} = 1\mu A$; the mirror operates with a reduced transconductance

$$g_{mm\Delta} \text{ and } \frac{I_0}{I_{in}} = 0.85.$$

16.6.2   DC measurements

16.6.2.1 Multiplying mirror of ratio 1:1, 1:2 and 1:4

Figure 16.24 represents the measured accuracy of a dynamic current mirror of ratio $I_{in} : I_{out1} : I_{out2} : I_{out4} = 1:1:2:4$ as a function of the input current I_{in}. The switching frequency f_{sw} is kept to 1kHz for the three different current ratios and the output is connected to an output voltage $V_{out} = 3V$. The output currents are normalised to x times the input current I_{in}, where x corresponds to the specific current ratio.

The circuit operates with only one external clock per cell and is designed for a nominal input current I_{in} of 0.75 μA. At low currents the accuracy is limited by the mismatch of the leakage currents of the reverse biased junctions, whereas at higher currents this effect becomes negligible. Therefore the measured results show a large spread at low currents. Furthermore the accuracy increases as the V_T mismatch and the charge injection is drowned in the higher gate voltage overhead.

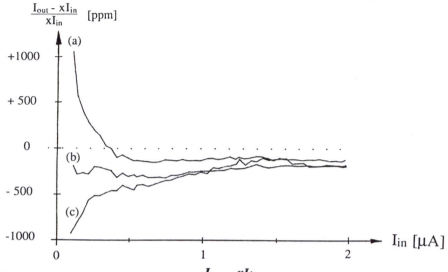

Figure 16.24 Measured error $\varepsilon = \dfrac{I_{outx}-xI_{in}}{xI_{in}}$ [ppm] for a self-biased n-type mirror; $f_{sw}=1\text{kHz}$.

(a) $\dfrac{I_{out}}{I_{in}} = 1{:}1,\ x{=}1$ (b) $\dfrac{I_{out}}{I_{in}} = 1{:}2,\ x{=}2$

(c) $\dfrac{I_{out}}{I_{in}} = 1{:}4,\ x{=}4$

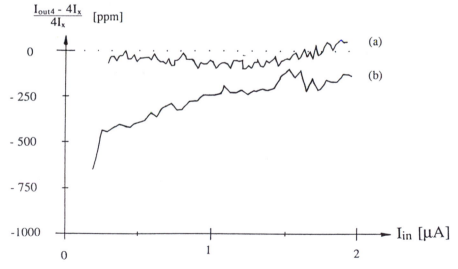

Figure 16.25 Measured error $\varepsilon = \dfrac{I_{out4}-4I_x}{4I_x}$ [ppm] for a "stacked" n-type mirror of ratio $I_{in}{:}I_{out1}{:}I_{out4} = 1{:}1{:}4$ as a function of I_{in} for $V_{out} = 3\text{V}$ and $f_{sw} = 1\text{kHz}$.
Curve (a) $I_x{=}I_{out1}$;   Curve (b) $I_x{=}I_{in}$

Figure 16.25 represents the measured error of a "stacked" n-type current mirror with two outputs of ratio $I_{in} : I_{out1} : I_{out4} = 1{:}1{:}4$, which is obtained by repeating six elementary cells. During the measurement, V_{out} is equal to 3V and f_{sw} to 1kHz. Curve (a) compares the two output currents I_{out1} and I_{out4}, hence $I_x = I_{out1}$, and curve (b) shows the current error while $I_x = I_{in}$. The accuracy is greater when two output currents are compared to each other (curve a) than if an output current is compared to an input current, because first-order errors cancel each other out.

16.6.2.2 Basic cell with a reduced transconductance $g_{mm\Delta}$

In Figure 16.26, the error $\varepsilon = \dfrac{I_{out}\text{-}I_{in}}{I_{in}}$ of a mirror operating with a normal basic cell (Figures 16.26 c, d) is compared to that of a mirror operating with an improved basic cell with a reduced transconductance $g_{mm\Delta}$ (Figure 16.26 a ,b). The ratio of the input current to the output current $\dfrac{I_{out}}{I_{in}} = 1$ and the ratio of the bypassing current to the input current $\dfrac{I_0}{I_{in}} = 0.85$. The output voltage is constant during the measurement: $V_{out} = 3V$ for the curves of Figures 16.26 a, c and $V_{out} = 2V$ for Figures 16.26 b, d. It is visible that the output conductance is not affected by using either the normal basic cell or the improved one. The improvement in accuracy is about a factor of six, corresponding to the ratio of the stored current in the improved structure and the normal configuration, respectively, in accordance with the theory.

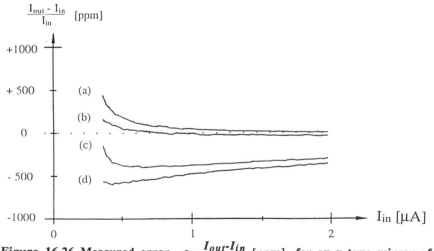

Figure 16.26 Measured error $\varepsilon = \dfrac{I_{out}\text{-}I_{in}}{I_{in}}$ **[ppm] for an n-type mirror of ratio 1:1 as a function of I_{in}; $V_{out} = 3V$ (a,c); $V_{out} = 2V$ (b,d). (a,b) basic cell with a reduced transconductance $g_{mm\Delta}$ ($\dfrac{I_0}{I_{in}} = 0.85$); (c,d) normal basic cell**

16.6.2.3 Influence of the clock frequency

Figure 16.27 shows the increase of the current error for very low switching frequencies (< 1kHz) for two chips taken from two different wafers (wafer 1 curves a,b; wafer 2 curves c,d). Two different effects superimpose in this graph: the current error due to the leakage currents; and the current error due to the glitches.

The error increase at low frequencies is due to the discharge of the storage capacitor C_j by the leakage currents of the sampling switch junctions (Section 16.3.2). Introducing typical numerical values for the example shown in Figure 16.27, the switching frequency f_{sw} needed to obtain an accuracy of $\approx$ 50ppm is found to be about 600Hz.

Furthermore it can be seen that the output current value can increase or decrease as a function of f_{sw}, i.e. that the slope may be larger or smaller than zero, depending on the sign of the voltage difference V_{dm1}-V_{dm0}. When f_{sw} is increased the number of glitches which occur during this period are also increased. If the common-gate transistor T_{c1} and the diode-connected transistor T_{c0} operate in weak inversion, the absolute current error due to transients is constant for a given clock configuration (Section 16.5.1.2) and the relative error varies proportionally to I_{in}. The slope of these curves and even their sign depends on the clock configurations and on the mismatch between the diode-connected transistor T_{c0} and the common-gate transistor T_{c1}[22].

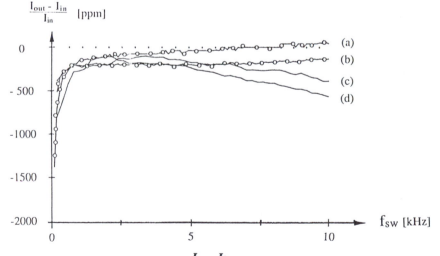

Figure 16.27 Measured error $\varepsilon = \dfrac{I_{out}-I_{in}}{I_{in}}$ [ppm] due to leakage currents and glitches for 2 self-biased n-type mirrors of ratio 1:1 as a function of f_{sw}.

wafer 1:	(a) $I_{in} = .25\mu A$	(b) $I_{in} = .5\mu A$
wafer 2:	(c) $I_{in} = .25\mu A$	(d) $I_{in} = .5\mu A$

16.6.3   Noise measurements

The noise of a dynamic current mirror was measured with the help of an FFT spectrum analyser. The output current noise was first converted into a voltage using a load resistance R_{load} and then amplified by 40dB with a low noise bandpass amplifier. The transconductance g_{mm} for a quiescent current I_{in} equal to 1 µA was found to be 3.2 µA/V and the total gain introduced in the noise path was 60dB. The value of the storing capacitor C was about 5.5pF and the parasitic capacitance C_{pi} was estimated to be about 0.3pF. The double-sided voltage white noise spectral power density $S_{VTw}(f)$ at the gate node of T_m before sampling was theoretically equal to 55 dBV/√Hz, which corresponds to -145 dBV/√Hz. The double sided voltage noise spectral power density $S_{VTw}(f)$ in the baseband, after sampling and bandpass amplification, was found to be - 63 dBV/√Hz and - 69 dBV/√Hz for f_s equal to 2kHz and 8kHz, respectively. As the aliasing effect of the white noise is dominating using the numerical values mentioned above, an increase of the sampling frequency f_s by four reduces the voltage noise spectral power density in the baseband by 6dB.

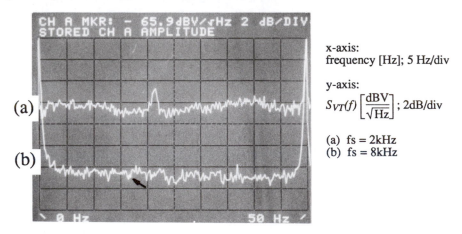

x-axis:
frequency [Hz]; 5 Hz/div

y-axis:
$S_{VT}(f) \left[\dfrac{dBV}{\sqrt{Hz}} \right]$; 2dB/div

(a) fs = 2kHz
(b) fs = 8kHz

Figure 16.28 Noise power spectral density $S_{VT}(f)$ in the baseband:
f_s in 1:4 ratio.

In Figure 16.28 the measured noise spectral density in the frequency band 0-50Hz is represented in dBV/√Hz and f_s is equal to 2kHz (16.28a) and 8kHz (16.28b). Note that the photograph represents the single-sided spectral power density which is twice as large as the double-sided spectral power density $S_{VT}(f)$, which corresponds to an increase of 3dB. The measured decrease of the baseband noise is about 6dB while the sampling frequency is increased by a factor of four in accordance with the theory. The value of the measured noise at the marker is about - 66 dBV/√Hz for f_s = 8kHz, hence it is in accordance with the above predicted values. At f_s = 2kHz the foldover terms of the white noise spectral power density

$S_{VTwfold}(f)$ made up the 94% of $S_V(f)$. The spike visible is due to the 50Hz supply network and even with the help of a grounded box it is not possible to cancel it out completely.

16.7 Conclusion

Based on a simple idea, the dynamic current mirror (or current copier) eliminates the main limitations of standard current mirrors, that are due to offset and 1/f noise. A new problem is created by the charge injected from the switches but the resulting error can be kept very low by adequate design procedures, especially when speed is not a limitation. Except in very special cases, the cascode configuration (or equivalent means) is needed to avoid spoiling the excellent intrinsic precision by errors due to the nonzero output conductance and to the gate-to-drain capacitance.

The basic cell, which provides a single one-to-one discontinuous copy of the input current, can be extended to obtain continuous multiple copies as well as current multiplication or division by integer numbers.

The principle of dynamic mirrors can probably be extended to a variety of different circuits, to create very precise analogue CMOS building blocks.

References

[1] J. B. Shyu, G. C. Temes and F. Krummenacher, "Random error effects in matched MOS capacitors and current sources," *IEEE J. Solid-State Circuits*, vol. SC-19, pp. 948-955, Dec. 1984.

[2] K. R. Lakshimikumar, R. A. Hadaway and M. A. Copeland, "Characterization and modeling of mismatch in MOS transistors for precision analog design," *IEEE J. Solid-State Circuits*, vol. SC-21, pp. 1057-1066, Dec. 1986.

[3] E. A. Vittoz, "MOS transistors operated in lateral bipolar mode and their applications in CMOS technology," *IEEE J. Solid-State Circuits*, vol. SC-18, pp. 273-279, June 1983.

[4] K. C. Hsieh, P. R. Gray, D. Senderowicz and D. Messerschmitt, "A low-noise chopper stabilized differential switched capacitor filtering technique," *IEEE J. Solid-State Circuits*, vol. SC-16, pp. 708-715, Dec. 1981.

[5] C.C. Enz, *High Precision CMOS Micropower Amplifiers*, PhD thesis No 802, Lausanne: Swiss Federal Institute of Technology EPFL, 1989.

[6] R. C. Yen, P. R. Gray, "A MOS switched-capacitor instrumentation amplifier," *IEEE J. Solid-State Circuits*, vol. SC-17, pp. 1008-1013, Dec. 1982.

[7] M. Degrauwe, E. A. Vittoz and I. Verbauwhede, "A micropower CMOS-instrumentation amplifier," *IEEE J. Solid-State Circuits*, vol. SC-20, pp. 805-807, June 1985.

[8] J. C. Candy and B. A. Wooley, "Precise biasing of analog-to-digital converters by means of auto-zero feedback," *IEEE J. Solid-State Circuits*, vol. SC-17, pp. 1220-1225, Dec. 1982.

[9]   R. J. Van de Plassche, "Dynamic element matching for high accuracy monolithic
        D/A converters,"*IEEE J. Solid-State Circuits*, vol. SC-11, pp. 795-800, Dec. 1976.

[10]   E. A. Vittoz, "The design of high performance analog circuits on digital CMOS
        chips," *IEEE J. Solid-State Circuits*, vol. SC-20, pp. 657-665, June 1985.

[11]   Y. S. Yee, L. M. Terman and L. G. Heller, "A 1mV CMOS comparator," *IEEE J.
        Solid-State Circuits*, vol. SC-13, pp. 294-297, June 1978.

[12]   E. A. Vittoz and G. Wegmann, "High precision current mirrors," final seminar on
        project "CMOS Functional Blocks" of the Swiss National Research Foundation PN
        13, May 1988.

[13]   G. Wegmann and E. A. Vittoz, "Very accurate dynamic current mirrors," *IEE
        Electronics Letters*, vol. 25, pp. 644-646, 11th May 1989.

[14]   S. J. Daubert, D. Vallancourt and Y. P. Tsividis, "Current copier cell," *IEE
        Electronics Letters*, vol. 24, pp. 1560-1562, 8th Dec. 1988.

[15]   D. Vallancourt, Y. P. Tsividis, S. J. Daubert, "Sampled current sources," in
        ISSCC Dig. Tech. Papers, pp. 1592-1595, Feb. 1989.

[16]   D.G. Nairn, *Current Mode Algorithmic Analog-to-Digital Converters*, PhD thesis,
        Dept. of Electrical Engineering, Toronto University, 1989

[17]   D. W. J. Groeneveld, H. J. Schouwenaars, H. A. H. Termeer and C. A. A.
        Bastiaansen, "A self calibration technique for monolithic high-resolution D/A
        converters," *IEEE J. Solid-State Circuits*, vol. SC-24, pp. 1515-1522, Dec. 1989.

[18]   J. H Shieh, M. Patil and B. J. Sheu, "Measurement and analysis of charge injection
        in MOS analog switches," *IEEE J. Solid-State Circuits*, vol. SC-22, pp. 277-281,
        April 1987.

[19]   G. Wegmann, E. A. Vittoz and F. Rahali, "Charge injection in analog MOS
        switches," *IEEE J. Solid-State Circuits*, vol. SC-22, pp. 1091-1097, Dec. 1987.

[20]   C. Eichenberger, *Charge Injection in MOS-Integrated Sample-and-Hold and
        Switched-Capacitor Circuits*, PhD thesis, Zurich: Swiss Federal Inst. of Tech.
        ETHZ, *Hartung-Gorre Series in Microelectronics*, vol. 3, 1989.

[21]   E. A. Vittoz and G. Wegmann, "Dynamic Current Mirrors" in C. Toumazou, F. J.
        Lidgey, and D. G. Haigh, Eds, *Analogue Integrated Circuit Design: The Current-
        Mode Approach,* London: Peter Peregrinus Ltd.,1990.

[22]   G. Wegmann, *Design and Analysis Techniques for Dynamic Current Mirrors*, PhD
        Thesis no. 890, Lausanne:Swiss Federal Institute of Technology EPFL, 1990.

[23]   G. Wegmann and E. A. Vittoz, "Analysis and improvements of accurate dynamic
        current mirrors",*IEEE J. Solid-State Circuits*, vol. SC-25, pp. 699-706, June 1990.

[24]   J. Robert, P. Deval and G. Wegmann, "Very accurate current divider", *IEE
        Electronics Letters*, vol. 25, pp. 912-913, 6th July 1989.

[25]   E. A. Vittoz, "MOS Transistors", Intensive Summer Course on CMOS VLSI
        Design: Analog & Digital, EPFL, Lausanne, 1989.

[26]   E. A. Vittoz, "Dynamic analog techniques" in Y. P. Tsividis and P. Antognetti,
        Eds,*VLSI circuits for Telecom-munications*, Prentice-Hall, 1985.

[27]   P. E. Allen and D. R. Holberg, *CMOS Analog Circuit Design*, New York: Holt, Rinehart & Winston, 1987.

[28]   P. R. Gray, R. G. Meyer, *Analog Integrated Circuits*, 2nd Edition, New York: J. Wiley & Sons, 1984.

[29]   C.-A. Gobet, *Modélisation et Calcul de Bruit des Circuits a Capacites Commutees*, PhD thesis No 582, Lausanne: Swiss Federal Institute of Technology EPFL, 1985.

[30]   J. H. Fischer, "Noise sources and calculation techniques for switched-capacitor filters," *IEEE J. Solid-State Circuits*, vol. SC-17, pp. 742-752, Aug. 1982.

[31]   E. A. Vittoz, "Micro Power Techniques, in " Advanced Summer Course on "Design of MOS-VLSI Circuits for Telecommunications," L'Aquila, Italy, June 1984.

[32]   G. Wegmann and E. A. Vittoz, "Basic principles of accurate dynamic current mirrors", *IEE Proc.*, Pt. G, vol. 137, no.2, April 1990.

[33]   R. E. Lüscher and J. Solo de Zaldivar, "A New Approach to a High Density CMOS Process: SACMOS," in *ISSCC Dig. Tech. Papers*, pp. 260-261, New York, 1985.

[34]   Y. P. Tsividis, *Operation and Modeling of the MOS Transistor*, New York: McGraw-Hill, 1987.

[35]   I. S. Gradshteyn and I. M. Ryzhik, *Table of Integrals, Series and Products*, New York: Academic Press, 1965.

Appendix 16A

Symbols and Definitions

The symbols and definitions that are used for n- and p-channel transistors are shown in Figure 16A.1. The analysis assumes a symmetrical device, hence source, V_S, and drain, V_D, voltages can be interchanged.

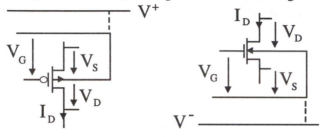

p-channel   n-channel

Figure 16A.1 Symbols and definitions for n- and p-channel MOS transistors.

V_S, V_D and gate voltage V_G are referred to the local substrate which is either the general substrate of the circuit or a separate well. The substrate will be omitted when it is connected to one rail of the power supply (V^+ for p-channel, V^- for n-channel).

The MOS transistor can be characterised by the following parameters [10,34]:

- the gate threshold voltage V_{T0} for $V_S = 0$
- the factor n which represents the reduction of the gate voltage effect due to fixed charges in the channel
- the specific current I_s
- the saturation current I_{Dsat}
- the transfer parameter β and defined as

$$\beta = \mu C'_{ox} \frac{W_{ef}}{L_{ef}} \qquad (16A.1)$$

where μ is the mobility of charge carriers and C'_{ox} the gate oxide capacitance per unit area. The effective channel length L_{ef} and channel width W_{ef} of the transistor are defined by the layout, the tolerances and the overlap given by process parameters.

The effective gate threshold voltage V_{TE} will be assumed to depend linearly on the source voltage V_S:

$$V_{TE} = V_{T0} + nV_S \qquad (16A.2)$$

The factor n stands for the effect of substrate modulation. It is usually between 1.5 and 2 for small gate voltage values and tends to 1 with increasing V_G [25]. The specific current I_s [5] determines whether the device operates in **weak inversion** ($I_{Dsat} < I_s$) or in **strong inversion** ($I_{Dsat} > I_s$):

$$I_s = 2n\beta U_T^2 = 2n\mu C_{ox}' \ U_T^2 \frac{W_{ef}}{L_{ef}} \qquad (16A.3)$$

$$\text{with} \qquad U_T = \frac{kT}{q} \qquad (16A.4)$$

The saturation voltage V_{Dsat} is defined as

$$V_{Dsat} = \frac{V_G - V_{T0}}{n} = V_s + 2U_T \sqrt{\frac{I_{Dsat}}{I_s}} \qquad (16A.5)$$

If the transistor operates in **strong inversion** it is modelled by the following relationships

triode or linear $\quad I_D = \beta \ (V_D - V_S) \ (V_G - V_{T0} - \frac{n}{2} (V_S + V_D))$

(conduction) $\qquad = \beta \ (V_D - V_S) \ (V_G - V_{TE} - \frac{n}{2} (V_D - V_S)) \qquad (16A.6)$

saturation $\qquad I_D = I_{Dsat} = \frac{\beta}{2n} \ (V_G - V_{T0} - nV_S)^2$

$$= \frac{\beta}{2n} \ (V_G - V_{TE})^2 \qquad (16A.7)$$

where (16A.6) is valid while $V_D < V_{Dsat}$ and (A.7) for $V_D > V_{Dsat}$.

The drain current in **weak inversion** is given by

$$I_{Dsat} = I_s \ e^{\frac{V_G - V_{T0}}{nU_T}} \ (e^{\frac{-V_S}{U_T}} - e^{\frac{-V_D}{U_T}})^2 \qquad (16A.8)$$

which reaches its saturation value when the term in V_D becomes negligible, thus for the minimum possible value of the saturation voltage

$$V_{Dsat} = V_s + (2 \ to \ 6)U_T \qquad (16A.9)$$

The small signal transconductance in saturation $g_m = \frac{dI_{Dsat}}{dV_G}$ is given by:

weak inversion $\qquad g_m = \frac{I_{Dsat}}{nU_T} \qquad (16A.10)$

strong inversion $\quad g_m = \sqrt{\dfrac{2\beta I_{Dsat}}{n}} = \dfrac{\beta}{n}(V_G - V_{TE})$ $\qquad$ (16A.11)

Furthermore, the residual drain-to-source conductance g_{ds} in saturation due to the channel shortening effect can be approximated by

$$g_{ds} = \frac{I_{Dsat}}{V_E} = \frac{I_{Dsat}}{\lambda L_{ef}} \qquad (16A.12)$$

where the Early voltage V_E is of the order of tens of volts, λ is a process parameter and its unit is $V/\mu m$. The Early voltage also increases slowly with the gate voltage.

Appendix 16B

Calculation of the equivalent white noise bandwidth for a second-order filter

The second-order filter transfer function is given by

$$B(j\omega) = \frac{1}{(1 - \omega^2 \alpha\beta) + j\omega\alpha} \qquad (16B.1)$$

hence

$$|B(\omega)|^2 = \frac{1}{|1 - \omega^2 \alpha\beta|^2 + |\omega\alpha|^2} = \frac{1}{1 + \omega^2 \alpha(\alpha - 2\beta) + \omega^4 \alpha^2 \beta^2} \qquad (16B.2)$$

Equations (16B.2) yield

$$\Delta f_w = \frac{1}{4\pi} \int_{-\infty}^{+\infty} \frac{d\omega}{1 + b\omega^2 + c^2\omega^4} = \frac{1}{8\pi\sqrt{c}} \int_{-\infty}^{+\infty} \frac{dx}{\sqrt{x}\,(1 + \frac{b}{c}x + x^2)} \qquad (16B.3)$$

with

$$b = \alpha(\alpha - 2\beta) \qquad c = \alpha\beta \qquad x = c\omega^2 \qquad (16B.4)$$

which according to [35] (p. 297) can be solved after replacing with $cos(t) = \dfrac{b}{2c}$, hence $1 + cos(t) = \dfrac{\alpha}{2\beta}$, and leads to

$$\Delta f_w = \frac{1}{4\pi\sqrt{c}} \, (-\pi) \, \frac{sin(-\frac{t}{2})}{sin(t)\, sin(\frac{\pi}{2})} = \frac{1}{4\sqrt{2c}} \, \frac{1}{\sqrt{1 + cos(t)}} = \frac{1}{4a} \qquad (16B.5)$$

which surprisingly depends only on the linear term α [31].

Switched-Current Cellular Neural Networks for Image Processing

Angel Rodríguez-Vázquez, Servando Espejo,
José L. Huertas and Rafael Domínguez-Castro

17.1   Introduction

Artificial Neural Networks (ANNs) consist of arrays of highly interconnected, elementary computing units and provide new, efficient paradigms for the development of massive data processing ICs [1,2]. Unlike conventional computers, whose capabilities rely on the sequential operation of a small number of highly accurate computing artefacts, making them unsuitable for analogue circuit techniques, the capabilities of ANNs emerge as a consequence of the close co-operation among many simple, low-precision units (called neurons). This low-precision feature and the possibility of versatile exploitation of small analogue circuits (formed by a few transistors) for a wide variety of low-level linear and non-linear signal processing tasks required in ANNs, led to the expected use of analogue techniques in the implementation of neural network ICs [3]. Also, since ANNs are much better suited than conventional computers for tasks of cognitive nature, the practical significance of these chips in future advanced signal processing applications should not be underestimated.

A major problem in the design of ANN chips is the complex and area consuming routing. In fully interconnected algorithms (for instance the Hopfield model [4]), in which each cell is connected to every other one, the routing area increases to N^3, where N is the neuron count; this infers that only small dimension networks, not complex enough to support practical applications, are readily realisable in monolithic form. One possible way to overcome this problem is to use multichip architectures [5] and/or introduce serialisation [6]. Another alternative is to devise paradigms requiring a smaller number of connections. Cellular Neural Networks (CNNs), proposed by *Chua* and *Yang* in 1988, explore this second possibility [7-9].

CNNs consist of arrays of elementary processing units (cells), each one connected only to a set of nearby cells (neighbours). This local connection property renders CNN's routing easy, allows increased cell density per silicon area, and makes their computation paradigms very suitable for VLSI implementation. This is more pertinent for the important class of translationally invariant CNNs, in which all cells are identical (with the exception of those on the net borders) and, consequently, the layout is very regular. Also, since the number of different weights is very small for this latter CNN class, programmability can easily be incorporated without significant extra routing cost, by just adding several global control lines, one per weight.

CNNs are non-linear dynamic systems described by differential equations. This chapter focus on Discrete-Time Cellular Neural Networks (DTCNN) whose architecture is similar to CNNs (local connections, uniform network etc.), but with discrete-time dynamics described by finite-difference equations. DTCNNs can be treated as numerical emulators of CNNs [10], particularisations of threshold neural networks, or as brain-state-in-a-box neural models for those cases allowing only local connections [11,12]. They can also be considered as the analogue counterpart of the cellular automata paradigm [13]. The relationships to these well-established models are quite interesting, from the point of view of analysis [12] as well as from the point of view of applications [14].

CNNs are especially suited for image processing since their two-dimensional structure and local interactions are typical features of many image processing algorithms [15]. As a result, a large number of applications have been developed in this field [8,16-22]. Other CNN applications cover fields such as motion detection [23], control [21], character recognition [24] or the simulation of physical systems described by lattice equations [25]. Also, the extension of the original single-layer, linear-interaction model to multi-layer [26] and non-linear-interaction [27] CNNs have opened many new application possibilities in fields of strong economical interest like artificial vision, robotics, pattern recognition, etc. Most of CNNs applications can be performed by DTCNNs too. There are also some image processing tasks, such as the transformation of objects to their concentric contours or the search for objects with minimal distance to a fixed point, which have been demonstrated for DTCNNs but not yet for CNNs [28].

The many potentials of the family of CNN computation paradigms can hardly be realised fully unless special-purpose hardware is devised for them. In particular this chapter explores the use of switched-currents for the design of DTCNN chips. Emphasis is on image processing applications for which switched-current circuits are especially suited because they can be directly interfaced to CMOS-compatible photosensor devices, and fabricated in a common silicon substrate in the standard CMOS digital technology. To keep the chapter to a reasonable size and, simultaneously, guarantee readability we concentrate on single-layer uniform structures and binary output images. For the sake of completeness we discuss elementary algorithmic issues (Section 17.2), circuit design techniques (Sections 17.3 and 17.4) and input/output strategies (Section 17.5). Possible model extensions are outlined in the conclusion.

17.2   The DT-CNN Model

17.2.1   Network architecture and dynamics

Basically, DTCNNs consist of collections of identical processing units (called cells) arranged regularly on a two-dimensional array, where each cell is connected only to a reduced number of neighbours. The concept is illustrated in Figure 17.1 where Figure 17.1(a) presents the cells arranged on a rectangular grid and Figure 17.1(b) presents a hexagonal grid; other grids (triangular, pentagonal etc.) can also be used. A comparison of theoretical issues in regard to the use of different grids can be found in [29].

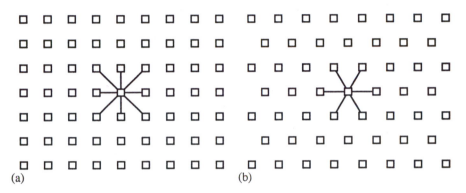

(a)   (b)

Figure 17.1 Basic CNN architecture
(a) Rectangular grid (b) Hexagonal grid.

In Figure 17.1 we assume for simplicity that cells are connected only to their nearest neighbours. In a more general case, each cell, denoted by the index c, is excited by a set of nearby cells located within a distance r in the grid. For this generic c-th cell we define the cell r-neighbourhood, $N_r(c)$, which consists of cell c itself and its interacting cells. For illustration purposes Figure 17.2 shows some r-neighbourhoods on a rectangular grid, for $r = 1$, 2 and 3.

A class of CNNs which is very well suited for VLSI is that of translationally invariant networks (also called uniform CNNs), in which all inner cells are identical. In this case, the neighbourhood radius, r, as well as the neighbour interactions, remain constant throughout the net. This chapter will implicitly assume that nets are translational invariant.

Each cell in a DTCNN has three associated variables[1]:

---

[1] Simple subscripts are used to denote variables for the different grid cells. In particular, the index c indicates a generic cell in the grid. The set of all possible inner grid locations is called the grid domain, represented by GD, so that $c \in GD$, $\forall c$. We will use the index d to indicate the neighbours of the generic c-th cell, so that $d \in N_r(c)$. Double subscripts are

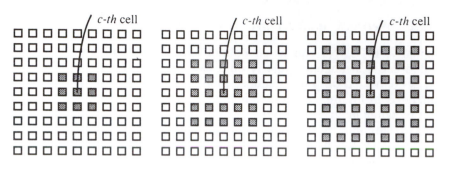

Figure 17.2 Neighbourhoods on a rectangular grid (r=1,2,3).

(a) Cell state: $x_c(n)$, which conveys cell energy information as a function of time, represented by the discrete-time integer variable n.

(b) Cell output: $y_c(n)$, obtained from the cell state through a non-linear transformation.

(c) Cell input: u_c, which, together with the initial cell state for $n = 0$, represents the external excitation of the cell.

The generic c-th inner cell is excited by the input and output of the cells included in its neighbourhood, giving the following equation for the transient evolution of the cell state:

$$x_c(n+1) = D_c + \sum_{d \in N_r(c)} [A_{cd} y_d(n) + B_{cd} u_d] \qquad (17.1)$$

which applies $\forall c \in GD$, where GD represents the set of all possible inner cell locations-the grid domain (see [1]). Also the cell outputs are obtained from the corresponding states through the cell non-linearity, which is ideally represented by the following piecewise-linear function:

$$y_c(n) = f(x_c(n)) = \frac{m}{2} \left(\left| x_c(n) + \frac{1}{m} \right| - \left| x_c(n) - \frac{1}{m} \right| \right) \qquad (17.2)$$

displayed in Figure 17.3. Bear in mind that the dynamic state equation applies only to inner cells. Net border cells remain static, with associated input and state fixed to constant values, characteristic of each application task.

Note that the slope of the cell non-linearity is m. In the literature two different values are considered for this parameter: $m = 1$, corresponding to

---

used in parameters which scale contributions of one cell (indicated by the second subscript) to another (indicated by the first subscript).

soft non-linearity [10] and $m \to \infty$, corresponding to hard non-linearity [11]. Smoother sigmoidal approximations of Figure 17.3 also qualify for practical implementations [30,31].

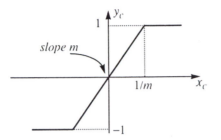

Figure 17.3 Generic DTCNN cell non-linearity.

17.2.2 *Network processing and operating modes*

CNNs are processing devices. Input information is conveyed by the initial state vector $x(0) = \{x_c(0), \forall c \in GD\}$ and the vector $u = \{u_c, \forall c \in GD\}$. Output information is conveyed by the output vector $y = \{y_c, \forall c \in GD\}$. Input-output mapping is determined by the network dynamic evolution following the transient initialised by $x(0)$, driven by u, and under the boundary conditions associated to border cells. This mapping, and hence the processing task performed by the net, depends on the actual time instant at which the net output is sampled, as well as on the parameters B_{cd}, A_{cd} and D_c of (17.1). These are called control, feedback and offset parameters, respectively. The control and feedback parameters can be arranged into matrices, which provide a pictorial view of the interactions within each cell's neighbourhood. For uniform networks these matrices are invariant throughout the grid domain-they are templates. The functionality of an uniform CNN is determined by its control (**B**) and feedback (**A**) templates, and by its offset parameter *(D)*.

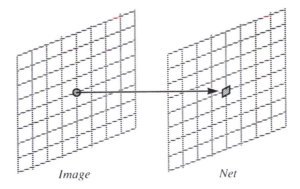

Figure 17.4 Pixel-to-cell correspondence in a DTCNNN.

CNNs are especially suited to processing signals from sensor arrays. In particular, for most image processing applications each inner cell in the CNN corresponds to an image pixel (see Figure 17.4). We will assume that the value 1 of the cell output corresponds to black in the pixel associated to the cell, while -1 corresponds to white.

There are two different modes of operation for DTCNNs: steady-state and transient mode. The steady-state mode is used for most applications, sampling the output once the net has converged to a stable equilibrium state. In transient mode the output is a snapshot at a given time instant after the net initialisation time instant. The net input for both operating modes (composed of vectors $x(0)$ and u) can be signals, although in this case we implicitly assume that their time constants are much larger than the internal net time constants (quasi-static approach).

Figure 17.5 illustrates these two operating modes for a 32x32 DTCNN. Figure 17.5(a) represents the steady-state mode and shows the application of a network with soft cell non-linearity ($m = 1$ in Figure 17.3) to extract the borders of a binary image. Figure 17.5(b) represents the transient mode and corresponds to a net with hard non-linearity ($m \rightarrow \infty$), used to increase objects step by step. The templates for each application are given in Table 17.1.

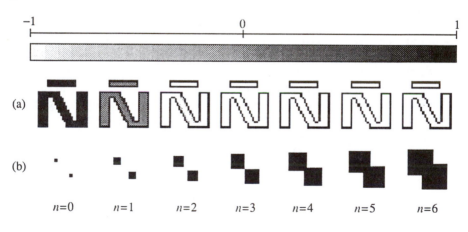

Figure 17.5 Illustrating image processing via a DTCNN
(a) Steady-state mode, (b) Transient mode.

The images presented in Figure 17.5 are snapshots of the net output transient evolution following the net initialisation at $n = 0$ with the input image. Pixel values are represented according to the grey scale on top of the images. For the border extraction process, the final output is available once the net has converged to a stable pattern, after $n = 3$ iterations. However, for Figure 17.5(b), each subsequent time instant represents a valid net outcome.

Application	NL	[A]	[B]	D	x(0)	u	x_s	u_s
Noise Filtering	S	$\begin{bmatrix}0&1&0\\1&2&1\\0&1&0\end{bmatrix}$	$\begin{bmatrix}0&0&0\\0&0&0\\0&0&0\end{bmatrix}$	0	M	DNC	0	DNC
Hole Filling	S	$\begin{bmatrix}0&1&0\\1&2&1\\0&1&0\end{bmatrix}$	$\begin{bmatrix}0&0&0\\0&4&0\\0&0&0\end{bmatrix}$	-1	$\begin{bmatrix}1&1&1\\1&1&1\\1&1&1\end{bmatrix}$	M	-1	DNC
Convex Corners Extraction	S	$\begin{bmatrix}0&0&0\\0&2&0\\0&0&0\end{bmatrix}$	$\begin{bmatrix}-1/4&-1/4&-1/4\\-1/4&4&-1/4\\-1/4&-1/4&-1/4\end{bmatrix}$	-3	M	M	DNC	-1
Borders Extraction	S	$\begin{bmatrix}0&0&0\\0&2&0\\0&0&0\end{bmatrix}$	$\begin{bmatrix}-1/4&-1/4&-1/4\\-1/4&4&-1/4\\-1/4&-1/4&-1/4\end{bmatrix}$	-2	M	M	DNC	-1
Connected Component Detection	S	$\begin{bmatrix}0&0&0\\1&2&-1\\0&0&0\end{bmatrix}$	$\begin{bmatrix}0&0&0\\0&0&0\\0&0&0\end{bmatrix}$	0	M	DNC	-1	DNC
Shadow Creation	S	$\begin{bmatrix}0&0&0\\0&2&2\\0&0&0\end{bmatrix}$	$\begin{bmatrix}0&0&0\\0&2&0\\0&0&0\end{bmatrix}$	0	$\begin{bmatrix}1&1&1\\1&1&1\\1&1&1\end{bmatrix}$	M	-1	DNC
Linear Thresholding	H	$\begin{bmatrix}0&0&0\\0&0&0\\0&0&0\end{bmatrix}$	ANY	ANY	DNC	M	DNC	ANY
Local Linear Transform	S	$\begin{bmatrix}0&0&0\\0&0&0\\0&0&0\end{bmatrix}$	ANY	ANY	DNC	M	DNC	ANY
Concentric Contours Extraction	H	$\begin{bmatrix}0&-1&0\\-1&3&-1\\0&-1&0\end{bmatrix}$	$\begin{bmatrix}0&0&0\\0&4&0\\0&0&0\end{bmatrix}$	-4	M	M	-1	DNC
Increassing Objects Step by Step	H	$\begin{bmatrix}1&1&1\\1&1&1\\1&1&1\end{bmatrix}$	$\begin{bmatrix}0&0&0\\0&0&0\\0&0&0\end{bmatrix}$	8	M	DNC	-1	DNC
Decreassing Objects Step by Step	H	$\begin{bmatrix}1&1&1\\1&1&1\\1&1&1\end{bmatrix}$	$\begin{bmatrix}0&0&0\\0&0&0\\0&0&0\end{bmatrix}$	-8	M	DNC	-1	DNC

Table 17.1 CNN templates.
M: Matrix of input pixels, NL: Non-linearity, H: Hard, S: Soft,
DNC: Do not care, x_s and u_s: State and input of border cells

An important issue for the steady-state operation mode is to determine conditions of the template parameters that guarantee convergence for any input. General necessary conditions have not yet been found for this, although convergence can be proven for important classes of templates. In particular, convergence to binary outputs, as required in most DTCNN image processing applications, is guaranteed provided that mA_{cc} is larger than unity and the templates fulfil some regularity conditions [10,12,28,32].

Since their first proposal in 1988, a large number of CNN applications have been suggested, for which the associated templates can be found in the references. For illustration purposes Table 17.1 summarises the templates and boundary conditions for some significant single-layer-DTCNN image processing functions.

17.3 Basic Building Blocks for SI DTCNNs

17.3.1 Cell operators

Figure 17.6 is an analogue computer conceptual-block diagram for a DTCNN cell, where circles indicate summation, triangles are used for signal scaling and the other two blocks represent delay and non-linear transformation. Equation (17.1) can be mapped into this figure; however, note that no weighting is performed at the cell input on neighbours' contributions, as would correspond to (17.1). Instead, each cell produces weighted replicas of its output and input for the different neighbours. To handle this, implementation templates following the concept of Figure 17.6 must be obtained by interchanging entries along all radial lines in the original template matrices, as illustrated below for a rectangular grid with unity neighbourhood radius parameter ($r = 1$),

$$\begin{bmatrix} a & b & c \\ d & e & f \\ g & h & i \end{bmatrix} \rightarrow \begin{bmatrix} i & h & g \\ f & e & d \\ c & b & a \end{bmatrix}$$

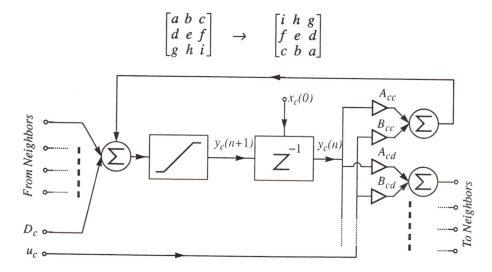

Figure 17.6 DTCNN conceptual cell diagram.

Summations in Figure 17.6 are performed in current-mode by rooting currents to a common node. Remaining operators for DTCNN implementation (signal scaling, non-linear transformation, delay and replication) are performed using current-mirrors, current switches [33,34] or a combination of these.

17.3.2 Current-mode replication and scaling

Figure 17.7(a) shows a generic multi-output current mirror, consisting of $N+1$ three-terminal abstract transconductors, represented by the symbol of Figure 17.7(b). We assume this generic three-terminal transconductor is characterised as follows:

$$i_2 = Pg(v_1,v_2)$$

$$i_1 \approx 0 \qquad\qquad (17.3)$$

where $g(.)$ is assumed invertable and P is a scale factor. We further assume that the impedance at port 2 is large ($di_2/dv_2 \approx 0$), so that variations of the voltage at this port has only a slight influence on the current i_2.

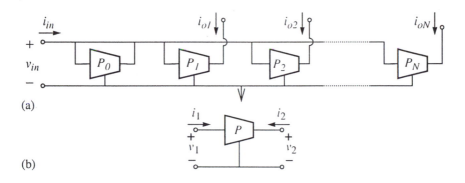

(a)

(b)

Figure 17.7 Current mirror (a) Concept (b) Primitive.

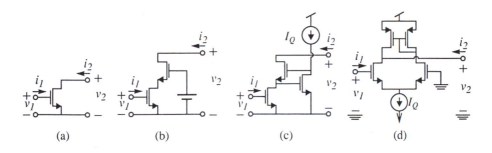

(a)   (b)   (c)   (d)

Figure 17.8 MOS transconductors for mirrors
(a) Single (b) Cascode (c) Regulated-cascode (d) Tunable circuit.

Figure 17.8 shows some MOS transconductors which apply as basic building blocks for current mirrors. Note that the use of an abstract block (represented by the symbol in Figure 17.7(b)) is convenient to render our presentation somewhat independent of transconductor topology. Also, most concepts discussed in the following, relating to the abstract transconductor, are valid for IC technologies other than CMOS.

Assume each transconductor in Figure 17.7(a) has a different parameter value, P_i $(0 \leq i \leq N)$; that all devices are matched (in the sense that they all have the same normalised non-linear characteristic) and that the influence of the voltage at port 2 is negligible. Under these assumptions, elementary analysis shows that output device non-linearities cancel (one by one) the non-linearity of the input device, yielding

$$i_{ok} = P_k g(v_{in}) = Pg\left[g^{-1}\left(\frac{i_{in}}{P_0}\right)\right] = \frac{P_k}{P_0} i_{in} \qquad 1 \leq k \leq N \qquad (17.4)$$

The replication operation is implemented in this manner. Also, as long as P_k/P_0 is designer-controlled (for instance, by changing the W/L of the transistors if the transconductor is a single MOS transistor), Figure 17.7(a) provides a simple way to scale currents.

Most elementary transconductors exhibit inherent rectification characteristics. For instance, when Figures 17.8(a)-(c) replace the generic transconductors in Figure 17.7(a), negative input currents, $i_{in} < 0$, drive the input voltage, v_{in}, below a cut-in value, so that the input and output devices become cut-off and hence the output currents are null, $i_{ok} = 0$[2]. Due to this, the scaled replication implemented by Figure 17.7(a) is unilateral; only positive currents are allowed. For bilateral operation, current-shifted biasing at the input and output nodes or complementary devices[3] must be used, as shown in Figures 17.9(a)-(b). An arrow has been added to the common terminal of the device symbols in Figure 17.9(b) to differentiate complementary devices.

Note that the bilateral current amplifiers of Figures 17.9(a)-(b) are of the inverting type: only negative scale factors are implemented. Non-inverting amplification (that is, positive scale factors) are achieved by cascading two bilateral mirrors, as shown in Figure 17.9(c).

17.3.3   Cell non-linearity

The non-inverting amplifier of Figure 17.9(c) is also used to implement the saturation non-linearity required for soft DTCNNs. It is achieved by using transconductor cut-off. Consider first $i_{in} < -I_Q$; then the mirror on the left cuts off and current i_{o1} is supplied only by the rail, making $i_{o1} = -I_Q$ and,

---

[2] To prevent inconsistencies when dealing with cut-off transistors, bear in mind that actual current sources are non-linear - their current becomes null for zero voltage drop.
[3] For instance, from drain to source, the NMOS draws positive currents, whilst the current drawn by the PMOS is negative.

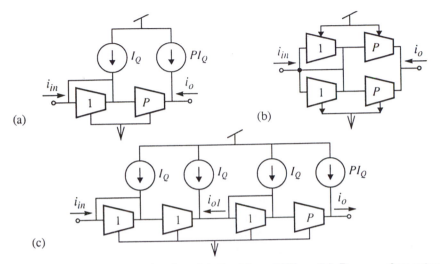

(a)

(b)

(c)

Figure 17.9 Bilateral weighting (a) By bias shifting (b) By complementary devices, (c) Non-inverting current amplifier.

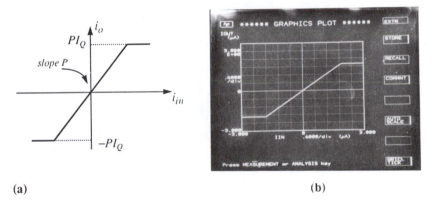

(a)

(b)

Figure 17.10 Soft non-linearity (a) Transfer characteristics of Figure 17.9(c) (b) Measurements from a silicon 1.6μm CMOS prototype (P=1).

hence, $i_o = -PI_Q$; in a similar way, the cut-off of the second mirror makes $i_o = PI_Q$ for $i_{in} > I_Q$. Thus, the saturation characteristics of Figure 17.10(a) are implemented. Note that the inner piece has a slope of value P. The non-linearity required for soft DTCNNs is readily implemented by making $P = 1$.

Figure 17.10(b) illustrates the practical performance of this circuit. It shows measurements from a silicon prototype in a 1.6μm n-well single-poly CMOS technology, using the transconductor of Figure 17.8(b) and $I_Q = 2\mu A$. Measurements are made via the HP4145B semiconductor parameter analyser, whose resolution is in the range of pA's. Cell linearity is very good (less than 1% deviation) over the whole current range; on the other hand, non-linearity is quite abrupt.

Figure 17.9(c) could also be used for hard DTCNNs if P were made large enough. In this case, a stage with $P = 1$ should be cascaded to limit the output current to $\pm I_Q$. A much simpler and more convenient circuit uses a current-switch-based comparator, like that represented in Figure 17.11(a), capable of yielding very high resolution and low offset [34]. In this structure, positive input currents force M_p ON and yield a voltage of about $-V_{Tp}$ at node v_{o1} so that node v_{o2} is at V_{DD} and M_s becomes ON, yielding $i_o = I_Q$. On the contrary, negative input currents force $v_{o1} \cong V_{Tn}$ turning M_s OFF and yielding $i_o = -I_Q$ (see [2]). Also, since in the quiescent point (for $i_{in} = 0$) neither M_p or M_n conduct, the input impedance in this quiescent point is purely reactive, and consequently the sign of the input current is detected with very high resolution. Figure 17.11(b) shows the v_{o2} - i_{in} characteristics measured from a silicon prototype, and illustrates the operation of this circuit. Inverters are realised in this prototype using a simple two-transistor complementary structure. Measured resolution and offset for this very simple structure are 10pA's.

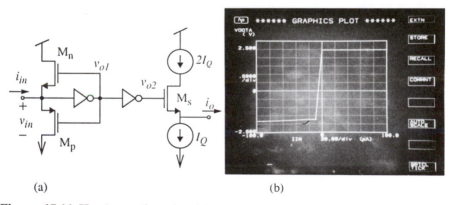

(a)   (b)

Figure 17.11 Hard non-linearity (a) Implementation using a current switch (b)Measurements from a silicon prototype.

17.3.4   Delay block

The delay operator is readily realised in current-mode by cascading two current track-and-hold circuits [35], as shown in Figure 17.12(a). This block requires two non-overlapping complementary clock signals and relies on the analogue memory operation performed by the capacitors (nominal or parasitics) at the transconductor inputs. An interesting observation concerning this circuit is that it allows concurrent implementation of the full delay operator and the cell non-linearity required for soft DTCNNs. This is shown by taking into account transconductor cut-off, which yields

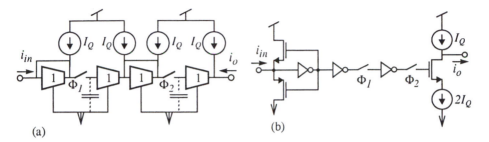

(a)   (b)

Figure 17.12 Delay/Non-linearity operators (a) Concurrent full-delay and soft limitation (b) Full-delay/hard limiter block.

$$i_o(n+1) = \frac{I_Q}{2}\left(\left|\frac{i_{in}(n)}{I_Q} + 1\right| - \left|\frac{i_{in}(n)}{I_Q} - 1\right|\right) \qquad (17.5)$$

Thus, since delay and saturation are realised in the same circuit, a separate non-linear block need not be considered in the cell architecture. In a similar way, delay and non-linearity can also be concurrently implemented for hard DTCNNs, using the circuit of Figure 17.12(b).

17.3.5 Current mode CNN conceptual cell schematics

Figure 17.13(a) shows a generic cell schematic diagram for current-mode soft DTCNNs. A corresponding schematic diagram for hard DTCNNs is represented in Figure 17.13(b). In each figure, the switches labelled *ST* and $\overline{ST}$ are used for initialisation.

Note that Figures 17.13(a) and (b) contain only the parts of the cell corresponding to the evaluation of the cell state (x_c) and the generation of the output variable replicas for neighbours $(A_{cd}y_c)$. Generation of scaled replicas of the input cell current (u_c) is straightforward using Figures 17.7 and 9. Note also that Figures 17.13(a) and (b) include only two outputs per cell: one for a generic positive weight and other for a generic negative weight. Additional positive and/or negative outputs are obtained by using the replication concept. Finally, for both figures, the bias currents displayed can be obtained by replicating a master current.

For the simplest case of single transconductors and switches and, for instance, for the application of connected component detection (see Table 17.1), a switched-current DTCNN cell can be implemented with about 20 transistors. This rather low complexity, together with the simple layout and, in particular, the possibility of interconnecting cells by abutment yields pixel densities ranging from 60 to more than 160 cells/mm² in a standard 1.6μm CMOS technology.

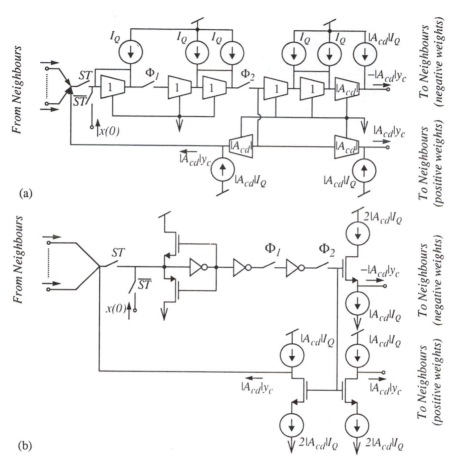

Figure 17.13 DTCNN cell schematic diagram (a) Soft cell non-linearity
(b) Hard cell non-linearity.

17.4 CMOS Current-Mode CNN Design Issues

17.4.1 *CMOS mirror circuits: static non-idealities and sizing equations*

Current mode DTCNNs operation is degraded by random and systematic error sources. Random errors are caused by statistical variations of the technological parameters across the die (mainly $K = \mu C_{ox}$ and V_T in CMOS) and can be attenuated using large devices [36], careful layout [37] and proper biasing. On the other hand, systematic errors are corrected by transconductor choice and sizing.

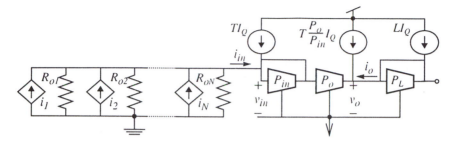

Figure 17.14 Generic mirror and boundary circuit.

Let us focus on systematic static current mirror errors. Figure 17.14 presents a generic mirror and driving and loading devices, and will be used to discuss these errors. We assume the device is nominally intended to yield output to input current scaling by a factor P_o/P_{in}. Two major error sources can be identified:

(a) Input-output voltage mismatch $(v_{in} \neq v_o)$ at the bias point, defined as the point at which transconductors sink only their bias currents, $(i_{in}=0)$.

(b) Finite R_o/R_{in} ratios, where R_o and R_{in} represent the current mirror output and input resistances (for an ideal mirror, R_o/R_{in} should be infinitely large).

Voltage mismatch produces current offset $(i_o \neq 0$ for $i_{in} = 0)$, originated by the transconductor current dependence on the output voltage. This offset is eliminated by forcing the current density to be the same in the mirror input transconductor and loading device. This is achieved in the case of Figure 17.14 by specifying

$$\gamma = \frac{L}{P_L} = \frac{T}{P_{in}} \qquad (17.6)$$

which must hold for all transconductors in the net.

A finite R_o/R_{in} ratio causes current gain error due to spurious current division at the mirror input and output nodes. For simplicity, we shall assume equal output resistance for all mirrors in Figure 17.14 (actually output and input resistance depend on the current level) obtaining

$$\frac{\dfrac{i_o}{N}}{\displaystyle\sum_{j=1}^{N} i_j} = \frac{P_o}{P_{in}} \frac{1}{\left[1 + \dfrac{NR_{in}}{R_o}\right]\left[1 + \dfrac{R_{in}}{R_o}\right]} \approx \frac{P_o}{P_{in}}\left[1 - \frac{(N+1)R_{in}}{R_o}\right] \qquad (17.7)$$

where N is the number of mirrors driving node vin. It is seen that the gain error is inversely proportional to R_o/R_{in}, and increases proportionally with N. A more precise evaluation, taking into account R_{in} and R_o variations with current level, yields the same proportionality.

Since N in (17.7) can be a large number (up to 11 for the corner and border detection templates on a rectangular grid net with $r = 1$), the importance of the current gain error cannot be underestimated. The error is especially significant if small dimension single-transistor transconductors are used, due to the very low Early voltages associated with short channel transistors. This can be corrected by increasing device size (channel length) but this does not yield optimum area and speed for DTCNN implementations. For improved R_o/R_{in} figures with short channel devices, cascode transconductors, feedback transconductors, or a combination of both must be used [38].

It is illustrative to discuss in some detail the area advantages associated to using these improved transconductors. Figure 17.15(a) shows a simple CMOS mirror and Figure 17.15(b) shows a mirror using a cascode transconductor. Each figure includes the complementary devices used for biasing. Design parameters are displayed in the figures: W, L_n and L_p for the simple mirror and W, L and V_{CAS} for the cascode mirror. Figure 17.15(c) shows a circuit to provide this cascode voltage.

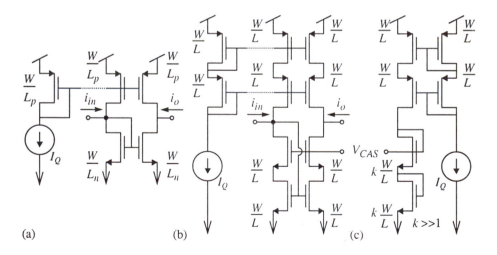

Figure 17.15 Bilateral CMOS mirrors (a) Simple structure (b) Cascode structure (c) Cascode voltage generation.

We assume that Early voltages are proportional to the channel length: $V_{An} = a_n L_n$, $V_{Ap} = a_p L_p$. In Figure 17.15(a) channel lengths are different for NMOS (L_n) and PMOS (L_p), to obtain equal Early voltages for both devices and hence optimise R_o/R_{in}. In the case of Figure 17.15(b), R_o/R_{in} is intrinsically much larger and all channel lengths are made equal $(L_n = L_p)$ for simpler design. For simplicity, we have also assumed that all transistors have the same channel width, W.

Table 17.2 gives sizing equations for Figures 17.15(a) and (b). These equations are intended to ensure that the mirrors handle the whole input current range with minimum distortion, and use the smallest possible devices. The W expressions given in the table correspond to a bias current I_Q; W values for larger currents (associated to scaled replication) are calculated taking into account the requirement for equal current density in all transconductors given in (17.6).

	Channel widths $(W = W_n = W_p)$	Lengths	Bias voltage
Single device mirror	$W = \dfrac{4I_Q L_n}{(V_{DD} - V_{SS} - V_{Tn})^2} \left[\sqrt{\dfrac{1}{K_n}} + \sqrt{\dfrac{\alpha_n}{2K_p \alpha_p}} \right]^2$	$L_p = \dfrac{\alpha_n}{\alpha_p} L_n$	
Cascode mirror	$W = \dfrac{16 I_Q}{K_n V_{TN}^2} L_n$	$L_p = L_n$	$V_{CAS} \approx V_{SS} + 2V_{Tn}$

Table 17.2 Sizing equations for CMOS mirrors.

Note that the sizing equations are parameterised by L_n, which is chosen by the designer to control R_o/R_{in} and the channel area. For Figure 17.15(a) the result of evaluation of these figures is

$$\frac{R_o}{R_{in}} = 1 + \sqrt{\frac{K_n W L_n}{2I_Q}} \, \alpha_n \qquad Area = 2WL_n \left[1 + \frac{\alpha_p}{\alpha_n} \right] \qquad (17.8)$$

and the following is obtained for the cascode mirror of Figure 17.15(b):

$$\frac{R_o}{R_{in}} = \frac{2WL_n}{I_Q} \left[\frac{\alpha_p^2 \alpha_n^2 K_n \sqrt{K_p K_n}}{\alpha_n^2 K_n + \alpha_p^2 \sqrt{K_p K_n}} \right] \qquad Area = 8WL_n \qquad (17.9)$$

Figure 17.16 shows the current gain relative error per cell versus the total cell area of a soft-DTCNN for connected component detection (see Table 17.1), with single and cascode mirrors. The technology is a standard digital n-well 1.6µm CMOS; the scale is logarithmic. Two curve families are shown: the top family is for the single transistor mirror, and the bottom family for the cascode. The parameter for each family is the rail current I_Q, which varies from 0.25µA (bottom curve in each family) to 128µA (top curves), and doubles from each curve to the next one. As shown, a simple current mirror requires large area to achieve an acceptable margin of error. On the other hand, cascode mirrors allow the use of shorter channel-length devices, resulting in much higher area efficiency and speed.

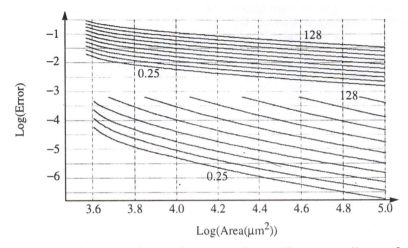

Figure 17.16 Current gain relative errors (per cell) versus cell area for a CCD soft-DTCNN and different rail currents.

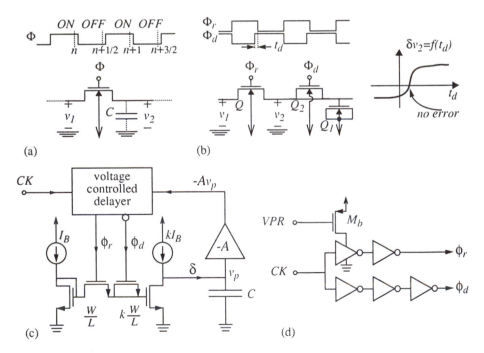

Figure 17.17 Feedthrough and cancellation techniques (a) Simple analogue switch (b) Analogue switch with extra MOS capacitor and dummy transistor (c) Adaptive feedthrough cancellation scheme (d) Delayer.

17.4.2   CMOS current mode dynamic operators non-idealities

Assume the analogue switch consists of a single MOS transistor, as shown in Figure 17.17(a). Correct operation of this circuit depends on the capacitor's ability to maintain the input voltage when the analogue switch turns off. Assume that voltage decay due to unavoidable leakage is negligible (equivalently that the clock frequency is large enough). The largest cause of error in the delay operation arises due to the necessity to evacuate the MOS channel charge during the switch turn-off transient process, giving

$$v_2(n+1/2) = v_1(n) + \frac{\Delta q}{C} \qquad (17.10)$$

where Dq represents the part of the evacuated charge delivered to the capacitor node. This error, generically called feedthrough error [39], is very large (up to 20% or more of the bias current) if small geometry transconductors are used. This effect can be attenuated with several techniques. A simple choice is to include a nominal capacitor at the switch output node. Since neither linearity nor accuracy is required in the capacitance for this purpose, a shorted transistor may be used, represented by Q_1 in Figure 17.17(b). This technique can significantly lower the error without a severe area penalty.

Much lower feedthrough error is achieved using a dummy transistor, represented by Q_2 in Figure 17.17(b), and exploiting the fact that the feedthrough is a monotone function of the delay between the control signals of the real switch and the dummy transistor and, in particular, that this function is null for some specific delay value. This property can be exploited to design an adaptive circuit for feedthrough cancellation, as shown in Figure 17.17(c). Output of the current-mode track-and-hold stage is either null or equal to the charge injection error (disregarding static errors). This error is integrated at capacitor C, whose voltage is amplified to drive a voltage controlled delayer. Implementation of this delayer is easily accomplished using a MOS transistor to control the biasing of a simple CMOS inverter during its transitions, as shown in Figure 17.17(d). This technique has achieved errors as little as 0.3% of the bias current on silicon prototypes with mirror transistor areas of $12\mu m^2$.

17.4.3   Programmability issues for current-mode blocks

Although fixed weight DTCNN chips can be useful as stand-alone units for image processing tasks, programmability is an important feature for general purpose applications. Programmability can either be discrete, where control signals are digital, or continuous, where controlling signals are analogue. Discrete programmability can be incorporated very simply, through analogue multiplexing of current contributions from different mirrors. These mirrors can either implement fixed templates (with applications, for instance, in cases where well-defined tasks must be performed sequentially [40,41]), or be binary-weighted (for more general applications). Binary programmability provides ease of controllability and accurate results, at the cost of a strong area penalty.

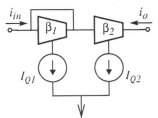

Figure 17.18 Programmable current-mirror using tunable transconductors.

A simple way to achieve analogue programmability is using tunable transconductors. Figure 17.18 shows a programmable current mirror using the transconductor of Figure 17.8(d). Two different situations arise, depending on whether transistors operate in weak or strong inversion,

$$\frac{i_o}{i_{in}}\bigg|_{strong} = \sqrt{\frac{\beta_2 I_{Q2}}{\beta_1 I_{Q1}}} \qquad \frac{i_o}{i_{in}}\bigg|_{weak} = \frac{I_{Q2}}{I_{Q1}} \qquad (17.11)$$

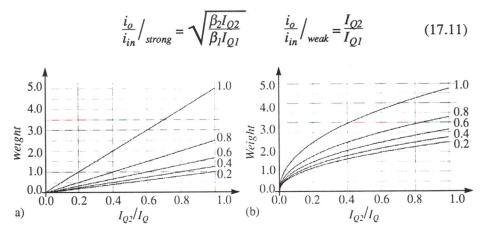

Figure 17.19 Weight variation with I_{Q2}/I_Q for different values of I_{Q1}/I_Q in Figure 17.18 (a) Weak inversion ($I_Q = 10$nA) (b) Strong inversion ($I_Q = 50\mu$A).

The dependence is linear for weak inversion, which provides larger weight adjustment ranges. This is presented in Figure 17.19, showing the current weight as a function of I_{Q2}/I_Q for different values of I_{Q1}/I_Q, where I_Q is a normalisation factor of value 10nA for weak inversion and 50μA for strong inversion. Non-linearity cancellation is exact in weak inversion due to the exponential nature of current-to-voltage characteristics, while it is only approximate for strong inversion: non-linearity in weak inversion is less than 1% up to $i_o = I_{Q2}$, while the corresponding value for strong inversion is $i_o = 0.13I_{Q2}$. Drawbacks of weak inversion are low accuracy due to mismatch and reduced speed. These can be overcome by using CMOS compatible lateral BJTs [42], which exhibit an exponential feature for larger current ranges and excellent matching properties [43]. Other alternatives for tunable CMOS current mirrors may be found elsewhere [44,45].

17.4.4 Bias current selection: area, power and reliability

Transistor geometry factors, static gain error due to finite R_o/R_{in} and power dissipation, increase with I_Q. Hence, a bias current as small as possible should be chosen. The issue is to identify the minimum feasible rail current value. The lowest limit is certainly established by leakage (about 10pA). However, a more restrictive bound exists due to MOS transistor mismatch [36,46] and Early voltage (V_A) degradation with channel length.

Mismatch is mainly produced by variations of the threshold voltage (V_T) and large signal transconductance $(\beta = KW/L)$ of equally designed transistors in the same chip. Standard deviations for these parameters have two major components: one inversely proportional to the square root of the channel area and the other proportional to the distance between devices. Results in [36] demonstrate that the distance-dependent component is negligible for devices with a channel area less than about $100\mu m^2$. Since for bias currents below about $50\mu A$ and cascode mirrors, the device area, calculated from Table 17.2 is well below this bound and so the distance dependent component need not to be considered for current mode CNNs.

Another important consideration is that, for a given $s(V_T)$ and $s(\beta)/\beta$, the ratio $s(I)/I$ in MOS transistors is inversely proportional to $v_{gs} - V_T$. This means that once W/L factors have been set to achieve acceptable mismatch levels, bias current cannot be decreased too far below the upper bound given by the equations in Table 17.2, since this would produce a low v_{gs} voltage at the bias point, with the corresponding high $s(I)/I$. Hence, mismatch considerations establish bounds for both minimum area and power trends.

For example, we have obtained 100% success (out of 30 trials) for full device level Monte Carlo simulation of a connected component detector (CCD) DTCNN with 16 cells in a row, using the cell diagrams of Figure 17.13(a). Unitary transistor geometries of $W/L = 4\mu m/3.2\mu m$ for both n and p-channel devices, cascode transconductors and a bias current $I_Q = 2\mu A$ were used. Further geometry reduction has not been considered, since minimum contact size ($4\mu m$ with surrounding diffusion in the 1.6μm n-well technology used) does not allow a significant area reduction anyway.

In these Monte Carlo simulations, global biasing voltages (the bias stage is also simulated) are used for current reference generation. Dispersion due to mismatch among transistors of different current sources did not produce critical results. Thus, global biasing is a fair approach.

However, in some cases, to prevent switching noise coupling, it will be convenient to include an independent current reference in each cell. Regarding this, Montecarlo simulation of an entire CCD system in which cell current references have a standard deviation larger than 5% have shown 100% success.

17.5 Input-Output Strategies

High input-throughput in image-processing DTCNN chips is achieved by implementing photosensor devices directly at the processing cells. Since photosensor output signals are in the form of current, they are easily interfaced to current-mode cells. Also, since in most application cases either $x_c(0)$ or u_c do not convey signal information, only one photosensor per cell is required.

Figure 17.20 shows the layout and measured performance of a CMOS photosensor, using the compatible pnp vertical transistor in a n-well 1.6μm technology. The characteristics in Figure 17.20(b) have been measured by making the emitter-to-collector voltage vary slowly from 5V to 0V. In a first swing, the light is constant under typical laboratory lighting. In a second swing, the light is decreased progressively while the voltage changes. Two sets of characteristics are shown for two different layout sizes. It is seen that the CMOS compatible photosensor exhibits good resolution, with output current changing from μAs (this level increases with the area of Figure 17.20(a)) to few pA's.

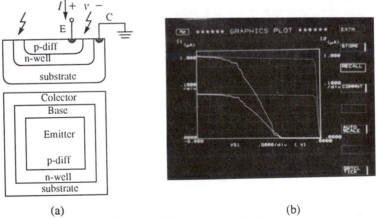

(a) (b)

Figure 17.20 CMOS phototransistor
(a) Conceptual layout, (b) i-v characteristics.

In case photosensors are not included, I/O strategies other than parallel loading and/or downloading through bonding pads must be considered. However, regardless of whether the initialisation process is parallel or serial, some control circuitry must be included at cells to isolate initialisation and computation processes. In the fully parallel input case this is handled by using one global signal, *ST*, to control the switches. Clocks signals must be kept at a high state while *ST* is high.

Serial cell loading requires more involved control circuitry: local logic must be included in each cell, and additional control signals must be employed. This logic should be implemented by serial/parallel switches, to avoid noise coupling from switching digital gates. Serial loading also confronts the designer with electrical issues related to the need to maintain each cell state and input, while remaining cells are initialised.

Net downloading processes must be performed serially in the more general case and, hence, local logic and control signals are also required. However, this additional circuitry can be basically the same than that used for initialisation. Also, since downloading can be performed while the network remains in operation (with the help of an additional output replication branch), leakage and charge injection errors are not of concern in this case.

As an example, Figure 17.21 depicts a high level diagram of a CNN chip with a cell by cell loading and downloading strategy. While *ST* is low, cells are initialised one by one through the global data path *DATA*, which must be driven by a current source. The row and column selection signals are generated by a digital $2P$ $(P = log_2N)$ binary counter and two $P:2^P$ binary to one-hot decoders. In a design following this strategy, only 8 external connections are required (6 if digital and analogue supplies share the same pin).

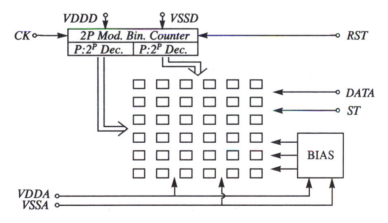

Figure 17.21 Architecture of a serial I/O CNN chip.

With these I/O techniques, and if input signals are binary, current mode CNN chips can be tested with digital equipment. For this purpose, input and output signals must be voltages. Figure 17.22(a) shows a simple binary voltage-to-current converter, which is used to interface digital test equipment with network input. This circuit is also used for the implementation of the offset terms and border cell contributions.

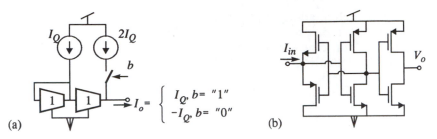

Figure 17.22 (a) One bit voltage-to-current converter with $\pm I_Q$ output
(b) High resolution current comparator [34].

The parasitic capacitance of the global data path is typically very high in large networks. Hence, the loading process may be very slow if low rail currents are used. The speed of this process can be increased by making the current rails of Figure 17.22(a) larger than the cell current rail, I_Q. This excessive current must be limited at the cell input by a saturation non-linearity.

Output data can also be transformed into voltage either by a simple CMOS inverter or, for low rail currents I_Q, by a faster and highly sensitive current comparator as the one shown in Figure 17.22(b) [34].

17.6 Discussion and Perspectives

Figure 17.23 shows experimental results obtained from a CMOS 1.6μm prototype of a 9x9 switched-current DTCNN chip which can be reconfigured via local logic to implement the templates for noise removal, feature extraction (borders and edges), shadow detection, hole filling, and CCD on a rectangular grid with unity neighbourhood radius. The rail current used in the prototype is 10μA. Switches are implemented using minimum dimension n-channel transistors. Minimum dimension dummy transistors are used for adaptive feedthrough cancellation, no feedthrough cancellation capacitor is used. The clock frequency is 5MHz. Serial loading/unloading process on a cell by cell basis is used. Convergence is achieved in about four clock cycles, which corresponds to about 1μs computation time. For representation convenience, the results in Figure 17.23 correspond to electrical simulations. Measurements taken from silicon prototypes are in full accordance to the simulation results.

To conclude, we see that the combination of switched-current techniques and cellular neural networks allows simple design of robust and fast analogue massive processing ICs for image processing tasks. Although the techniques discussed in the chapter are mainly oriented to binary output images, the use of the soft-non-linearity in transient mode of operation produces non-binary output images as well. Also, the computational capabilities of these massive processing switched-current chips can be

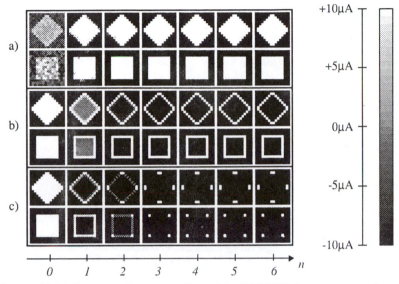

Figure 17.23 Results of a reconfigurable DTCNN image processing chip.

significantly enhanced by incorporating the newest CNN model improvements: (a) Time-variant and non-linear templates [27], (b) Multilayer DTCNNs [26], (c) Multipath systems [41]. Just to mention a few, this would allow applications such as radon transform, halftoning [28], automatic vehicle guidance [40], skeletonisation [26] and motion detection and estimation [21].

References

[1]   E. Sanchez-Sinencio and C. Lau, Eds, *Artificial Neural Networks*, New York: IEEE Press, 1992.

[2]   J. M. Zurada, *Introduction to Artificial Neural Networks*, St. Paul: West Publishing, 1992.

[3]   E. A. Vittoz, "Future of analog in the VLSI environment," in *Proc. IEEE International Symposium on Circuits and Systems*, pp. 1372-1375, May 1990.

[4]   J. J. Hopfield, "Neural networks and physical systems with emergent computational capabilities," in *Proc. Natl. Acad. Sci. USA*, vol. 79, pp. 2254-2258, 1982.

[5]   B. Linares-Barranco et al, "A modular T-mode design approach for analog neural network hardware implementations," *IEEE J. Solid-State Circuits*, vol. 27, pp. 701-713, May 1992.

[6]   R. Dominguez-Castro et al, "Analog neural programmable optimizers in CMOS VLSI technologies," *IEEE J. Solid-State Circuits*, vol. 27, pp. 1110-1115, July 1992.

[7]   L. O. Chua and L. Yang, "Cellular neural networks: theory," *IEEE Trans. Circuits and Systems*, vol. 35, pp. 1257-1272, Oct. 1988.

[8]   L. O. Chua et al, "Signal processing using cellular neural networks," *J. of VLSI Signal Processing*, vol. 3, pp. 25-51, Jan. 1991.

[9]   J. A. Nossek et al, "Cellular neural networks: Theory and circuit design," *Int. J. Circuit Theory and Applications*, vol. 20, pp. 533-554, Sept. 1992.

[10]   A. Rodriguez-Vazquez et al., "Accurate design of analog CNN in CMOS digital technologies," in Proc. *IEEE Int. Workshop on Cellular Neural Networks and Their Applications,* pp. 273-280, Dec. 1990.

[11]   H. Harrer and J.A. Nossek, "Discrete-time cellular neural networks," *Int. J. Circuit Theory and Applications*, vol. 20, pp. 453-468, Sept. 1992.

[12]   S. Hui and H. Zak, "Dynamical analysis of the brain-sate-in-a-box (BSB) neural model," *IEEE Trans. Neural Networks*, vol. 3, pp. 86-94, Jan. 1992.

[13]   T. Toffoli and N. Margolus, *Cellular Automata Machines: A New Environment for Modeling*, Cambridge: MIT Press, 1987.

[14]   L. O. Chua and B. Shi, *Exploiting Cellular Automata in the Design of Cellular Neural Networks for Binary Image Processing*, UCB/ERL M89/130, University of California at Berkeley, Nov. 1989.

[15]   J. C. Russ, *The Image Processing Handbook*, Boca Raton: CRC Press, 1992.

[16]   L. O. Chua and L. Yang, "Cellular neural networks: applications," *IEEE Trans. Circuits Syst.*, vol. 35, pp. 1273-1290, Oct. 1988.

[17]   T. Matsumoto et al, "CNN cloning template: connected component detector," *IEEE Trans. Circuits Syst.*, vol. 37, pp. 633-635, May 1990.

[18]   T. Matsumoto et al, "CNN cloning template: hole filler," *IEEE Trans. Circuits Syst.*, vol. 37, pp. 635-638, May 1990.

[19]   T. Matsumoto et al, "CNN cloning template: shadow detector," *IEEE Trans. Circuits Syst.*, vol. 37, pp. 1070-1073, Aug. 1990.

[20]   T. Roska (ed.), *Proc. IEEE Int. Workshop on Cellular Neural Networks and Their Applications*, Budapest, Dec. 1990.

[21]   J.A. Nossek (ed.), *Proc. of the IEEE Int. Workshop on Cellular Neural Networks and Their Applications*, Munich, Oct. 1992.

[22]   K. Slot, "Cellular neural network design for solving specific image-processing problems," *Int. J. Circuit Theory and Applications*, vol. 20, pp. 629-637, Sept. 1992.

[23]   T. Roska et al, "Detecting moving and standing objects using cellular neural networks," *Int. J. Circuit Theory and Applications*, vol. 20, pp. 613-628, Sept. 1992.

[24]   H. Suzuki et al, "A CNN handwitten character recognizer," *Int. J. Circuit Theory and Applications*, vol. 20, pp. 601-612, Sept. 1992.

[25] S. Paul et al, *Mapping Nonlinear Lattice Equations onto Cellular Neural Networks*, UCB/ERL M92/42, University of California at Berkeley, May 1992.

[26] L. O. Chua and B. Shi, "Multiple layer cellular neural network: A tutorial," in *Algorithms and Parallel VLSI Architectures*, North Holland, 1991.

[27] T. Roska and L. O. Chua, "Cellular neural networks with non-linear and delay-type template elements and non-uniform grids," *Int. J. Circuit Theory and Applications*, vol. 20, pp. 469-482, Sept. 1992.

[28] H. Harrer, *Discrete-Time Cellular Neural Networks*, PhD dissertation, Technical University of Munich, Aachen: Verlag Shaker, 1992.

[29] G. Seiler, *Symmetry Properties of Cellular Neural Networks on Square and Hexagonal Grids*, TUM/LNS/TR/90/2, Technical University of Munich, May 1990.

[30] J. M. Cruz and L. O. Chua, "A CNN Chip for Connected Component Detection," *IEEE Trans. Circuits Syst.*, vol. 38, pp. 812-817, July 1991.

[31] K. Halonen et al, "Programmable analogue VLSI CNN chip with local digital logic," *Int. J. Circuit Theory and Applications*, vol. 20, pp. 573-582, Sept. 1992.

[32] N. Fruhauf et al, "Convergence of reciprocal time-discrete cellular neural networks with continuous nonlinearities," in *Proc. IEEE Int. Workshop on Cellular Neural Networks and Their Applications*, pp. 106-111, Oct. 1992.

[33] Z. Wang, "Novel pseudo RMS current converter for sinusoidal signals using a CMOS precision current rectifier," *IEEE Trans. on Instrumentation and Measurement*, vol. 39, pp. 670-671, Aug. 1990.

[34] R. Dominguez-Castro et al., "High resolution CMOS current comparators," in *Proc. European Solid-State Circuits Conference*, pp. 242-245, Sept. 1992.

[35] J. B. Hughes, "Switched currents - A new technique for analog sampled-data signal processing," in *Proc. IEEE International Symposium on Circuits and Systems*, pp. 1584-1587, May 1989.

[36] M. J. M. Pelgrom et al, "Matching properties of MOS transistors," *IEEE J. Solid-State Circuits*, vol. 24, pp. 1433-1440, Oct. 1989.

[37] E. A. Vittoz, "The design of high performance analog circuits on digital CMOS chips," *IEEE J. Solid-State Circuits*, vol. 20, pp. 657-665, June 1985.

[38] E. Sackinger and W. Guggenbuhl, "A high-swing high-impedance MOS cascode circuit," *IEEE J. Solid-State Circuits*, vol. 25, pp. 289-298, Feb. 1990.

[39] C. Eichenberger, *Charge Injection in MOS-Integrated Sample-and-Hold and Switched-Capacitor Circuits*, Zurich: Hartung-Gorre Series in Microelectronics, vol. 3, 1989.

[40] G. Eros et al, "Optical tracking system for automatic guided vehicles using cellular neural networks," in *Proc. IEEE Int. Workshop on Cellular Neural Networks and their Applications*, pp. 216-221, Oct. 1992.

[41] L. O. Chua et al., "Some novel capabilities of CNN: Game of life and examples of multipath algorithms," in *Proc. IEEE Int. Workshop on Cellular Neural Networks and their Applications*, pp. 276-281, Oct. 1992.

[42]   E. A. Vittoz, "MOS transistors operated in the lateral bipolar mode and their application in CMOS technology," *IEEE J. Solid-State Circuits*, vol. 18, pp. 273-279, June 1983.

[43]   T. Pan and A. A. Abidi, "A 50dB variable gain amplifier using the parasitic bipolar transistors in CMOS," *IEEE J. Solid-State Circuits*, vol. 24, pp. 951-961, Aug. 1989.

[44]   K. Bult and H. Wallinga, "A class of analog CMOS circuits based on the square-law characteristics of an MOS transistor in saturation," *IEEE J. Solid-State Circuits*, vol. 22, pp. 357-365, June 1987.

[45]   E. A. Klumperink and E. Seevinck, "MOS current gain cells with electronically variable gain and constant bandwidth," *IEEE J. Solid-State Circuits*, vol. 24, pp. 1465-1467, Oct. 1989.

[46]   C. Michael and M. Ismail, "Statistical modeling of device mismatch for analog MOS integrated circuits," *IEEE J. Solid-State Circuits*, vol. 27, pp. 154-166, Feb. 1992.

Chapter 18

Test for Switched-Current Circuits

Paul Wrighton, Gaynor Taylor, Ian Bell
and Chris Toumazou

18.1 Introduction

The concept of test for electronic circuits originated with mass production circuit boards, progressed into digital ICs and is now becoming established in analogue and mixed analogue/digital ASIC's.

Digital test has enjoyed (and is still enjoying) many years of intensive research providing test and design engineers with tools and techniques for the generation of near optimum and minimal length test sets, to excite and exhibit erroneous behaviour in faulty devices. The field of analogue (and more recently mixed analogue/digital) device test is however still very much in its infancy and remains somewhat of a 'black art'.

Until recently digital device test has relied upon voltage monitoring techniques, that is stimuli are chosen to display an incorrect logic level at a circuit output in case of a fault [1]. These methods have been shown to provide less than 100% fault coverage, in certain cases, with some faults remaining undetectable. An alternative approach, power supply quiescent current (Iddq) monitoring [2,3] has been shown to provide better coverage. The principle of Iddq monitoring relies on the complementary design of CMOS devices, which have nano amp quiescent state (logical 1 or 0) currents, but which draw currents several magnitudes greater, (tens of micro amps,) when in transition. The presence of a fault is indicated by a raised Iddq level [4]. Figure 18.1 shows this principle for the case of a CMOS inverter.

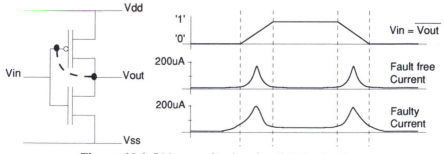

Figure 18.1 Iddq monitoring in digital circuits.

The concept of current monitoring has been extended from digital devices into the analogue and mixed analogue/digital field [5,6,7,8]. Unlike digital circuits, analogue devices have no fixed voltage levels, but are merely restricted by the power supplies thus supply current is instantaneous. Current changes exhibited due to faulty behaviour are not as pronounced as those in the digital case, but are none the less significant. They are dynamic in nature and thus require analogue current sampling. There are problems associated with the current testing of mixed analogue/digital devices, the principle one being that digital switching noise masks the delicate analogue signal. Partitioning digital and analogue blocks for test is one method of circumventing this, but introduces further problems, such a silicon overhead (for the partitioning and associated circuitry) and considerable increases in the overall test time as each block must be tested separately and reconfigured. A far more elegant solution is the introduction of Built In Current (BIC) sensors, placed in the analogue cells between the device itself and a power line, These provide a perfect copy of the current drawn from the analogue cell, free from digital noise [9,10,11].

The switched current (SI) devices which form the subject of this book are a subset of analogue devices. Test of SI circuits has, thus far, seen little interest [12]. It is the purpose of this chapter to suggest a method by which it is hoped that the basis of switched current test can be established. The basic design concept lends itself readily to test by current monitoring as described above and it will be shown that the addition of a relatively small amount of test circuitry can improve testability and simplify test strategy.

18.2 Basic concepts for the test of SI cells

The basis of any switched current-mode environment is essentially an array of current sensors (copiers and memory cells - Chapter 3) and little additional test circuitry is required. Figure 18.2 shows a typical second generation cascode memory cell element from which SI devices can be developed and comprises simply five transistors and two current sources. The principle of test relies essentially on that of its operation; that of charge storage and transportation from gate source capacitances of the first memory transistor M_1 to the next in the circuit. Each memory transistor is biased with M_2 & M_3, and switched to subsequent cells with M_4 & M_5. Since SI devices are essentially sequential, a fault exhibited anywhere in the circuit should be carried sequentially through the circuit and hence should be observable at the output.

Analogue circuit test based upon current monitoring essentially relies on comparison of current signatures from a known good source and the device under test. The good signature can be obtained from a "golden chip", (a device that has passed stringent parameter and functionality checks,) or via simulation. Signal processing errors will limit the accuracy and sensitivity of fault detection, which in SI cells is mainly due to finite conductance and charge injection (see Chapter 4). This can be taken a stage further where both signatures are generated together on chip, giving a form of Built-In Self Test (BIST).

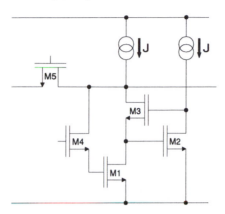

Figure 18.2 Second Generation Cascode Memory Cell.

18.2.1 *Injection and simulation of faults in SI cells*

Faults in integrated circuits can manifest in a variety of ways. Pinhole faults in oxides or polycrystaline silicon causing bridges, bridges from one metal track to another or gaps (opens) in the tracks, floating gates, and ion implantation defects, comprise a few of the major causes of faulty behaviour. Thus there are many fault models defined [13,14] from the abstract basic stuck at fault in a digital circuit, to the physical defects of gate oxide shorts, floating gates and pinhole faults.

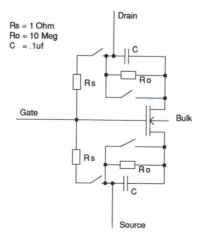

Figure 18.3 Fault model.

A distinction is made between a catastrophic fault which causes a full break or short/bridge on a transistor, and a non-catastrophic fault which denotes, not complete failure in manufacture, but a 'bad spot' which may or may not cause functional problems initially, but is likely to fail after burn in. This form of fault has not been investigated to date with SI cells, work has concentrated on purely catastrophic faults. (The fault model used shown in Figure 18.3 [15] and is capable of developing four distinct catastrophic faults by means of opening and closing the switches.)

These are :

      - Gate Source Short   (GSS)

      - Gate Drain Short   (GDS)

      - Drain Open   (DOP)

      - Source Open   (SOP)

In this model, an open does not necessarily denote a break in the conductance path, as it is modelled as a high impedance imperfection, such as a point defect in the oxide layer of a MOSFET channel. The reason for this is that the SPICE simulator requires at least two connections at each node and a path to ground from every node, hence opens must be simulated using large resistances. If there is only a single path then an injected open under simulation will be exhibited as a high impedance node, with a reduced current flow and or increased associated voltage levels.

It is assumed that only one fault will be present in any circuit at one time, this makes for simpler analysis of faulty behaviour, and provides a minimal saturated fault set. That is to say each possible fault is injected, in turn, into every transistor of the circuit under test, simulated with the prescribed test stimuli and the results analysed to see if that fault is detectable.

Simulation of all circuits was carried out on HSPICE using process parameters typical of a 2.5μm P-Well Process.

18.2.2 *Fault detection and detectability measures*

Assuming that a 'good signal', in response to some particular stimulus is generated from a known fault free source, whether this is simulation, or a known good ("golden") chip, a 'test signal' can be found for each device under test by exciting it with the same stimulus. The sampled test signal can then be subtracted from the good signal leaving a sampled 'error signal'. In theory should this signal have a non zero value then the Device Under Test (DUT) is faulty, clearly this is over simplistic, process variations, inaccuracies in simulation models, variations in the ambient conditions affecting test equipment adversely, etc. all potentially cause non zero values of error signal. Hence a thresholding must be used. The value of threshold chosen is critical, too low and faulty chips will pass, yet too high and good chips that have slightly different process parameters from the good signal source will be failed. Knowledge of a particular process is required to determine such a threshold. Several methods for the 'successful' detection of faults have been investigated, these are :

i)   Pre/post-subtraction digitisation [16] - Where the good and faulty signatures are digitised into a predefined number of levels before being subtracted (pre) or after subtraction (post), giving a digitised error signal, providing discrete detectability levels. The quantity and size of the digitised steps gives a measure for the quality of fault detection.

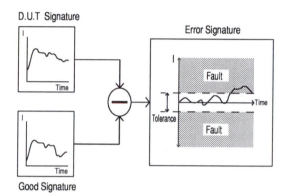

Figure 18.4 Linear fault detection.

ii)   Linear tolerance band definition [22] - A variation of post subtraction digitisation, with only one quantised band and as such readily practically implementable (as shown in Figure 18.4). In this instance there are two factors determining the detection of a fault, the straying of the error signal outside the tolerance region and its duration. The level of the tolerance band is defined as a factor of the average current drawn by the device (typically 10 - 20%), and is dependent upon the characteristics of the DUT. This band can be set to cope with the accuracy of the test equipment as well as process variations and noise. In this case a fault is said to be detected when at least a certain percentage of the total number of samples is greater than the predefined linear band.

iii)   Frequency domain variance [17] - Calculates a weighted distribution for the error signal in the frequency domain, and is used not just for fault detection, but for location and identification. Using an FFT algorithm the major and the some side lobes from the frequency responses of the faulty and good devices are compared with an RMS threshold level as detection criterion. The detectability of a fault decreases as the observance of variation occurs with increasing distance from the major lobe. It has the obvious functional overhead of complex arithmetic in the form of an FFT.

iv)   Histographic analysis - This is a second method for the calculation of weighted distribution, it does not rely on complex mathematical procedures, but calculates a normalised value or weight for the distribution of the histogram of the error signal, the higher the value the larger and greater the number of non zero samples.

A logarithmic histogram of the error signal, in 9 decades from 1e-12 to 1e-4A, is generated. Then,

$$Weight = \frac{[\ \Sigma(No\ of\ samples\ per\ decade)(decade)]}{(Max\_decade)(No\ of\ sample\ points)} \qquad (18.1)$$

Providing the following weighted distribution profile (Figure 18.5), giving a method of comparison of test vectors on injected faults, simply by comparing the weighted numbers generated, the larger the number the greater excitation produced.

For the purposes of the investigations under taken with SI cells the detectability measures of Linear Tolerance Band and Histographic Analysis were implemented. Linear tolerance provides a readily practicable implementation of fault detection, and histographic analysis a simple efficient method for the comparison of test vectors. It is also possible with histographic analysis to perform a thresholded detection similar to the linear tolerance band method, but the former has a significant grey region, because in linear tolerance band >10% of all the samples must be in error (above the threshold) for detection, with histographic analysis this is not necessary, as shown in Figure 18.5 as the horizontal dotted line. if a significant proportion of the samples are just below the threshold then the fault can be said to be detectable. The threshold shown is calculated to be equivalent to 10% of all samples with a threshold of 1μA.

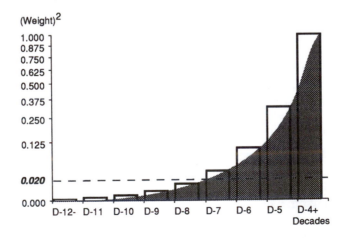

Figure 18.5 Weighted threshold distribution.

18.3   Test vector analysis

Initial work on the potential of current monitoring for test addressed a simple circuit comprising two cascaded cascode memory cells (Figure 18.6) and investigate variation of fault coverage with different test stimuli. In particular:

i)   -   what effect has the period of the test vector with respect to the clock ?

ii)   -   what form of signals will elicit the greatest response ?

iii)   -   Lastly which faults are hardest to detect, can they be grouped or categorised? and how are they best executed?

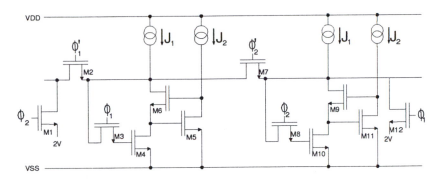

Figure 18.6 Initial test circuit - 2 cascaded cascode memory cells.

The circuit comprises twelve transistors and, with the fault model introduced in Section 18.1.1 a possible 48 faults. Current sources J_1 and J_2 supplied bias currents of 100μA and 15μA respectively. Normal operation for this circuit (a delay of two clock cycles) is displayed in Figure 18.7. The top trace is the phase of the clock controlling the output of the second cell, the second and third traces respectively show a non sampled (linear) input square wave and the resultant output wave delayed by a clock cycle.

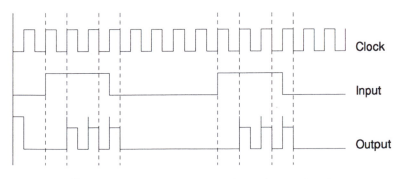

Figure 18.7   Normal operation of test circuit.

18.3.1 Effect of test stimulus period

To analyse the effect of test stimulus period, the test circuit was fully fault simulated (all 48 faults) for test vectors of 2λ, 4λ, 8λ and 16λ where λ is the clock period of the circuit, shown in Figure 18.8. The profile of the test vectors was a positive square wave of amplitude $30\mu A$.

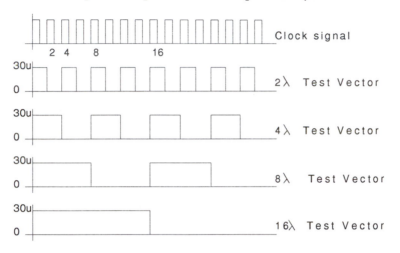

Figure 18.8 Period variation - test vectors.

The weighted distribution generated with histographic analysis was used in generating the results as the profile of the error signal holds more importance than a linear level.

The following four graphs (Figures 18.9 - 18.12) show the coverage obtained with each test vector in respect of the four fault groups, GDS, GSS, DOP and SOP.

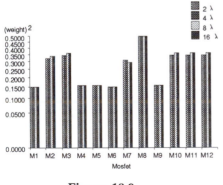

Figure 18.9
Effect of detection on GDS faults.

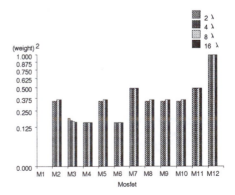

Figure 18.10
Effect of detection on GSS faults.

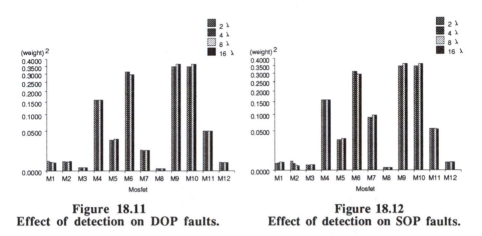

Figure 18.11
Effect of detection on DOP faults.

Figure 18.12
Effect of detection on SOP faults.

On first inspection there appears to be very little change in coverage with the variation of test vector period. In analysing the graphs and data it is found that there are two types of fault: Static and Transient, usually found in conjunction, with the first being far more prevalent than the latter. A static fault occurs over a time period far greater than the clock signal, to all intents it is time invariant, this is shown in Figure 18.13b where the faulty case saturates irrespective of the test signal. A transient fault is exhibited either; only at the commencement of the test signal and decays within a clock cycle or appears on the edge of a clocked transition as an over size pulse, Figure 18.13a, since with a larger test vector period there are fewer transitions per cycle then the less detectable transient faults become. This effect can be seen clearly in Figure 18.11 MOSFET M_3 with coverage falling slightly as the period of the TV increases.

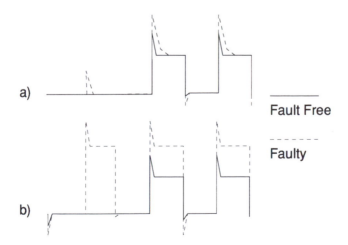

Figures 18.13(a,b) Transient and static faults.

To summarise the results; The largest change in fault distribution was in the region of 1.5% (0.015 weight) which is just less than the linear detection level of 10% of faulty samples greater than 1e-6A (0.02 weight). It can be noted that the larger the test vector period the greater the detection of the bulk of faults, 41% of faults exhibited an increase in their weighted distributions being static in nature, and a worsening distribution on 8% of faults denoted as having transient behaviour, as test vectors increase in period from 2λ, to *16λ*.

18.3.2 *Different profile test stimuli*

To investigate the effect of different test stimuli on circuits, and to see whether unipolar or bipolar signals will produce the greatest excitation, four profiles of test vector have been investigated :

i)   Positive pulse square wave.   '+' $0 > X$ Amps.

ii)   Negative pulse square wave.   '-' $-X > 0$ Amps.

iii)   Bi-polar pulse square wave.   '+/-' $-X > X$ Amps.

iv)   Sinusoid.   'Sin' $-X > 0 > X$ Amps.

Each wave was fully fault simulated with amplitudes of 10, 30, 50, and 70 micro amps with a period 16 times that of the clock signal. There were two basic aims behind this :

•   Investigating which form of test vector elicited the greatest coverage.

•   The optimum amplitude for each test vector.

It is also possible to go a stage further and analyse which test vector provides the best performance, in gaining responses from the four types of injected fault, and which faults are inherently difficult to test.

18.3.2.1 *Overall coverage*

Figures 18.14(a,b) show respectively the total weighted and percentage coverage obtained for all test vectors at all four respective amplitudes (10, 30, 50, 70μA). The criterion for the successful detection of a fault was set at 10% of all samples having an error greater than 1μA, for the weighted distribution this translates to a weighted value of 0.02.

Comparing the two methods of fault detection, higher coverage is obtained with weighted distribution thresholding, having approximately 80-86% coverage for '+', '+/-', '-' TV's. Compared to linear thresholding, where less than 10% of faulty samples fail to exceed 1μA, but with a non the less significant and detectable error signal, with coverage in the region

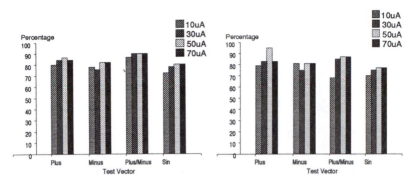

Figure 18.14(a,b) Total percentage coverage, Linear and Weighted thresholds.

of 70-77%. Suggesting that in the region of 5-10% of faults are marginally detectable with the applied test vectors.

The general trend to be seen with increasing amplitude is for an increase in overall coverage, as erroneous activity is proportional to the magnitude of the stimulus. Highest coverage is observed with '+' and '+/-' test vectors, although it can be observed in both graphs that for the '+' stimuli coverage falls with a test vector of 70μA. When investigated further it is found that '+' fault coverage is approximately Gaussian in nature, having a mean of 50μA, which is half the bias current J_1. This is because with amplitudes up to 50μA increased fault excitation with larger magnitude test stimuli is displayed, above the mean the fault may be excited further but it cannot be seen as the maximum current is 100μA, as such the erroneous behaviour is effectively restricted.

18.3.2.2 Transistor level analysis

Investigating in greater depth the effect of the different profile test vectors and amplitudes, with respect to individual transistors, gives a greater insight into the testability of the SI circuit. To achieve this requires the weighted error signal distributions to be calculated for every transistor with all four faults, then assuming equi-probability of occurrence, the arithmetic mean of the four values is calculated. The graphs shown in Figures 18.15 to 18.18 show the average weighted distribution of the error signal, for each MOSFET of the test circuit, with all four injected fault types, at amplitudes of 10, 30, 50, & 70μA respectively. MOSFETs M_1 - M_6 comprise the first cell and M_7 - M_{12} the latter as shown in the circuit diagram Figure 18.6.

Again it can be observed that the trend in increasing excitation occurs with greater magnitude stimuli for all MOSFETs up to 50μA, and although it was not visible with the overall thresholded fault coverage in the previous section, the weighted distributions at 70μA, are for all test stimuli

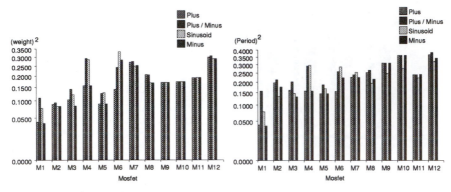

Figure 18.15
Weighted distribution of MOSFETs 10µA test vectors.

Figure 18.16
Weighted distribution of MOSFETs 30µA test vectors.

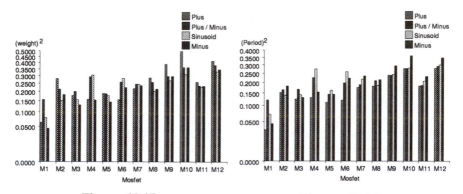

Figure 18.17
Weighted distribution of MOSFETs 50µA test vectors.

Figure 18.18
Weighted distribution of MOSFETs 70µA test vectors.

reduced from that of 50µA, due to the physical limitation imposed by the bias currents and the test vector. Indicating a Gaussian distribution for fault excitation against test stimuli amplitude, for all test stimuli, with peak or optimum fault excitation observable at the mid-point of the memory transistor bias current.

Turning to consider the effect of the test vector profiles on fault excitation, what is immediately obvious is that, individual MOSFETs respond differently to the test stimuli. In terms of the overall excitation achieved, up to 50µA, the '-' stimulus provides the least performance, along with a sinusoid stimulus with transistors M_4 & M_6 being the most notable exceptions. The greatest response is obtained with '+' at 50µA, although at 70µA the '-' stimulus is less restricted than signals with positive components. In short, what can be stated is that from the vectors applied and from the information obtained to date, no one profile is to any significant degree superior to the others. All illicit responses, some better than others, follow the same Gaussian trend but provide different levels of fault excitation for different faults.

Considering individual MOSFET's in the circuit. The weighted distributions for the first six transistors to the latter six, show that fault detection is greater in the second cell, this is most prominent at $10\mu A$, decreasing in significance towards the optimum of $50\mu A$, as would be anticipated with a Gaussian distribution. A transistor showing little erroneous behaviour in the presence of a fault, does so for all test the stimuli, with minimal differences. The exception to this rule is transistor M_1 from which a far greater level of fault detection is found with the '+/-' test stimuli. In general as would be expected the twin transistor in the second cell responds similarly to that in the first. This effect is most observable with the transistors M_4, M_{10} and M_6, M_9 where they all display large erroneous responses, these transistor duals are the memory transistors, and the cascode regulator respectively, and is hardly surprising since they are the main focus of the circuit. What is unexpected is that the cascode transistors M_5 & M_{11} are severely less detectable than M_6 or M_{12}, the reason for this it is that this transistor is not directly in the flow of current passing from cell to cell, its purpose is to regulate the gate of M_6 (or M_9), providing the high gain required to minimise channel length modulation of the memory transistor. So when a fault occurs in this transistor all that is affected is the source voltage of the memory transistor, leaving the circuit in all probability functionally correct but with a greater error due to channel length modulation.

18.3.2.3 Individual faults groups and failure modes

In the previous section it was noted that optimum coverage with the test stimuli applied was achieved with a positive square wave of amplitude $50\mu A$, the Figure 18.19 shows the weighted distribution of the four injected faults for this test stimuli. The striking feature of the graph is the 100% (weight = 1.00) detectability of M_{12} with a gate source short, the reason behind this is that, with SI circuits being dual phase synchronous in nature they therefore have a 50% dead zone, when the output from cell 'n' is isolated from cell 'n+1' whilst memorising the current from cell 'n-1'. This isolation is provided with pass transistors, in this case M_2, M_7 and M_{12}, where M_{12} provides the output switch for cell 2 on phase ϕ_1. It is the current emerging from M_{12} that provides the device signature, thus with a gate source short the output is always in error.

Faults injected into the test circuit fall into four distinct bands, this can be seen clearly in the graph of Figure 18.19. Analysis of the other test vectors shows these same bands with very little variance, suggesting that all faults lie in one of four failure modes which can only generate faulty

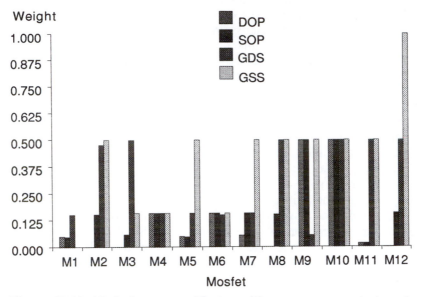

Figure 18.19 All fault groups, 50µA positive square wave test vector.

behaviour characterised by that band, the magnitude within which is determined by the test vector. These four modes exhibit faults as shown in Figure 18.20 and as follows:

i) Saturation failure - The device saturates at the biasing current level, in this case 100µA, giving a weighted distribution of approximately between 0.450 and 0.500 (in one case 1.000).

ii) Non-saturation failure - This occurs when the circuit exhibits faulty behaviour for of a constant non-saturated level. This gives rise to the largest weighted distribution from 0.125 to 0.350.

iii) Transient failure - Pulse lengthening of spikes and usually a minimal level of static fault, with a weighted distribution of not more than 0.060.

iv) Undetectable (failure) - For an injected fault no erroneous behaviour is noticeable in the weighted distribution above fourth decimal place.

The table shown in Figure 18.21 shows the percentage of each of the four faults in the four detection bands, as an average for the four test vectors simulated at 50µA. Highlighting the most detectable faults as being gate source shorts, then gate drain shorts, with the classes of opens being far less observable, with the bulk of faults in transient and undetectable failure modes.

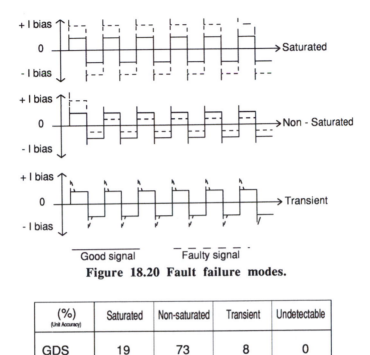

Figure 18.20 Fault failure modes.

(%) (Unit Accuracy)	Saturated	Non-saturated	Transient	Undetectable
GDS	19	73	8	0
GSS	36	56	0	8
DOP	4	35	40	21
SOP	4	50	29	17
Total	16	54	19	11

Figure 18.21 Table - Detection band percentages.

18.4 Built In Self Test for SI cells

Switched current circuits, in particular (Chapter 8), are comprised entirely from memory cells with feedback loops and some additional (biasing) transistors providing specific functionality. It was demonstrated above that cascades of such cells are highly testable with standard patterns, and thus there is a clear immediate gain in testability if global feedback loops and can be disconnected for test. Rather than comparing the signature of the device under test with simulated values of a 'good' board, two equal length cascades of cells on the same device can be compared one against the other as shown schematically in Figure 18.22. This minimises error due to process variation. To achieve a nominally zero error current one path must be inverted with respect to the other - this inversion may be available as part of normal circuit function as demonstrated in the following example:

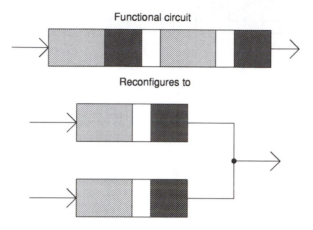

Figure 18.22 Block reconfiguration.

18.4.1   *Functional inversion test circuit*

In this example the functionality of the circuit provides the necessary inversion to generate the error signal at the summing node. Modifying the circuit given in Chapter 8 (Figure 8.2) for a lowpass 2nd order filter, to give self testing capabilities requires five extra in-circuit test switches, as shown in Figure 18.23, 'a' severs the unswitched feed back path, 'b' and 'c' two connect the output of the first integrator's current copier to the output of the second with the same transistor ratios forming a summing node and giving that summing node access to the outside world. 'd' and 'e' switches toggle between the scaled input to the second integrator and an input equal to that of the primary input. Other modifications to the circuit are the logical combination of the test line and the clock signals, and extra routing for the combined clock/test lines. The inserted test switches are minimal size transistors and therefore introduce a negligible overhead. A block diagram of the circuit in normal and test modes is shown in Figure 18.24.

The circuit will now function as either a biquadratic filter section or a self testing integrator section, depending upon the setting of the test switches. When configured for testing the circuit will not initially have the required zero level fault free error signal, this must be generated, since at switch on an unknown level of charge is stored. The two integrators must be configured to provide equal and opposite current levels at their outputs. Obtaining opposite responses is not a problem as this is provided by the functionality of $H[z]$ and $-H[z]$, but obtaining equal levels is not so easy, there are only two definite levels for an SI circuit element, these being the positive and negative saturation levels but with the implemented class A

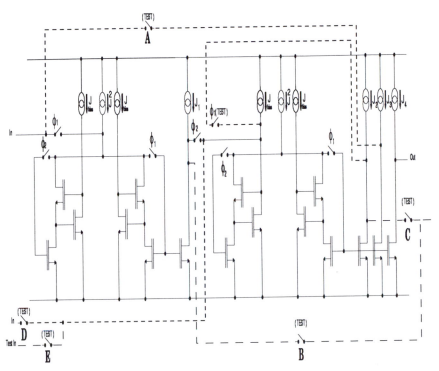

Figure 18.23 Self testing biquadratic filter.

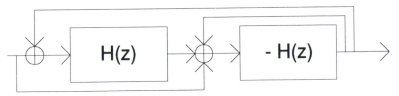

Integrator based bi-quadratic SI filter

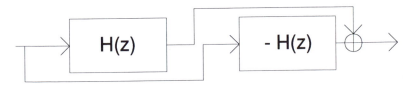

Self testing SI Integrator sections

Figure 18.24 Circuit functionality, normal and test operation.

cells the absolute positive and negative saturation levels are unequal. So although it is feasible to saturate the two devices by the application of a suitably large amplitude current, the error signal remains non zero, but is now of a known difference:

$$Err[n] = (h[n]_{+SAT}) + (-h[n]_{-SAT}) \qquad (18.2)$$

Removing this difference is achieved simply by changing the input stimulus half way through a clock cycle, since $H[z]$ reads on ϕ_2 and $-H[z]$ on the opposite phase, a change in the input signal between $H[z]$ and $-H[z]$ reading causes an effective subtraction step. Hence providing a step equal in magnitude to the error difference for half a clock cycle negates the error leaving the required zero level fault free signal. At this point the device is ready to accept the test stimuli, providing two restrictions on maintaining the zero error signal are observed: Firstly at no time must either integrator be allowed to saturate. Secondly input the must only change on the beginning of a clock cycle with a step, hence linear ramps and sinusoids are not permitted unless quantised. Figure 18.25 gives an example test stimuli, and shows the fault free envelope for $Err[Z]$. $Err[Z]$ must be defined in an envelope because of integrator drift δ:

$$Err[Z] = H[Z] + (-H[Z]) = \delta \neq 0 \qquad (18.3)$$

Thus the zero level fault free error signal degrades with time, this can be recovered to a degree by applying a large change in input stimulus forcing one integrator up to (but not into) saturation.

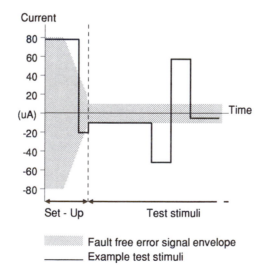

Figure 18.25 Example test stimuli.

18.4.1.1 Fault coverage

Using the test vector shown in Figure 18.25, the circuit of 18.23 was fault simulated in test configuration, with all four fault models. Figure 18.26 shows the fault coverage obtained for all of the MOSFET's directly in current flow, ie. the current copiers isolated with the severance of feedback were not fault injected. The detection envelope about *Err[Z]* was defined as +/-10μA, just bracketing the fault free error signal, thus any point outside this envelope can be regarded as erroneous, and hence a fault has been detected, partial detection of a fault is denoted as <5% of the total sample points being in error.

(%) (Unit Accuracy)	Detectable	Partially Detectable	Non - Detectable
GDS	69	22	9
GSS	50	37	13
DOP	50	11	39
SOP	60	0	40
Total	57	18	25

Figure 18.26 Reconfigurable biquadratic section fault coverage.

From the table it can bee seen that overall coverage including partially detectable faults is 75%, with the bulk of these faults being readily detectable. It is the case that a different profile test stimulus will highlight different faults, thus with several such stimuli the coverage will be increased. The earlier work on test vector analysis although applicable with this circuit is restricted, as saturation levels cannot be breached, thus a compromise between stimuli wavelength and amplitude must be reached, for example a +50/-50μA stimulus of wavelength 16 clock cycles is inappropriate as it will drive the integrators into saturation.

Figure 18.27 shows several failure modes of the test circuit observed, along with the fault free zero error signal, the vertical dashed line represents the point when the error signal falls within the 10μA detection envelope, after 2.7μs of the 10μs test stimuli. These failures do not cover all of the failure modes but represent three areas where the bulk of faults are observed. Fault detection is only possible if before the error signal settles, hence if faulty behaviour is displayed only in the settling period then that fault will be classed as undetectable.

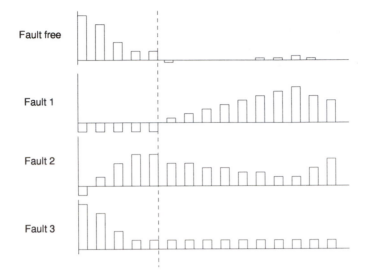

Figure 18.27 Failure modes in reconfigurable biquadratic sections.

Should the current copiers be included in the fault coverage then this falls as only 40% of faults are detectable 26% of which must be declared as marginally detectable, this is an area that requires further work as noted in the conclusions of this chapter.

18.5   Conclusions and future work

The material presented here shows that there exists the possibility of testing much of th basic transistor circuitry of switched current devices by simple reconfiguration. The technique offers the advantage of in-circuit test signature generation with no special test equipment needed for go/no go testing. One cascade of cells is tested against a second on the same device. This reduces errors due to process variation and also means that special test generation for new devices is no longer required - in test mode all devices essentially look the same.

Clearly a great deal of further work is needed. In particular use of different test patterns made possible by the inclusion of a third (test only) clock phase is being investigated in order to increase coverage. The optimum length of cascades needs to be determined and work is also ongoing using soft fault models - For example evidence from digital CMOS [13] suggests current monitoring is able to predict failures due to gate oxide failures prior to functional failure of the device. High coverage for this type of fault has been demonstrated in analogue circuits [14]. Testing the transistors does not ensure correct functioning of the usual feedback paths or complete coverage of the transistors of the current sources and so this also needs to be considered.

Acknowledgements

The authors would like to acknowledge Dr Nick Battersby of Imperial College very useful discussions related to this work.

References

[1]   B. R. Wilkins, *Testing digital circuits an introduction*, Van Nostrand Reinhold (UK) 1986.

[2]   D. Wu, "Can Iddq Test replace conventional stuck at fault test ?," in *Proc IEEE Custom Integrated Circuits Conf.*, Paper 13.2, 1991.

[3]   T. Storey and W. Maly, "Current test and stuck-at fault comparison on a CMOS chip," *Electronic Engineering*, pp. 89-95, Nov. 1991

[4]   L. Baldo, J. Figueras, A. Rubio, V. Champaca and R. Rodriguez, J. Segura, "Quiescent current estimation for current testing," in *Proc. IEEE Test Conf.*, 1991, pp. 543 - 548.

[5]   R. L. Ruey-wen, Ed., *Testing and diagnosis of analog circuits and systems*, New York: Van Nostrand, 1991.

[6]   D. Camplin, I. Bell, G. Taylor and B. Bannister, "Can Supply current monitoring be applied to the testing of analogue and digital portions of mixed asics?," in *Proc. IEEE Test Conf.*, pp. 538 - 542, 1991.

[7]   D. Camplin, G. Taylor, B. Bannister and I. Bell, "Supply current testing of mixed analogue and digital IC's," *Electronics Letters*, vol. 27, Aug. 1991.

[8]   D. Camplin. I. Bell. G. Taylor and B. Bannister, "Investigations into current sensing strategies," *IEE colloquium No. 1992/118*.

[9]   M. Patyra and W. Maly, "Circuit design for built in current testing," in *Proc. IEEE Custom Integrated Circuits Conf.*, Paper 13.4, 1991.

[10]   M. Roca and A. Rubio, "Selftesting CMOS operational amplifier," *Electronics Letters*, vol. 28, July 1992.

[11]   D. Camplin, K. Eckersall and P. Wrighton, "MIXIT 2nd year deliverable 1992," VLSI group, Dept. Electonics, Hull University.

[12]   G. Taylor, P. Wrighton, C. Toumazou and N. Battersby, "Mixed signal test considerations for switched current signal processing," in *Proc. Midwest conference on Circuits and Systems*, Aug. 1992.

[13]   C. Hawkins and J. Soden, "Reliability and electrical properties of gate oxide shorts IC's," in *Proc. IEEE International Test Conf.*, pp.443, paper 13.4, 1986.

[14]   K. Eckersall, P. Wrighton, G. Taylor, I. Bell, B. Bannister, "Testing Mixed signal ASICs through the use of supply current monitoring," in *Proc. European Test Conference*, 1993.

Chapter 19

Analysis of Switched-Current Filters

Antônio Carlos Moreirão de Queiroz

19.1   Introduction

Ideally, switched-current filters operate as periodically switched linear networks, in which the circuit reaches a static steady state between the switching instants. This mode of operation is essentially the same as that of switched-capacitor filters, and results in an analogue circuit that processes signals in the sampled-data domain, as in a digital filter. In switched-capacitor filters, the signal is represented by capacitor voltages, and the computations are done by charge balancing at the switching instants. In switched-current filters, the signal is represented by currents and the computations are done by current balancing between the switching instants.

The similarities between switched-current filters and switched-capacitor filters allow the simulation of most structures using their switched-capacitor equivalents [1,3], by simply translating transistor currents into capacitor voltages and arranging the switches and operational amplifiers to obtain the same operation. The idealised operation of the filter can then be studied with the aid of an ideal switched-capacitor filter simulator. The problem with this approach is that it is not obvious how to model imperfections in the switched-current filters, such as finite R_{ds} resistances and C_{gd} capacitances in MOS transistors and non-zero resistances in current-conducting switches. These imperfections do not change the fundamental assumption that the circuit stabilises between the switching instants, but effects the current balancing in a way that has no correspondence in usual switched-capacitor circuits.

Time-domain analysis can always be carried out using general circuit simulators, and the simulation can take into account all the important imperfections. This approach is the only one available to study most of the non-linear effects, and is also useful in the study of switching transients. Brute-force frequency-domain analysis can be done by analysing the circuits in the time domain and extracting the frequency response from the waveforms obtained. This may be useful for final project verification, but is too inefficient for investigations related to structures and synthesis methods. It is also somewhat inaccurate, because the time-domain analysis of non-linear networks is always approximate and, if non-linear effects are considered, the response depends on the signal level.

Direct frequency-domain simulations, ignoring non-linear effects, can be obtained with the techniques used to analyse general periodically switched linear circuits [8]. This approach is useful in the evaluation of the switching frequency limitations of a filter. It is, however, unnecessarily complex when effects such as incomplete current balancing between switching instants are not significant.

This chapter discusses an efficient technique for the frequency-domain analysis of switched-current filters, similar to those used for ideal switched-capacitor circuits [11-13]. The method is described as it was implemented in a computer program named ASIZ, that analyses multi-phase switched-current filters; obtaining transfer functions in z-transform, poles and zeros, frequency responses, transient responses and sensitivities [9,14]. The method can also be used for the analysis of ideal switched-capacitor circuits.

19.2 Frequency-Domain Nodal Analysis of Switched-Current Filters

In switched-current filters, nodal voltages are generally non-linear functions of the input (a current), and only AC currents are linear functions of the input (if proper design techniques are used). However, if the circuit can be considered as operating with small signals, the non-linearity of the transconductances and other elements can be neglected and AC voltages and currents can be considered as linear functions of the input. In small signal AC analysis, the transistor currents are linear functions of their V_{gs} voltages, that represent the signals inside the filter as well as the currents. Nodal analysis can then be used to analyse the circuit at transistor level, using the basic elements that constitute a switched-current filter, shown in Figure 19.1:

- Ideal switches: Their function is to short-circuit two nodes when closed, leaving them unconnected when open.

- Constant current sources: Their function is to bias the MOS transistors. The value of their DC currents has no effect on the small-signal AC behaviour of the filter.

- Signal current sources: These feed AC signals to the filter, and so their currents are significant.

- Resistors: These model resistive losses, as output resistances of current sources and MOS transistors, and can be used to transform currents into voltages, at the filter output, for example.

- Transconductors: Used in the modelling of several elements.

- MOS enhancement mode transistors: For small-signal AC analysis, the two kinds of MOS transistors can be represented by the same model. The output circuit is modelled by a transconductance controlled by V_{gs}, and a possible non-infinite output resistance. The input circuit consists of a C_{gs} capacitance. A parasitic C_{gd} capacitance can also be included.

·   Capacitors: These model the inputs of the MOS transistors, parasitic capacitances and are basic elements in switched-capacitor filters.

·   Ideal operational amplifiers: These are basic elements in switched-capacitor filters, and can also be used in precision switched-current filters [2].

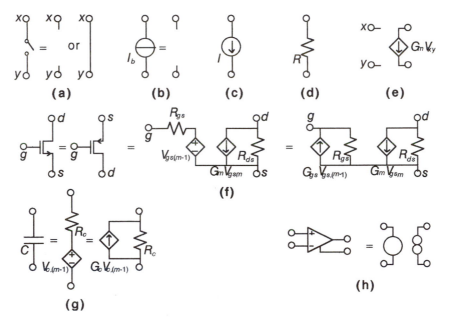

Figure 19.1   Elements of switched-current filters and their AC small signal models used in the analysis method.
(a) Switch (b) Biasing current source (c) Signal current source, (d) Resistor (e) Transconductor (f) MOS transistor, including the C_{gs} capacitance, where $G_{gs}=1/R_{gs}$ (g) Capacitor, where $G_c=1/R_c$ (h) Ideal operational amplifier; a nullator-norator pair.

Capacitances must be modelled by an equivalent circuit that presents no load (infinite resistance) when fed by low-impedance circuits and acts as a voltage memory when disconnected from the remaining circuit. Additionally, the effects of charge redistribution, essential for switched-capacitor filters and circuits with parasitic capacitances, must also be correctly modelled.

These functions can be realised, in principle, by representing a capacitor by a large resistor R_c in series with a voltage source that remembers the value of V_c just before the last change in the configuration of the switches ($V_{c,(m-1)}$ in Figure 19.1). For the correct computation of the charge redistribution effects, the R_c resistances must be inversely proportional to the capacitance values. In the z-transform domain, the value of the memory voltage source can be expressed by

$$V_{c,m}(z) = V_{c,(m-1)}(z)z^{-1/f} \qquad (19.1)$$

where m is the number of the present phase, and it is assumed that the switches operate periodically with a period T, subdivided into f phases of equal duration[1]. At the end of each phase, the configuration of the switches is changed. Note that $m-1$ is defined as f if $m=1$. In the models for the capacitor and MOS transistor in Figure 19.1, this voltage source is considered as being controlled by the capacitor voltage at the previous phase, or as a transconductance with input and output at different phases in the Norton equivalents shown.

The basic unknowns in the circuit are the partial nodal voltages $E_{i,mk}(z)$, for the $i=1,...,n_t$ nodes, for the $m=1,...,f$ output phases, and for the $k=1,...,f$ input phases. Added for all k they form the nodal voltages at the output phases m, $E_{i,m}(z)$, and also added for all m they form the global nodal voltages $E_i(z)$. The output signal of a switched-current filter can be obtained as the global voltage over a load resistor.

Considering initially all the switches open, by modelling the elements as in Figure 19.1, the system of linear equations (19.2) can be constructed.

$$\begin{bmatrix} \mathbf{G}_n & 0 & \cdots & 0 & -\mathbf{G}_c z^{-1/f} \\ -\mathbf{G}_c z^{-1/f} & \mathbf{G}_n & \cdots & 0 & 0 \\ \vdots & \vdots & \ddots & \vdots & \vdots \\ 0 & 0 & \cdots & \mathbf{G}_n & 0 \\ 0 & 0 & \cdots & -\mathbf{G}_c z^{-1/f} & \mathbf{G}_n \end{bmatrix} \begin{bmatrix} \mathbf{E}_{1k} \\ \mathbf{E}_{2k} \\ \vdots \\ \mathbf{E}_{(f-1)k} \\ \mathbf{E}_{fk} \end{bmatrix} = \begin{bmatrix} \mathbf{I}_1 \\ 0 \\ \vdots \\ 0 \\ 0 \end{bmatrix} \begin{bmatrix} 0 \\ \mathbf{I}_2 \\ \vdots \\ 0 \\ 0 \end{bmatrix} \cdots \begin{bmatrix} 0 \\ 0 \\ \vdots \\ \mathbf{I}_{f-1} \\ 0 \end{bmatrix} \begin{bmatrix} 0 \\ 0 \\ \vdots \\ 0 \\ \mathbf{I}_f \end{bmatrix} \qquad (19.2)$$

$\mathbf{G}_n$ is the nodal conductance matrix of the entire circuit, with the exception of the $\mathbf{G}_c$ transconductances for all the capacitances. $\mathbf{G}_c$ is the nodal conductance matrix of the $\mathbf{G}_c$ transconductances only. $\mathbf{E}_{mk}$ are the unknown nodal voltage vectors in the z domain for the $m=1,...,f$ output phases and for the $k=1,...,f$ input phases. The signal inputs at the $k=1,...,f$ input phases are represented by f $\mathbf{I}_k$ input current vectors, initially equal, arranged as shown. For frequency response calculations, unity values are used for the input currents. The order of the system is $n_t \times f$ and all the submatrices and subvectors are of order n_t.

The effect of a closed switch is to short-circuit two nodes during the phases in which it is closed. In the system of (19.2), the effect of a closed switch between nodes x and y, closed at phase m, is to add the rows and the columns x and y in the submatrices and subvectors corresponding to the

---

[1] The generalisation for the case of phases with arbitrary duration is simple, but will not be discussed here, to maintain the expressions in the simplest form.

m^{th} phase (if x or y is zero, representing a connection to the signal ground, the other column and row are eliminated). The unknowns $E_{x,mk}$ and $E_{y,mk}$, $k=1,...,f$, are reduced to single ones (or eliminated if $x=0$ or $y=0$). The order of the system is reduced by the number of switches.

Ideal operational amplifiers can be processed in a similar way: For each phase, the columns corresponding to the input terminals are added, eliminating one unknown, and the rows corresponding to the output terminals are also added, eliminating one equation (for single-input or single-output amplifiers, the input column or the output equation is removed from the system). As this is done for all the phases, each operational amplifier eliminates f equations and $f \times f$ unknowns. The order of the system is reduced by $f \times$(number of op-amps).

The processing of the switches and op-amps is actually carried out by a preprocessor that generates two sets of pointers; one pointing to columns and the other to equations in the reduced version of (19.2), that indicate where a node, unknown or equation of the original system will be placed in the final system.

The reduced system can be solved, for a particular numerical value of z, by the LU factorisation of the system matrix and f substitutions, one for each of the $k=1,...,f$ input phase excitation vectors, being obtained the $f \times f$ partial nodal voltage vectors $\mathbf{E}_{m,k}(z)$. Assuming that input and output signals remain constant for the duration of each phase, the frequency response of a global nodal voltage E_i relative to an unitary input can be computed by

$$E_i(j\omega) = \frac{1 - e^{j\omega T/f}}{j\omega T} \sum_{m=1}^{f} \sum_{k=1}^{f} E_{i,mk}(e^{j\omega T}) \qquad (19.3)$$

where $z=e^{j\omega T}$ is used. The term multiplying the summations is a "sampling function" corresponding to the duration T/f in a period T of each phase. The phase difference between the different phases is already taken into account by the way the system of (19.2) was built.

There is a problem with this formulation: The use of "large" resistors to represent the capacitors is necessary to maintain the accuracy of the basic algorithm, but can lead to numerical problems due to the near singular systems that result, and it also introduces spurious artefacts in the analysis if the resistors are not "large" enough.

19.3   Refinements to the Basic Algorithm

The objective of using large R_c resistors is to avoid their influence on the equations of the nodes where the voltage is forced by low-impedance circuits. *The R_c resistors only have a significant role in an equation if there are no low-impedance circuits connected to the corresponding node at the corresponding phase.* Assuming that there are no subcircuits connected to the remaining circuit only through capacitors and current sources at some phase (Figure 19.2), the equations in which R_c resistors are significant are those that remain empty when the "stamps" of all the elements are mounted in the final system, with the exception of those with R_c resistors and G_c transconductances. Normalised values can then be used for the R_c resistances, if only the parts of their "stamps" that are in these "high-impedance" equations are mounted. Independent normalising factors can then be used for resistive and capacitive elements, allowing the formation of a well-conditioned system of equations. The circuit nodes can be classified, possibly depending on the phase, as "high-impedance" nodes and "low-impedance" nodes, corresponding to their nodal equations.

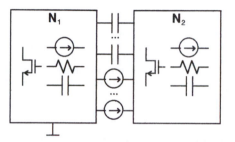

Figure 19.2 Forbidden circuit at any phase in the refined algorithm. The resulting system of equations is singular, even when there are no current sources connecting N_1 and N_2, a case where the circuit has a solution. A node connected to capacitors only is an allowed exception.

This analysis method corresponds to the analysis of a periodically switched RC-active network, where the circuit is allowed to stabilise between the switching instants. It can be used with switched-current circuits, switched-capacitor circuits, or any combination of the two.

19.4 Voltage Sources in Nodal Analysis

Switched-capacitor filters need a voltage source input. As in these circuits this voltage source is connected to a capacitive circuit, its output resistance is not important. A voltage source with a series resistance can be simulated by its Norton equivalent, as shown in Figure 19.3a. The same idea can be used to create a voltage amplifier with finite gain that is ideal in switched-capacitor circuits, this is shown in Figure 19.3b.

Ideal voltage sources and voltage amplifiers that can be connected to any circuit can be efficiently simulated by the circuits shown in Figures 19.3c and 19.3d in nodal analysis. These circuits do not increase the number of equations in the nodal system, because the operational amplifier compensates for the inclusion of the extra node x. Note that the common circuit between the two realisations is an ideal current-controlled voltage source (that reduces the order of the system of equations by f). The last type of controlled source, the current-controlled current source, can be obtained by using this circuit controlling a transconductor.

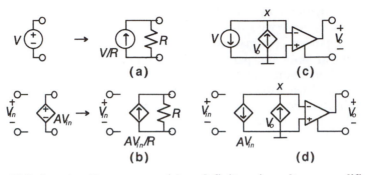

Figure 19.3 Input voltage source (a) and finite-gain voltage amplifier (b) for use with switched-capacitor filters (or with any capacitive load). Voltage source (c) and voltage amplifier (d) that can be used with any circuit, in implementations that do not increase the order of the system of equations.

19.5 Interpolation of z-transforms

The systems of equations obtained by this method for switched-current filters are almost always larger than those for equivalent switched-capacitor filters, because switched-current filters do not use operational amplifiers and also usually have more nodes and use fewer switches. The systems are sparse, and so the problem is not serious if a sparse solver is used. In the ASIZ program, the method used to speed up the analysis was the interpolation of z-transforms.

A very efficient and powerful technique for frequency and time domain analysis of linear lumped-parameter circuits, is the interpolation of transforms in the form of the ratio of polynomials using the Fast Fourier Transform (FFT) [5]. The technique can be efficiently implemented in the following way:

- It is more convenient to work with positive powers of z by multiplying all the equations by $z^{1/f}$. In the refined algorithm, only the high-impedance equations need to be multiplied, because there are no terms in z in the low-impedance equations.

- Let n be the number of high-impedance equations. The degree for the FFT interpolations, N, is the next higher power of 2 above n/f, that is the highest power of z that can appear in the $E_{mk}(z)$.

- The circuit is analysed for $z_l = Ke^{j2\pi l/N}$, $l=0,...,N/2$ (see Figure 19.4), using the determinants $D(z_l)$ of the system matrix and the nodal voltages $E_{mk}(z_l)$, for all the $m=1,...,f$ phases, for all the $k=1,...,f$ input phases, and for all the $l=0,...,N/2$ z_l frequencies. K is a constant close to 1, chosen so that no z_l coincides with a natural frequency of the circuit.

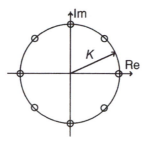

Figure 19.4 Values of z used in the FFT interpolation, in the case of order 8. As the polynomials being interpolated are real, their values in the lower half of the circle are the complex conjugates of the values in the upper half. The circuit must be analysed only in the upper half and real axis points.

· The numerators can now be computed. For the FFT interpolation, denominators and numerators must be adjusted to make their values correspond to values of polynomials with only integer powers of z. The increment between successive powers of z in the denominator and in all the partial numerators is always 1, but n/f can be non-integer and the delay between an input in one phase and an output in another causes the $E_{mk}(z)$ to be multiplied by a fixed fractional power of z when $m \neq k$. The necessary adjustments are:

$$D(z_l) = D(z_l) z_q^{1-\operatorname{frac} n/f}, \text{ if frac } n/f \neq 0 \tag{19.4a}$$

$$\alpha_{mk} = (m - k + f) \operatorname{mod} f \tag{19.4b}$$

$$N_{mk}(z_q) = D(z_l) E_{mk}(z_l) z_l^{-\alpha_{mk}/f} \tag{19.4c}$$

The operator frac x returns the fractional part of x, and x mod y returns the remainder of the division of x by y. The remaining values for the $D(z_l)$ and $N_{mk}(z_l)$, for $l=N/2+1...N-1$, are the complex-conjugates of these.

· The direct FFT algorithm with order N can now be used to recover the denominator $D(z)$ (the same for all) and the numerator polynomials of all the $E_{mk}(z)$, frequency normalised by the factor K. The denormalisation of the polynomials, some "cleaning" of numerical residues (terms of excessive degree in the computed polynomials, ideally null) and the increase of the powers of z in the numerators $N_{mk}(z)$ by $_{mk}/f$ give the correct z-transforms of the $E_{mk}(z)$.

Only $N/2+1$ analyses are needed for the interpolation[2]. From the z-

_____

[2] Actually, it is possible to save one more analysis by rotating the analysis points, turning all them into complex-conjugate pairs. The correct polynomials can be recovered after the FFT interpolations by a simple frequency transformation.

transforms obtained, frequency response graphs can be quickly plotted using (19.3), poles and zeros can be computed by polynomial root calculation, and time responses can be obtained from the corresponding difference equations.

Example 19.1

The circuit for a "second generation" integrator [4] is shown in Figure 19.5.

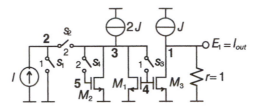

Figure 19.5 Second generation integrator circuit. The output current corresponds to the voltage over the unitary resistor. The switch s_1 is necessary to avoid the singularity that would occur in phase 1 if the input current source were left in open circuit.

With C_{gs} capacitances represented by G_g conductances, parasitic resistors $R=1/G$ considered in parallel with the current sources and from the drain to the source of the transistors and, ignoring parasitic capacitances, the system of equations takes the form shown in (19.5), with all the stamps of the elements in place, and the multiplications by $z^{1/f}$ still not performed.

$$
\begin{array}{c}
\\
\begin{array}{cccccccccc}
1 & 0 & 2 & 2 & 3 & 4 & 5 & 5 & 6 & 5
\end{array}\\
\begin{array}{c}1\\0\\2\\2\\3\\4\\5\\5\\6\\5\end{array}
\left[\begin{array}{cccc|cccc}
1+2G & & Gm_3 & & & & & & & \\
& G & & & & & & & & \\
& & 3G & Gm_1 & Gm_2 & & & & & \\
& & & G_{g1}+G_{g3} & & & & & \left(-G_{g1}-G_{g3}\right)z^{-1/2} & \\
& & & & G_{g2} & & & & & -G_{g2}z^{-1/2}\\
\hline
& & & & & 1+2G & & & Gm_3 & \\
& & & & & & G & & & \\
& & & & & & & 3G & Gm_1 & Gm_2\\
& & \left(-G_{g1}-G_{g3}\right)z^{-1/2} & & & & & & G_{g1}+G_{g3} & \\
& & & & -G_{g2}z^{-1/2} & & & & & G_{g2}
\end{array}\right]
\end{array}
$$

$$
\begin{bmatrix}
E_{1,11} \\ E_{2,11} \\ E_{3,11} \\ E_{4,11} \\ E_{5,11} \\ \hline E_{1,21} \\ E_{2,21} \\ E_{3,21} \\ E_{4,21} \\ E_{5,21}
\end{bmatrix},
\begin{bmatrix}
E_{1,12} \\ E_{2,12} \\ E_{3,12} \\ E_{4,12} \\ E_{5,12} \\ \hline E_{1,22} \\ E_{2,22} \\ E_{3,22} \\ E_{4,22} \\ E_{5,22}
\end{bmatrix}
=
\left[\;\cdots\;\middle|\,
\begin{array}{c} \\ I_1 \\ \\ \\ \\ \hline \\ I_2 \\ \\ \\ \\ \end{array},
\begin{array}{c} \\ \\ \\ \\ \\ \hline \\ \\ \\ \\ \end{array}
\right]
\tag{19.5}
$$

The unknowns and current source vectors corresponding to the two input phases are shown separately. The switches cause the addition of several rows and columns, causing the columns and equations to be mapped as indicated into the final system (the second equation and column are eliminated by s_1). The final system is shown in (19.6), where the terms in G_g are maintained only in the high-impedance equations (marked). The input during phase 1 is null, and so all the $E_{i,m1}=0$. Six unknowns remain to be computed. The others are null or equal to one of them.

$$
\begin{bmatrix}
1+2G & Gm_3 & & & & \\
 & 3G+Gm_1 & Gm_2 & & & \\
* & & G_{g2}z^{1/2} & & -G_{g2} & \\
\hdashline
 & & & 1+2G & Gm_3 & \\
 & & & 4G+Gm_2 & Gm_1 & \\
* & -G_{g1}-G_{g3} & & & (G_{g1}+G_{g3})z^{1/2} &
\end{bmatrix}
\begin{bmatrix}
E_{1,12} \\ E_{3,12} \\ E_{5,12} \\ \hdashline E_{1,22} \\ E_{2,22} \\ E_{4,22}
\end{bmatrix}
=
\begin{bmatrix}
\\ \\ \cdots \\ \\ I_2 \\
\end{bmatrix}
\quad (19.6)
$$

Using the FFT interpolation method discussed above, just one analysis of the circuit is needed. For the ideal circuit, with all the $Gm=1$, $G=0$, $I=1$, and $G_g=1$, the four parts of $E_1(z)$ are obtained as

$$
E_{1,11}(z) = E_{1,21}(z) = 0; \quad E_{1,12}(z) = \frac{z^{1/2}}{z-1}; \quad E_{1,22}(z) = \frac{1}{z-1} \quad (19.7)
$$

$E_{1,22}(z)$ corresponds to the ideal forward Euler lossless integrator function, with $E_{1,12}(z)$ being an anticipated copy. The other parts are null because the input is sampled only in phase 2. Taking into account the non-infinite resistances in parallel with the current sources, assumed to have a normalised value of 100 ($G=0.01$), the outputs are interpolated as

$$
E_{1,11}(z) = E_{1,21}(z) = 0; \quad E_{1,12}(z) = \frac{0.915z^{1/2}}{z-0.934}; \quad E_{1,22}(z) = \frac{0.915}{z-0.934} \quad (19.8)
$$

where the effect of the losses can be observed to be the formation of a lossy integrator, with some attenuation.

19.6 Time-Domain Analysis

Transient responses in the time-domain can be easily obtained from the difference equations corresponding to the z-transforms of the output signal. The output $E_{i,m}(z)$, at the output phase m, is given as a function of the input signals at each input phase k by

$$
E_{i,m}(z) = \sum_{k=1}^{f} \frac{(a_N z^N + a_{N-1}z^{N-1}+\cdots+a_0)z^{a_{mk}/f}}{z^M + b_{M-1}z^{M-1}+\cdots+b_0} I_k(z) \quad (19.9)
$$

where the numerators are the $N_{i,mk}(z)$ (degree N and coefficients a_l dependent on i, m, and k), and the common denominator is $D(z)$ (degree M, the same for all i, m and k) obtained by the FFT interpolation. M is the smallest integer greater or equal to n/f, due to the adjustment carried out by (19.4a) and N is an integer due to (19.4c). α_{mk} is given by (19.4b). By transforming (19.9) into the corresponding difference equation, the values of $E_{i,m}(t)$ as a function of the input $I(t)$ (sampled and held for the duration of each phase), the zero-state response, can be obtained as

$$E_{i,m}(t) = \sum_{k=1}^{f} \sum_{l=0}^{N} a_l I_k \left(t + \left(l - M + \frac{\alpha_{mk}}{f} \right) T \right) - \sum_{l=0}^{M-1} b_l E_{i,m} (t + (l - M)T) \quad (19.10)$$

where $I_k(t)=0$ and $E_{i,m}(t)=0$ for $t<0$. The output voltage is obtained as a combination of the $m=1,...,f$ $E_{i,m}(t)$, each one valid during its own phase. Note that the polynomial summation in k can be precomputed, as it depends only on the input.

19.7   Sensitivity Analysis

Sensitivity analysis in the frequency domain is a simple extension of the algorithm. The method described in this section uses the adjoint network method [6] directly, avoiding any reformulation as, for example, that used in [7] for the analysis of ideal switched-capacitor filters. Along with the normal system of equations $\mathbf{G}(z)\mathbf{E}(z)=\mathbf{I}(z)$, the adjoint system $\mathbf{G}^T(z)\hat{\mathbf{E}}(z)=\hat{\mathbf{I}}(z)$ is also solved. The adjoint system corresponds to an adjoint network, with the same topology of the normal network, but with the input and output branches of the transconductances and operational amplifiers interchanged, the input current sources removed and currents of value $-z^{1/f}$ (not -1 because all the system was multiplied by $z^{1/f}$) applied at the output nodes in each phase, forming one $\hat{\mathbf{I}}_k(z)$ for each input phase, as in the normal system. From the $\mathbf{E}(z)$ and $\hat{\mathbf{E}}(z)$, the sensitivities of a chosen nodal voltage in relation to any circuit parameter can be computed.

In the switching period an element acts f times, once in each phase and, in the case of the R_c resistances, with two different roles; as normal conductances and as transconductances with input and output branches in different phases. Because of this, the derivative of a nodal voltage relative to an element value is the sum of the derivatives of the nodal voltage relative to all these occurrences of the element. Considering that a nodal voltage is a sum of its $f \times f$ partial values, the expression for the sensitivity of a nodal voltage $E_i(z)$ relative to a parameter $x(z)$ is given by

$$S_x^{E_i} = \frac{x}{E_i} \sum_{m=1}^{f} \sum_{k=1}^{f} \sum_{l=1}^{f} \frac{\partial E_{i,mk}}{\partial x_l} \quad (19.11)$$

where $x_l(z)$ is the parameter $x(z)$ acting in phase l. Note that the term indicating the sampling operation, present in (19.3), does not appear in the sensitivities, but must be included if only the partial derivatives are considered (see next example).

The most general element is a transconductance with input and output branches active in different phases. The expression that gives the derivative of a nodal voltage $E_{i,mk}(z)$ relative to a transconductance $Gm_{ab,f_1f_2}(z)$ (control voltage in phase f_1 over branch a and output current in phase f_2 at branch b) is similar to the usual expression for a transconductance in a continuous-time circuit [6]:

$$\frac{\partial E_{i,mk}}{\partial Gm_{ab,f_1f_2}} = V_{a.f_1k}\hat{V}_{b.f_2m} \qquad (19.12)$$

where $V_{l,mk}$ is the voltage over branch l in phase m, with the input active in phase k, and $\hat{V}_{l,mk}$ is the same in the adjoint network.

There are two details that must be taken into account in the analysis of the adjoint network as a consequence of the refinements made in the basic algorithm:

· To multiply by $z^{1/f}$ only the high-impedance equations in the normal system is equivalent to multiplying by $z^{1/f}$ the unknowns in the adjoint system corresponding to the remaining low-impedance equations (voltages at the low-impedance nodes in the adjoint network). The solutions for the adjoint system must be divided by $z^{1/f}$ in these cases.

· The elimination of the terms in R_c in the low-impedance equations of the normal system is equivalent, in the adjoint system, to considering as null the voltages in all the low-impedance nodes when compared (added or subtracted) to the voltages in the high-impedance nodes. As seen below, this effects only the sensitivities with respect to G_c conductances (capacitances).

A resistor in branch a can be considered as a transconductance with input and output in the same branch and phase, with $Gm=1/R_a$:

$$S_{R_a}^{E_i} = -\frac{1}{R_aE_i}\sum_{m=1}^{f}\sum_{k=1}^{f}\sum_{l=1}^{f}V_{a.lk}\hat{V}_{a.lm} \qquad (19.13)$$

A normal transconductance, with its input in branch a and its output in branch b, in the same phase, results in

$$S_{Gm_{ab}}^{E_i} = \frac{Gm_{ab}}{E_i}\sum_{m=1}^{f}\sum_{k=1}^{f}\sum_{l=1}^{f}V_{a.lk}\hat{V}_{b.lm} \qquad (19.14)$$

An input current source in branch a can be considered as a transconductance controlled by a unity voltage that exists only in the phases in which the source is active:

$$S_{I_a}^{E_i} = \frac{I_a}{E_i} \sum_{m=1}^{f} \sum_{k=1}^{f} \hat{V}_{a,km} \qquad (19.15)$$

Sensitivities to voltage sources and voltage gains are directly equivalent to sensitivities to current sources and transconductances, if the models of Figure 19.3 are used. Note that if the models employing operational amplifiers are used, although the voltage E_x is eliminated as unknown in the normal system (its value is 0), the unknown $\hat{E}_x$ is retained in the adjoint system, and the values required by (19.14) and (19.15) are all available. For a voltage amplifier with its input in branch a and its output in branch b (not used in the formula), modelled as in Figure 19.3d with internal node x, the sensitivity is obtained from (19.14) as

$$S_{A_{ab}}^{E_i} = \frac{A_{ab}}{E_i} \sum_{m=1}^{f} \sum_{k=1}^{f} \sum_{l=1}^{f} V_{a,lk} \hat{E}_{x,lm} \qquad (19.16)$$

For a voltage source in branch a, modelled as in Figure 19.3c with internal node x, from (19.15)

$$S_{V_a}^{E_i} = \frac{V_a}{E_i} \sum_{m=1}^{f} \sum_{k=1}^{f} \hat{E}_{x,km} \qquad (19.17)$$

Similar expressions can be derived for the other two types of controlled sources.

For a G_c conductance (or C capacitance) at branch a, the two roles must be taken into account: At each phase l, it acts as a conductance of value G_c and as a transconductance of value $-G_c z^{-1/f}$, controlled by the branch voltage at the preceding phase (if $l=1$, $l-1=f$). The branch voltages $\hat{V}_a^*$ are measured considering as null the voltages at low-impedance nodes in the adjoint network. In an ideal switched-current circuit these sensitivities (to the C_{gs} capacitances) must be null. It is not true if parasitic capacitances (such as C_{gd}) are present:

$$S_{C_a}^{E_i} = S_{G_{ca}}^{E_i} = \frac{G_{ca}}{E_i} \sum_{m=1}^{f} \sum_{k=1}^{f} \sum_{l=1}^{f} (V_{a,lk} - V_{a,(l-1)k} z^{-1/f}) \hat{V}_{a,lm}^* \qquad (19.18)$$

FFT interpolation can be used also in the sensitivity analysis, with the $\hat{E}(z)$ also interpolated. The interpolations can be performed exactly as described above. The denominator polynomial is the same as that of the normal system. Note, however, that the adjustment factors (19.4b) for the adjoint numerators must be used in the transpose form, with m and k interchanged (The phase sequence in the adjoint network is opposite to that in the normal network, due to the reversal of the direction of the G_c transconductances). The z-transforms of the sensitivities can be easily obtained by the use of (19.13)-(19.18).

Example 19.2

For the circuit used in example 19.1, the adjoint system of equations is

$$
\begin{bmatrix}
1+2G & & & & & & & \\
Gm_3 & 3G+Gm_1 & & & & & & \\
 & Gm_2 & G_{g2}z^{1/2} & & & & & \\
\hline
 & & & 1+2G & & & -G_{g1} & -G_{g3} \\
 & & -G_{g2} & & 4G+Gm_2 & & & \\
 & & Gm_3 & & Gm_1 & (G_{g1}+G_{g3})z^{1/2} & &
\end{bmatrix}
\begin{bmatrix}
\hat{E}_{1,11}z^{1/2} \\
\hat{E}_{3,11}z^{1/2} \\
\hat{E}_{5,11} \\
\hat{E}_{1,21}z^{1/2} \\
\hat{E}_{2,21}z^{1/2} \\
\hat{E}_{4,21}
\end{bmatrix}
,
\begin{bmatrix}
\hat{E}_{1,12}z^{1/2} \\
\hat{E}_{3,12}z^{1/2} \\
\hat{E}_{5,12} \\
\hat{E}_{1,22}z^{1/2} \\
\hat{E}_{2,22}z^{1/2} \\
\hat{E}_{4,22}
\end{bmatrix}
=
$$

$$
=
\begin{bmatrix}
-z^{1/2} \\
\\
\hline
\\
\\
\end{bmatrix}
,
\begin{bmatrix}
\\
\\
\hline
-z^{1/2} \\
\\
\end{bmatrix}
\tag{19.19}
$$

The sensitivity of $E_1(z)$ in relation to Gm_1 is computed by (19.14), with $V_a=E_4$ and $\hat{V}_b=\hat{E}_3$, as

$$
S^{E_1}_{Gm_1} = \frac{Gm_1}{E_1}\left(E_{4,11}\hat{E}_{3,11} + E_{4,21}\hat{E}_{3,21} + E_{4,12}\hat{E}_{3,11} + E_{4,22}\hat{E}_{3,21} + \right.
$$
$$
\left. + E_{4,11}\hat{E}_{3,12} + E_{4,21}\hat{E}_{3,22} + E_{4,12}\hat{E}_{3,12} + E_{4,22}\hat{E}_{3,22} \right)
\tag{19.20}
$$

The required voltages are obtained from (19.6) and (19.17). In the ideal case, the denominator for all the nodal voltage z-transforms is $D(z)=z-1$, and the required numerators are

$$
\begin{array}{lll}
N_{4,11} = 0 & \hat{N}_{3,11} = z & N_1 = z^{1/2}+1 \\
N_{4,12} = -z^{1/2} & \hat{N}_{3,12} = z^{1/2} & \\
N_{4,21} = 0 & \hat{N}_{3,21} = -z^{1/2} & \\
N_{4,22} = -1 & \hat{N}_{3,22} = -1 &
\end{array}
\tag{19.21}
$$

Note that some values are not listed as unknowns in (19.6) and (19.19), but they are equal to other unknowns, as indicated by the column pointers shown in (19.5) in the case of the E_4, and by the equation pointers in the case of the $\hat{E}_3$. For example: $E_{4,12}=E_{3,12}$ (column pointer 2) and $\hat{E}_{3,22}=\hat{E}_{2,22}$ (equation pointer 5). Substituting these values in (19.20), the sensitivity value of -1 is obtained. Gm_1 controls only the gain of the integrator, and so the sensitivity is frequency-independent.

The sensitivity of $E_1(z)$ in relation to C_{gs2} is given by (19.18), with $V_a = E_5$ and $\hat{V}_a = \hat{E}_5$, as

$$
\begin{aligned}
S^{E_1}_{C_{gs2}} = \frac{C_{gs2}}{E_1} \Big(& \left(E_{5,11} - E_{5,21} z^{-1/2} \right) \hat{E}_{5,11}{}^* + \left(E_{5,21} - E_{5,11} z^{-1/2} \right) \hat{E}_{5,21}{}^* + \\
& \left(E_{5,12} - E_{5,22} z^{-1/2} \right) \hat{E}_{5,11}{}^* + \left(E_{5,22} - E_{5,12} z^{-1/2} \right) \hat{E}_{5,21}{}^* + \\
& \left(E_{5,11} - E_{5,21} z^{-1/2} \right) \hat{E}_{5,12}{}^* + \left(E_{5,21} - E_{5,11} z^{-1/2} \right) \hat{E}_{5,22}{}^* + \\
& \left(E_{5,12} - E_{5,22} z^{-1/2} \right) \hat{E}_{5,12}{}^* + \left(E_{5,22} - E_{5,12} z^{-1/2} \right) \hat{E}_{5,22}{}^* \Big)
\end{aligned}
\tag{19.22}
$$

As C_{gs2} is between node 5 and ground; $\hat{E}_5{}^*$ is simply $\hat{E}_5$. The required numerators for the $E(z)$ and $\hat{E}(z)$ are

$$
\begin{aligned}
E_{5,11} &= 0 & \hat{E}_{5,11} &= -z \\
E_{5,12} &= 0 & \hat{E}_{5,12} &= -z^{1/2} \\
E_{5,21} &= z^{1/2} & \hat{E}_{5,21} &= -z^{1/2} \\
E_{5,22} &= z & \hat{E}_{5,22} &= -1
\end{aligned}
\tag{19.23}
$$

With these values, (19.22) reduces to 0, which will always occur with the C_{gs} capacitances in an ideal filter.

Sensitivity analysis can also be used to predict offset effects caused by mismatches in the biasing current sources. Equation (19.15) cannot be used directly because the AC value of the biasing sources is 0. However, it can be observed that

$$
\frac{\partial E_i}{\partial I_{bias}}(1) = \frac{\hat{V}_{bias}(1)}{f}
\tag{19.24}
$$

meaning that the effect of an error in a current source is proportional to the global AC voltage over the source in the adjoint network. The expression is evaluated at DC ($z=1$) for a biasing source. The division by f appears because signals during different phases (summation in m in (19.15)) must be added as in (19.3), where the term preceding the summations reduces to $1/f$ as $\omega \to 0$. Note that (19.24) is meaningless if applied to high-impedance nodes, in which the connection of a current source would violate the restrictions shown in Figure 19.2.

For example: An error in the current $2J$ entering node 3 would cause the error

$$
\frac{\Delta E_1}{\Delta(2J)} = -\frac{1}{2}\frac{\hat{N}_3(1)}{D(1)} = -\frac{1}{2}\frac{z-1}{z-1}\bigg|_{z=1} = -0.5
\tag{19.25}
$$

where the values listed in (19.19) were used. The integrator pole is cancelled, meaning that the biasing current is not integrated.

Example 19.3

As a more complex example, consider the analysis of a third-order elliptic low-pass filter, with a passband ripple of 1dB, a minimum stopband attenuation of 30dB and a switching frequency to passband border frequency ratio of 10. A normalised circuit realising the filter, obtained by the application of the bilinear transformation to a passive prototype [10], is shown in Figure 19.6. No attempt was made to optimise the dynamic range or transconductance value spread, and the circuit was considered in the simplest form; without any circuitry to reduce parasitic effects. The filter presents unity gain at DC, due to the use of the 2Ω normalised load resistor.

The global z-domain transfer function for the ideal circuit is obtained by the algorithm as

$$T(z) = z^{1/2}\left(1 + z^{1/2}\right)\frac{0.04632z^3 + 0.00689z^2 + 0.00689z + 0.04632}{z^4 - 2.13139z^3 + 1.78125z^2 - 0.54344z} \quad (19.26)$$

The term $1+z^{1/2}$, that apparently doubles the filter gain, is due to the sampling of the output once at each phase assumed by the algorithm. When this term is multiplied by the $1/f$ sampling function, as in (19.3), the correct one-period sampling function is obtained, and the gain is also corrected. The extra pole at $z=0$ is due to the sample-and-hold circuit made by the input transistors, that introduce a $z^{1/2}$ term multiplying the denominator, completed by the application of (19.4a) to round up its z powers. The multiplying term $z^{1/2}$ is due to the output formulation during phase 2.

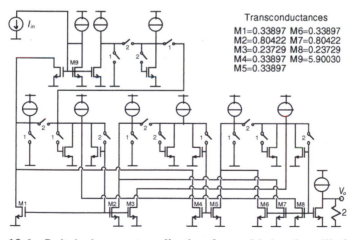

Figure 19.6 Switched-current realisation for a third-order elliptic filter. The unidentified transistors have unitary transconductances, and the ratios of the other transconductances to theirs are as listed. The biasing currents are proportional to the sum of the transconductances of the transistors connected to them.

Figure 19.7 shows the gain frequency response obtained with a 10kHz switching frequency. The transmission zeros at 5kHz, where the *s*-domain infinity is mapped by the bilinear transformation, can be observed. The small amount of distortion due to the sampling function is also observable. The expected error margins when 5% random errors are assumed in the transconductance values are shown, added and subtracted from the nominal curve. The error was computed as statistical deviation, using sensitivity analysis, by

$$\Delta\left|E_o(j\omega)\right| = 0.05 \times 8.686 \sqrt{\sum_i \left(\operatorname{Re} S_{Gm_i}^{E_o(j\omega)}\right)^2} \qquad (19.27)$$

The error is small and mostly frequency-independent, as a consequence of the use of the low-sensitivity second-generation integrators. In the small window, the poles and zeros, roots of the numerator and denominator of (19.26), considered in powers of $z^{1/2}$, are plotted.

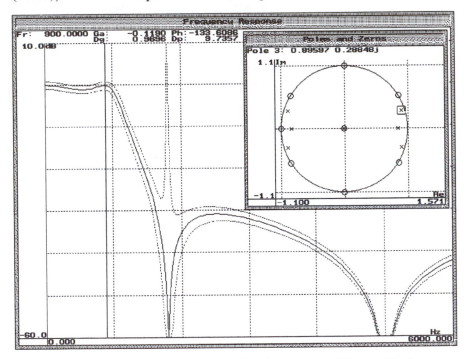

Figure 19.7 Gain frequency response for the example filter, with the error margins due to 5% random errors in the transconductances shown (ASIZ screen). The line at 900 Hz is a cursor, and at the top are listed the gain, phase, and gain and phase errors at this frequency. In the small window are the poles and zeros in $z^{1/2}$.

In Figure 19.8 the ideal gain curve (a) is compared with those obtained when some imperfections are taken into account. Curve (b) shows the effect of R_{ds} and current source resistances. A G_{ds} conductance equal to

$0.02G_m$ was added to each transistor. The input current source was assumed to be ideal. Curve (c) shows the effect of C_{gd} parasitic capacitances. Normalised C_{gs} capacitances with values equal to G_m were assumed, and G_{gd} capacitances with values equal to $0.01C_{gs}$ were added to each transistor.

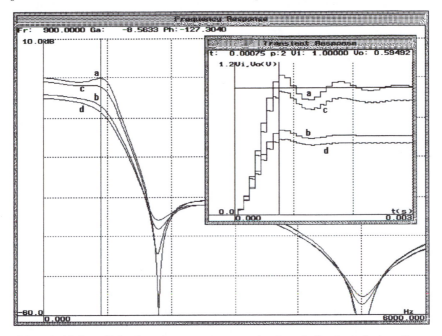

Figure 19.8   Gain frequency responses (main window) and step time responses (small window): (a) for the ideal filter (b) with the effect of G_{ds} conductances, (c) with the effect of C_{gd} capacitances and (d) with both.

These proportions take account of the fact that different transconductances and biasing current sources are obtained by changing the channel width of the MOS transistors (or placing fixed size transistors in parallel), and $C_{gs} \alpha G_m$, $C_{gd}=0.01C_{gs}$, $G_{ds}=0.01G_m$, and $G_{bias}=0.01G_m$. Curve (d) shows the combined effect. In the small window, the corresponding step responses are shown. The effects of G_{ds} and C_{gd} on the frequency response are similar, and they are a severe source of errors in these filters. In the transient response it can be observed that one effect of the C_{gd} capacitances is to cause the output signal to change at each phase. The example circuit resulted in a system of 18 equations, and 5 analyses were needed to generate the results for each curve.

19.8   Conclusions

The small-signal AC analysis of ideal switched-current filters, at transistor level, can be done in a similar way to those used for ideal switched-capacitor filters. In switched-current filters, the algorithm described can model precisely the effects of drain resistances of MOS transistors and other resistive losses, and also the effects of parasitic capacitors, such as C_{gd} capacitances. The interpolation of z-transforms by applying the FFT is not only a fast way to obtain frequency responses; a great deal of information can is also available from the interpolated transforms, their poles and zeros, and time-domain responses. Methods for computing sensitivities in relation to all the element values in a simple way were presented, directly derived from the methods used for continuous-time circuits. The algorithm described is not able to evaluate non-linear effects, and cannot model precisely (although some effects can be approximated by proper linearised models) effects caused by incomplete circuit stabilisation during a switching interval. It is, however, a powerful tool for the evaluation and verification of structures intended for switched-current or switched-capacitor filters.

References

[1]   J. B. Hughes, N. C. Bird and I. C. Macbeth, "Switched currents – a new technique for analog sampled-data signal processing," in *Proc. IEEE International Symposium on Circuits and Systems*, Portland, USA, pp. 1584–1587, May 1989.

[2]   D. Vallancourt and Y. P. Tsividis, "Sampled-current circuits," in *Proc. IEEE International Symposium on Circuits and Systems*, Portland, USA, pp. 1592-1595, May 1989.

[3]   J. B. Hughes, I. C. Macbeth and D. M. Pattullo, "Switched current filters," *IEE Proceedings*, vol. 137, Pt. G, pp. 156–162, April 1990.

[4]   J. B. Hughes, I. C. Macbeth and D. M. Pattullo, "Second generation switched-current signal processing," in *Proc. IEEE International Symposium on Circuits and Systems*, New Orleans, USA, pp. 2805–2808, May 1990.

[5]   K. Singhal and J. Vlach, "Symbolic analysis of analog and digital circuits," *IEEE Trans. Circuits Syst.*, vol. CAS-24, pp. 598–609, November 1977.

[6]   G. C. Temes and J. W. LaPatra, *Circuit Synthesis and Design*, McGraw-Hill, 1977.

[7]   J. Vandewalle, H. De Man and J. Rabaey, "The adjoint switched capacitor network and its application to frequency, noise and sensitivity analysis," *Int. J. Circuit Theory and Appl.*, vol. 9, pp. 77–88, 1981.

[8]   A. Opal and J. Vlach, "Analysis and sensitivity of periodically switched linear networks," *IEEE Trans. Circuits Syst.*, vol. 36, pp. 522–532, April 1989.

[9]   A. C. M. de Queiroz, P. R. M. Pinheiro and L. P. Calôba, "Systematic nodal analysis of switched-current filters," in *Proc. IEEE International Symposium on Circuits and Systems*, Singapore, pp. 1801–1804, June 1991.

[10] A. C. M. de Queiroz and Paulo R. M. Pinheiro, "Exact design of switched-current ladder filters," in *Proc. IEEE International Symposium on Circuits and Systems*, San Diego, USA, pp. 855–858, May 1992.

[11] E. Hökenek and G. S. Moschytz, "Analysis of multiphase switched-capacitor (m. s. c.) networks using the indefinite admittance matrix (i. a. m.)," *IEE Proceedings*, vol. 127, Pt. G, pp. 226–241, October 1980.

[12] J. Vlach, K. Singhal, and M. Vlach, "Computer oriented formulation of equations and analysis of switched-capacitor networks," *IEEE Trans. Circuit Syst.*, vol. CAS-31, pp. 753–765, September 1984.

[13] A. C. M. de Queiroz and L. P. Calôba, "CAPZ – A PC program for switched-capacitor filter analysis in the z-transform domain," in *Proc. 34th Midwest Symposium on Circuits and Systems*, pp. 134-137, Monterey, USA, May 1991.

[14] A. C. M. de Queiroz, P. R. M. Pinheiro, and L. P. Calôba, "Nodal analysis of switched-current filters," *IEEE Trans. Circuits Syst.*, vol. 39, CAS-II, Dec. 1992.

Chapter 20

Non-linear Behaviour of Switched-Current Memory Circuits

Gordon W. Roberts and Philip J. Crawley

20.1 Introduction

The Switched-Capacitor (SC) circuit technique has been successfully employed in the realisation of sampled-data filters, analogue-to-digital and digital-to-analogue converters in fully monolithic form since the early 1980s [1]. Their performance is recognised to be exceptionally good for most applications. Nevertheless, using SC techniques to implement analogue functions on mixed-signal ICs consisting largely of digital electronics is not without its price. A second poly-silicon layer must be added to an otherwise standard digital process to allow for the realisation of floating linear capacitors. This therefore adds more complexity to the fabrication process and increases the overall cost of the IC. With a smaller percentage of the mixed-signal IC occupied by analogue circuits today, this additional cost is under increasing pressure to be eliminated or, at least, reduced.

More recently, a new method of realising analogue sampled-data circuits has been proposed that circumvents the requirement for floating linear capacitors. This new method is called the Switched-Current (SI) or Current Copier technique [2,3]. In contrast to the SC approach which operates on voltage samples of the input signal, the SI technique operates on current samples of the input signal. Although charge storage is an essential operation for both circuit techniques, as it provides the means of realising signal delay, it is handled very differently in the two technologies. In the case of SC circuits, charge is moved from one capacitor to another, usually with the assistance of an op-amp, in such a way that the voltage transfer function of the SC circuit is determined by the ratio of the two capacitors. By contrast, the operation of SI circuits does not depend on the exact value of any capacitance in the circuit, but instead, relies on the ratio of transistor areas of a current mirror circuit. A capacitor is used only to remember the state of the transistor.

Although SI circuits are intended to be used in linear applications, their linear behaviour is questionable. According to some recent experimental evidence, simple SI circuits generate large levels of harmonic distortion when excited by a simple sinusoidal signal [4]. It is therefore hard to conclude from these experimental results whether SI circuits can meet the same performance requirements that are presently being met by SC circuits. It therefore seems fitting at this time to investigate the non-linear behaviour of SI circuits. In the past, most, if not all, SI circuit design techniques and analysis published in the open literature have dealt strictly with the linear behaviour of SI circuits. However, since no linear components are used in SI circuits, one wonders how well a linear analysis of a SI circuit can reveal its true behaviour, especially in light of the experimental evidence disclosed by [4]. With this in mind, this chapter will outline the non-linear operation of the most basic element of SI circuits, the current memory cell and, in turn, identify the major source of distortion in these circuits. A formula that places an upper bound on the total harmonic distortion generated by these SI circuits will also be given. The method used to derive this bound is general and applies to any sampled-data system, regardless of its implementation. The results of this new theory are supported by both HSPICE simulation and experimental results.

20.2 Basic Memory Cell Operation

Figure 20.1 shows the most basic circuit that illustrates the SI technique [3,2,5]. Herein we shall refer to it as the current memory cell. It has one principal transistor T_1, three switches, and a current source I_{bias}. The non-linear gate-to-source capacitance of T_1, denoted C_{GS}, is explicitly indicated in the figure, as this capacitance plays an important role in the operation of SI circuits.

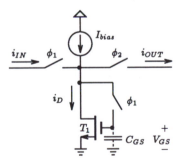

Figure 20.1 Simple switched-current memory cell.

The operation of the cell is as follows: during ϕ_1 the input current i_{IN} is added together with bias current I_{bias} and is fed into diode-connected transistor T_1. A corresponding voltage is then established between the gate and source of transistor T_1 and thus a proportional charge is stored on C_{GS}. At the moment the switches controlled by clock phase ϕ_1 are opened, the present value of the gate-to-source voltage on T_1 will be held constant by the charge stored on C_{GS}. In turn, the drain current of transistor T_1 is maintained constant for the entire duration of clock phase ϕ_2. The current that appears at the output of the current memory cell is then the difference between the bias current I_{bias} and the drain current i_D. Assuming non-overlapping complementary clocks with period T, we can write

$$i_{OUT}(n) = - i_{IN}(n-1/2) \qquad (20.1)$$

Equivalently, we can express this relationship in terms of an input-output z-transform current transfer function as follows:

$$\frac{I_{out}(z)}{I_{in}(z)} = - z^{-1/2} \qquad (20.2)$$

Here $I_{out}(z)$ and $I_{in}(z)$ represent the z-transform of $i_{OUT}(n)$ and $i_{IN}(n)$, respectively.

To demonstrate the operation of the SI memory cell we simulated the circuit in Figure 20.1 using HSPICE. The circuit was clocked at an arbitrary rate of 512kHz and a 16kHz sampled-and-held sinusoidal signal having a 10µA amplitude was applied as the input. Memory transistor T_1 was designed with a width and length of 400µm by 60µm, respectively, and biased with a current source of 20µA. The switches were essentially modelled as ideal switches.

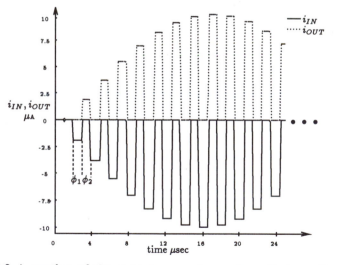

Figure 20.2 A portion of the output current waveform that results when a sampled sinusoidal signal is applied as input to the simple SI memory cell.

On completion of the HSPICE analysis, we plotted a portion of the input and output current waveforms associated with the current memory cell in Figure 20.2. The time interval is limited to 24µs in order to clearly illustrate the behaviour of the current memory cell. From this figure we can see the sample-and-hold action of both the input and output current signals associated with the current memory cell. We see that during ϕ_2 the output current is a constant (but inverted) version of the input current waveform one half-cycle previously. Conversely, during ϕ_1, the output current goes to zero as the output node has been disconnected from the memory cell.

Although it is clear from Figure 20.2 that the input-output transfer function does appear to be that described by (20.2), it is not immediately obvious whether or not there is any deviation from this ideal operation. One way to determine whether any deviation from the ideal behaviour exists is to compute the power spectral density (PSD) of the output current waveform using FFT techniques and compare it against the PSD of the input signal. Ideally the input and output PSDs should be equal. Adopting this approach, we plot the PSD of both the input and output signals in Figure 20.3. This was simply obtained by plotting the magnitude of the FFT results and expressing them in dBs. As is clearly evident, tones harmonically related to the input frequency of 16kHz are present in the output current waveform. As a point of reference, the second harmonic, which is the largest harmonic, is calculated to be 52.3dB below the fundamental tone. The fact that these harmonic tones are present in the output signal indicates that the circuit in Figure 20.1 has some unexplained non-linear behaviour. Furthermore, the non-linearity appears quite significant owing to the level of the harmonic tones present in the output spectrum. Also, as was previously stated, we expect that the magnitude of the input and output tones at 16kHz to be equal. However, we found that they differed by about 0.009dB (i.e. by a factor of 0.9989). Although one may regard this difference as part of some numerical error, we shall show below that this error is the result of a non-zero settling time error that occurs during the memorising phase of the memory cell. Furthermore, we shall show that the distortion appearing in the output current is the result of variation in the settling error from one sampling instant to the next.

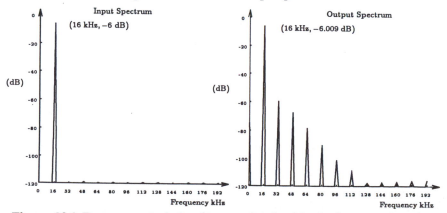

Figure 20.3 Power spectral density associated with the input and output current waveforms of the simple SI current memory cell.

20.3   Small-Signal Behaviour of the Memory Cell

The current memory cell of Figure 20.1 has two modes of operation. On clock phase ϕ_1, the voltage across the gate of diode-connected transistor T_1 is charged to a level which corresponds to the input current. On clock phase ϕ_2 the output current is held constant at a level which is determined by the charge that was stored on the gate of transistor T_1 on the previous clock phase. However, the charging process that occurs during clock phase ϕ_1 is not instantaneous, i.e. the greater the time that the circuit is given to settle the greater the accuracy of the charge transfer process. Thus the settling behaviour of the circuit, which determines the time needed to obtain a particular accuracy in the charge transfer process, limits the frequency of operation of the circuit. In this section we shall use small-signal methods to predict the settling behaviour of the current memory cell. In turn, we shall derive another input-output transfer function for the current memory cell that accounts for its settling behaviour. This formula will allow us to explain the difference between the magnitude of the fundamental of the input and output signals from the current memory cell uncovered by the HSPICE simulation in the last section. We should note that a similar formula was recently presented by *Hughes* [6].

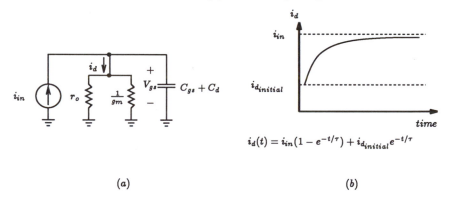

$$i_d(t) = i_{in}(1 - e^{-t/\tau}) + i_{d_{initial}} e^{-t/\tau}$$

(a)   (b)

Figure 20.4 (a) Small-signal model of current memory cell on phase ϕ_1, (b) The drain current response to a step change in the input current.

Figure 20.4(a) shows the small-signal model for the current memory cell during the memorising phase ϕ_1. In this figure we have replaced the diode-connected transistor by an incremental resistance $1/g_m$ in parallel with the output resistance r_o and a gate-to-source capacitance C_{gs} combined with the capacitance connected to the drain node denoted as C_d. The switches are assumed ideal. Although this is not true in practice, it simplifies the analysis significantly without sacrificing much accuracy. It is also assumed that the operating point around which the small-signal model is derived corresponds to a drain current level of I_{bias}. Thus all the currents referred to in the figure represent small deviations from this bias current level. In keeping with standard notation we have labelled the small-signal currents in lower case. Let us also make the assumption that the level of the signal drain current at the start of the memorising phase ϕ_1 is equal to $i_{dinitial}$ and that the input current is denoted as i_{in}.

Given these assumptions we expect that during ϕ_1 the drain current of transistor T_1 will move from its initial level of $i_{dinitial}$ towards the final level of i_{in}. The function that describes how the drain current moves from the initial level towards the final level over time can be derived from the small-signal model. Solving for the drain current as a function of time $i_d(t)$ we find that

$$i_d(t) = (i_{in} - i_{dinitial})[[1 - e^{-t/\tau}] + i_{dinitial} \qquad (20.3)$$

where τ can be approximated by $(C_{gs} + C_d)/g_m$ because $r_o \gg 1/g_m$. Thus we find that the drain current is described by an exponentially decaying function, as is depicted in Figure 20.4(b). Of course, we hope that given sufficient time the drain current of transistor T_1 would exactly equal the input current. Indeed, as $t \to \infty$, we see from (20.3) that $i_d(t)$ approaches i_{in} in the limit. We can therefore conclude from this result that, under the assumption of small-signal conditions, the settling behaviour of the current memory cell follows a single time-constant response.

Let us now ask ourselves what effect this settling behaviour has on the charge transfer process of the current memory cell, i.e. what current errors result from incomplete settling. Consider that at the end of clock phase ϕ_1, the switch that diode-connects T_1 is opened. At that instant, the voltage across the gate-source region of the transistor is held constant and, in turn, keeps the drain current constant. If we assume that the current memory cell is driven by two complementary clocks of equal duration with period T, then we can conclude that the drain current at the end of clock phase ϕ_1, which we shall denote as i_{dfinal}, will be

$$i_{dfinal} = (i_{in} - i_{dinitial})[1 - e^{-T/(2\tau)}] + i_{dinitial} \qquad (20.4)$$

Alternatively, we can rearrange the above equation and express the ratio of the actual change in drain current over the ideal change in the drain current, i.e. $i_{in} - i_{dinitial}$. On doing so, we can write

$$\frac{i_{dfinal} - i_{dinitial}}{i_{in} - i_{dinitial}} = [1 - e^{-T/(2\tau)}] \qquad (20.5)$$

Ideally this ratio should be one as it indicates the effectiveness of the charge transfer process during the memorisation phase. Thus, subtracting (20.5) from unity provides us with an error measure which we shall denote as the **settling error** γ given by the following:

$$\gamma \triangleq e^{-T/(2\tau)} \tag{20.6}$$

Now that we have identified the incomplete charge transfer occurring on the memorising clock phase, we would like to quantify this effect on the overall input-output discrete-time transfer function of the current memory cell. Once again we shall work with the drain current of the transistor. Consider the plot of a portion of an arbitrary input and transistor drain current waveform shown in Figure 20.5. The input signal current is assumed piece-wise constant and the transistor drain current follows a single time-constant response to each step change in the input current level. Let us further assume that the sampling instant occurs at the end of clock phase ϕ_2. We shall denote each of these sampling instants with an integer relative to the nth sample. In contrast, we shall denote the instant of time occurring at the end of clock phase ϕ_1 with a fraction $(1/2)$ corresponding to the half-period difference.

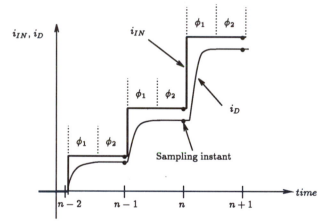

Figure 20.5 Typical drain current behaviour with an input current step.

Let us consider the following situation: The $(n\text{-}1)$ sampling instant has just ended and that the current memory cell has memorised a current of $i_D(n\text{-}1)$. During the next clock phase ϕ_1 a new input $i_{IN}(n\text{-}1/2)$ is applied that causes the drain current of transistor T_1 to change towards the new input current level from its previous current level of $i_D(n\text{-}1)$. Due to the settling behaviour of the current memory cell, this drain current will differ from the intended input current of $i_{IN}(n\text{-}1/2)$ by the settling error γ times the difference between the $i_{IN}(n\text{-}1/2)$ and $i_D(n\text{-}1)$. Thus at the end of clock phase ϕ_1 we have a difference equation that describes the new value of the drain current, namely

$$i_D(n) = (1 - \gamma)[i_{IN}(n\text{-}1/2) - i_D(n\text{-}1)] + i_D(n\text{-}1) \tag{20.7}$$

The above result can then be used to compute the output current $i_{OUT}(n)$ by noting that the drain current i_D is held constant over the full duration of clock phase ϕ_2. Thus, we know

$$i_{OUT}(n) = - i_D(n) \tag{20.8}$$

which allows us to further state

$$i_{OUT}(n-1) = - i_D(n-1) \tag{20.9}$$

Therefore, combining (20.8) and (20.9) with (20.7), allows us to write

$$i_{OUT}(n) = - (1 - \gamma)[i_{IN}(n-1/2) + i_{OUT}(n-1)] + i_{OUT}(n-1) \tag{20.10}$$

Noting that the above equation is linear, i.e. γ is assumed to be constant, we can take the z-transform of this equation and write the ratio of output current to input current, and get

$$\frac{I_{out}(z)}{I_{in}(z)} = - \frac{(1 - \gamma)}{1 - \gamma z^{-1}} z^{-1/2} \tag{20.11}$$

In contrast to the ideal transfer function stated above for this current memory cell, i.e.

$$\frac{I_{out}(z)}{I_{in}(z)} = - z^{-1/2}$$

we see that the inclusion of transistor settling behaviour gives rise to an additional multiplicative error term in the overall current transfer function which is $(1 - \gamma)/(1 - \gamma z^{-1})$.

It is interesting to compare the above theory with the results obtained from the HSPICE simulations of the last section. In the previous section we found that the output tone corresponding to the input tone of 16kHz had an unexpected 0.009dB drop in magnitude (see Figure 20.3). Based on the above theory, this loss in magnitude is almost fully accounted for by (20.11). To see this, consider expressing the transfer function given in (20.11) in terms of physical frequencies by substituting for the variable z with $e^{j\omega T}$. Thus (20.11) becomes

$$\frac{I_{out}(e^{j\omega T})}{I_{in}(e^{j\omega T})} = - \frac{(1 - \gamma)}{1 - \gamma e^{-j\omega T}} e^{-j\omega T/2} \tag{20.12}$$

With the input tone located at 16kHz, and the period of the clock set at 1/512msec, we can evaluate the above transfer function as follows and write

$$\left|\frac{I_{out}(e^{j2\pi.16/512})}{I_{in}(e^{j2\pi.16/512})}\right| = \frac{|1 - \gamma|}{|1 - \gamma e^{-j2\pi.16/512}|} \tag{20.13}$$

The only unknown is the settling error γ. To obtain γ we look back at its definition in (20.6) and note that it depends on τ which is equal to $(C_{gs} + C_d)/g_m$ and the settling period $T/2$. Using the DC operating point information generated from HSPICE during our initial simulation we find that $g_m = 85.16\mu A/V$, $C_{gs} = 22.08pF$ and $C_d = 0.05pF$. Thus we find that

$$\gamma = e^{-(85.16\mu/512k)/[2.(22.08p+0.05p)]} = 0.0233$$

Finally, by substituting this result back into (20.13), we can write

$$\left[\frac{I_{out}(e^{j16/512})}{I_{in}(e^{j16/512})}\right] = 0.9995$$

When expressed in dB's, this is equivalent to -0.004dB. Thus, comparing it with the result obtained from simulation (i.e. -0.009dB or 0.9989) we see that we are in reasonably close agreement.

If we take a moment and reflect on the current transfer function for the memory cell given by (20.12), we see that for increasing input frequency the magnitude of this transfer function decreases. This implies that the input and output signal will differ by a larger amount as the frequency increases. This suggests that linear errors are minimised by maintaining a high sampling to input signal frequency ratio.

20.4 Large-Signal Behaviour of the Memory Cell

As long as small-signal conditions apply, (20.12) can be used to describe the input-output current behaviour of the current memory cell under non-ideal settling conditions. Unfortunately, SI circuits are generally not used in a small-signal manner (i.e. the ratio of input current level to transistor bias level is not kept small). Instead, one finds SI circuits designed such that the input current level varies between 30% to 70% of the bias level [7,4]. This comes about largely from dynamic range considerations; if the input current level is small, the corresponding gate voltage on transistor T_1 of the current memory cell will be indistinguishable from the noise voltage generated by the transistor. Alternatively, if the input current level is too high relative to the bias level, excessive distortion will result.

The settling behaviour of the current memory cell described by the small-signal model fails to accurately predict the behaviour of the current memory cell when large input currents are applied. The reason for this is that the underlining assumption of a small-signal analysis: namely, that the circuit has a constant operating point or that the variation in the operating point is small is no longer true. This stems from the fact that large input currents applied to SI circuits cause equally large drain current swings to occur within the memory transistor. This, in turn, causes large changes in the transconductance of the transistor. This variation is evident from the definition of the instantaneous transconductance of a MOSFET, i.e.

$$g_m(t) = \frac{\partial i_D}{\partial v_{GS}} = \sqrt{\mu C_{OX} \frac{W}{L} i_D(t)}$$

Obviously, for large changes in $i_D(t)$, $g_m(t)$ will see correspondingly large changes. The impact of this on the behaviour of the current memory cell can be seen by reviewing the equation for the settling error γ given in (20.6). Here we see that the settling error depends on τ, which in turn depends on the gate-to-source capacitance C_{gs} and the transconductance g_m of the transistor. Thus the variations in transconductance will lead to exponential variations in the settling error. Also, since the gate capacitance C_{gs} does not depend very much on the drain current level, provided the transistor stays within the saturation region, we can say that the variation of the settling error depends solely on the variation of the transconductance g_m.

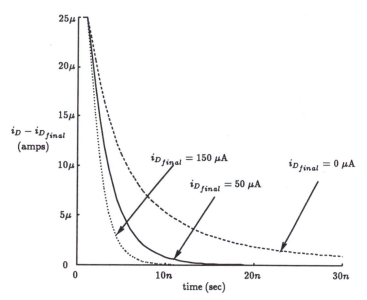

Figure 20.6 Comparing the relative settling behaviour of the current memory cell under three different current levels.

To illustrate the large-signal settling behaviour of the current memory cell, we have simulated the transient behaviour of the memory cell depicted in Figure 20.1 during clock phase ϕ_1 using HSPICE. The drain current of transistor T_1 was initialised at three different current levels: 175µA, 75µA, and 25µA. These levels are thought to typify the drain current of a current memory cell subject to a wide-ranging input current signal. After the start of each simulation (i.e. $t = 0+$) a 25μA current step signal was pulled from the input node. This, in turn, should cause the drain current over the course of each simulation to decrease by about 25μA. On completion of the HSPICE simulation, the drain current of transistor T_1 was observed and the difference between the drain current and its intended final current value (i.e. for the three cases stated above, 150μA, 50μA, and 0μA, respectively) was plotted in Figure 20.6. This was done so that all three curves could be compared on the same graph. Clearly the curve depicting the change in the drain current corresponding to an initial current level of 175μA has the fastest settling behaviour. The next fastest corresponds to a current level of 75μA, and the slowest corresponds to the lowest current level of 25μA. This behaviour is expected as we know that g_m is proportional to $\sqrt{i_D}$, resulting in the settling behaviour being fastest when the current level is highest. As far as the settling error is concerned, the opposite is true. We see that the settling error is largest when the initial drain current is smallest.

While it is clear from above that the settling behaviour of the current memory cell is dependent on its instantaneous drain current, what might not be so obvious is why variations in the drain current leads to harmonic distortion. As an example, let us look at the case where the input to the current memory is a sinusoidal signal whose amplitude is large relative to the bias current level. Recall that the operation of the current memory cell is to track-and-hold the input signal, thus it is reasonable to assume that the output signal will bear a close resemblance to the input signal, albeit with a phase inversion and possibly some distortion. Thus for a sinusoidal input signal having, say, 10μA amplitude, one would expect the input and output waveforms shown in Figure 20.7. We can see how the distortion arises by focusing our attention on the peak and trough of the output signal. At its peak, we know that the drain current of the memory transistor is at a minimum. As a result, the settling error at this instant is the largest it will be over the full cycle of the input signal. This implies that the difference between the input and output signals (ignoring the phase inversion) is largest at this instant, as can be seen in the figure. Conversely, at the trough of the output signal, the settling error would be smallest which, in turn, results in a smaller difference between the input and output signals (once again ignoring the phase inversion). As a result of all this, a perfectly symmetrical sinusoidal input is transformed into an asymmetric output waveform. Obviously, the asymmetric nature of this output waveform gives rise to harmonic distortion.

Based on the above discussion, it makes intuitive sense that the larger the variation in the settling error, the larger the harmonic distortion will be. In the following we shall develop a formula for quantifying the level of harmonic distortion present in the output signal from a current memory cell when excited by a sinusoidal signal. This was originally presented in [8].

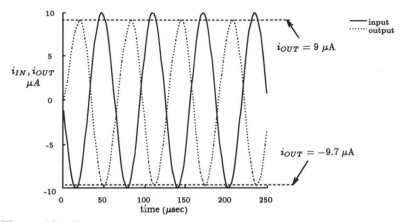

Figure 20.7 Example of the asymmetric nature of the output current waveform. The magnitude of the positive peak is less than the magnitude of the negative peak of the output current waveform.

20.4.1 A total harmonic distortion bound

The distortion associated with the output signal of a SI circuit excited by a pure sinusoidal signal can be quantified by a Total Harmonic Distortion (THD) measure. It is defined as the square-root of the sum of the harmonic powers of the output signal divided by the power of the fundamental signal. Since the input is a periodic signal, the output is also periodic and thus can be expressed in terms of a Fourier Series having the following form:

$$i_{OUT}(t) = a_0 + \sum_{i=1}^{\infty} a_i cos(i\omega_o t + \phi_i) \qquad (20.14)$$

where a_0, a_i and ϕ_i are the corresponding Fourier coefficients. If we make the substitution

$$i_{OUT_i}(t) = a_i cos(i\omega_o t + \phi_i) \qquad (20.15)$$

we can express the output signal as

$$i_{OUT}(t) = a_0 + \sum_{i=1}^{\infty} i_{OUT_i}(t) \qquad (20.16)$$

and on doing so, we can define the THD measure as

$$THD = \left[\frac{\sum_{i=2}^{\infty} \| i_{OUT_i} \|_2^2}{\| i_{OUT_1} \|_2^2} \right]^{1/2} \qquad (20.17)$$

We are using the short-hand L_2-norm notation $\|x\|_2^2$ to represent the power associated with the signal $x(t)$ to simplify our presentation, i.e.

$$\|x\|_2^2 \triangleq \frac{1}{T}\int_0^T x^2(t)dt$$

Recognising that the sum of the powers associated with all the harmonic components is bounded by the power of all deviations from the fundamental component of the output signal (Schwartz inequality [9]), i.e.

$$\Sigma_\infty^{i=2} \|i_{OUT_i}\|_2^2 \leq \|i_{OUT} - i_{OUT_1}\|_2^2$$

we can bound the THD measure given in (20.17) by the following more convenient expression:

$$THD \leq \left[\frac{\|i_{OUT} - i_{OUT_1}\|_2^2}{\|i_{OUT_1}\|_2^2}\right]^{1/2} \tag{20.19}$$

Given this definition, we know that if we can find a measure for $i_{OUT} - i_{OUT_1}$ (or equivalently Δi_{OUT}) and i_{OUT_1}, we can find a bound on the THD. To find a measure of these components, we refer back to the difference equation that relates the input and output current signals of the current memory cell in (20.10) derived in the previous section and modify it to include a time-varying settling error $\gamma(n)$. Here $\gamma(n)$ represents the settling error that occurs on the nth sample. The result is the following time-varying difference equation:

$$i_{OUT}(n) = -(1-\gamma(n))[i_{IN}(n-1/2) + i_{OUT}(n-1)] + i_{OUT}(n-1) \tag{20.20}$$

Recognising that the settling error $\gamma(n)$ varies between two extreme values of γ_{max} and γ_{min}, we can write

$$\gamma_{min} \leq \gamma(n) \leq \gamma_{max} \tag{20.21}$$

Furthermore, as a matter of convenience, we can divide $\gamma(n)$ into two components: a nominal settling error, denoted γ_{nom}, and a time-varying settling error representing the difference denoted $\Delta\gamma(n)$. Thus,

$$\gamma(n) = \gamma_{nom} + \Delta\gamma(n) \tag{20.22}$$

As a result of this designation, we can consider the output current i_{OUT} as consisting of two components: a linear component i_{OUT_1} associated with the nominal settling error term γ_{nom} and an error component Δi_{OUT} associated with $\Delta\gamma(n)$, allowing us to write

$$i_{OUT}(n) = i_{OUT_1}(n) + \Delta i_{OUT}(n) \qquad (20.23)$$

Substituting the above two equations back into (20.20), we can write

$$\begin{aligned} i_{OUT_1}(n) + \Delta i_{OUT}(n) &= -(1 - \gamma_{nom} - \Delta\gamma(n))[(i_{IN}(n-1/2) + i_{OUT_1}(n-1) \\ &\quad + \Delta i_{OUT}(n-1)] + [i_{OUT_1}(n-1) + \Delta i_{OUT}(n-1)] \qquad (20.24) \end{aligned}$$

If we consider that $\gamma(n) = \gamma_{nom}$, then we know that no distortion will result at the output of the SI circuit when stimulated by a sinusoidal input. We can therefore write $i_{OUT_1} = i_{OUT}$ because $\Delta i_{OUT} = 0$. Under this assumption we immediately see from (20.24) that

$$i_{OUT_1}(n) = -(1 - \gamma_{nom})[i_{IN}(n-1/2) + i_{OUT_1}(n-1)] + i_{OUT_1}(n-1) \qquad (20.25)$$

Furthermore, we can isolate Δi_{OUT} in (20.24) by subtracting off the equation for i_{OUT_1} given by (20.25), resulting in

$$\Delta i_{OUT}(n) = (\Delta\gamma(n)+\gamma_{nom})\Delta i_{OUT}(n-1) + \Delta\gamma(n)[i_{OUT_1}(n-1)+i_{IN}(n-1/2)] \qquad (20.26)$$

While (20.25) and (20.26) give us an expression for $i_{OUT_1}(n)$ and $\Delta i_{OUT}(n)$, respectively, we cannot use either of these expressions to determine a bound on the THD because $\gamma(n)$ is still unknown. However, we know that this term is bounded between γ_{min} and γ_{max} allowing us to observe the worst-case conditions imposed on the THD measure.

Since the THD specification has been bounded by the ratio of the signal power associated with Δi_{OUT} over the power associated with i_{OUT_1}, we know that in order to ensure that we have the largest possible ratio, we must look to find the largest Δi_{OUT} and the smallest i_{OUT_1}. This will guarantee that we have created an upper bound on the THD because the right hand-side of (20.19) will be bounded.

Based on the difference equation for i_{OUT_1} given in (20.25), we see that i_{OUT_1} is smallest when γ_{nom} is largest. This will occur when $\gamma_{nom} = \gamma_{max}$. The resulting difference equation then becomes

$$i_{OUT_1}(n) = -(1 - \gamma_{max})[i_{IN}(n-1/2) + i_{OUT_1}(n-1)] + i_{OUT_1}(n-1) \qquad (20.27)$$

Taking the z-transform of the above equation and rearranging it in the form of $I_{out_1}(z)/I_{in}(z)$, we get

$$\frac{I_{out_1}(z)}{I_{in}(z)} = -\frac{(1 - \gamma_{max})}{(1 - \gamma_{max}z^{-1})} z^{-1/2} \qquad (20.28)$$

Also, the largest Δi_{OUT} results when the γ coefficients in front of the currents on the right hand side of (20.26) are largest, namely

$$\Delta i_{OUT}(n) = \gamma_{max}\Delta i_{OUT}(n\text{-}1) + (\gamma_{max} - \gamma_{min})[i_{OUT1}(n\text{-}1) + i_{IN}(n\text{-}1/2)] \quad (20.29)$$

Since the above equation is time-invariant, we can take the z-transform of this equation to yield

$$\Delta I_{out}(z) = \gamma_{max}z^{-1}\Delta I_{out}(z) + (\gamma_{max} - \gamma_{min})[z^{-1}I_{out1}(z) + z^{-1/2}I_{in}(z)] \quad (20.30)$$

We can eliminate $I_{out1}(z)$ from this expression by substituting a rearranged form of (20.28) into the above equation, and get

$$\frac{\Delta I_{out}(z)}{I_{in}(z)} = (\gamma_{max} - \gamma_{min})\frac{(1 - z^{-1})}{(1 - \gamma_{max}z^{-1})^2}z^{-1/2} \quad (20.31)$$

From the theory of linear discrete-time systems, we know that the amplitude of the output signal from a circuit excited by a sine-wave with amplitude A is equal to A times the magnitude of the transfer function relating the input and output signals evaluated at the same frequency as the input signal. The power associated with the output signal is therefore equal to $A^2/2$ times the squared magnitude of the circuit's transfer function, again evaluated at the same frequency as the input signal.

Using the above principle, let's consider the power associated with the two fictitious output signals, i_{OUT1} and Δi_{OUT}, from the current memory cell. Based on an input sine-wave with amplitude A and frequency ω_o, the power associated with these two signals can be found from the following two equations:

$$\|i_{OUT1}\|_2^2 = \frac{A^2}{2}\left|\frac{I_{out1}(z)}{I_{in}(z)}\right|_{z=ej\omega_oT}^2 \quad (20.32)$$

and

$$\|i_{OUT}\|_2^2 = \frac{A^2}{2}\left|\frac{I_{out}(z)}{I_{in}(z)}\right|_{z=ej\omega_oT}^2 \quad (20.33)$$

Both $I_{out1}(z)/I_{in}(z)$ and $\Delta I_{out}(z)/I_{in}(z)$ can found from (20.28) and (20.31), respectively.

We are now in a position in which to impose a bound of the THD. Substituting the expression for $\|i_{OUT1}\|_2^2$ and $\|i_{OUT}\|_2^2$ found above into (20.19) with the appropriate transfer functions substituted, and after cancelling common terms, we can write our final result as

$$THD(\omega_o) \le \left|\frac{\gamma_{max} - \gamma_{min}}{1 - \gamma_{max}}\frac{1 - e^{-j\omega_oT}}{1 - \gamma_{max}e^{-j\omega_oT}}\right| \quad (20.34)$$

As is evident from the above expression, the THD bound changes with the frequency of the input signal. In fact, owing to the terms in the numerator dominating the denominator terms, as γ_{max} and γ_{min} are normally quite small, we see that the THD increases with increasing input frequency because of the high-pass nature of the term $1 - e^{-j\omega_o T}$. The implication of this is that the least amount of distortion occurs when the sampling (or clock) rate of the current memory cell is much higher than the frequency of the input signal. One should note that under such conditions, the linear error associated with the memory cell also decreases, as was noted in the previous section. Thus, the linear magnitude and phase errors and non-linear distortion associated with the current memory cell can be reduced by increasing the sampling to input signal frequency ratio.

20.4.2 Comparing THD bound with simulation results

To demonstrate the accuracy of the THD bound derived above, let us return to the example of Section 20.2. In that section, a current memory cell with a 20μA bias current was excited by a 16kHz tone having a 10μA amplitude. The PSD associated with the output signal was then calculated and plotted in Figure 20.3. By summing the power associated with each harmonic and dividing this total by the power of the fundamental and taking the square-root, we find that the THD for this memory cell is 0.26%. To compare this result to that predicted by the bound given in (20.34), we require estimates of the maximum and minimum settling errors of this circuit. Since the settling error is dependent on the circuit time-constant, which in turn is dependent on the instantaneous drain current of the memory transistor, we can deduce the extremities in the settling error from the maximum and minimum levels in the drain current of the memory transistor. With the memory transistor biased with a 20μA current source and a 10μA sinusoidal signal applied as input, the drain current in the memory transistor will vary between 10μA and 30μA. Alternatively, we can state that the maximum and minimum drain currents will be 1.5 and 0.5 times the nominal drain current of 20μA, respectively. These two factors can simply be written as $(1+m)$ and $(1-m)$, where m is the drain current modulation factor defined as the ratio of the amplitude of the input signal to the bias current level. Of course in this case $m = 0.5$. Since the time constant of the memory transistor is inversely proportional to the square-root of the drain current, we can immediately state that the minimum and maximum time constants are related to the nominal time constant τ_{nom} as follows:

$$\tau_{min} = \frac{\tau_{nom}}{\sqrt{1+m}} \qquad (20.35)$$

and

$$\tau_{max} = \frac{\tau_{nom}}{\sqrt{1-m}} \qquad (20.36)$$

Substituting these two expression into the equation for the settling error given in (20.6), and defining the nominal settling error as $\gamma_{nom} = e^{-T/(2\tau_{nom})}$, we can write

$$\gamma_{min} = (\gamma_{nom})\sqrt{1+m} \tag{20.37}$$

and

$$\gamma_{max} = (\gamma_{nom})\sqrt{1-m} \tag{20.38}$$

We are now in a position in which to compute the minimum and maximum values of γ. Recall from Section 20.3 that γ_{nom} was found to be 0.0233. Thus, with $m = 0.5$, we find $\gamma_{min} = 0.01$ and $\gamma_{max} = 0.071$. Therefore, according to (20.34) we find the upper bound to the THD at 1.38%. When compared to the result that was obtained from the HSPICE analysis, i.e. 0.26%, we see that it is clearly within the bound predicted by (20.34).

To highlight the fact that the THD of a current memory cell depends on frequency, we repeated the HSPICE analysis of Section 20.2 seven times under exactly the same conditions except that the frequency of the input tone was changed from 1kHz to 64kHz logarithmically by octaves. The PSD of the output signal was then calculated using an FFT from which the THD was then determined. The results of this analysis are summarised in Figure 20.8. Also shown superimposed on this graph of THD versus frequency is the bound predicted by (20.34). As is clearly evident, the actual THD measure found from simulation is always within the bound determined by (20.34).

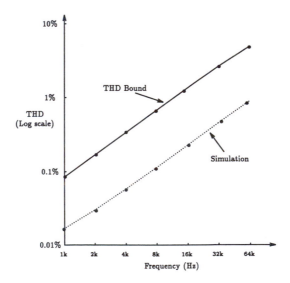

Figure 20.8 THD bound predicted by equation and verified by simulation.

One can observe from the above example that the bound predicted by (20.34) is about five times as great as that obtained from simulation. Although one may consider this bound to be too conservative, other examples demonstrate that this is not always the case.

20.4.3 *Experimental confirmation of the THD bound*

To provide experimental confirmation of the bound on THD provided by (20.34) we will make use of some experimental data recently made available in the literature. *Hughes* and *Moulding* [4] (see also Chapter 10) fabricated a delay line composed of a cascade of five current memory cells. Each of the memory cells were constructed in differential form. In essence, each differential memory cell consists of two single-ended current memory cells of the type shown in Figure 20.1. Additional transistors were included to provide a virtual ground at the input to the memory cell, as well as provide for some common-mode feedback to cancel common-mode effects. The memory behaviour of the cell remains essentially the same as that described for the simple current memory cell above. Therefore the equation for the bound on THD given by (20.34) is applicable to this case.

The circuit consisting of a cascade of five memory cells was clocked at 12.7MHz. A 100kHz tone with an amplitude equal to 70% of the bias level was applied as input to the circuit. The output PSD was then measured using a spectrum analyser. From the photograph of the PSD provided in [4] (also Figure 10.25), we calculated that the THD would be approximately 2%. Since there are five identical memory cells in cascade each assumed to be contributing the same amount of distortion, we can divide the above THD measure by a factor of $\sqrt{5}$ and estimate that the THD for one memory cell will be 0.89%.

To check this value of THD against the bound provided by (20.34), we use the same approach described above in the last subsection. First of all, we require information about the nominal time constant of the memory cell. We are told [4] that the bandwidth of the current memory cell is nominally 12MHz. This implies that the nominal time constant of a single memory cell is $1/(2\pi.12\text{MHz})$ or 13.26nsec. Therefore, from (20.6) we compute the nominal settling error to be 0.0514. With an input tone amplitude of 70% of the bias level implies that the drain current modulation factor m will be 0.7. Therefore, using (20.37) and (20.38), we find $\gamma_{min} = 0.021$ and $\gamma_{max} = 0.002$. Finally, using the THD bound provided by (20.34), we find that the THD should be less than 1.43%. This clearly bounds the measured THD of 0.89% and again demonstrates the usefulness of the formula provided by (20.34).

20.5 Conclusions

The non-linear behaviour of the SI memory circuit has been described in this chapter. This, in turn, was used to identify the major source of distortion in these circuits. Specifically, it was shown that distortion in SI circuits is the result of variations in the settling behaviour of the current memory cell from one sampling instant to the next. These variations are a direct result of the operating point of the current memory cell being a function of the time-varying input current signal.

As a means of quantifying the amount of distortion generated by a current memory circuit under specific input conditions, an upper bound on the THD was developed. Several examples demonstrated the simplicity of using the formula for the THD bound. Moreover, the bound in each example was checked against the THD computed from a HSPICE simulation and, in one particular case, compared to some experimentally observed results. For all these cases, the actual THD measure was within the proposed bound confirming our theoretical predictions. We conclude that SI circuits will operate with the least amount of distortion when the sampling (or clock) rate is made much larger than the frequency of the input signal. This is also true for its linear behaviour (i.e. magnitude and phase errors) and suggests that for best performance, a highly oversampled situation should be used.

Finally, it should be noted here that the THD bound and its development are valid for all sampled-data systems, regardless of its circuit implementation.

Acknowledgement

The work presented in this chapter was supported by NSERC and by the Micronet, a Canadian federal network of centres of excellence dealing with Microelectronic Devices, Circuits and Systems for Ultra Large Scale Integration.

References

[1]   G. S. Moschytz, Ed., *MOS Switched-Capacitor Filters: Analysis and Design*, New York: IEEE Press, 1984.

[2]   J. B. Hughes, N. C. Bird, and I. C. Macbeth, "A new technique for analog sample data signal processing," in *Proc. IEEE International Symposium on Circuits and Systems*, pp. 1584-1587, May 1989.

[3]   S. J. Daubert, D. Vallancourt and Y. P. Tsividis, "Current copier cells," *Electronics Letters*, vol. 24, pp. 1560-1562, Dec. 1988.

[4]   J. B. Hughes and K. W. Moulding, "Switched-Current Video Signal Processing," in *IEEE Custom Integrated Circuits Conf.*, pp. 24.4.1-24.4.4, May 1992.

[5]   G. Wegmann and E. A. Vittoz, "Basic principles of accurate dynamic current mirrors," *IEE Proceedings*, vol. 137, Pt. G, pp. 95-100, April 1990.

[6]   J. B. Hughes, *Analogue Techniques For Very Large Scale Integrated Circuits*, Ph.D. Thesis, Faculty of Engineering and Applied Science, University of Southampton, March 1992.

[7]   T. S. Fiez and D. J. Allstot, "CMOS switched-current ladder filters," *IEEE J. Solid-State Circuits*, vol. 25, pp. 1360-1367, Dec. 1990.

[8]   P. J. Crawley and G. W. Roberts, "Predicting harmonic distortion in switched-current memory circuits," in *Proc. IEEE International Symposium on Circuits and Systems*, May, 1993.

[9]   S. Haykin, *Communication Systems*, New York: John Wiley & Sons, Second Edition, pp. 634, 1983.

Chapter 21

GaAs MESFET Switched-Current Circuits

Chris Toumazou and Nicholas C. Battersby

21.1   Introduction

So far, this book has concentrated exclusively upon the implementation of switched-current techniques using CMOS technology, with potential applications in the audio and even video arenas (Chapter 10). To extend the application of these techniques to frequencies beyond the capability of CMOS technology, the use of GaAs MESFET (Gallium Arsenide MEtal Semiconductor FET) technology is promising. This chapter considers the feasibility and basic implementation of switched-current techniques in GaAs MESFET technology for very high speed applications. Although many of the ideas that follow are relatively immature, they certainly fall under the heading of future perspectives and directions.

Since this book has been primarily concerned with CMOS technology, this chapter begins with a general introduction to GaAs technology to clarify some of the basic differences between GaAs and Silicon. A review of continuous-time GaAs MESFET current mirrors is described in order to set the scene for the development of the required sampled-data GaAs circuits.   The chapter then briefly describes the particular difficulties imposed by the use of GaAs MESFET technology on switched-capacitor circuit design. A GaAs MESFET switched-current memory cell is then proposed and its application to an integrator discussed. Factors affecting the performance of the memory cell are then described and some enhanced circuits to improve performance presented. Finally, simulated results of a memory cell, an integrator and a bandpass filter are presented to confirm the functionality of the proposed circuits.

21.2 GaAs Versus Silicon Technology For Analogue Design

The Gallium Arsenide MESFET technology is very suitable for realising analogue functions at both radio and microwave frequencies. The high resistivity of the GaAs substrate allows passive components to be realised on the same chip as active devices, leading to fairly sophisticated, single chip MMICs (Monolithic Microwave Integrated Circuits).

The first GaAs MESFET based MMIC was a single stage monolithic X-band amplifier, reported by Plessey in 1976. This amplifier combined both GaAs MESFETs, for active elements, and transmission lines and MIM capacitors for the low parasitic passive elements, on the same semi-insulating substrate. Since then, a substantial number of MMICs that include more than one system function on the same chip have been realised. Sophisticated system chips performing multiple functions, such as radar transmit/receive modules, receivers and frequency synthesisers have also been developed. The level of maturity that GaAs has now reached is such that it a serious contender to Silicon technology in specific application areas. In particular, the radiation hardness of the GaAs MESFET and lower parasitic capacitances, due to the semi-insulating substrate, are perhaps the overriding reasons for the choice of GaAs over Silicon technology, for very high frequency applications in the 100s of MHz and GHz bands.

Silicon MOSFET technology has always dominated low frequency analogue design due to high, predictable transconductance and the low offset (random variations of input thresholds between nominally identical devices) and also complementary devices are available, making it easier to realise efficient push-pull and differential circuits. Furthermore, the lower current density and smaller device feature sizes of the Silicon FET give it a significantly lower power consumption than the GaAs MESFET. GaAs MESFETs on the other hand have a relatively poor transconductance and offset, and the gain per stage can be very low. In addition, the $1/f$ noise is high in GaAs at frequencies below 100MHz. In the frequency range up to 100MHz Silicon would thus appear to be more attractive than GaAs.

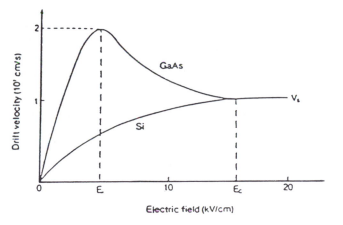

Figure 21.1 Electron drift velocity as a function of electric field strength.

For RF applications GaAs technology becomes more attractive than Silicon because it offers lower noise performance and the semi-insulating substrate of the GaAs MESFET gives fewer and lower values of parasitic capacitances. Minority carrier mobility is higher in GaAs than in Silicon which leads to higher electron drift velocity for GaAs as shown in Figure 21.1.

As can be seen clearly from the graph, the speed advantage of using GaAs is most apparent for low field strengths. This feature coupled with the semi-insulating properties of the GaAs substrate, namely high resistivity ($10^7\Omega$-cm compared to 30 Ω-cm for Silicon), together with associated low parasitics, gives the GaAs MESFET a tremendous speed advantage over Silicon BJT and FET technologies. Although the maximum f_t of BJT Silicon technology is steadily creeping towards that of GaAs, for precision sampled-data techniques such as the switched-capacitors and switched-currents FET technology is the most useful. Generally Silicon MOSFETs have maximum f_ts about an order of magnitude lower than that of GaAs MESFETs. Although predictions in Chapter 10 (10.1), suggest maximum clockrates of $\approx$1GHz for Silicon switched-current techniques, the relationship did not consider the large parasitic embedment that will be associated with these Silicon MOSFETs at such high frequencies. Unless special processing steps are taken, then this will severely limit the speed of operation to well below the device f_t.

A major disadvantage of GaAs is that its hole mobility is about one fifth that of Silicon and for this reason p-channel devices are not made. This is a serious limitation to analogue design flexibility and an unfortunate missing part in the designers tool kit.

The most generally adopted active device in GaAs technology is the depletion mode n-channel MESFET which can be produced with a relatively high degree of uniformity and which has a typical gate length in the range 0.5 to 1µm. Gate widths vary between a minimum of about 10µm and several hundred microns. Enhancement mode devices are not so popular for analogue design in GaAs, the main reason being that the gate forms a Schottky diode with the channel and so if the gate bias is greater than about 0.5V above the source, conduction will occur. Thus critical control of the gate voltage is essential with the enhancement mode FET. Depletion-mode devices also suffer from a restricted operating range of gate-source voltage between the threshold voltage (typically -1V) and the Schottky breakdown at about 0.5V, but the working range is more useful.

Analogue design is also quite seriously hindered by the low inherent device voltage gain (transconductance/output conductance) of the GaAs MESFET, typically as low as 20. This is due to the channel length modulation parameter λ which is frequency dependent. This frequency dependence [1] is associated with channel length frequency dispersion effects caused by traps in the semi-insulating substrate. At DC, MESFET λ is typically 0.06, however above several kHz MESFET λ typically falls to 0.3, leading to unacceptably high levels of output resistance as shown in Figure 21.2.

21.3 MESFET Modelling

The work presented in the following sections is based on simulations and measured results of two foundry processes, namely the Anadigics 0.5µm

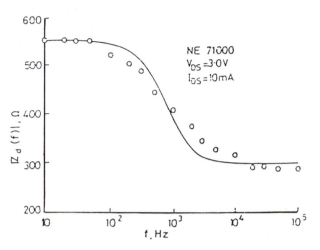

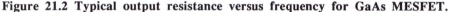

Figure 21.2 Typical output resistance versus frequency for GaAs MESFET.

gate length, -1V threshold process and the Plessey 0.5µm, -2V threshold process.

Over the years much effort has been expended on the development of accurate GaAs MESFET CAD models for simulation. In most cases fairly major compromises have had to be made between model complexity and accuracy [2]. A fairly accurate model for the drain current of a short channel depletion mode MESFET was presented by *Curtice* in 1980 [3], as a modification to the SPICE JFET model, and is given below,

$$I_d = \beta(V_{GS} - V_T)^n \tanh{(\alpha V_{ds})(1 + \lambda V_{ds})} \qquad (21.$$

Where β is the transconductance factor, V_T the device threshold voltage, λ the channel length modulation factor, and α is a parameter that describes early saturation of the drain current versus V_{ds}. The power index n lies typically between 1.9 and 2.1 and depends upon doping profiles.

Equation (21.1) is essentially the SPICE MESFET Level 3 model, and its main advantage over the simple SPICE Level 1 Model (JFET model) is that it models the early saturation effect. If the process exhibits early saturation then the device saturation voltage is $V_{min} < V_{GS} - V_T$. In fact this is a very advantageous feature for the analogue designer, particularly when designing for low power supply operation, since the head room required to saturate the device can be low.

Of particular importance for analogue design is the power index n which should ideally be $n = 2$, if a mature analogue design methodology based upon Silicon FETs is to be utilised. In fact a novel linear analogue circuit design synthesis algorithm has been developed [4], which exploits the FET square law output current versus V_{gs} characteristic in a suite of linear tunable transconductor and multiplier circuits.

A recently measured [5] I_d versus V_{gs} curve for a 32µm GaAs MESFET is shown in Figure 21.3 and confirms the validity of the square law assumption for the MESFET.

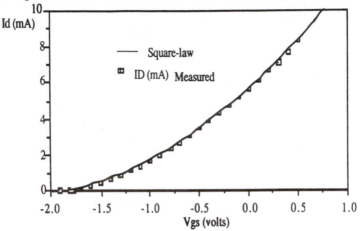

Figure 21.3 I_d **versus** V_{gs} **characteristic for a GaAs MESFET.**

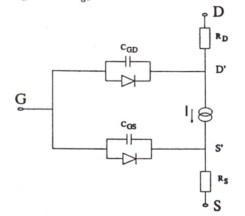

Figure 21.4 The SPICE level 1 MESFET model.

In fact for the technology utilised in our work, there was no evidence of early saturation, excellent conformity to a square-law was achieved and so the simple SPICE level 1 JFET model shown in Figure 21.4 could be employed.

The non-linear current $I(V_{gs}, V_{ds})$ is assumed to have the same form as (21.1). Only parasitic drain and source resistors are linear elements (do not change with bias conditions). The capacitors $C_{gs}(V_{gs})$ and $C_{gd}(V_{gd})$ represent the depletion layer and fringing capacitances associated with source and drain ends of the channel respectively and the diodes represent the respective Schottky barrier junctions. For our work, key level 1 SPICE MESFET parameters are shown in Table 21.1. Note that since this model does not predict λ changes with frequency, two different models, one for high frequency (HF) and one for low frequency (LF), should be used. It is in fact quite common to shunt a series RC circuit across the constant drain current generator in the model to mimic the change in output resistance with frequency.

Parameter	Value	Units
V_{TO}	-1	Volt
β	$0.067*10^{-3}$	mA/V^2
λ (LF)	0.06	V^{-1}
λ (HF)	0.3	V^{-1}
R_d	2920	Ω
R_s	2920	Ω
C_{gs}	$0.39*10^{-15}$	Farad
C_{gd}	$0.39*10^{-15}$	Farad
P_b	0.79	Volt
I_s	$0.075*10^{-15}$	Ampere
Parameters are for a typical 1μm gate length GaAs Process		

Table 21.1 Key parameters for SPICE JFET model of GaAs MESFET.

21.4 GaAs MESFET Current-Mirror Techniques

21.4.1 Cross-coupled MESFET pair

The principal building block for GaAs switched-current techniques is the current mirror. It is therefore appropriate to consider several current mirror circuits implemented in GaAs technology.

The very simple, elegant cross-coupled MESFET pair [6] shown in Figure 21.5 is responsible for the birth of a completely new generation of analogue design in GaAs technology. In particular it has been shown to be an important building block for active dynamic loads in push-pull amplifier stages, and negative current-mirrors, both applications compensating for the lack of p-channel devices.

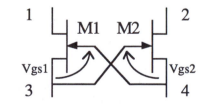

Figure 21.5 Cross-coupled MESFET pair.

In Figure 21.5 the pair of n-channel GaAs MESFETs have their gates and sources cross-coupled. The arrangement is biased such that $V_{GS1} = V_{GS2} = 0$

with both transistor saturated. By connection, the input device gate source voltage is equal but opposite in sign to the output device gate source voltage and so the small signal output current will be the inverse of the input current, and the output voltage will directly follow the input voltage. The connection effectively creates a voltage-following, inverting current-mirror which realises a form of positive impedance converter. Thus, if the input to M_1 (node 3) is connected to ground, the source of M_2 will be at virtual ground. Similarly, if node 3 is connected to a high impedance current source then, ideally, the output of M_2 will be also be a high impedance. Obviously, the above assumes MESFETs with infinite output resistance, which is not the case in practice, as will be discussed later. This elegant impedance transforming property of the cross-coupled current-mirror connection is of crucial significance for the design of high speed switched-current circuits. Although the input and output nodes of the current-mirror are physically separate, the fact that parasitic capacitance at one node may be equivalently transferred to the other node means that they can be regarded in some respects as a single node. When the current-mirror is used in a feedback loop, as is frequently required, this single-node feature minimises the number of poles introduced, leading to increased stability margins and optimum dynamic response.

Under the assumption of (21.1), and a square-law characteristic ($n = 2$), with negligible λ and early saturation effects, then the drain current of the MESFET given in (21.1) simply reduces to

$$I_d = \beta(V_{GS} - V_T)^2 \qquad (21.2)$$

Using (21.2) it can be shown that the circuit in Figure 21.5 has a current transfer characteristic into a short-circuit load of [7]

$$\frac{I_{d2}}{I_{dss2}} = \frac{I_{d1}}{I_{dss1}} + 4\left[1 - \sqrt{\frac{I_{d1}}{I_{dss1}}} \right] \qquad (21.3)$$

where $I_{dss1} = \beta_1 V_T^2$ and $I_{dss2} = \beta_2 V_T^2$. For this first order analysis we have assumed λ effects to be negligible. Although (21.3) shows that the cross-coupled MESFET-pair has a non-linear current transfer characteristic, it should be noted that the gate-source voltage of M_2 is the negative of the gate-source voltage of M_1 (i.e. $V_{gs2} = -V_{gs1}$). This property of perfect voltage inversion allows the MESFET pair of Figure 21.5 to be utilised as the core block of positive current-mirrors (feeding out of nodes 3 and 4), negative current-mirrors (feeding out of nodes 1 and 2), and a voltage buffer between nodes 1 and 2. What is also interesting about this structure is that it is reminiscent of the 1st generation current-conveyor [8].

21.4.2   Linear current-mirror

The major difference between the cross-coupled connection of Figure 21.6 and a standard MOSFET current-mirror pair is that it is possible to tie the gate-source connections of two MOSFETs directly together. Both MOSFETS will operate on a similar point on the I_d-V_{GS} characteristic of the FET ensuring similar drain source current over virtually the entire current-dynamic range. In contrast, a depletion mode MESFET pair of Figure 21.5 requires the gate to operate at a more negative potential than the source. Thus to create a similar gate-source connection to the MOSFET would require additional level-shift biasing circuitry with feedback control of the gate node [9]. While such a solution is feasible it does lead to fairly complex circuitry and potential instability due to the local negative feedback loop. The cross-coupled connection of Figure 21.5 ensures that when the V_{GS} of M_1 is positive then that of M_2 is negative. This does limit the V_{GS} operating range to about ±0.5V (0.5V being the forward Schottky diode conduction voltage), but gives a small signal linear current transfer characteristic about the operating I_{dss} (V_{GS} = 0) of the transistor. For large signals, the current transfer is non-linear as in (21.3) since the MESFETs are operating on essentially opposite points on the I_d vs V_{GS} characteristic of the device.

A GaAs equivalent to the linear MOSFET current-mirror can be formed by combining two cross-coupled MESFET-pairs. This will give an additional voltage inversion, and so ensure that the input and output gate source voltages are of the same polarity, as required for linear large signal current transfer (as with MOSFETs). The dual connection is shown in Figure 21.6a [7].Assuming a MESFET drain current given by (21.2) then the input drain current of MESFET M_1 in Figure 21.6 is given by I_{d1} = $\beta_1(V_{GS1} - V_T)^2$. The V_{GS} of M_1 is then inverted and applied to M_2 giving I_{d2} = $\beta_1(-V_{GS1} - V_T)^2$.

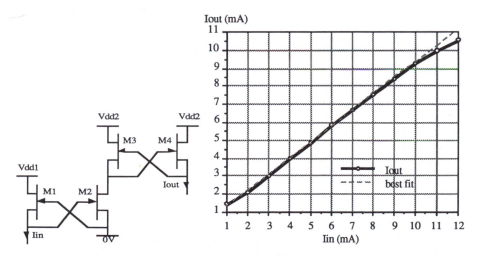

Figure 21.6 (a) Linear current mirror (b) Measured transfer characteristics.

Since the drain current of M_2 is equal to that of M_3 then the V_{GS} of M_3 is $(-V_{GS1})$. The V_{GS} of M_3 is then similarly inverted and applied to M_4, giving $V_{GS4} = V_{GS1}$ and therefore $I_{d4} = \beta_4 (V_{GS1} - V_T)^2$. Thus Figure 21.6a possesses a short-circuit current transfer characteristic given by

$$I_{out}/I_{in} = I_{d4}/I_{d1} = \beta_4/\beta_1 = W_4/W_1 \qquad (21.4)$$

corresponding to a linear non-inverting current amplifier (W is the gate width of MESFET). Note that the linearity is completely independent of the nominal MESFET model and so does not rely on a square-law MESFET characteristic as with a single cross-coupled pair.

Unfortunately, finite λ reduces current-transfer accuracy in a similar way to MOSFET current-mirrors, and also introduces some non-linearity [10]. Measured transfer characteristics of an integrated linear current mirror are shown in Figure 21.6b, confirming excellent linearity for the range $2\text{mA} < I_{in} < 10\text{mA}$ [5].

21.4.3 Cascoding

To improve current-mirror accuracy, linearity and precision, the development of suitable cascoding techniques is essential. The small-signal output (drain) conductance of a single MESFET, or basic current-mirror, is of the order of g_o. For a single cascode current-mirror, this conductance is reduced to the order of g_o^2/g_m, and for a double cascode current-mirror, reduced to g_o^3/g_m^2. Without cascoding, the low device voltage gain for GaAs MESFET's (g_m/g_o) of typically 20 leads to unacceptable current-mirror transfer accuracy. Transistors connected as current-sinking elements can be cascoded in the conventional manner (bottom cascoding) using a chain of stacked common-gate transistors. However, for devices connected as current sources, such as the current mirror of Figure 21.6a, cascoding is a lot more complex since voltages on the gates of cascode transistors must be defined in such a way that the current-mirror is maintained in the saturation region (top cascoding). In general, the gate voltage bias can be provided in two ways, by using level shifting diodes or by using a 'double level-shifting, self bootstrapping' technique which relies upon the ratioing of device gate widths [11]. The double level shifting approach has the advantage that both the voltage level shift introduced and the voltages required to keep the devices in saturation scale proportionally to the device threshold voltage V_T making circuit performance less sensitive to process variations. A full treatment of the single and double cascoding double level-shifting techniques can be found in [7]. An example of the cross-coupled mirror of Figure 21.5, incorporating single and double cascoded is shown Figure 21.7. In the Figure 21.7a, M_4 cascodes M_2. M_1, M_3 and I_{IN} essentially mimic a floating voltage source which provides a double level shift to ensure M_2 has sufficient drain source voltage to keep it saturated and operating in its high gain region. The V_{GS} of M_4 alone is insufficient to keep M_2 saturated. The technique can be extended to a double cascode mirror as shown in Figure 21.7b with additional cascode device M_6 and level shifter M_5.

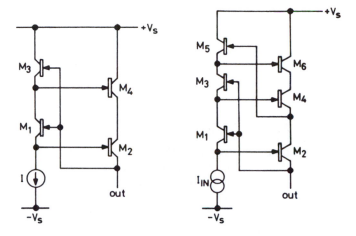

Figure 21.7 Cascode current-mirrors (a) Single (b) Double.

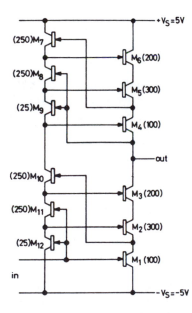

Figure 21.8 High gain GaAs op-amp.

21.5   Previous Work on a GaAs Switched-Capacitor Filter

Before utilising some of the previously developed continuous-time current-mirror techniques to realise GaAs switched-current cells, it is worthwhile reviewing some of what has actually been achieved to date with GaAs switched-capacitor (SC) techniques.

For SC filters the settling time of operational amplifiers is a very important parameter since it determines the maximum speed of circuit operation. Op-amp dc gain must also be above specified limits, typically 60dB. An example of an amplifier which achieves this performance is shown in Figure 21.8 [11].

The design is based on a single stage architecture with double cascodes to achieve high gain, a current-mirror to produce a push-pull output and a double-level shifting bias technique. The minimum settling time of 450ps for the amplifier in Figure 21.8 permits its use in sampled data systems with maximum switching rates up to 500MHz.

State of the art in GaAs SC filters is that of a 2nd order switched capacitor bandpass filter designed to realise a Q-factor of 16, midband gain unity and midband frequency of 1/25th of the switching frequency reported in [12]. The design uses the single-stage, double-cascode, double-level-shifting operational amplifier design of Figure 21.7 At a clock rate of 500MHz, excellent centre frequency performance at 20MHz and Q accuracy at $Q = 15$ was obtained. The $1/f$ noise floor at a measuring bandwidth of 10kHz was -75dB down from the reference signal [12].

Now, while the feasibility of switched-capacitor filters in GaAs MESFET technology has certainly been demonstrated, there are still several features of the technique that can make their design particularly unattractive, i.e.

- When SC circuits are implemented in CMOS technology, the gates of the MOSFETs can be driven between the negative and the positive power supply rails since the gate is insulated from the channel. Unfortunately, the gate electrode of a GaAs MESFET forms a Schottky diode with its source and drain. The gate voltage cannot therefore be more than about 0.5V more positive than the source, if conduction through the parasitic gate-source diode is to be avoided. In a switched-capacitor circuit, the switch terminals can experience wide voltage swings and thus to avoid parasitic gate-source diode conduction, the clock voltage must track the switch's source voltage [12]. It is thus necessary to introduce a switch control circuit. A typical switch control circuit is shown in Figure 21.9 [12].

- The digital input is fed to an inverter comprising M_3 and M_2, the output of which drives the gate of the switch MESFET M_1. The output of the inverter switches between the inverter negative supply voltage $-V_s$ which opens the switch, and the inverter positive supply voltage, closing the switch. The inverter positive supply is derived from the output of a voltage buffer comprising M_4 and M_5 which ensures that the positive supply of the inverter and hence the switch gate terminal during conduction tracks the analogue input, as required to prevent the junction diode forward bias. Many more sophisticated control circuits can be found in the literature [12], to reduce clock-feedthrough effects etc. The need for 'switch control' circuits to track the source voltage of the switch adds significantly to the overall circuit complexity.

- The high quality floating capacitors required by switched-capacitor circuits are normally provided by a Silicon nitride layer. Typically a minimum sized 0.1pF Silicon nitride capacitor has a via size of about 8x8 microns and so capacitors can in fact consume a tremendous amount of chip area. Furthermore, this Silicon nitride layer is not required by digital circuits and consequently switched-capacitor circuits are not fully compatible with digital GaAs MESFET technology.

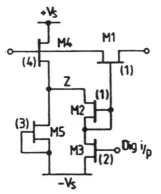

Figure 21.9 Typical switch control circuit.

Despite the aforementioned drawbacks and significant challenges that still remain for GaAs SC filter design, previous work demonstrates the engineering feasibility of using GaAs for sampled data analogue signal processing in radio frequency applications [12]. This previous work on SC filters is also confirmation of the realistic modelling, simulation and of the specially developed analogue design techniques, used throughout the filter realisation. This coupled with the encouraging measured performance, means that we can now confidently apply similar design techniques to realistic GaAs switched-current circuits and systems, and so in the following sections appropriate design techniques will be developed.

21.6   Towards GaAs MESFET Switched-Current Techniques

In this section we explore ideas for the realisation of switched-current (SI) circuits using GaAs MESFET technology. The main advantage of the SI technique over the SC technique and why we believe that it has tremendous potential for GaAs technology is that it fully exploits digital technology without the requirement for special analogue processing options, such as high quality linear floating capacitors (Silicon Nitride). Also, it does not require the realisation of operational amplifiers and all switches operate close to ground, hence avoiding the use of complex switch control circuits [12].

It is not the intention of this work on GaAs SI techniques to even attempt to realise the level of maturity and sophistication encountered with previous Silicon SI filters, since there is still a major gap between the levels of integration possible with GaAs compared to Silicon. However, it is intended that this work may help bridge some of this gap, principally by designing with smaller device dimensions than is typical to classical microwave design, and incorporating more active devices since the signal wavelengths at the frequency of operation (between 100MHz to about 1GHz) would allow lumped design approaches. In this section, we will introduce the basic switched-current memory cell in GaAs MESFET technology, its application to the realisation of a linear non-inverting memory an integrator and a simple bandpass filter discussed. In addition to demonstrating the feasibility of the technique, the main factors limiting the performance of the basic cells are also described.

21.6.1 Basic memory cell

The basic GaAs MESFET switched-current memory cell [13,14] is shown in Figure 21.10b and it is derived from the versatile continuous time cross-coupled MESFET pair discussed earlier (Figure 21.5), and shown connected as a current-mirror in Figure 21.10a.

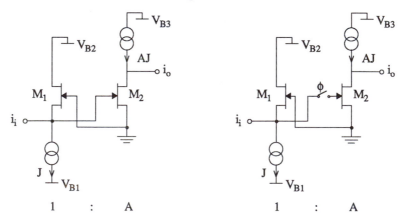

(a) Continuous time current mirror   (b) Switched-current memory cell

Figure 21.10 Basic GaAs MESFET switched-current memory cell.

The operation of the memory cell can be explained as follows. Assume initially that MESFETs M_1 and M_2 (Figure 21.10b) have the same size (i.e. $A=1$) and are operating in saturation. Then with the switch closed and $i_i = 0$ both MESFETs are set up with an equal I_{DSS}, giving quiescent $V_{gs} = 0$. On application of an input i_i the V_{gs} of M_1 will incrementally decrease whilst that of M_2 will increase by the same amount due to the cross coupling. Now, since $V_{gs1} = -V_{gs2}$ the circuit has the property of an inverting current mirror, as discussed earlier, and so $i_o = -i_i$ and the source of M_1 is held close to ground. When the switch is opened the cross coupling link is disconnected and the V_{gs} of M_2 is held on its gate capacitance and so as i_i continues to change i_o remains constant, hence memorising the previous input current. The circuit of Figure 21.10b thus implements a switched-current memory cell with the same functional characteristics as the first generation CMOS cell mentioned briefly in Chapter 3. This functional equivalence allows the GaAs MESFET switched-current memory cell to utilise the filter synthesis techniques already developed for its CMOS counterpart (see Chapter 8).

The basic memory cell can easily be extended to generate a scaled output current, by scaling the aspect ratio of M_2 with respect to M_1, as shown in Figure 21.10b. Extra scaled output currents can also be obtained by adding additional parallel connected output stages.

The voltages seen at the terminals of the switch, in Figure 21.10b, are equal in magnitude to the gate-source voltages of M_1 and M_2, which are held close to ground and not subject to large voltage excursions. Since these terminal voltages are held close to ground, 'switch control' circuits are not required and the switch can be implemented by a single MESFET. This is a significant advantage over switched-capacitor circuits, where large voltage excursions necessitate the use of 'switch control' circuits to avoid conduction through the switch's parasitic gate-source diode as discussed earlier.

21.6.2   *Linear non-inverting memory cell*

Under small signal conditions the transfer characteristic of the simple cross-coupled current mirror (Figure 21.10a), and the switched-current memory cell based upon it (Figure 21.10b), is sufficiently linear. However, for large signals, the quadratic dependence of the drain current on the gate-source voltage in the saturated MESFET device, results in the non-linear transfer characteristic given by (21.3). However, this large signal non-linearity is cancelled in a cascade of two mirrors as discussed earlier in Section 21.4.2. Applying this to the memory cell of Figure 21.10b gives the linear non-inverting memory cell shown in Figure 21.11.

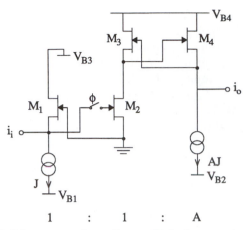

1 : 1 : A

Figure 21.11 Linear non-inverting switched-current memory cell.

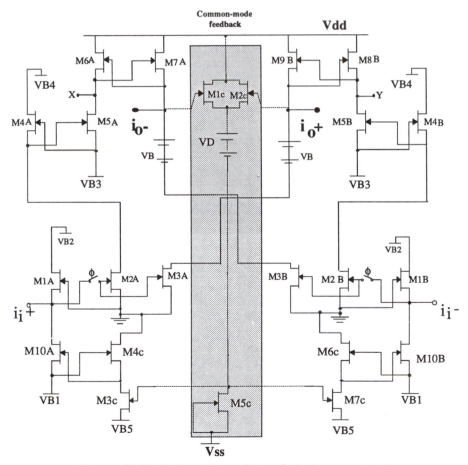

Figure 21.12 Fully differential switched-current cell.

21.6.3 Fully differential, linear transconductance GaAs switched-current cell

The aim of differential structures is to improve performance by cancelling unwanted effects that are common to both inputs and outputs. These include charge injection, offsets and power supply noise. In this section, a fully differential switched-current memory cell is presented. This is achieved by using two of the linear non-inverting SI memory cells with dual outputs, and linear negative current-mirrors to give the output difference as shown in Figure 21.12. M_1 - M_3 (A and B) form a pair of inverting SI cells (similar to Figure 21.10). Each cell has a dual output so that a final fully differential output can be obtained. The linear negative current-mirrors M_4 - M_7 (A and B) combine the differential signals at a single high impedance current output node. As with the basic cell, the large signal current transfer is non-linear. However, we will now show that the large signal transconductance gain of the differential memory is in fact linear. For simplicity assume the relationship

$$I_d = F(V_{GS}) \qquad (21.5)$$

where $F()$ is a non-linear function in this case a square-law according to (21.2). We will also assume negligible finite λ, and early saturation.

Consider operating the cell with a fully differential input current such that $(i_i^+ - i_i^-) = 0$. From (21.5)

$$i_i^+ = I_{d1A} = F(V_{GS1}) \qquad (21.6)$$

$$i_i^- = I_{d1B} = F(-V_{GS1}) \qquad (21.7)$$

Since M_2 and M_3 are cross-coupled it follows that $V_{GS2} = V_{GS3} = -V_{GS1}$ and so

$$I_{d2A} = I_{d3A} = F(-V_{GS1}) \quad and \quad I_{d2B} = I_{d3B} = F(V_{GS1}) \qquad (21.8)$$

now since M_2 drives linear current-mirror M_4 to M_7 then the output drain current of M_7 is

$$I_{d7} = I_{d2} \qquad (21.9)$$

Thus it can be readily shown from (21.8) and (21.9) that

$$i_o^- = I_{d2A} - I_{d3B} = F(-V_{GS1}) - F(V_{GS1}) \qquad (21.10)$$

and

$$i_o^+ = I_{d2B} - I_{d3A} = F(V_{GS1}) - F(-V_{GS1}) \qquad (21.11)$$

which for a square-law MESFET saturation-mode characteristic gives

$$i_o^- = - i_o^+ = -4\beta V_T V_{GS} \tag{21.12}$$

From (21.12) it is clear that the differential transconductance gain is a constant $-2\beta V_T$ and is positive since V_T is negative for a depletion mode device. A linear transconductance is a very desirable feature of this cell since changes in the stored gate-source voltage, due to charge injection effects, are only multiplied by a constant and thus only appear as a dc offset at the output, and do not produce non-linear distortion.

A common-mode feedback circuit has also been introduced to the cell and comprises M_{1C} to M_{7C}. The circuit essentially forms a negative feedback loop which regulates the input bias current when a common-mode output current is present. For example, assume an output common-mode current such that output voltages of the cell rise causing the gates of M_{1C} and M_{2C} to rise equally by an incremental common-mode voltage. This change will modulate the bias current set by current source M_{5C} causing it to incrementally increase and so raise the gate voltages of M_{3C} and M_{7C} causing their drain current to increase. Since M_{3C} and M_{7C} are driving M_{4C} and M_{6C}, then their gate source voltages will also incrementally increase. M_{4C} and M_{6C} are cross-coupled to M_{10A} and M_{10B} and so their gate source voltage will incrementally reduce. The net result is an incremental increase in the drain currents of M_{3A} and M_{3B}, pulling down the output voltage to counteract the initial common-mode increase. Such common-mode circuitry does add quite significantly to the complexity of the circuit, but is essential if high input and output CMRR performance is to be obtained so that common errors can be minimised. In fact a highly symmetrical cell with excellent common-mode rejection can be obtained by taking the outputs from nodes X and Y of Figure 21.12. This output stage coupled with the recently proposed linear differential input transconductor reported in [15], modified in the same way as the differential CMOS transconductor reported in Chapter 6, would yield a highly area efficient alternative. However, these ideas have yet to be fully explored.

21.6.4   Generalised integrator

A generalised GaAs MESFET switched-current integrator can be realised using the same architecture as employed by the first generation CMOS switched-current integrator and the resulting circuit is shown in Figure 21.13.

The diode level shifts, shown in Figure 21.13, are used to maintain the signal voltage level close to ground and thus enable the cells to be cascaded.

The transfer function of the integrator is given by

$$i_o(z) = A_1 \frac{z^{-1}}{1 - Bz^{-1}} i_1(z) - A_2 \frac{1}{1 - Bz^{-1}} i_2(z) - A_3 \frac{1 - z^{-1}}{1 - Bz^{-1}} i_3(z) \tag{21.13}$$

where A_1, A_2, A_3 and B are scaling factors determined by transistor aspect ratios.

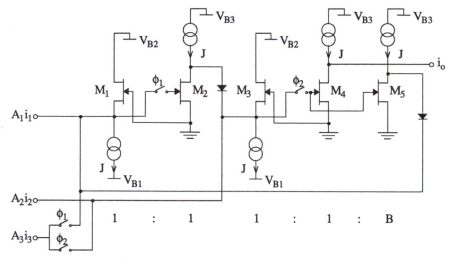

Figure 21.13 Generalised switched-current integrator.

21.6.5   Performance limitations

The performance attainable using GaAs MESFET switched-current circuits is limited by the same factors that limit the performance of their CMOS counterparts, but the magnitude of the errors caused is different since the characteristics of the two technologies differ significantly.   The performance limitations of CMOS switched-current circuits were discussed in detail in Chapters 4 and 5, and in this section these limitations are briefly reconsidered in the context of GaAs MESFET technology, with reference to the memory cell shown in Figure 21.10b.

(i)   Charge injection errors

Charge injection errors result from errors in the value of the stored gate-source voltage of the memory transistor M_2, caused by the flow of charge from the parasitic capacitance of the switch transistor during its turn-off transient (described in Chapter 4). In CMOS switched-current circuits this error, although significant, is manageable because the area of the memory transistor is generally much larger (typically 2 to 3 orders of magnitude) than that of the switch transistor. In GaAs MESFET technology, the design rules demand the minimum transistor width to be much larger than its length and only a single transistor length is available; it is therefore not possible to obtain large memory transistor to switch transistor area ratios and large charge injection errors result.

(ii) Finite conductance ratio errors

An ideal switched-current memory cell would present an infinite input conductance during the sampling phase and zero output conductance during the retrieval phase. In reality, this cannot be achieved using either the CMOS or GaAs MESFET technologies and the resulting finite input to output conductance ratio causes errors in the magnitude of the transmitted signal current. Unfortunately, the output resistance of the GaAs MESFET device is low (typically less than 1kΩ for frequencies above several kHz) and this gives rise to large conductance ratio errors in the basic memory cell.

(iii) Settling errors

If the memory cell is given insufficient time to settle properly (i.e. the clock frequency is too high) then an error in the stored current will result. The effect of this on the GaAs MESFET switched-current cell is the same as for the CMOS cell and can be minimised by optimising the settling characteristic of the cell for the intended clock frequency.

(iv) Mismatch errors

Mismatch errors are a consequence of the small variations in transistor characteristics resulting from manufacturing tolerances. The GaAs MESFET switched-current memory cell has the same basic structure as the first generation CMOS cell and therefore inherits its mismatch characteristics and sensitivities.

(v) Noise errors

Noise in the GaAs MESFET switched-current cell originates from the flicker and thermal noise generated by the MESFETs used in the circuit implementation and it manifests itself in two forms, sampled and direct noise. The way in which the sampling operation of the memory cell shapes the noise spectrum is the same as that occurring in the second generation CMOS cell or current-copier cell, as discussed in Chapter 5.

(vi) Gate leakage

The gate electrode of a MESFET is not as well isolated from its source and drain contacts as the gate of a MOS device and a very small leakage current will flow from the gate during the hold phase. The effect of this leakage current is to cause an error in the stored current if it is stored for a long enough time period. However, for the very high clock frequencies of interest, this leakage has a negligible effect on the stored current value.

(vii) Large signal non-linearity

As mentioned earlier, the basic GaAs MESFET memory cell (Figure 21.10b) exhibits a degree of non-linearity under large signal conditions, which will contribute offset, gain error and distortion components to the signal [4]. The impact of this non-linear behaviour can be minimised by operating the cell with a large ratio of bias current to signal current, using the cell in cascaded pairs of two and fully-differential structures.

21.7 Circuit techniques for enhanced performance

In the above section the main factors limiting the performance of GaAs MESFET switched-current cells have been highlighted. Circuit techniques to improve performance are currently at an early stage, but two techniques, one to improve charge injection performance and the other the conductance ratio of the basic cell, are described here.

21.7.1 Dummy switch compensation

The simplest possible implementation of the switch, required by the GaAs MESFET switched-current cell, is the single MESFET switch shown in Figure 21.14a. However, in practice, using this simple switch results in unacceptable levels of charge injection error.

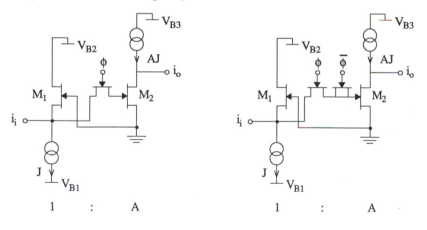

(a) Single transistor switch   (b) Dummy switch compensation

Figure 21.14 Switch implementations.

The dummy switch charge injection cancellation technique, previously described in Chapter 6 for the CMOS switched-current cell, can also be applied to the GaAs MESFET cell. The technique simply consists of adding a half size source-drain connected transistor between the main switch transistor and the gate of the memory transistor M_2, and clocking it with the inverse of the main switch's clock, as shown in Figure 21.14b.

21.7.2 Cascoded memory cells

The high output conductance of the GaAs MESFET gives rise to unacceptable errors in the basic memory cell (Figure 21.10b) and therefore it is necessary to introduce single or double cascoding techniques to improve performance.

In Figure 21.15a, M_1 and M_2 are single cascoded in the conventional way by M_3 and M_4. The addition of single cascoding to the basic memory cell typically reduces the output conductance of the cell by $\approx 20\text{dB}$. Figure 21.15b shows an alternative novel cascoding arrangement [10], which has two significant advantages over Figure 21.15a. Consider the case when switch ϕ is closed and all devices are assumed to be matched. By connection $V_{gs2} = -V_{gs1}$ and since M_1 is driving M_3, V_{gs3} is equal to V_{gs1}. Also since M_2 is driving M_4, then V_{gs4} is $-V_{gs1}$ and so the drain-source voltage of M_2 is simply $V_C - V_{gs3} - V_{gs4} = V_C$. Similarly the drain-source voltage of M_1 is $V_C - V_{gs3} + V_{gs1} = V_C$. Hence the drain-source voltages of both M_1 and M_2 are independent of input and output signals. This result leads to higher output resistance and improved current transfer accuracy. However, one disadvantage of this arrangement is that when ϕ is open, the drain-source voltage of M_2 will vary due to changes in input current and this will cause some signal feed through in certain configurations.

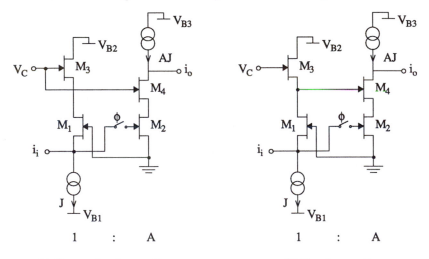

(a) Conventional cascoding   (b) Novel cascoding

Figure 21.15 Cascoded memory cells.

The current sources shown so far have been represented as being ideal, however, in practice they will be cascoded current sources. The lower current source cascode can be simply realised by stacking transistors. However, to realise the top cascode source requires the use of double level shifting cascode techniques [11]. A fully cascoded memory cell is shown in Figure 21.16 employing a top cascoded double level shifting bias chain.

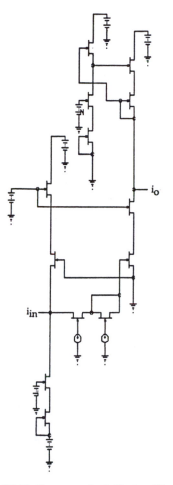

Figure 21.16 Fully cascoded GaAs SI memory cell.

21.8 Simulation results

To verify the operation of the proposed GaAs MESFET switched-current cells, a memory cell and a damped forward Euler integrator were simulated, both employing the single cascoded cell (of Figure 21.15b) and dummy switch cancellation (Figure 21.14b). The simulations used the HSPICE JFET model and the process parameters of an N-channel depletion mode GaAs MESFET process with a 1µm gate length and -1V threshold (Table 21.1).

21.8.1 Fully cascoded memory cell

The simulated transfer characteristic of the single cascoded memory cell, with dummy switch charge injection cancellation, is shown in Figure 21.17.

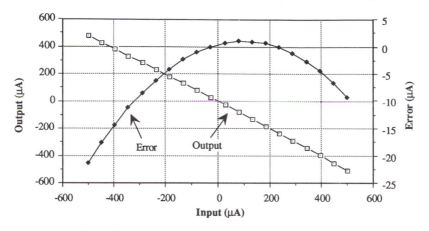

Figure 21.17 Simulated transfer characteristic of memory cell (clock=1GHz).

From Figure 21.17 it can be seen that the error in the output current of the memory cell is relatively small for small input currents, but increases progressively as the input current magnitude increases because of the large signal non-linearity of the single cell, mentioned earlier. This large signal non-linearity can be cancelled by operating the memory cells in cascaded pairs or kept small by maintaining a large bias current to signal current ratio.

The output resistance of single and double cascoded memory cells was also measured by simulation. For the single cascoded cells, the output resistance using the conventional cascoding technique (Figure 21.15a) was 14kΩ whilst using the novel cascode (Figure 21.15b) a value of 19kΩ was obtained. Using double cascoding, the output resistance's were boosted to 166kΩ for the conventional cascode and 228kΩ for the novel cascode.

So that all transistors remained saturated, the bias voltages were chosen to give a quiescent source-drain voltage of 1.5V for each transistor, giving power supply voltages in the range -3V to +6V and a quiescent $I_{DSS} \approx$ 4.3mA for the nominal sized device. The clock frequency was 1GHz and the clock voltage swing -5V to 0V.

A settling time characteristic of the fully cascoded memory cell of Figure 21.16 was simulated by measuring the settling time of the cell for various memory transistor gate widths. The simulated settling time (to 0.01%) curve for the cell is shown in Figure 21.18. The minimum settling time for positive and negative steps is 140ps with transistor widths of 72μm (gate length fixed at 1μm). The width of the main switch transistor was 30μm and the dummy switch transistor 15μm (minimum dimensions). This value settling time is lower than the a state of the art op-amp design [11], which gave a minimum settling time of 215ps.

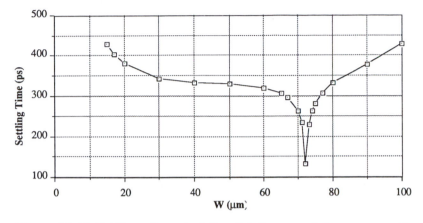

Figure 21.18 Settling characteristic of fully cascoded memory cell.

It is however, clear from Figure 21.18 that the settling minima is fairly sharp which indicates multiple poles in the system. An operating point should be chosen slightly to the right of the minimum to account for process tolerances and to ensure a certain amount of settling flexibility. With careful design and optimisation and layout we would expect a 200ps settling time to be more realistic.

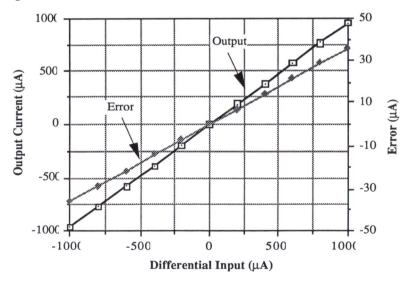

Figure 21.19   Differential mode transfer characteristic of differential cell.

The transfer characteristic of the fully differential cell of Figure 21.12 was also simulated by applying a differential current at the inputs. The basic memory cells were cascoded, but the current-mirror active loads remained as shown in Figure 21.12. The net power supply voltage was ± 5V with each transistor biased to operate with a minimum saturation voltage of 1.5V. The resulting transfer characteristic is show in Figure 21.19.

A slight improvement in current-transfer accuracy was obtained with the differential cell (factor of approx. 3) and Figure 21.18 shows that for very small signals errors are very low (less than 1%). However, excellent CMRR performance of the order of -70dB was simulated. The CMRR can be improved by top cascoding or by employing common-mode feedback. Much work has still to be done to develop the cell further, in particular to exploit the linear g_m feature of the cell.

21.8.2 Damped integrator

A damped forward Euler integrator was simulated using the circuit shown in Figure 21.13, where $A_1 = 1$, $A_2 = A_3 = 0$ and $B = 0.375$, corresponding to a first order lowpass section with a -3dB frequency of 100MHz for a clock frequency of 1GHz. In this configuration a large signal linear characteristic can be expected because there is a cascade of two memory cells between the input and output of the circuit.

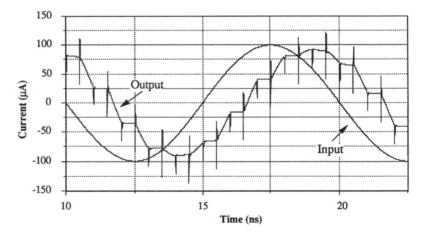

Figure 21.20 Simulated integrator waveforms (clock=1GHz).

The simulated input and output current waveforms of the damped integrator, for an input current of 100μA (peak) at 100MHz, are illustrated in Figure 21.20 and demonstrate the expected function of the circuit. The frequency response of the integrator was simulated for a 100μA (peak) input current and a 1GHz clock frequency and the results are plotted as Figure 21.21. The frequency response (Figure 21.21) confirms the correct operation of the damped integrator, although errors are present in both the low frequency gain and the -3dB frequency.

The simulated results, shown in this section, have confirmed the correct functionality of the proposed GaAs MESFET switched-current circuits. The maximum simulated clock frequency at which the memory cell functioned correctly was 3GHz, representing a considerable improvement over the maximum clock frequency so far reported for a CMOS switched-current circuit of 13.3MHz (see Chapter 10).

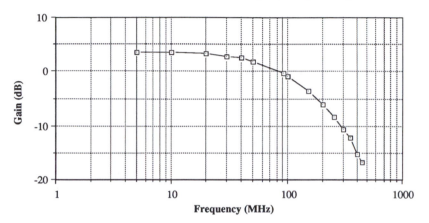

Figure 21.21 Simulated frequency response of damped integrator (clock=1GHz).

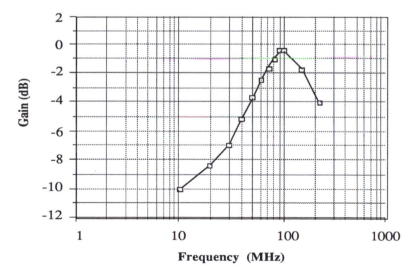

Figure 21.22 Frequency response of bandpass biquad filter.

21.8.3 Biquadratic filter example

The generalised integrator, of Figure 21.13, can be used to develop filters. As a simple example consider a bandpass biquadratic filter designed to have a centre frequency of 100MHz and a Q-factor of $1/\sqrt{2}$. The filter design was based on the biquadratic filter synthesis method developed in Chapters 3 and 8.

The filter was simulated with a 1GHz clock and the resulting frequency response is shown in Figure 21.22.

The responses obtained were close to predictions, demonstrating the feasibility of GaAs switched-current filters. However, much further work is still required to achieve accurate performance.

21.9 Discussion and Conclusion

In this chapter a new family of switched-current circuits, using GaAs MESFET technology, has been presented for very high speed applications. The new switched-current memory cell offers a number of potential advantages over existing attempts to realise switched-capacitor filters using GaAs MESFET technology, specifically it does not require high gain operational amplifiers, 'switch control' circuits and is fully compatible with digital processing technology. Perhaps the latter two advantages are the most significant.

Having described the basic memory cell it was shown how the basic memory could be used to develop a generalised integrator, and a biquadratic filter based on the synthesis techniques originally developed for the first generation CMOS memory cell. Using integrators as a starting point it is possible to develop a wide range of filters, again based on first generation CMOS switched-current synthesis techniques. Unfortunately, filters based on the first generation CMOS switched-current technique exhibit poor sensitivity characteristics, and the power consumption will be very high due to the need to maintain large bias currents. Attempts have been made to realise second generation GaAs SI cells. These attempts have generally lead to fairly complex switching arrangements because of the negative threshold and the forward junction diode of the MESFET.

The factors limiting the performance of the memory cell were then highlighted and it was seen that obtaining good accuracy from the GaAs MESFET cell is more difficult than in CMOS, because of the high output conductance of the MESFET, large charge injection errors and large signal non-linearity problems. However, the main sources of error are the same for both GaAs and CMOS technologies and the analyses presented in Chapters 4 and 5 can be applied to the GaAs MESFET circuits, with modifications to account for the differences in technologies. The use of dummy switch compensation to reduce charge injection errors and cascoding to increase the input to output conductance ratio were described.

To verify the functionality of the proposed GaAs MESFET switched-current circuits a memory cell, damped integrator and biquad filter were simulated. The results confirmed that the circuits operate as expected, but the accuracy obtained was poor. There is clearly a need to develop new enhanced circuit structures to improve performance, specifically the development of fully differential structures to enhance performance is encouraging.

The most promising future application of the SI technique is probably sigma-delta modulators. Although, attempts have been made to realise sigma-delta modulators with GaAs MESFETs using SC op-amp techniques [16], the switched-current circuits required are relatively simple and their digital process compatibility is particularly advantageous (see Chapter 14).

In many of the cell examples described throughout this chapter, the final complexity required to realised a practical design with low errors may well exceed the complexity of op-amps employed in SC techniques. However, the potential for higher speed operation still exists because of the virtual ground operating environment. This, coupled with the use of pure digital GaAs technology are perhaps good reasons why the technique should be considered as a compliment to SC structures in GaAs [17]. In fact much can be learnt from techniques developed to enhance SC performance, and these may be mapped into the SI domain. Of particular interest for a future phase of research is to investigate the application of SC gain and offset compensated techniques [18] to SI equivalents.

In conclusion, the implementation of the switched-current technique using GaAs MESFET technology has been proposed in this chapter, however, further circuit design innovation is still required before the technique can be practically exploited. We believe that the work is an encouraging step towards LSI in GaAs design, specifically for realising the analogue parts such as modulators, integrators, filters, and comparators for mixed mode RF front-end interfaces on pure digital technology.

Acknowledgements

The authors would like to acknowledge Dr David Haigh for his collaborative work on some of the background work in this chapter. We also acknowledge the British Science and Engineering Research Council for financial support.

References

[1]   C. Camocho-Penalosa and C. S. Aitchison, "Modelling frequency dependence of output impedance of a microwave MESFET at low frequencies," *Electronics Letters*, vol. 21, pp. 528-529, 1985.

[2]   D. G. Haigh and J. Everad, *GaAs technology and its impact on circuits and systems*, London: Peter Peregrinus Ltd, 1989.

[3]   W R Curtice, " A MESFET model for use in the design of GaAs integrated circuits," *IEEE Trans. MTT*, vol. 28, pp. 448-456, May 1980.

[4]   D. G. Haigh and C. Toumazou "Synthesis of Transconductor/Multiplier Circuits for Gallium Arsenide Technology", *IEEE Trans. Circuits and Syst.*, vol. 39, pp. 81-92, Feb. 1992.

[5]   D. G. Haigh, C. Toumazou and S. J. Newett, "Measurements on Gallium Arsenide Building Blocks and implications for Analog IC design," *Analog Integrated circuits and Signal Processing Kluwer Academic publishers*, vol. 1, pp. 137-149, Dec. 1991.

[6]   C. Toumazou and D. G. Haigh, "GaAs analogue IC design," in C. Toumazou, F. J. Lidgey and D. G. Haigh, Eds, *Analogue IC design: the current-mode approach*, London: Peter Peregrinus Ltd, 1990.

[7]   C. Toumazou and D. G. Haigh, "Cross-coupled GaAs MESFET circuits for potential MMIC applications," *Proc. IEEE International Symp. Circuits Syst.*, pp. 1853-1856, June 1991.

[8]   K. C. Smith and A. S. Sedra, " The current-conveyor - a new circuit building block," *IEEE Trans. on Circuit Theory, Proc IEEE*, vol. 56, pp. 1369-1369, Aug. 1968.

[9]   N. Scheinberg, "Design of high speed operational amplifiers with GaAs MESFET's," in *Proc. IEEE International Symp. Circuits Syst.*, pp 193-198, May 1987.

[10]   C. Toumazou, D. G. Haigh and J. M. Fopma, "Transconductor/multiplier circuits for Gallium Arsenide technology, Part 2- Practical Circuit Design," in *Proc. IEEE International Symp. Circuits Syst.*, pp. 2991-2994, June 1991.

[11]   C. Toumazou and D. G. Haigh, "Design of GaAs operational amplifier for analogue sampled data applications," *IEEE Trans. Circuits and Systems*, vol. 37, pp. 922-935 July 1990.

[12]   D. G. Haigh, C. Toumazou, S. J. Harrold, K. Steptoe and J. I. Sewell, "Design, optimisation and testing of a GaAs Switched Capacitor filter," *IEEE Trans. Circuits Syst.*, vol. 38, pp. 825-837, Aug 1991.

[13]   C. Toumazou and N. C. Battersby, "High speed GaAs switched-current techniques for analogue sampled-data signal processing," *Electronics Letters*, vol. 28, pp. 689-690, 26 March 1992.

[14]   C. Toumazou, N. C. Battersby and M. Punwani, "GaAs switched-current techniques for front-end analogue signal processing applications," *Proc. IEEE Midwest Symp. Circuits Syst.*, Aug. 1992.

[15]   C.Toumazou and D. G Haigh, "Integrated Microwave Continuous time active filters using fully tunable GaAs Transconductors," *Proc. IEEE International Symposium on Circuits and Systems,* Singapore, pp. 2569- 2572, June 1991.

[16]   K. T. Chan and K. W. Martin, "Components for a GaAs delta-sigma modulator oversampled analog-to-digital converter," *Proc. IEEE International Symp. Circuits Syst.*, pp. 1300-1303, May 1992.

[17]   N. C. Battersby and C. Toumazou, "Towards high frequency switched-current techniques," *Proc. IEEE International Symp. Circuits Syst.*, May 1993.

[18]   A. K. Betts, D. G. Haigh and J. T. Taylor, "Design issues for switched capacitor filters using GaAs technology," *Electronics Letters,* pp. 2216-2219, 1990.

Chapter 22

Switched-Currents:
State-of-the-Art and Future Directions

Chris Toumazou, Nicholas C. Battersby and John B. Hughes

22.1   Introduction

The switched-current technique is an analogue sampled-data signal processing technique which can be implemented on any digital CMOS or GaAs MESFET process without the requirement for special analogue process options. By virtue of this digital process compatibility, the technique seeks to replace switched-capacitors in mixed analogue and digital applications where provision of the additional process steps, needed to implement switched-capacitors, increases overall cost.

This book has developed the idea of switched-currents, providing a detailed description of the technique, proposing several ideas which advance the capabilities of the technique and illustrating several state-of-the art application areas. This final chapter, summarises state-of-the-art material in each chapter of the book and presents promising future prospects and avenues of research.

22.2   Switched-currents: an analogue technique for digital technology

The book began, with **Chapter 2** by Nicholas Battersby and Chris Toumazou, which placed the switched-current technique into a historical and technology perspective. Trends in processing technology and their impact on analogue circuits were then described to illustrate the need that switched-currents fulfil and to set the technique in perspective for the rest of the book. The evolution of analogue sampled-data signal processing was traced from early work, through the development of bucket brigade devices to charge-coupled devices and switched-capacitors. The emergence of mixed-mode analogue and digital signal processing and the subsequent need to integrate analogue sampled-data circuits on a digital process was then identified. Unfortunately switched-capacitors are not fully compatible with pure digital processing technology and advances in technology were seen to result in degraded performance. The switched-current technique was then introduced as a potential replacement for switched-capacitors for digital technology.

In **Chapter 3** a review of the fundamental concepts of switched-currents, the basic switched-current memory cell and its principles of operation were given by John Hughes, Ian Macbeth and Neil Bird. The authors also describe how basic filter building blocks such as delay cells, integrators and differentiators, having identical sensitivities to switched-capacitor counterparts, could be constructed. A worked example of a 6th order low-pass filter synthesised from biquadratic sections is explained.

Just as with the switched-capacitor technique, MOS transistor imperfections resulted in deviations from the ideal performance described by the algorithmic properties of the signal processing modules. For the basic switched-current cell to be of practical use in applications a good understanding of limitations and non-ideal behaviour is essential. John Hughes and William Redman-White give an exhaustive analysis of the origin and impact of non-ideal circuit behaviour on the performance of switched-current circuits, in **Chapter 4.**

Noise is one of the most fundamental limitations of any electrical circuit. In sampled-data systems such as the switched-current cells the noise is generated by the cells transistors extends to high frequencies and the sampling operation extends this noise into the lower frequency operating band. In **Chapter 5** Steven Daubert gives a first class analysis of the various sources of noise in the switched-current memory cell. The chapter starts with a noise analysis of the continuous-time current-mirror and then gives a detailed analysis of the noise performance of the switched-current copier cell for comparison. Some useful tips for noise reduction are given throughout the chapter.

In **Chapter 6**, John Hughes, Kenneth Moulding and Douglas Pattullo consider circuit techniques to improve on the performance attainable with basic switched-current cells to give higher levels of analogue performance. Two main approaches were adopted in the chapter, techniques to reduce conductance errors and fully differential structures to reduce charge injection errors and cross-talk from the digital environment. The work presented in this chapter is a positive step towards cells which may subsequently achieve performances close to that obtainable with state-of-the-art switched-capacitor filter techniques.

To appropriately conclude the section of the book on cells and enhancements, **Chapter 7** by Nicholas Battersby and Chris Toumazou considered class AB techniques which significantly reduced the quiescent bias current demands of 'Class A' memory cells by allowing signal currents to be much larger than the bias currents to be used. Techniques such as these are desirable when power dissipation is at a premium.

Once the basic cells for switched-current processing were developed they were employed in a number of applications. Probably the application area which has had most impact on switched-current processing is that of precision sampled-data filters. As with early developments of switched capacitor techniques, switched-currents by virtue of their charge memorising capabilities most aptly realise filter building blocks such as delay cells, integrators and differentiators. Many of these filter building blocks are becoming the bench marks for future switched-current performance.

Nicholas Battersby and Chris Toumazou in **Chapter 8** give a complete overview of the switched-current filter design approach. Basic filter building blocks are discussed employing many of the enhanced cells developed in earlier chapters. Both biquadratic and bilinear ladder filter synthesis methods are proposed. The design and fabrication of a class A biquadratic filter and a fifth order bilinear elliptic filter are presented with measured results from a 1.2μm digital CMOS process. The elliptic filter example can be clocked at 5.75MHz and represents the fastest high order switched-current filter of its kind to date.

A final section in **Chapter 8** looks towards fully tuneable switched-current filters and presents a new concept relating to a method for facilitating the accurate and continuously tuneable processing of switched-current circuits in a fully integrable way. The concept is called switched-transconductance techniques and is a step towards bridging the gap between the precision of the sampled-data processor and the tunability of the continuous-time processor.

Designing switched-current building blocks does not mean re-inventing the wheel! This is the message given by Gordon Roberts and Adel Sedra in **Chapter 9**, who have pioneered a signal flow graph transformation which converts from the ubiquitous switched-capacitor element to the switched-current equivalent, while still maintaining similar sensitivity properties. A multiple input switched capacitor network is readily converted into a multiple output switched-current network using this powerful signal flow graph analysis. Such signal flow graph methods are essential if switched-currents are to exploit the wealth of circuit functions already developed for switched capacitors.

The continuing trend of communication systems to migrate towards higher operating frequencies and the quest for fully integrated solutions has created the need for high frequency integrated filters with good inherent accuracy. In **Chapter 10**, John Hughes and Kenneth Moulding push the frequency barriers of state-of-the-art switched-current processing with switched-current building blocks operating in the video frequency band. The components include memory cells and delay lines particularly suited to the implementation of FIR filters. Performance is enhanced by the use of so-called grounded-gate negative feedback and fully differential topologies. Design formulas and graphs were included in the chapter to facilitate engineering of transmission errors and S/N ratio, and also to ensure monotonic settling behaviour.

It is conceded that while switched-current techniques have advantages allowing the implementation of mixed analogue/digital systems on digital technology, the feasibility of realising switched-current structures as good as state-of-the-art switched-capacitor techniques is still a challenge. Probably where this will be made possible is in techniques which become less dependent upon the accuracy of the raw analogue component and manufacturing variations in the process technology. Several such techniques were presented in Chapters 11 to 15.

In **Chapter 11** Adoración Rueda, Alberto Yúfera and José Huertas take a parallel approach to classical integrator based filter synthesis, by resurrecting the digital wave active filter approach and identifying the advantages of a switched-current analogue implementation. The wave active filter approach was discovered some 20 years ago. The idea is to simulate the equations for an incident and reflected wave in an LC ladder filter, similar in concept to the signal propagation in transmission lines. Early analogue implementations of wave active filters used active RC techniques, with operational amplifiers and performance was generally limited to low frequency operation. Since wave filters rely on the simulation of the behaviour of passive filters through the use of wave quantities instead of port voltages or currents, a principle advantage of the technique is that it does not require the use of high quality precision integrators. Instead the corresponding discrete-time multiport 'adaptor' is realised. The net advantages are that (i) lossless filter topologies can be utilised (reducing the strain on accuracy requirements of switched-current cells), (ii) filter synthesis is based upon the bilinear transformation allowing clock frequencies up to the Nyquist limit and (iii) the necessary summation and multiplication functions are easily implemented using switched-current processing. A further feature of the wave filter approach demonstrated in the chapter is the ease of programmability and the modular style of the cells. By overcoming many of the fundamental limitations of classical switched-current filter approaches, the pioneering work in **Chapter 11** is yet another significant step towards achieving comparable performance to state-of-the-art switched capacitor filter techniques.

The requirement for small size, high resolution, high speed analogue to digital and digital to analogue converters will always be of paramount importance in mixed analogue/digital VLSI systems. In fact the demands for the linearity of high-resolution D/A and A/D converters for measurement and digital audio equipment are currently so high that the achievable accuracy based upon component matching alone is far from sufficient. This book comprises a very comprehensive section on data converters with three distinct methods proposed in switched-currents to overcome the 'analogue inaccuracy' constraints and yield, in some cases, state-of-the-art A/D and D/A performances. All approaches seek to minimise the impact of manufacturing variations on the converters performance but in different ways.

In **Chapter 12**, David Nairn and in **Chapter 13**, Philippe Deval consider data conversion approaches based on algorithmic and pipelined A/D converters. The advantage of these types of converter is that high performance switched-current cells can be exploited in converters which are intended to operate close to the Nyquist sampling rate. Another very important feature of the algorithmic approach is the small chip area consumption. The approach is similar to the successive approximation technique but uses much simpler hardware, leading also to reduced power consumption. In both Chapters 12 and 13 a number of novel switched-current building blocks are presented including bi-directional current samplers, divide-by-two switched-current cells, current-comparators and current multiplexers. Throughout both chapters the basic principles of the algorithmic converter approach are described and complementary material covering cell enhancements, errors and improvements is given.

Converter performance demonstrating 14 bit resolution with only 2.5 mW of average power dissipation is demonstrated in Chapter 13. Also in Chapter 13 are some important messages due to the effect of technology scaling on switched-current memory performance. In particular the basic memory's time*injection product reduces as technology shrinks.

In **Chapter 14**, Gordon Roberts and Philip Crawley consider the feasibility of high resolution data converters based upon the over-sampling sigma-delta technique. The techniques developed in the chapter are for use in voice band telecommunication applications, with particular emphasis upon low power operation. The chapter presents the design and implementation of several different building blocks for switched-current double-integrator sigma-delta modulators. Although, far from state-of-the-art in resolution, the low power consumption and low chip area of measured test chip examples are very encouraging. With improved circuit techniques the switched-current sigma delta modulator looks set to tackle the high performance analogue interface requirements for future voice band communication systems on pure digital technology.

An alternative technique to over sampling, which achieves high resolution data conversion is the self calibration technique. A team of researchers from Philips Research Laboratories in Eindhoven, namely, Wouter Groeneveld, Hans Schouwenaars, Corné Bastiaansen and Henk Termeer have pioneered the current storage technique for data conversion and demonstrate in **Chapter 15** outstanding performance with world leading results for data converters based upon the self-calibrating switched-current approach.

The idea behind self calibration is to keep a memory cell calibrated to a constant reference by constantly refreshing its memory. Techniques to achieve this are presented in the chapter. The authors show that using the switched-current technique, an array of current sources which are equal to each other to within 0.02% can be realised. Two fabricated test chip examples are reported. A 16 bit self calibrating CMOS D/A converter for digital audio is designed showing measured linearity of 0.0025% within a 3 volt power supply. In a second example a multibit over sampling D/A converter is described which achieves an impressive 115 dB of dynamic range. The work in this chapter is a clear example of how the switched-current technique can compete favourably with state-of-the-art switched capacitor performance.

So far the applications of the switched-current technique have been their use as storage elements to create either filter functions or data converters. It is in fact the discontinuous storage mode of the cell which is exploited. However, in **Chapter 16** by George Wegmann two memory cells are combined to realise the sampled-data version of a continuous-time current-mirror. Based upon this idea the main limitations of standard CMOS current-mirrors such as offsets, matching inaccuracy and $1/f$ noise are almost eliminated. The work in Chapter 16 is an extension of previous work by the author and gives a very detailed account of dynamic mirror operation, performance accuracy limitations and improvements, different types of current mirror such as multiple outputs and current-dividers are also discussed. There is also an excellent treatment of errors and limitations, in particular the noise analysis of dynamic mirrors. Measurements on practical chips confirm theoretical expectations with impressive current-mirror transfer accuracy measured to within 1250 ppm.

To date much research effort has been devoted to signal processing applications of neural networks such as pattern recognition etc. A new exciting development is the electronic circuit implementation of cellular neural networks for image processing. In **Chapter 17**, Angel Rodríguez-Vázquez, Servando Espejo, José Heurtas and Rafael Domínguez-Castro describe a novel application of switched-currents to realise cellular neural networks for image-processing applications. The high levels of integration required by the technique is most aptly suited to switched-currents on digital technology. The performance features of switched-current neural networks includes improved frequency performance and linearity. Techniques discussed in the chapter include noise reduction in an imaging neural network, a re-configurable image processing chip and techniques for feature extraction etc. The application and achievable performance is promising. The work is a necessary step towards robust and fast analogue massive processing for cellular neural networks.

Throughout the book many innovative applications of switched-current cells have been discussed and cell enhancements described to maximise performance. However, up until this stage, little attention had been devoted to the important area of test, simulation and analysis of switched-current structures, and so this was the subject of Chapters 18, 19 and 20.

It is well recognised that it is easier to monitor hard faults in electronic circuits by monitoring short circuits and open-circuits in the current-domain. The concept of current-monitoring has recently been extended from digital devices into the analogue and mixed analogue/digital field. Current changes exhibited due to faulty behaviour are not as pronounced as those in the digital case, but are none the less significant.

Current-monitoring self test of switched-current systems appears to be natural to the switched-current cell operation as discussed by Paul Wrighton, Gaynor Taylor, Ian Bell and Chris Toumazou in **Chapter 18**.

The ease at which currents can be summed and compared during the functional cycles means that switched-current cells can be self-configured for test during 'dummy' cycles. A full treatment of current-monitoring for test, test vector generation and ideas for built in self test for switched-current cells are described in the chapter. In a final example a biquadratic switched-current filter is re configured in test mode and fault coverage of up to 75% is achieved. The results are far from ideal, but the chapter does demonstrate the ease at which structures can be re configured in a logical way for built-in self-test. This work is an encouraging step forward and a major advance for test on large scale or even future wafer scale integrated circuits employing the switched-current technology.

In **Chapter 19**, Antônio Carlos Moreirão de Queiroz concentrates exclusively upon the simulation of switched-current circuits in the frequency domain at the transistor level. The algorithm described, precisely models losses such as finite drain conductance and parasitic capacitance. The algorithm is implemented in a computer programme named ASIZ and analyses multi-phase switched-current filters, obtaining transfer functions in both the z-domain and s-domain. Such research investment in CAD tools for switched-current analysis is essential if the switched-current technique is to be readily accepted.

The algorithm described in Chapter 19 is unable to evaluate non-linear effects, and in fact most of the analysis given for switched-current cells throughout the book, thus far, assumed linear operation. It was therefore fitting that Gordon Roberts and Philip Crawley, in **Chapter 20**, give a detailed analysis of the non-linear behaviour of switched-current-memory cells. An interesting observation from their work is that the major source of distortion is the result of settling behaviour of the current memory cell from one sampling instant to the next.

Chapter 21 by Chris Toumazou and Nicholas Battersby proposes a completely new implementation of the switched-current technique using GaAs MESFET technology for Gigahertz clock-rate applications. Gallium Arsenide (GaAs) technologies, with their higher electron mobility and intrinsically low parasitics, have traditionally proved superior to silicon based technologies at very high frequencies. The GaAs MESFET technology is well suited to the realisation of precise sampled-data structures using the switched-current technique. The techniques presented offers a number of advantages over GaAs switched-capacitor techniques, specifically it does not require high gain operational amplifiers, switch-control circuits and of course is fully compatible with digital processing technology. Perhaps these advantages are even more significant than with equivalent CMOS cells, since it is well recognised that designing high gain, high speed op-amps in GaAs technology is a non-trivial task. Furthermore, the abolition of switch control circuits, essential in GaAs SC techniques due to avoid junction diode conduction of the MESFET, dramatically reduces overall complexity. This work is on GaAs SI circuits is an encouraging step towards LSI in GaAs design, specifically for realising analogue parts such as modulators, integrators, filters, and comparators for mixed mode RF front-end interfaces on pure digital GaAs technology and with switching frequencies in excess of 1GHz

22.3 Future Research Directions

In addition to the direct continuation of the research described in this book, there are many other promising avenues of further research to be pursued in the switched-current field and some of these will be detailed in this section.

It was noted earlier that switched-current filter performance is not yet competitive with that of switched-capacitors. Although this book has presented new ideas for enhanced memory cell performance, more work to improve performance further is essential if switched-currents are to become widely accepted. Many of the measured results presented in this book have demonstrated good performance when compared to previous switched-current filters, but the levels of performance obtained still fall short of those attainable using switched-capacitor techniques. So, whilst switched-currents may be a good choice for applications in which a basic digital process must be used, performance is not yet sufficient to displace switched-capacitors entirely.

Temes et al give an excellent overview [1] of the state-of-the-art in SC techniques and compare limitations of SC circuits to SI circuits. In many cases performance limitations are very similar in the two techniques. Where SC techniques seem to currently have the edge is in SNR, speed and low power consumption, while the switched-current technique has the edge if low chip area and low cost is imperative [1]. In fact much has been, and still can be, learnt from techniques already developed to enhance SC performance.

22.4 Towards total memory cell error cancellation schemes

A major proportion of the first few sections of this book was devoted to memory cell improvement via methods to combat charge injection and finite output conductance errors. In summary two distinct ways of tackling these errors were presented. There were the approaches that employed charge cancellation tricks such as dummy switches, and gain enhancement tricks such as cascoding, to tackle the errors separately. There were also the global correction techniques such as the self-calibration methods of Chapter 15 and over sampling techniques of Chapter 14 where amplitude errors are converted to the time domain. However, there exists one more category of error cancellation scheme not explicitly described in the book, where the lumped error source can be sampled and then subtracted from itself. The approach is similar to the self calibrating technique but does not require a separate reference current sources. In fact the approach is reminiscent of the classical gain and offset compensated SC techniques [1], sometimes referred to as Finite-Gain Insensitive (FGI) techniques because they have the effect of desensitising the op-amp gain at low frequency. An attempt at realising an SI equivalent to the SC FGI technique was the algorithmic memory cell reported by Toumazou, Battersby and Maglaras in 1990 [2], and briefly described below.

22.4.1 Algorithmic memory cell

The algorithmic memory cell [2] uses an algorithmic means to cancel total memory cell errors in one lump without relying upon any component matching. The algorithmic memory cell is shown in Figure 22.1a and its timing diagram in Figure 22.1b. The cell comprises 3 basic switched-current memory cells and requires 5 clock phases (phases ϕ_2' and ϕ_3' are the same as ϕ_2 and ϕ_3 but have slightly delayed trailing edges) and for the purposes of explanation each clock cycle has been divided into 12 equal periods.

The idea behind the circuit is to access the error of a memory cell, invert it and then pass it through the same cell again, thus cancelling the error without the need for matching.

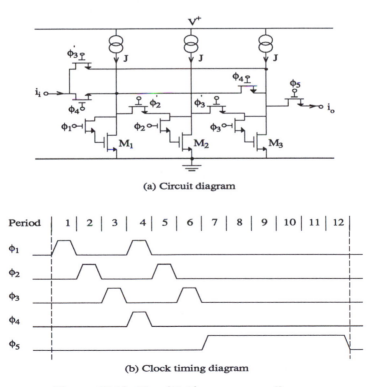

(a) Circuit diagram

(b) Clock timing diagram

Figure 22.10 Algorithmic memory cell.

The cell operation is as follows. M_3 is the main memory transistor. The aim is to introduce the input current to accumulate a total error in the memory, subtract the input current from the cells output leaving just the error and then feed back the error to subtract it with itself. In order to achieve the required subtraction an odd number of cells are required, hence the additional two transistors M_1 and M_2. Note M_1 and M_2 only ever sample the error current, and not the input current! A more detailed analysis is given below.

During periods 1 and 2, the input current is not present and the charge injection errors of the first two cells are accumulated. During period 3, the total error (including conductance error) of the third cell is added in the presence of the input current (i_i). In period 4, the output of the third cell is summed with the input (i_i), leaving just the accumulated charge errors to be sampled by the first cell. The uneven number of signal inversions around the loop ensures that the error of the first cell in period 1 cancels with its error in period 4. The same cancellation occurs in the second cell and then again in the third cell, where the input signal is reintroduced. Finally, during periods 7-12, the stored input current is outputted together with any residual error. Thus using the proposed cell, errors have been cancelled, according to the following relationship

$$i_o(z) = - z^{-1/2}i_i(z) - (\delta_{14} - \delta_{11}) + (\delta_{25} - \delta_{22}) - (\delta_{36} - \delta_{33}) \qquad (21.1)$$

where δ_{nj} is the error current of the n-th cell during period j.

Under ideal conditions (21.1) indicates that complete error cancellation is possible. However, if the memory transistors M_1, M_2 and M_3 are operating in saturation, there will be small variations in their transconductances, due to the presence of the error currents, which will prevent complete cancellation. By operating the memory transistors M_1, M_2 and M_3 unsaturated their transconductances will become approximately constant and the error cancellation improved. In fact simulations have shown error reductions of some two orders of magnitude.

The use of a more complex clocking scheme reduces the speed of the algorithmic cell to one sixth of the speed of a single cell using the same size transistors. However, because such a large improvement in error reduction has been achieved, the speed/accuracy ratio of the cell has still been significantly improved. To assess this speed/accuracy trade-off, the proposed algorithmic cell was simulated with a 10µA input current and a settling tolerance of 0.01%. Using a 2µm CMOS technology, the maximum sampling frequency of the algorithmic cell was found to be 650kHz whilst that of the simple cell was 3.9MHz. In this instance, using the algorithmic cell reduced the total error by a factor of 46, giving a speed/accuracy improvement of 7.7 times.

22.4.2 S^2I memory cell

Recently, John Hughes and Kenneth Moulding have proposed an elegant and simpler scheme [4] compared to the algorithmic cell, which detects and cancels the combined memory cell error but with virtually no detriment in any other performance. The cell is termed an S^2I Memory Cell which has a 'coarse step' in which an input sample is memorised, and a 'fine step' in which the error current is sampled. The principle advantage of this technique is that it uses a single PMOS device to sense and subtract the error, as opposed to two additional NMOS devices employed in the algorithmic scheme [3]. The basic S^2I Memory Cell arrangement together with clock waveforms is shown in Figure 22.2.

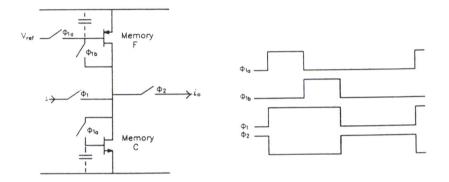

Figure 22.2 S^2I Memory Cell [3] circuit diagram and clock waveforms.

The operation of the cell is easily described by considering the clock-wave forms. The basic idea is that the coarse memory M_1 samples the input whilst M_2 provides the bias. The difference between the input and the memorised input (i.e. the error) is then sampled by fine memory M_2. The input is then disconnected and the output current is formed by the difference in current between the two memories, which to a first order will be identical to the input current since the error has been subtracted. Since M_2 only handles the error currents, the cell produces an effective 'virtual earth' at its input which reduces conductance errors and signal dependant charge injection errors.

   The inherent simplicity of this switched-current memory cell holds the promise of considerably higher performance and bandwidth than is possible with the previous algorithmic cell, and all the aforementioned error cancellation schemes. Furthermore the need for dummy switch compensation schemes is also avoided in both algorithmic [2] and S^2I Memory Cell [3]. In fact if such techniques hold up to expectations in practice then a completely new generation of very high performance switched-current processing could soon be underway.

22.3.3 Noise performance

Probably, the major limitation on the performance of switched-current systems will be SNR. Noise introduced into switched-current circuits by the physical behaviour of transistors is sampled, resulting in noise components above half the clock frequency being aliased back into the baseband, whilst low frequency components are attenuated by the correlated double sampling effect. The theoretical background and nature of noise in switched-current circuits is similar to that of switched-capacitor circuits, in which the noise spectrum is also shaped by the sampling process. The noise properties of the switched-current integrator stem from its method of integration and so it cannot be improved significantly by using enhanced circuits; rather other circuit structures, such as differentiators or FIR filters, should be used if integrator noise is a limiting factor. The aliasing of high frequency noise components back into the baseband by the sampling operation results in the dominant noise contribution in many cases and the most effective ways to reduce this are to ensure that the bandwidth of the circuit is no larger than that required in order to settle correctly at the desired clock frequency and to operate the memory transistors with the maximum allowable saturation voltage $(V_{GS} - V_T)$.

Probably, the ultimate limitation on the noise floor of SI cells is the KT/C noise in the gate voltage of the memory device [1]. KT/C can be reduced in SC filters by the use of large unit capacitors, and by large memory transistors in SI cells (see Chapter 5).

22.3.4 Towards low power

One of the fastest growing sectors of the electronics industry is that of portable computer and communications equipment. To increase battery life and product functionality, the design of fully integrated low power filters and data converters is necessary. The class AB memory cell described in Chapter 7 is a step towards this goal, but further work in this area is needed. A possible alternative approach is to try to use class A memory cells in which the memory transistor is operated in the weak inversion region, however it is not yet clear how feasible this will be in practice. The benefits of weak inversion switched-current cell operation are enormous for micropower analogue computational systems for large and wafer scale integration

22.3.5 Tuning schemes

Adaptive filters are widely used in communications systems and the realisation of fully tuneable switched-current filters could have wide application (Chapter 11). Currently, tuneable switched-current filters can be obtained by varying the clock frequency or by introducing an array of different sized current mirrors that can be switched in or out of the circuit.

A new technique that uses variable transconductance ratios to tune switched-current filters is currently being investigated (see Chapter 8, [4]).

22.3.6 Filter building blocks

The synthesis of switched-current filters described in this book has emphasised the use of integrator based structures and proposed a bilinear integrator based synthesis approach. The application of alternative differentiator based filters has been demonstrated successfully using switched-capacitors, where it was claimed to have advantages in certain applications [5], and equivalent SI cells were described in chapter 3, but further investigation is needed to confirm the expected benefits. The application of switched-current techniques to the implementation of wave-active filters is also a very promising alternative to conventional approaches (Chapter 11, [6]) requiring further investigation.

22.3.7 Data converters

Impressive data converter performance has been achieved using the SI technique, and this is indeed consolidated by the work in chapters 12 to 15. A combination of self calibration techniques and sigma delta over sampling techniques for extremely high resolution, coupled with algorithmic converter techniques for area efficiency , low power consumption and high speed has a tremendous scope for achieving the very high specifications warranted by today's telecommunications industry.

22.3.8 GaAs technology

For very high frequency applications the proposed implementation of switched-current circuits using GaAs MESFET technology is very promising (see Chapter 21). In particular, the design of sigma-delta modulators seems to be a particularly suitable application of the technique. The work presented in this book has laid the foundations for the development of GaAs MESFET switched-current circuits, although much work is still needed before a commercially viable circuit can be integrated.

22.3.9 CAD, simulation and test

If switched-current techniques are to gain wide acceptance, not only do they need to provide levels of performance that are competitive with switched-capacitors, but they also require a suite of CAD (Computer Aided Design) tools to automate and simplify the design process. At present, low level simulation tools are frequently used in the design of switched-current filters, since there are no commercial switched-current specific simulation tools available. However, some existing switched-capacitor simulators can be used (such as SCALP from Philips) and switched-current specific tools are now being developed (see Chapter 19, [7]). The modular nature of switched-current circuits makes them amenable to design automation and the development of such tools is just starting to be pursued.

Finally, once switched-current filters start to be designed and integrated into complex mixed analogue and digital systems, the issue of how to test the resulting integrated circuit must be tackled and this is the topic of a current research programme (see Chapter 18, [8]).

22.4 Conclusions

This book has demonstrated that the future of switched-currents is very 'dynamic'! This list of possible future research avenues has shown that there is considerable scope for the development of the switched-current technique in forthcoming years. The quotation given in the preface of the book perhaps most aptly concludes the book:

"This is not the end. It is not even the beginning of the end. But it is perhaps the end of the beginning."

Winston Churchill

References

[1]   G. C. Temes, P. Deval and V. Valencia "SC Circuits : State of the Art Compared to SI Techniques," in *Proc. IEEE International Symp. Circuits Syst.*, May 1993.

[2]   C. Toumazou N. C. Battersby and C Maglaras, " High-Performance Switched-Current Memory Cell", *Electronics Letters*, vol. 26, pp 1593-1595, 13th Sept 1990.

[3]   J. B. Hughes and K. W. Moulding, "S^2I; A two-Step Approach to Switched-Currents," in *Proc. IEEE International Symp. Circuits Syst.*, Chicago, May 1993

[4]   C. Toumazou and N. C. Battersby, "Switched-transconductance filters: A new approach to switched-current processing," to be published in *Proc. IEEE International Symp. Circuits Syst.*, May 1993.

[5]   T.-C. Yu, C.-H. Hsu and C.-Y. Wu, "The bilinear-mapping SC differentiators and their applications in the design of biquad and ladder filters," in *Proc. IEEE International Symp. Circuits Syst.*, pp. 2189-2192, May 1990.

[6]   A. Yúfera, A. Rueda and J. L. Huertas, "Switched-current wave analog filters," *Proc. IEEE International Symp. Circuits Syst.*, pp. 859-862, May 1992.

[7]   A. C. M. de Queiroz, P. R. M. Pinheiro and L. P. Calôba, "Systematic nodal analysis of switched-current filters, *Proc. IEEE International Symp. Circuits Syst.*, pp. 1801-1804, June 1991.

[8]   G. E. Taylor, C. Toumazou, P. Wrighton and N. Battersby, "Mixed signal test considerations for switched-current signal processing," *Proc. IEEE Midwest Symposium on Circuits and Systems*, Washington D. C., August 1992.

Index